DEEPER PATHWAYS
IN
HIGH-ENERGY PHYSICS

Studies in the Natural Sciences

A Series from the Center for Theoretical Studies
University of Miami, Coral Gables, Florida

Recent Volumes in this Series

A Continuation Order Plan is available for this series. A continuation order will bring
delivery of each new volume immediately upon publication. Volumes are billed only upon
actual shipment. For further information please contact the publisher.

ORBIS SCIENTIAE

DEEPER PATHWAYS IN HIGH-ENERGY PHYSICS

Chairman
Behram Kursunoglu

Editors
Arnold Perlmutter
Linda F. Scott

Scientific Secretaries
Mou-Shan Chen
Joseph Hubbard
Michel Mille
Mario Rasetti

Center for Theoretical Studies
University of Miami
Coral Gables, Florida

SPRINGER SCIENCE+BUSINESS MEDIA, LLC

Library of Congress Cataloging in Publication Data

Orbis Scientiae, University of Miami, 1977.
  Deeper pathways in high-energy physics.

  (Studies in the natural sciences; v. 12)
  "Held by the Center for Theoretical Studies, University of Miami, Coral Gables,
Florida, January 17-21, 1977."
  Includes index.
  1. Particles (Nuclear physics)—Congresses. I. Kursunoğlu, Behram, 1922-   II.
Perlmutter, Arnold, 1928-   III. Scott, Linda F. IV. Miami, University of, Coral
Gables, Florida. Center for Theoretical Studies. V. Title. VI. Series.
QC793.07 1977   539.7   77-9628
ISBN 978-1-4757-1567-5   ISBN 978-1-4757-1565-1 (eBook)
DOI 10.1007/978-1-4757-1565-1

Proceedings of Orbis Scientiae 1977 on Deeper Pathways in High-Energy Physics
held by the Center for Theoretical Studies, University of Miami,
Coral Gables, Florida, January 17-21, 1977

PREFACE

These proceedings contain the papers presented during the 14th annual High Energy Physics meeting convened under the aegis of Orbis Scientiae 1977. The title "Deeper Pathways in High Energy Physics" was adopted to indicate either further penetration into the nature of the structure of the elementary particle or the deepening of the continuously trodden path that gets gradually deeper and deeper, evolving into a trench. In the latter instance, the visibility of the real nature of elementary particles may be getting dimmer and dimmer. It is hoped that some of the papers in these proceedings do, indeed, contain further revelations on the "deeper" nature of elementary particles. We must not be overly charmed with "charm" even if it may fit the data of the current generation of e^+ and e^- experiments. It would be much less than prophetic to say that a complete and totally satisfactory theory comprising the entire physical reality is yet to be discovered, and there is presumably no paper of that kind in these proceedings. Despite this short-coming, the editors do not wish to hide their admiration for the caliber of the papers contributed by the participants of Orbis Scientiae 1977.

Appreciation is extended to Mrs. Helga S. Billings, Mrs. Elva Brady, and Ms. Yvonne L. Leber for their skillful typing of the proceedings, which they have performed with great enthusiasm and dedication.

Orbis Scientiae 1977 received some support from the Energy Research and Development Administration.

The Editors

ORBIS SCIENTIAE 1977 PARTICIPANTS

CONTENTS

THE DYNAMICS OF STREAMS OF MATTER

P.A.M. Dirac

Florida State University

Tallahassee, Florida

If one builds up a theory based on point particles
interacting with fields, the particles cause singularities in the fields and lead to awkward mathematics. One
is led to doubt whether the basis is really suitable for
a description of nature. One can evade the difficulties
by getting away from the point particle model. One can
have instead a continuous stream of matter, each element
of which interacts with the field. One can arrange for
the field to act on each element of matter in the same
way as it acts on a point particle, while each element
of matter disturbs the field like a point particle would,
but owing to the smoothing out coming from the continuous distribution there is no singularity in the
field.

In my talk at this conference last year, I showed
how the method could be applied to a continuous distribution of charged matter interacting with the electromagnetic field. Our problem now is to develop the
method to bring in Pb's (Poisson Brackets) so that the
equations of motion can be put in the Hamiltonian form.

The continuous matter is considered as a dust, each speck of which interacts with the field like a particle would. With a somewhat different notation from last year, let u_0 denote the density of the dust, the mass per unit volume, and let $u_r (r = 1,2,3)$ denote the flow, the mass crossing unit area per unit time. Conservation requires

$$\frac{\partial u_0}{\partial x_0} + \frac{\partial u_r}{\partial x_r} = 0 \ . \tag{1}$$

Let $v_\mu (\mu = 0,1,2,3)$ denote the velocity 4 - vector, satisfying

$$v_0 v_0 = 1 + v_r v_r \ . \tag{2}$$

Then
$$\frac{u_r}{u_0} = \frac{v_r}{v_0} \ . \tag{3}$$

We are writing the equations with all the suffixes downstairs, which is convenient when one wants to discuss the dynamics and not emphasize the relativity.

Let p_r denote the momentum per unit mass, so $u_0\, p_r$ is the momentum per unit volume. We shall have $u_0\, p_r$ as an ordinary field function, but p_r itself, like v_r, is a field function defined only where $u_0 \neq 0$.

If the matter is charged and e denotes the charge per unit mass, we get from the equations for a point particle

$$p_r = v_r + e A_r \ . \tag{4}$$

Similarly, if p_0 is the energy density per unit mass,

$$p_0 = v_0 + e A_0 \ . \tag{5}$$

The basic variables describing the stream of matter
at a certain time can be taken to be u_0, p_r for all
values of x_1, x_2, x_3. The other variables, namely v_r,
u_r, v_0, p_0, can be expressed in terms of these and the
A_μ by means of the equations (2), (3), (4), (5). Our
problem is to find the Pb's between the basic variables.
Those between the others will then follow.

Let us label each speck of dust by three para-
meters ξ_1, ξ_2, ξ_3 chosen so that there is unit mass per
unit $d\xi_1\, d\xi_2\, d\xi_3 = d^3\xi$. Then the mass per unit volume
is

$$u_0 = \frac{\partial(\xi_1,\xi_2,\xi_3)}{\partial(x_1\ x_2\ x_3)} \ . \tag{6}$$

Each speck of dust has three coordinates $x_r(\xi)$ and
three momenta $p_r(\xi)$ per unit mass. The standard Pb relations between them are, if we use the abbreviations x_r'
for $x_r(\xi')$, x_r'' for $x_r(\xi'')$ and so on,

$$[x_r',\ x_s''] = 0 \ , \quad [p_r',\ p_s''] = 0 \ ,$$

$$[x_r',\ p_s''] = \delta_{rs}\ \delta(\xi' - \xi'') \ . \tag{7}$$

These relations are the starting-point of our theory.
We have to eliminate all reference to the ξ's, which are
non-physical quantities, to get the relations we want.

The three x's are functions of the three ξ's.
These functions may be inverted to give us the three ξ's
as functions of the three x's, for regions of space
where the density of the matter does not vanish. In
such regions, if we have a field quantity V, we may
locate it either by ξ-values, so as to get $V(\xi)$, or by
x-values, so as to get $V(x)$. $V(x)$ is a linear function

of the $V(\xi)$, with coefficients which are determined by
the functions $x(\xi)$, so the coefficients are functionals
of the $x(\xi)$ for all ξ. Similarly $V(\xi)$ is a linear
function of the $V(x)$, with coefficients which are
functionals of the $\xi(x)$ for all x, or functionals of
the $x(\xi)$ for all ξ.

We shall use the following notation. Specific
values for the ξ's will be denoted by ξ' or ξ'', and
$V(\xi')$, $V(\xi'')$ will be written V', V''. This is in agreement with the notation of (7). Specific values for
the x's will be $x_r = a_r$ or $x_r = b_r$, and $V(a)$, $V(b)$ will
be abbreviated to V^a, V^b. One must keep a sharp
distinction between them and V', V''.

The Pb relations that we are seeking must not refer
to ξ-values, but to x-values. Thus they are

$$[u_0^a, u_0^b], \qquad [u_0^a, p_r^b], \qquad [p_r^a, p_s^b]. \qquad (8)$$

The first of these is easily worked out. The quantities u_0^a, u_0^b are determined by the functions $x(\xi)$, so
they are functionals of the x_r' for all ξ'. Since the
latter all have zero Pb, from the first of the relations
(7), we see that

$$[u_0^a, u_0^b] = 0. \qquad (9)$$

For the other Pb's (8) we need a more extensive calculation.

If we make a small change Δx_r in the x's as
functions of the ξ's, what is the corresponding change
$\Delta \xi_r$ in the ξ's as functions of the x's? To answer this
question, it is convenient first to consider the ξ's as
functions of the x's and the x's as functions of some

new variables ξ^*. Let us change each ξ_r as a function
of the x's by the amount $\Delta\xi_r$. Let us also change each
x_s as a function of the ξ^*'s by the amount Δx_s. This
leads to a change in the ξ_r of amount $(\partial\xi_r/\partial x_s)\,\Delta x_s$.
The total change in ξ_r is thus

$$\Delta\xi_r + (\partial\xi_r/\partial x_s)\Delta x_s. \tag{10}$$

If we now couple the two function changes so as to maintain $\xi^* = \xi$, the total change (10) must vanish and we
get

$$\Delta\xi_r = - (\partial\xi_r/\partial x_s)\Delta x_s. \tag{11}$$

Note the - sign here. A more careless argument would
not give it.

With our regular notation, when we vary the
functions x_r of the ξ's, the change at ξ' is $\Delta(x_r')$.
It is also $(\Delta x_r)'$. So we may write it simply $\Delta x_r'$.
This variation corresponds to a shift of the distribution of matter, and thus to a certain variation of the
dynamical variables.

One can generate a small change in the dynamical
variables by taking their Pb's with some definite
quantity. Thus if we set up

$$G = \int p_r'\,\Delta x_r'\,\ d^3\xi', \tag{12}$$

in which the quantities $\Delta x_r'$ are just numbers and so have
zero Pb with everything, we get from (7)

$$[\xi_r', G] = \Delta x_r', \qquad [p_r', G] = 0,$$

which is just the above shift. The change in the ξ_r^a is then

$$\Delta\xi_r^a = [\xi_r^a, G]$$

$$= \int [\xi_r^a, p_s'] \, \Delta x_s' \, d^3\xi'. \qquad (13)$$

Now the value of $\Delta\xi_r$ where $x = a$ is just the value of the right-hand side of (11) where $x = a$, which is

$$- (d\xi_r^a/da_s) \int \Delta x_s' \, \delta(\xi^a - \xi') d^3\xi'. \qquad (14)$$

Since (13) and (14) must be equal for any $\Delta x_s'$, we get

$$[\xi_r^a, p_s'] = - (d\xi_r^a/da_s) \, \delta(\xi^a - \xi'). \qquad (15)$$

Let ρ be some function of the ξ - variables. The total ρ for the whole distribution of matter is

$$\int \rho' \, d^3\xi' = \int \rho^a \, u_0^a \, d^3a.$$

Now this total is unaffected by the shift, so

$$[\int \rho^a \, u_0^a \, d^3a, G] = 0$$

for any $\Delta x_r'$, and hence

$$[\int \rho^a \, u_0^a \, d^3a, p_s'] = 0. \qquad (16)$$

We have

$$[\rho^a, p_s'] = \frac{\partial\rho^a}{\partial\xi_r^a} [\xi_r^a, p_s'] = - \frac{\partial\rho^a}{\partial\xi_r^a} \frac{\partial\xi_r^a}{\partial a_s} \delta(\xi^a - \xi')$$

$$= - \frac{\partial\rho^a}{\partial a_s} \delta(\xi^a - \xi'). \qquad (17)$$

(16) gives $\displaystyle\int \{[\rho^a, p_s'] u_0^a + \rho^a [u_0^a, p_s']\} d^3a = 0,$

and hence, by partial integration

$$\int \{\rho^a \frac{\partial}{\partial a_s} (u_0^a \delta(\xi^a - \xi')) + \rho^a [u_0^a, p_s']\} d^3a = 0. \qquad (18)$$

This holds for any function ρ^a, so

$$[u_0^a, p_s'] = - \frac{\partial}{\partial a_s} (u_0^a \delta(\xi^a - \xi')).$$

We have

$$p_s^b = \int p_s' \delta(\xi' - \xi^b) d^3\xi'. \qquad (19)$$

Now $\delta(\xi' - \xi^b)$ is a functional of the x_r'' for all ξ'' and so has zero Pb with u_0^a. Thus

$$[u_0^a, p_s^b] = \int [u_0^a, p_s'] \delta(\xi' - \xi^b) d^3\xi'$$

$$= - \frac{\partial}{\partial a_s} \{u_0^a \delta(\xi^a - \xi^b)\}.$$

Note the formula

$$u_0^a \delta(\xi^a - \xi^b) = \delta(a - b), \qquad (20)$$

which one checks by using (6) and seeing that each side gives 1 when multiplied by d^3a and integrated. Thus

$$[u_0^a, p_s^b] = - \frac{\partial}{\partial a_s} \delta(a - b). \qquad (21)$$

All reference to the ξ variables has disappeared from this result.

It remains to evaluate $[p_r^a, p_s^b]$. Using equations

like (19), we have

$$[p_r^a, p_s^b] = \iint [p_r' \delta(\xi' - \xi^a), p_s'' \delta(\xi'' - \xi^b)] d^3\xi' \, d^3\xi''$$

$$= \iint \{[\delta(\xi' - \xi^a), p_s''] p_r' \delta(\xi'' - \xi^b)$$

$$+ [p_r', \delta(\xi'' - \xi^b)] p_s'' \delta(\xi' - \xi^a)\} d^3\xi' \, d^3\xi''.$$

The first term in the integrand here gives, by an application of (17) with $\rho^a = \delta(\xi' - \xi^a)$ and p_s' replaced by p_s'',

$$-\iint \frac{\partial \delta(\xi' - \xi^a)}{\partial a_s} \delta(\xi^a - \xi'') p_r' \delta(\xi'' - \xi^b) d^3\xi' \, d^3\xi''$$

$$= - \delta(\xi^a - \xi^b) \frac{\partial}{\partial a_s} \int \delta(\xi^a - \xi') p_r' \, d^3\xi'$$

$$= - \frac{1}{u_0^a} \delta(a - b) \frac{\partial p_r^a}{\partial a_s}.$$

The second term may be evaluated in the same way, and we get finally

$$[p_r^a, p_s^b] = \frac{1}{u_0^a} \left(\frac{\partial p_s^a}{\partial a_r} - \frac{\partial p_r^a}{\partial a_s}\right) \delta(a - b). \qquad (22)$$

It is rather surprising that the momentum variables do not have zero Pb's. The reason is that the present momentum variables refer to definite locations in space instead of definite pieces of matter.

We have now obtained all the basic Pb's (8), namely (9), (21) and (22). The Hamiltonian is the total energy expressed in terms of the variables u_0^a, p_r^a and is

$$H = \int p_0^a \, u_0^a \, d^3 a \tag{23}$$

$$= \int \{[1 + (p_r^a - eA_r^a)(p_r^a - eA_r^a)]^{1/2}$$

$$+ eA_0^a\} \, u_0^a \, d^3 a. \tag{24}$$

The A_μ here may be either the potentials of a given external field or the potentials in terms of the field dynamical variables according to quantum electrodynamics, in which case one must add on to H a term representing the energy of the electromagnetic field to get the total Hamiltonian for the system.

It is interesting to see how the unusual Pb's we have obtained lead to the correct equations of motion. From (21) and (4)

$$[u_0^a, v_r^b] = - \frac{\partial}{\partial a_r} \delta(a - b).$$

With the help of (2),

$$[u_0^a, v_0^b] = - \frac{v_r^b}{v_0^b} \frac{\partial}{\partial a_r} \delta(a - b).$$

With (3) and (5)

$$[u_0^a, p_0^b] = - \frac{u_r^b}{u_0^b} \frac{\partial}{\partial a_r} \delta(a - b).$$

Thus from (23)

$$[u_0^a, H] = \int [u_0^a, p_0^b] \, u_0^b \, d^3b$$

$$= -\int u_r^b \frac{\partial}{\partial a_r} \, \delta(a - b) \, d^3b$$

$$= -\frac{\partial}{\partial a_r} \int u_r^b \, \delta(a - b) \, d^3b = -\frac{\partial u_r^a}{\partial a_r}.$$

Thus the equation of motion

$$\frac{\partial u_0^a}{\partial x_0} = [u_0^a, H]$$

leads to the conservation equation (1).

By similar arguments we find

$$[p_r^a, p_0^b] = \frac{v_s^b}{v_0^b} \frac{1}{u_0^b} \left(\frac{\partial p_s^a}{\partial a_r} - \frac{\partial p_r^a}{\partial a_s} \right) \delta(a - b)$$

and so

$$[p_r^a, H] = \int \{ [p_r^a, p_0^b] \, u_0^b + [p_r^a, u_0^b] \, p_0^b \} \, d^3b .$$

$$= \frac{v_s^a}{v_s^a} \left(\frac{\partial p_s^a}{\partial a_r} - \frac{\partial p_r^a}{\partial a_s} \right) - \frac{\partial p_0^a}{\partial a_r} .$$

The equation of motion $\partial p_r^a / \partial x_0 = [p_r^a, H]$ thus leads to

$$v_0^a \left(\frac{\partial p_r^a}{\partial x_0} + \frac{\partial p_0^a}{\partial a_r} \right) = v_s^a \left(\frac{\partial p_s^a}{\partial a_r} - \frac{\partial p_r^a}{\partial a_s} \right).$$

If one now uses (4) and (5) to express the p's in terms of the v's and A's one gets just Lorentz's equation of motion for each element of matter.

One can easily extend the theory to several streams of matter, that may be overlapping. Each stream will

have its own dynamical variables u_0, p_r, satisfying the
Pb relations (9), (21), (22), and two variables referring
to different streams will of course have zero Pb. The
total Hamiltonian will be the sum of the Hamiltonians
for the various streams plus the Hamiltonian for the
electromagnetic field, and the different streams will
interact through each one interacting with the field.

   The theory can also deal with streams of monopoles,
each element of monopole having a string attached to it.
It is now necessary that the monopoles and the strings
do not overlap with charged matter, otherwise there is
no action principle for the motion and the Hamiltonian
formalism fails.

ORIGIN OF SPIN

Behram Kursunoglu

Center for Theoretical Studies

University of Miami, Coral Gables, Fla. 33124

I. INTRODUCTION

Last August, Alan Krisch invited me to the symposium on High Energy Physics with Polarized Beams and Polarized Targets in Argonne National Laboratory to present a paper on my work concerning the structure of elementary particles and its relation to spin. Experiments by Krisch, et al.[1] with polarized proton beams scattered from polarized targets pointed to a strong possibility that protons do have an inner structure with different interaction regions. Their experimental data provided a basic support for the significant role of spin-spin forces at high energy scattering.

The total cross sections which measure the spin dependence are related to two types of quantities:

(i) Transversity cross sections (beam and target polarization are both transverse to the incident momentum) in which the quantity

$$\Delta\sigma_{tot}^{T} = \sigma_{tot}(\uparrow\downarrow) - \sigma_{tot}(\uparrow\uparrow) \quad , \tag{1}$$

13

can be surprisingly large.

(ii) Helicity cross sections (beam and target polariza-
      tion are both parallel to the incident momentum)
      in which the quantity

$$\Delta\sigma_{tot}^{H} = \sigma_{tot}(\rightarrow\leftarrow) - \sigma_{tot}(\rightarrow\rightarrow) \quad , \tag{2}$$

also has a behavior similar to transversity case.
According to the theory, to be presented shortly,
the quantities defined by (1) and (2) may be expected to increase as well as decrease with increasing energy.

Spin effects have also been observed in neutron-
proton elastic scattering and, in particular, they find
(presumably because of the suppression of Coulomb force)
totally different behavior between the n-p and p-p spin
orbit interactions. We shall see that the fundamental
reason for this difference of behavior may be found in
the fact that the n-p scattering is a four-body process,
compared to p-p scattering which is only a two-body
process. Thus, based on these considerations, the nucleon
or the concept of isospin is not necessarily an important internal symmetry for high energy scattering
but, perhaps, it is only for nuclear physics where isospin plays a major role. It will be seen that the
existence of spin is a consequence of a particle's
layered magnetic structure. At high energy scattering,
the Coulomb coupling can partially inhibit the effect
of the structure and hence the spin effect.

According to the theory for particles with layered
magnetic structure we must, at sufficiently high energies, such as those in the Argonne experiments, expect
for all p-p cross sections - elastic or inelastic - large

spin effects. The expectation is due to a close relationship between spin and structure of all elementary particles.[2,3]

The basic premises of the theory are quite novel and are not, at all, based on the current favorites of the establishment. I am, thus, appearing before you as an activist against the establishment. However, in this lecture I cannot give you more than an infinitesimal amount on the fundamentals of the theory. Please bear with me for a few moments to just outline a brief history, and then I shall quickly come to the physics of the theory. It was begun in 1950 by Einstein, and also by Schrödinger, who proposed a set of 16 field equations for the fundamental field quantities

$$\hat{g}_{\mu\nu} = g_{\mu\nu} + q^{-1}\, \Phi_{\mu\nu} \quad , \tag{3}$$

instead of the usual 10 field equations of general relativity for the symmetric tensor $g_{\mu\nu}$ describing gravitational field, where the field $\Phi_{\mu\nu}$ is measured in units of the constant q. The symmetric and antisymmetric parts $g_{\mu\nu}$ and $\Phi_{\mu\nu}$ of $\hat{g}_{\mu\nu}$ represent generalizations of gravitational and electromagnetic fields, respectively. A few months after the publication of these theories, they were challenged by W. Pauli and M. Born; the former on the basis of the absence of the "quantum action $\hbar$" in these theories, and the latter who assigned to them the status of "programs" rather than physical theories. The real stumbling block was later found to be the absence of a "fundamental length" in the Einstein and Schrödinger theories. In fact, such a property (i.e., a fundamental length r_o relating gravitational and electromagnetic fields) was intrinsically present in the

formalism and was missed only because of historical
prejudices (as related to me by Einstein in an October
19, 1953 conversation in his house at Princeton) experienced in the formulation of general relativity
where the so-called cosmological constant had led to
some undesirable cosmological consequences. In a
paper published in Physical Review (1952) I have,
uniquely, established a third and more general version
of Einstein's nonsymmetric theory where a fundamental
length r_0 appeared as a natural geometrical consequence
of the Bianchi-Einstein differential identities. This
length r_0, besides not being a cosmological constant,
formed the basis of a correspondence principle of the
theory, since for $r_0 = 0$ the new version of the
generalized theory of gravitation reduced to the field
equations of general relativity. The lack of a correspondence principle in the Einstein and Schrödinger
theories did not allow for a physical interpretation of
their theories and were, thus, quickly cast aside as another futile attempt for the construction of a unified
field theory.[2,3]

II. ORBITONS

By brute force and hard work I have at last
succeeded, in the past two years, for the case of
spherically symmetric fields, to obtain a few exact
solutions.[3] Naturally, at this time I shall not bother
you with the details, but will tell you about the results.
For the spherically symmetric time-independent space
time, I obtained four different massive solutions corresponding to proton, electron, e-neutrino, μ-neutrino.
The latter two are identified through the known appropriate symmetries. I named these solutions as <u>orbitons</u>

since each one of them has a structure consisting of
stratified layers with decreasing amounts of magnetic
charges, as we recede from the origin, with alternating
signs. The total magnetic charge in an orbiton is conserved according to

$$\sum_{0}^{\infty} g_n = 0 \quad , \tag{4}$$

where the partial magnetic charge g_n, with increasing
distance from the origin, obeys the limiting process

$$\lim_{n \to \infty} g_n = 0 \quad , \tag{5}$$

and where

$$g_n \, g_{n+1} < 0 \quad , \quad n = 0,1,2,\ldots \quad . \tag{6}$$

    All four orbitons (p,e,ν_e,ν_μ) and the correspon-
ding antiorbitons $(\bar{p},e^+,\bar{\nu}_e,\bar{\nu}_\mu)$ carry the same spectrum
of magnetic charges g_n. However, the densities of mag-
netic charge in the n-th layer differ for each orbiton.
The stability of an orbiton results from the equality
of the sum of the gravitational attraction (measured by
Gm^2 where $m^2 \sim \frac{\hbar c}{G}$), plus magnetic attraction of the layers
to the magnetic repulsion in each layer with continuous
magnetic charge distribution in the case of charged
orbitons; like p and e, the balance for stability includes the electrostatic repulsion of the orbiton. In
the absence of all g_n (n=0,1,2,...), the orbitons reduce
to point particles of the classical theory and all the
divergences of the various physical processes reappear.
Thus, the origin regains its classical and quantum field
theoretic sins where, instead of being zero as in the

present theory, all the fields assume infinite values.

III.   SPIN ANGULAR MOMENTUM

The energy density contains contributions of the form

$$4\pi \ r_o^2 \ [\tfrac{1}{2} \ (s^\mu s^\nu + \tfrac{1}{2} \ g^{\mu\nu} \ s^\rho s_\rho)] \quad ,$$

where

$$r_o^2 = \frac{2G}{c^4} \ (e^2 + g_n^2) \quad , \tag{7}$$

and where

$$s^\mu = \frac{1}{c} \ g \ \delta^\mu \quad , \tag{8}$$

is the conserved spin angular momentum density. The existence of the conserved magnetic current density δ^μ is due to the antisymmetric part of the affine connection $\Gamma^\rho_{\mu\nu}$. Thus, the spin angular momentum

$$S = \frac{(-1)^s}{c} \int g \ \delta^\mu \ d\sigma_\mu = (1)^s \ \frac{1}{c} \ \sum_{n=o}^{\infty} \ g_n^2 = (-1)^s \ \frac{1}{2} \ \hbar \tag{9}$$

is entirely a consequence of the existence of a torsion tensor $\Gamma^\rho_{[\mu\nu]}$ of the space time in the magnetic structure. In the relation (9) we have used

$$g_n^2 = \gamma_n \ \hbar c \quad , \tag{10}$$

where

$$\lim_{n\to\infty} \gamma_n = 0, \ \sum_o^{\infty} \gamma_n = \frac{1}{2}, \ s=0,1 \ \text{(spin up, spin down)}. \tag{11}$$

Hence, for the existence of spin angular momentum we

must have a density of matter. The latter, as will be
seen later, is the case for the spin angular momentum
of the photon.

Since all orbitons contain the same spectrum g_n
the addition of spin angular momenta for any number of
orbitons is easily obtained according to the rule

$$S_z = \pm \left[\frac{1}{2} \hbar \left(\sum_0^\infty \frac{1}{c} (g_n + g_n' + \ldots)^2 \right) \right]^{\frac{1}{2}} , \qquad (12)$$

and yields the same results as in quantum theory of
angular momentum. From the ongoing, we see that not
only spin angular momentum, but the magnetic structure
of the orbiton is due to a nonzero $\hbar$ and, conversely, the
existence of quantum action $\hbar$ is a consequence of the
magnetic structure of matter in general.

The fundamental length r_0 in (7) was computed from
the spherically symmetric solutions of the field equations in terms of the two constants of integration λ_0
and ℓ_0 according to

$$r_0^2 = \sqrt{(\lambda_0^4 + \ell_0^4)} , \qquad (13)$$

where the lengths ℓ_0 and λ_0 are given by

$$\ell_0^2 = \frac{2G}{c^4} g \sqrt{(e^2 + g^2)} \quad , \quad \lambda_0^2 = \frac{2G}{c^4} e \sqrt{(e^2 + g^2)} . \qquad (14)$$

In the fundamental field tensor

$$\hat{g}_{\mu\nu} = g_{\mu\nu} + q^{-1} \phi_{\mu\nu} , \qquad (15)$$

the constant q has the dimensions of an electric field
and is related to r_0 by the relation

$$r_o^2 \, q^2 = \frac{c^4}{2G} \quad , \tag{16}$$

where we further have

$$\ell_o^2 = g \, q^{-1} \, , \quad \lambda_o^2 = e \, q^{-1} \quad . \tag{17}$$

Thus, in the correspondence limit, $r_o = 0$ or $q = \infty$, for λ_o and ℓ_o to tend to zero e and g must remain finite. However, if we intend to obtain Nördstrom solutions in the limit of asymptotically large distances, then the electric charge e must assume very large values so that the observed electric charge is then the "renormalized value" of the one corresponding to asymptotic limits.

Now both of the relations (7) and (14), for given values of ℓ_o and λ_o, yield a connection between electric charge and magnetic charge. In particular, the lengths ℓ_o as well as λ_o can, in principle, be computed numerically from the solutions of the field equations inside the magnetic layers of the orbitons. In the latter case, the solutions of the first equation in (14) for the magnetic charge yields the relation

$$g_n^2 = \frac{1}{2} \, e^2 \, [\sqrt{(1+\omega_n^2)} - 1] \quad , \tag{18}$$

where

$$\omega_n = \frac{c^4 \ell_{on}^2}{e^2 \, G} \quad . \tag{19}$$

Hence, the infinite sum

$$\frac{1}{c} \, \Sigma \, g_n^2 = \frac{1}{2} \, \frac{e^2}{c} \, \sum_{n=o}^{\infty} \, [\sqrt{(1+\omega_n^2)} - 1] \quad ,$$

yields for the fine structure constant the series

relation

$$\sum_{o}^{\infty} [\sqrt{(1+\omega_n^2)}-1] = (\frac{\hbar c}{e^2}) \quad , \qquad (20)$$

where the infinite series involve elliptic functions of the first and second kind. It is hoped that the series in question will, in fact, sum up to the observed value of the fine structure constant.

For the case of neutrinos, setting e = 0, we obtain

$$g_n^2 = \frac{1}{2} \frac{c^4}{G} \ell_{on}^2 \quad . \qquad (21)$$

Hence, for the neutrino spin angular momentum to obtain the correct value we must have

$$\sum_{n=o}^{\infty} \ell_{on}^2 = \frac{\hbar G}{c^3} \text{ (Planck length)}^2. \qquad (22)$$

IV.   VACUUM

A novel and surprising result of the theory comes from the existence of two electric and two magnetic currents which are conserved according to

$$\int j_e^\mu \, d\sigma_\mu = \pm e \quad , \quad \int \delta^\mu \, d\sigma_\mu = 0 \quad , \qquad (23)$$

$$\int j_o^\mu \, d\sigma_\mu = 0 \quad , \quad \int \zeta^\mu \, d\sigma_\mu = 0 \quad , \qquad (24)$$

where the two currents j_o^μ and ζ^μ in (24) refer to neutral electric and neutral magnetic current densities of the vacuum. The two currents j_e^μ and δ^μ in (23) refer to ordinary electric current and the magnetic current of the orbiton structure. The energy density of the field can be split up into three parts in the form

$$T^{\nu}_{\mu} = T_{o\mu}^{\nu} + T_{oo\mu}^{\nu} + T^{\nu}_{\mu} \text{ (interaction) } , \qquad (25)$$

where T^{ν}_{μ} is conserved covariantly as

$$T^{\nu}_{\mu}|_{\nu} = 0 \qquad , \qquad (26)$$

and where

$$\int T^{\nu}_{oo\mu} \, d\sigma_{\nu} = 0 \qquad . \qquad (27)$$

Thus, the energy tensor $T^{\nu}_{oo\mu}$ represents vacuum energy density. All the currents and also the energy density are determined only in terms of the solutions of the field equations.

The vacuum consists of equal amounts of positive energy and negative energy orbitons. The total energy, total spin angular momentum, total electric charge, total magnetic charge of the vacuum is zero. The vacuum properties are invariant and therefore an external field can perturb the vacuum to produce pairs with positive energy. Each pair must carry zero total spin angular momentum and total zero electric charge and zero magnetic charge.

V. THE THREE FUNDAMENTAL INTERACTIONS

There exist only three fundamental interactions of fields and orbitons: (i) Long range gravitational inter-actions whose coupling strength at the orbiton core is proportional to Gm^2 where m is of the order of the Planck mass $\sqrt{(\frac{\hbar c}{G})}$; (ii) Long range electromagnetic inter-actions whose coupling strength in the neighborhood of an orbiton is proportional to e^2; (iii) Short range magnetic interactions extend through the stratified

magnetic layers and their coupling strengths are
measured by g_n^2 (n=0,1,2...). In the orbiton core, the
coupling strength is measured by $e^2 + g_n^2$. Thus, all
three fundamental interactions are of the similar orders
of magnitude.

All other interactions, strong, weak, atomic,
molecular, etc., are expressible in terms of the three
fundamental interactions. Because of the magnetic
couplings, it is seen that orbitons have different interaction regions. These interaction regions can be
classified in terms of the magnetic number n and its
energy dependence. For neutrinos the electric charge
density is zero, but for electron and proton the
electric charge density distribution extends beyond
magnetic charge density, but falls off sharply as $\frac{\lambda_o^4}{r^5}$.
The coupling of two orbitons say an electron and
proton with a relative energy E of the electron involves
an electric and magnetic interaction measured to first
order by $e^2+g_n^2$. In general, the coupling is nonlinear
in their fields and in the coupling constant $e^2+g_n^2$. At
low energies, only the high magnetic number $n_o(E)$ and,
hence, small g_n^2 participate in the interaction, and the
dominant coupling is of electrical character. However,
the electrostatic attraction, despite many alternating
magnetic attractions, is depressed by the corresponding
alternating magnetic repulsions. Thus, the net state is
an oscillating electron bound to the proton. For
higher energies, i.e., for $n(E) < n_o(E)$, the thicker
magnetic layers come into play and, in fact, at threshold
energies of a few kev we may expect the creation from
the vacuum of a ν_e and $\bar{\nu}_e$ pair. The resulting reaction
can proceed as

$$p+e \rightarrow p+e + \nu_e+\bar{\nu}_e = (p+e+\bar{\nu}_e)+ \nu_e$$

$$= n + \nu_e \quad .$$

Hence, we see that a neutron can be described as a deeply bound state of p, e and $\bar{\nu}_e$, viz.,

$$n = (p+e+\bar{\nu}_e) \quad , \tag{28}$$

where the arrows indicate spins.

Other simple examples can be obtained from storage ring reactions. Thus, from

$$e^+ + e^- \rightarrow e^+ + e^- + \nu_e + \bar{\nu}_e = (e^+ \nu_e)+(e^- \nu_e)= \pi^+ + \pi^-,$$

where the pions $\pi^\pm$ are described as deeply bound states of electrons and neutrinos according to

$$\pi^+ = (e^++\nu_e) \quad , \quad \pi^- = (e^-+\nu_e) \quad . \tag{29}$$

At still higher energies, we may consider creation of the both pairs ν_e and ν_μ and obtain the reactions

$$e^+ + e^- = e^+ + e^- + \nu_e + \bar{\nu}_e + \nu_\mu + \bar{\nu}_\mu$$

$$= (e^+ \nu_e \bar{\nu}_\mu) + (e^- \bar{\nu}_e \nu_\mu) = \mu^+ + \mu^- \quad ,$$

where now the muons are described as deeply bound states of two orbitons and one antiorbiton for μ^- and two anti-orbitons and one orbiton for μ^+; viz.,

$$\mu^+ = (e^+ + \nu_e + \bar{\nu}_\mu) \quad , \quad \mu^- = (e^- + \bar{\nu}_e + \nu_\mu) \quad . \tag{30}$$

Thus, in the collision of the antiparallel polarized electron and positron beams the branching ratio for the production of pions to the production of muons is greater than 1.

In the case of an encounter of parallel polarized electron-positron beams, in addition to the above reactions, the annihilation of the pair e^+ and e^- is, because of the correlation of spin and sequence of magnetic charge layers, inhibited. This allows many deeply bound states of e^+ and e^- where the magnetic coupling (mostly magnetic repulsion) delays the annihilation of the pair through their electric attraction. These 1^- states of this theory correspond to ψ-particles. The actual calculation of the decay rates of the above particle states may, in principle, be based on the geometry of space-time. The concept of general covariance which pertains to gravitational field in general relativity can, in the present theory, be extended to the unified field describing all the fundamental interactions. The problem of scattering of particles, bound states of particles, decays and lifetimes of particles, can be described as solutions of the field equations for special geometrical symmetries of space-time. The latter can be achieved by a choice of coordinate systems. The spherically symmetric space-time of the four different orbitons can be replaced by, for example, spheroidally or ellipsoidally symmetric space-time of neutron (three-body process) or of muon (three-body process). In the same way, a cylindrically symmetric space-time may be describing integral spin particles like, for example, pions, kaons etc. The scattering cross section may be obtained as solutions of the field equations in a cylindrically symmetric coordinate system.

VI.   MASS OF ORBITONS

From the derivation of the equations of motion, we obtain both positive and negative inertial mass defined as

$$\pm\ Mc^2 = \frac{1}{2}(\mp mc^2) - (\mp E_s)\ ,\qquad (31)$$

where m is of the order of Planck mass and where E_s is the finite self-energy and differs very little from the first term. In the correspondence limit $r_o = 0$ the self-energy E_s tends to infinity and yields the classical or quantum field theoretic mass. In the expression (31), mc^2 plays also the role of binding energy of an orbiton which is of the order of 10^{21} Mev. Thus, the observed mass is obtained as the difference of two unobserved large masses.

VII.   PHOTON

From the field equations of the theory[2,3]

$$G_\mu^\nu - \frac{1}{2}\ \delta_\mu^\nu\ G = \frac{8\pi G_o}{c^4}\ T_\mu^\nu\ ,\ \Phi_{\mu\nu} + r_o^2\ R_{[\mu\nu]} = F_{\mu\nu}\ ,\ \hat{g}^{[\mu\nu]}_{\ \ \ ,\nu} = 0\ ,\qquad (32)$$

and the definitions

$$[\sqrt{(-g)}\phi^{\mu\nu}]_{|\nu} = 4\pi j_e^\mu\ ,\ [\sqrt{(-g)}r_o^2\ R^{[\mu\nu]}]_{|\nu} = 4\pi j_o^\mu\qquad (33)$$

for electric currents,

$$(\sqrt{(-g)}f^{\mu\nu})_{,\nu} = 4\pi\ \delta^\mu\ ,\ (\sqrt{(-g)}\psi^{\mu\nu})_{,\nu} = 4\pi\ \zeta^\mu\qquad (34)$$

for magnetic currents, we can determine all the interactions of particles and fields, where

$$F_{\mu\nu} = \partial_\mu A_\nu - \partial_\nu A_\mu \; , \quad \Psi_{\mu\nu} = \frac{f_{\mu\nu} + \Lambda \; \Phi_{\mu\nu}}{\sqrt{(1+\Omega-\Lambda^2)}} = \partial_\mu B_\nu - \partial_\nu B_\mu \; . \quad (35)$$

We further have the definitions

$$\hat{g}^{\mu\nu} = \sqrt{(-g)} \; \frac{(1+\tfrac{1}{2}\Omega) g^{\mu\nu} + T^{\mu\nu} + \phi^{\mu\nu} - \Lambda f^{\mu\nu}}{\sqrt{(1+\Omega-\Lambda^2)}} \; ,$$

$$\hat{g}^{\mu\nu} = \frac{(1+\tfrac{1}{2}\Omega) g^{\mu\nu} + T^{\mu\nu} + \phi^{\mu\nu} - \Lambda f^{\mu\nu}}{1+\Omega-\Lambda^2} \; . \quad (36)$$

The equations (33) and (34), in terms of potentials, can be written as

$$\sqrt{(-g)} [G_\nu^\mu A^\nu - g^{\rho\sigma} A^\mu|_{\rho\sigma}] = 4\pi (j_e^\mu + j_o^\mu) \; , \quad A^\mu|_\mu = 0 \; , \quad (37)$$

$$\sqrt{(-g)} [G_\nu^\mu B^\nu - g^{\rho\sigma} B^\mu|_{\rho\sigma}] = 4\pi \; \zeta^\mu \; , \quad B^\mu|_\mu = 0 \; . \quad (38)$$

The vacuum field is represented by $\Psi_{\mu\nu}$ where the potentials B_μ have short range character. The field $\Psi_{\mu\nu}$ arises from the coupling between $f_{\mu\nu}$ and $\Phi_{\mu\nu}$, which, for $\Omega=0$ and $\Lambda=0$, decouple and we obtain $\Psi_{\mu\nu} = f_{\mu\nu}$.

The energy tensor T_μ^ν is conserved according to

$$T_\mu^\nu|_\nu = 0 \; , \quad (39)$$

where

$$T_\mu^\nu = q^2 [(\tfrac{1}{2} b_\rho^\rho - 1)\delta_\mu^\nu - b_\mu^\nu] + r_o^2 \; M_\mu^\nu \; . \quad (40)$$

Hence,

$$\Psi_{\mu\nu} \; s^\nu + \frac{r_o^2}{4\pi} \; (\sqrt{(-g)} M_\mu^\nu)|_\nu = 0 \; , \quad (41)$$

where the first terms represent magnetic force density.

Thus, for $r_o = 0$ magnetic force density vanishes and the field equations (32) reduce to the field equations of general relativity where the classical electromagnetic field (or waves) act as sources of the gravitational field. In that case, matter is represented as the singularities of the gravitational and electromagnetic fields.

In the equation (41), the magnetic current δ^μ interacts with the vacuum short range field $\Psi_{\mu\nu}$ and gives rise to creation of particle antiparticle pairs represented by the force density $\frac{1}{4\pi} r_o^2 \left(\sqrt{(-\rho)} M_\mu^\nu \right) \big|_\nu$ which conserves the magnetic force density $\Psi_{\mu\nu} \delta^\nu$. Thus, the force density $\Psi_{\mu\nu} \delta^\nu$ is counterbalanced by the "depletion rate" of the vacuum represented by the second term in eq. (41). The quantum character of the field is induced by the "orbiton" structure of the elementary particles. For example, in the momentum 4-vector

$$c \, P_\mu = \int \Psi_{\mu\nu} \, \delta^\nu \, d^4 x \ , \tag{42}$$

the total energy

$$E = \int \Psi_{4j} \, \delta^j \, d^4 x \ , \tag{43}$$

should be obtained as an infinite discrete sum with terms referring to states of polarization.

Photon or electromagnetic waves may be characterized by the two invariant statements

$$\Omega = 0 \ , \quad \Lambda = 0 \ . \tag{44}$$

The field equations (32) and the definitions (33) and

(34) become[2,3]

$$G^\nu_\mu - \frac{1}{2}\delta^\nu_\mu G = \frac{8\pi G_o}{c^4}\, \overset{o}{T}{}^\nu_\mu, \Phi_{\mu\nu} + r^2_o\, \overset{o}{R}_{[\mu\nu]} = F_{\mu\nu}, (\sqrt{(-g)}\Phi^{\mu\nu})_{,\nu} = 0 \ , \tag{45}$$

and

$$j^\mu_e = 0 \ , \ (\sqrt{(-g)}F^{\mu\nu})_{,\nu} = [\sqrt{(-g)}r^2_o\, \overset{o}{R}{}^{[\mu\nu]}]_{,\nu} = 4\pi j^\mu_o \ , \tag{46}$$

$$s^\mu = \zeta^\mu \ , \tag{47}$$

where

$$4\pi\, \overset{o}{T}{}^\nu_\mu = -\phi_{\mu\rho}\phi^{\nu\rho} + r^2_o [g^{\nu\alpha}S^\sigma_{\mu\rho}S^\rho_{\alpha\sigma} - \frac{1}{2}\delta^\nu_\mu g^{\alpha\beta}S^\sigma_{\alpha\rho}S^\rho_{\beta\sigma} - g^{\nu\alpha}\Gamma^\sigma_{\mu\rho}\Gamma^\rho_{\alpha\sigma}$$

$$+ \frac{1}{2}\delta^\nu_\mu g^{\alpha\beta}\Gamma^\sigma_{\alpha\rho}\Gamma^\rho_{\beta\sigma} - g^{\nu\alpha}S^\rho_{\mu\alpha|\rho}] \ . \tag{48}$$

The neutral vacuum electric current j^μ_o is the only source of the electromagnetic field and the total charged electric current j^μ_e ($= \overset{+}{j}{}^\mu_e + \overset{-}{j}{}^\mu_e = j^\mu_e - j^\mu_e = 0$) vanishes. Furthermore, the total magnetic current s^μ reduces to the vacuum magnetic current ζ^μ. Hence, we infer that the invariant statements (44) imply the annihilation from the vacuum particle-antiparticle pairs into pairs of photons or pairs of massive integral spin particles with opposite signs of electric charge. However, the latter might require less stringent conditions than the statement (44).

For the present case, the conservation laws (41) are replaced by

$$f_{\mu\nu}\, s^\nu + \frac{r^2_o}{4\pi}(\sqrt{(-g)}\overset{o}{M}{}^\nu_\mu)_{|\nu} = 0 \ , \tag{49}$$

where the first term represents magnetic force density
of the annihilated pair. The equation (49) may be
interpreted as the rate of creation of pairs plus the
rate of annihilation of pairs (photons). In this theory,
pair-annihilation could proceed by discrete annihilation of each of the corresponding magnetic layers in
particle and antiparticle. We may, thus, expect the resulting total energy to consist of an infinite number
of discrete frequencies, decreasing with increasing n
(layer number). In view of the opposite signs of the
corresponding layers, it is natural to create pairs of
photons with two states of polarization. We may, because of the implied localization of a short range
magnetic force density, regard photons as localized
"lumps" of energy. The massless character of these
lumps can be inferred from the vanishing of the ex-
tremum action function

$$S_{ext} = - \frac{q^2}{4\pi} \int [\sqrt{(-\hat{g})} - \sqrt{(-g)}] d^{\mu}x = 0 \ . \quad (50)$$

The four momentum vector of the photon pairs is given
by

$$c \, p_{\mu} = \int f_{\mu\nu} \, \zeta^{\nu} \, d^4x \qquad . \qquad (51)$$

The dependence of p_{μ} on the integral of the magnetic
force density does clearly exhibit the fact that
photons are created in the presence of matter and that,
like any other particle, it carries an angular momentum.
The integral nature of the photon spin is contained in
the appearance of ζ^{μ} in (51) since it does contain pairs
of orbitons and anti-orbitons.

REFERENCES

1. K. Abe et al., Phys. Lett. <u>63B</u>, 239 (1976).

2. B. Kursunoglu, Phys. Rev. <u>D9</u>, 2723 (1974).

3. B. Kursunoglu, Phys. Rev. <u>D13</u>, 1538 (1976).

"THE CHAMPIONS"

POLARIZATION EXPERIMENTS - A THEORETICAL REVIEW*

Francis Halzen

Rutherford Laboratory, Chilton, Didcot, Oxon.

and

University of Wisconsin, Madison, Wisconsin 53706

I would like to attempt to get two important points across in this necessarily superficial and incomplete theoretical review[1] of polarization experiments:

a) I would like to emphasize their versatility and their relevance to a large variety of aspects of hadron physics. I will discuss polarization measurements as

- tests of basic symmetries (parity, time reversal ......)
- a probe of strong interaction dynamics (e.g. in inclusive reactions)

---

*Work supported in part by the U.S. Energy Research and Development Administration, and in part by the University of Wisconsin Research Committee with funds granted by the Wisconsin Alumni Research Foundation.

      - a tool for doing hadron spectroscopy (N*
        spectra, baryonium and di-baryon resonances,
        multiquark (?) states .....).

b) I would like to present a qualitative picture
of the wealth of experimental data on polarization
parameters in pp and np scattering in the Regge
language and in the diffraction language, empha-
sizing the overlap in basic assumption of the two
approaches once the veil of their different
language has been lifted. More important, both
approaches agree that polarization data at <u>medium
energy</u> do reveal the structure of the <u>high energy</u>
diffractive amplitudes.

    This, as well as their power to do hadron spectro-
scopy in energy regions where straightforward bump
hunting fails, makes these experiments worthwhile and
indeed very exciting and <u>not</u> just a complicated way of
doing "6 GeV chemistry".

    The development of a polarized proton beam at the
Argonne National Laboratory[1] has completely changed the
concept of "polarization measurements" in high energy
physics. Whereas in the past these measurements were
usually limited to the determination of the asymmetry
(P parameter) in the cross-section between scattering
of hadrons on a target with its spin up and down relative to the interaction plane, now a completely new
avenue of spin measurements has been opened up. A
glossary to the experimental possibilities is schematically shown in Fig. 1a for the case of elastic
scattering. One can ask a completely novel list of
questions:

      - what is the correlation between the spins of the
        beam and target particles (C parameter)?

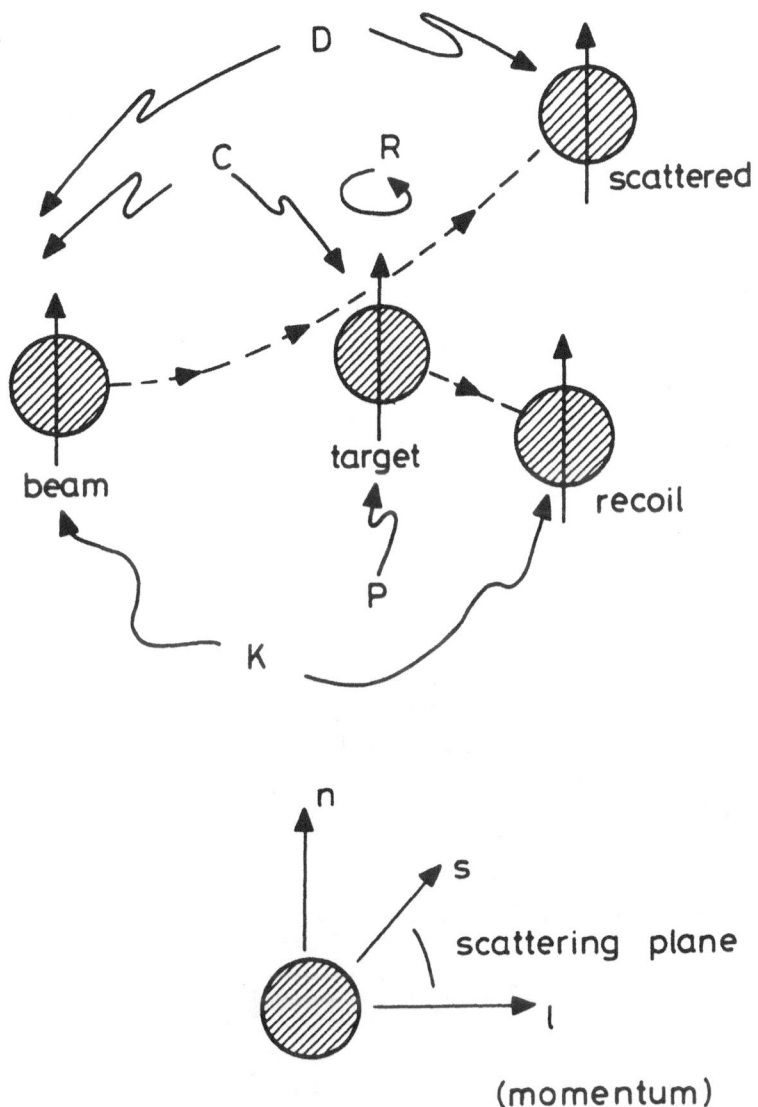

Figure 1 a,b: A road map to polarization parameters in elastic proton-proton scattering.

- is the spin of the beam particle affected (or
  depolarized) when scattered off the target (D
  parameter)?
- is the spin of the beam particle transferred to
  the recoil particle (K parameter)?
- is there a rotation of the spin of the beam
  particle in the collision (R parameter)?

. . . .

Moreover all these measurements can be made for three
independent orientations of the spin of each particle:
the normal to the scattering plane ($\vec{n}$), the direction
of its momentum ($\vec{\ell}$) and the third direction $\vec{s} = \vec{n} \times \vec{\ell}$
(see Fig. 1b).

(a1) BASIC SYMMETRIES

Rotational invariance, parity, time reversal and
charge conjugation relate many of these measurements, a
fact that has been used to test these symmetries with
higher precision ($\pm$ 6%) and in a wider kinematical range
($p_T^2 \simeq .6 - 1$ (GeV/c)^2) than before. We have no reason
to expect any violations, but if one decides to check
wide angle scattering is indeed the best bet. Time
reversal has been checked[2] by reversing both spins in
the initial and final state

$$\frac{d\sigma}{dt} (\uparrow\uparrow \rightarrow \downarrow\downarrow) \overset{?}{=} \frac{d\sigma}{dt} (\downarrow\downarrow \rightarrow \uparrow\uparrow) \ . \tag{1}$$

Vertical arrows refer to the n direction of beam, target,
scattered and recoil particle spins respectively. Also
parity and rotational invariance have been confirmed[2]
to the same 6% accuracy in the measurement

$$\frac{d\sigma}{dt} (\uparrow\downarrow \rightarrow \quad o\downarrow) \overset{?}{=} \frac{d\sigma}{dt} (\downarrow\uparrow \rightarrow \quad o\uparrow) \ . \tag{2}$$

Here the scattered particle's spin is unmeasured.

(a2) DYNAMICS OF STRONG INTERACTIONS

This subject will be expanded in the second part of the talk. I would like to discuss briefly however a polarization measurement in an inclusive process as just one example of a wealth of puzzling data[3], despite the early optimism of theorists that the summation involved in the definition of an inclusive reaction would make only the more important dynamic features of the process surface and make them, therefore, dynamically more tractable than exclusive reactions.

Let us consider Λ production in pp collisions. Its decay allows us to measure its spin orientation. Theory predicts the polarization to be small at high energy s and high missing mass M^2 with $x \cong 1 - \frac{M^2}{s}$ close to 1. Indeed, forward Λ's are described by K* exchange (Fig. 2a) and in the kinematic range under consideration we can Reggeize the K*p Reggeon-particle amplitude and obtain the celebrated triple Regge diagrams like the one shown in Fig. 2b.

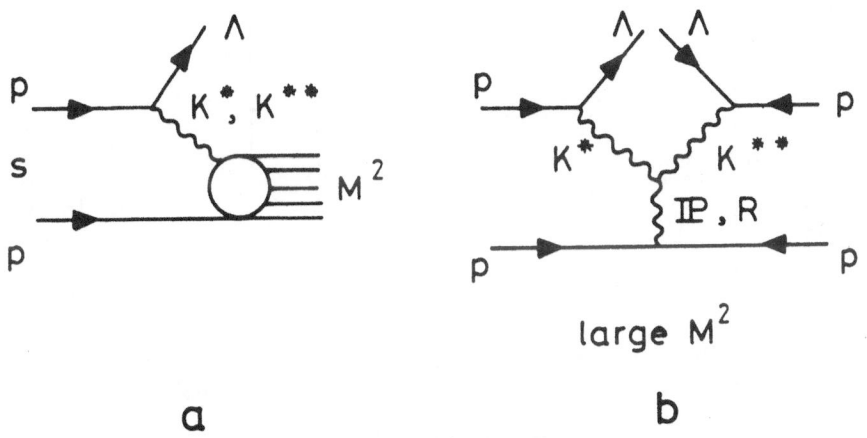

a   b

Figure 2 a,b: Triple Regge diagrams describing the polarization of inclusive Λ's.

Polarization of the Λ should result from the interference diagrams of Fig. 2b because of the different
phase of the vector K* and the tensor K**. But the
effect is at the one percent level[4]. Moreover, both
diagrams of Fig. 2b K*K**P and K*K**R, where P and R
represent the Pomeron and ρ, ω, f,A$_2$ exchanges respectively, are expected to vanish for symmetry reasons.
The K*K**P is zero by SU(3) because of generalized C
invariance. As the K*$^+$p channel is exotic even the
K*K**R diagrams will not produce any polarization in the
pole limit. Despite both corroborating arguments the
Λ polarization is observed[5] to be large (Fig. 3). The
data is FNAL data and well into the kinematic domain of
validity of above approximations.

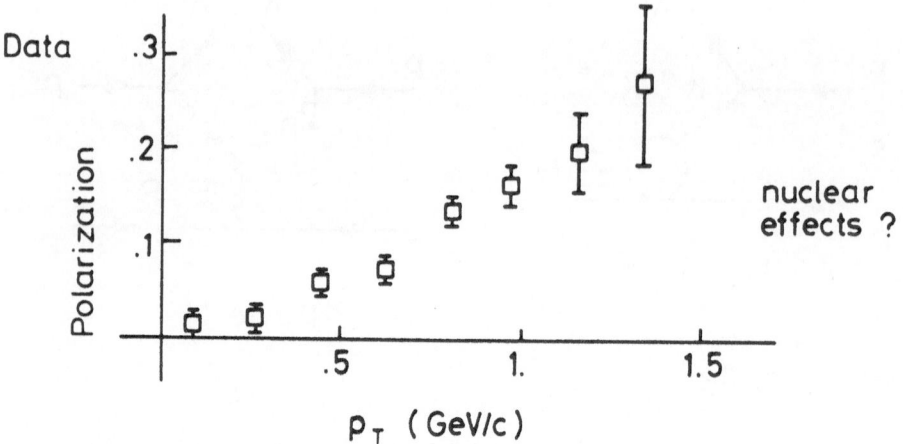

Figure 3:   Inclusive Λ polarization at Fermilab energies.

This data constitutes a real challenge, unless one argues that the effect is nuclear in origin because of the use of a Be target in the experiment. That would make the data interesting for a completely different reason. The rise in p_T of the Λ polarization (see Fig. 3) would then suggest that the effect is related to the anomalous behavior[6] of particle production at high p_T in nuclei. The presence of polarization would rule out some explanations of the anomalous A dependence of particle production, e.g. multiple quark scattering, a model in which all amplitudes are real and no polarization is expected. Either way this puzzle awaits explanation.

(a3) SPECTROSCOPY

Polarization measurements are the prime tool to probe resonances where the straight-forward bump hunting fails. We will discuss applications ranging from the old problem of the N* mass patterns in πN scattering to baryonium states and finally to the exciting discovery of an exotic di-baryon resonance, wanted by theorists since 1968[7].

It is well known that πN polarization experiments in the backward direction are very sensitive to the partial wave structure of the amplitudes. If one partial wave ℓ dominates at a given energy the spin independent amplitude f and the spin flip amplitude g have angular dependence

$$f \propto P_\ell(\cos\theta) \quad , \tag{3}$$

$$g \propto P_\ell'(\cos\theta) \quad . \tag{4}$$

Suppose the spin independent amplitude f dominates the differential cross-section, then

$$\frac{d\sigma}{d\theta} \propto P_\ell^2 \quad . \quad .$$ (5)

Polarization is produced by the interference of f and g,

$$P \frac{d\sigma}{d\theta} \propto P_\ell P_\ell' \quad . \quad$$ (6)

From Eqs. (5) and (6)

$$P \frac{d\sigma}{d\theta} \propto \frac{\partial}{\partial\theta} \left(\frac{d\sigma}{d\theta}\right) \quad .$$ (7)

Therefore, dominance of the scattering by a single partial wave is reflected in the direct relation of Eq. (7) between the polarization and the derivative in angle of the differential cross-section*.

Data on πp scattering satisfy Eq. 7 to an amazing degree of accuracy. Despite sign changes and variations in structure from energy to energy in both P and $\frac{d\sigma}{d\theta}$, Eq. (7) remains valid as can be seen in Fig. 4 for just a sample of tests performed in reference 8. The implication is that successive increasing ℓ values dominate the πN system with increasing energy. I.e. N*'s of the same ℓ will be degenerate in mass, or equivalently N*'s will appear in degenerate-pairs of same parity but different by a single unit in spin $J = \ell \pm \frac{1}{2}$. This pattern emerged indeed from a phase shift analysis[9], probably for the reason pointed out

---

*Above derivation is a hoax; it looks a lot like the correct, but more complicated, derivation given in reference 8.

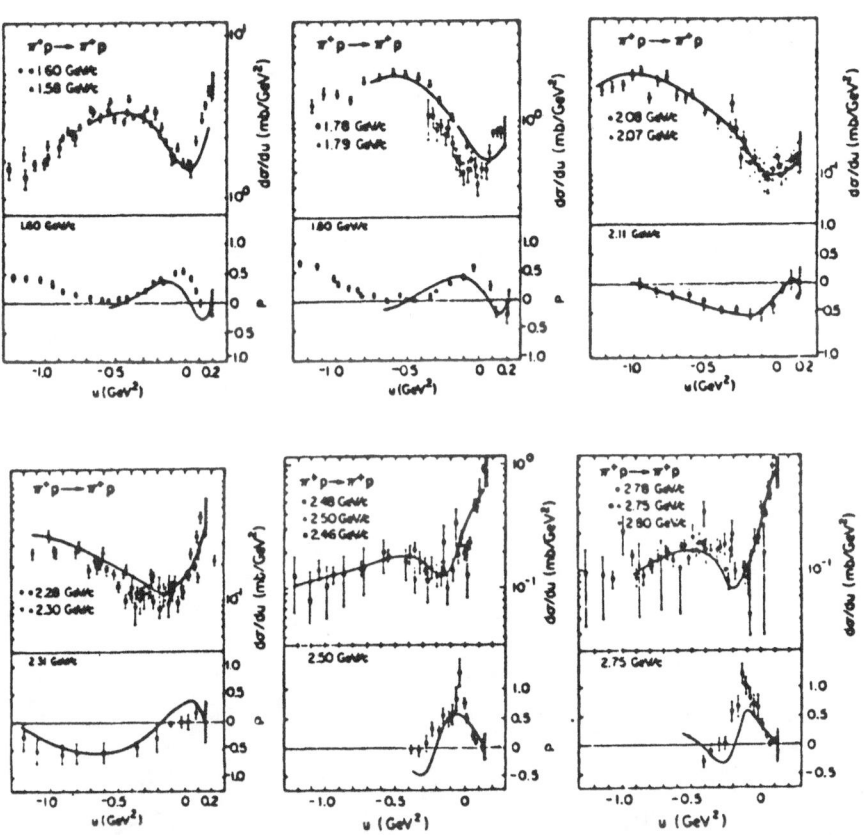

Figure 4:   Comparison of the derivative relation of Eq. 7
with a sample of π^+p elastic scattering data.

above, and has been checked up to $\ell = 10$ for the $I = \frac{3}{2}$ case[9]. This N* pattern is completely at variance with parity doubling arguments and is an open challenge to spectroscopists.

The same spectroscopists have, however, taken the initiative on the problem of baryonium and di-baryon resonances, as we will discuss, experiment obliged. Rosenzweig and Chew[10] have recently emphasized that the $N\bar{N}$ states might contain important hints on the dynamical origin of Zweig's rule. As is the case for ϕ and ψ on the quark flavor level, a Zweig rule would be operative at the particle level, coupling these states mainly to $N\bar{N}$ and not, say, to mesons. The most effective ways of probing them would be by formation in $\bar{p}p$ or through the effect of their exchange on the energy dependence of pp scattering (see Fig. 5). Chew[11] tentatively identified the present candidates of $N\bar{N}$ states with two baryonium trajectories (Fig. 6).

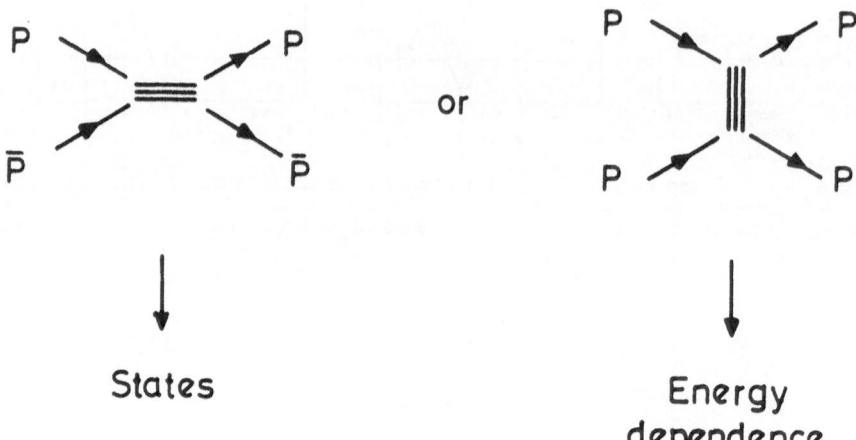

Figure 5:   Experimental probes of baryonium states.

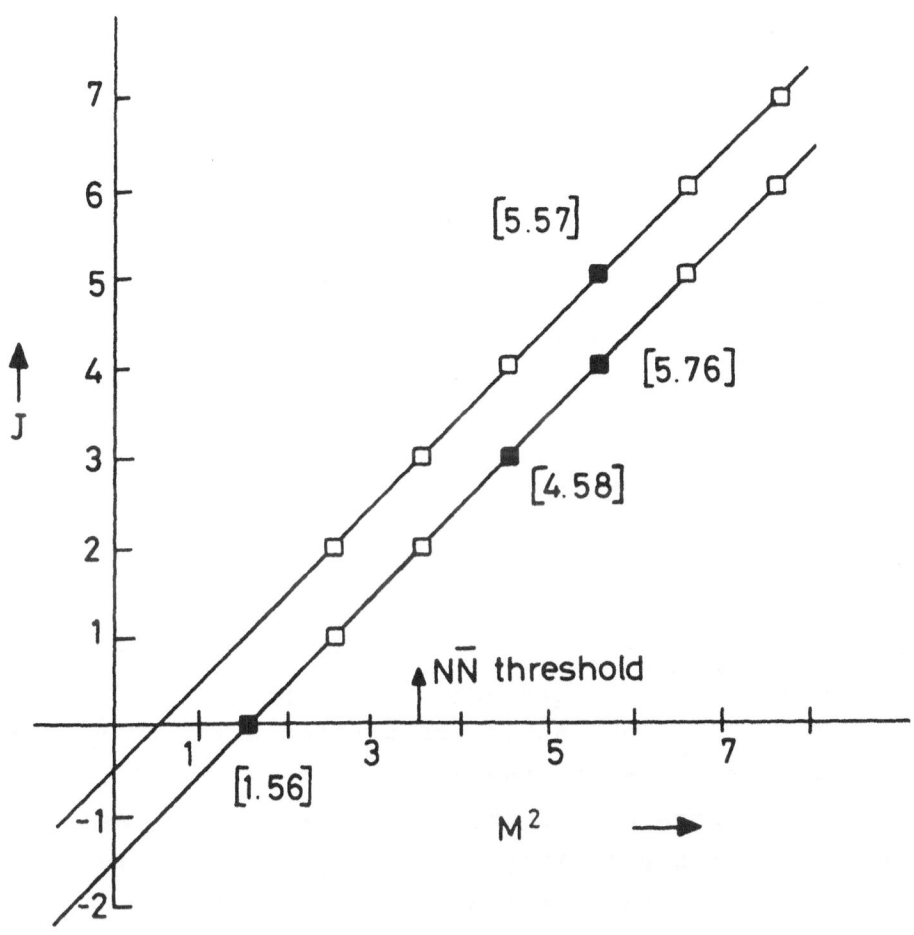

Baryonium

Figure 6: Baryonium trajectories.

Due to the low intercept their exchange should lead to dramatic energy dependence in pp scattering at low energies[12], typically 2 - 6 GeV/c. As can be seen from Fig. 7 this anomalous energy dependence has indeed been observed in measurements of P and the correlation parameter C_{nn}. Their magnitude varies from the 50% level at 2 GeV/c to a few percent at 6 or 12 GeV/c.

In Fig. 8a, we exhibit another striking low energy effect observed[13] in recent measurements of the pp total cross section in pure spin states. Arrows refer to beam and target particle spins in the n (vertical arrow) and ℓ (horizontal arrow) direction. The 16 mb peak in $\sigma(\overset{\rightarrow}{\leftarrow}) - \sigma(\overset{\rightarrow}{\rightarrow})$ around 1.5 - 2 GeV/c is even more striking than the 4 mb structure in $\sigma(\uparrow\downarrow)-\sigma(\uparrow\uparrow)$. In Fig. 8b we plot the unpolarized total cross-section, exhibiting the fact that the observed structure appears near π production threshold, presumably dominated by the $\Delta^{++}n$ channel. It is very tempting to associate the structure of Fig. 8a with the π production threshold via a Deck effect[14] or as a resonance[15] ($^1D_2$ state) driven by the ΔN inelastic channel. The latter explanation might make it similar to the S_{11} at ηπ threshold in πN scattering, the S* at $K\bar{K}$ threshold in the ππ channel or the Z* at KΔ threshold in K^+p scattering. The versatility of the polarized beam-target combination should allow experimentalists to decide the resonant status of this effect by looking at it on different backgrounds in different spin states, if necessary in exclusive channels. The solution might in fact already be there. Contrary to any of the above expectations Yokosawa[16] argued that the data show evidence against any dynamic relation of this bump with inelastic threshold, instead the effect seems to be completely associated with the

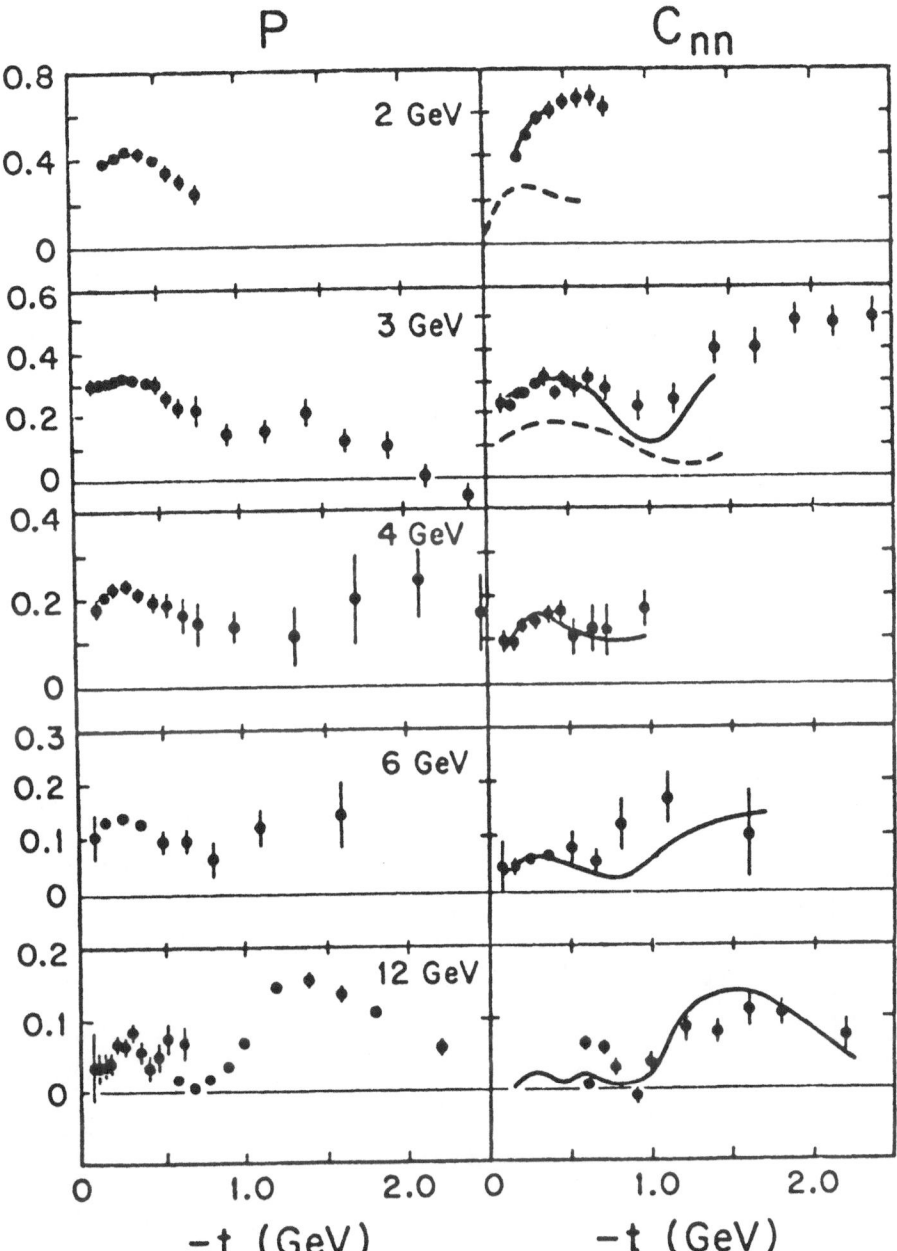

Figure 7: Illustration of the energy dependence of polarization parameters in the 2 ~ 12 GeV/c energy range.

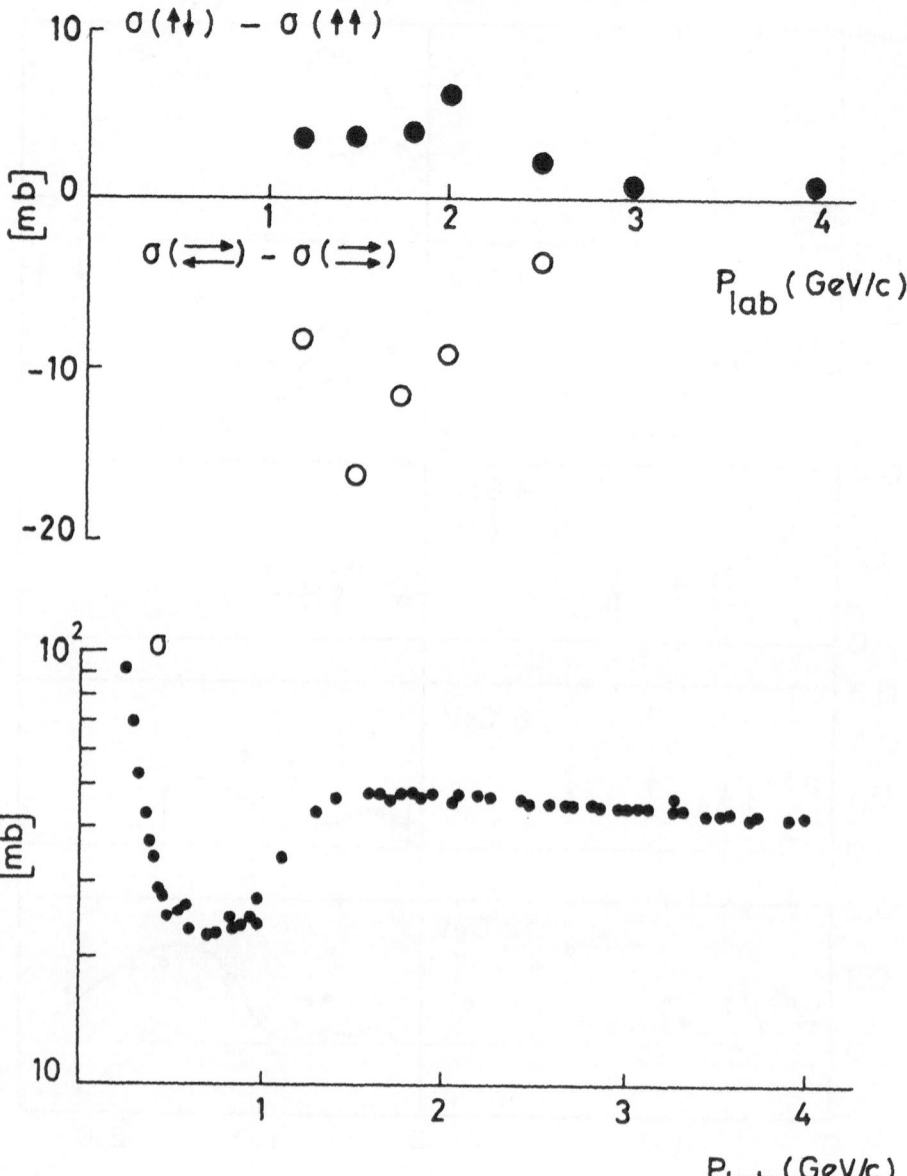

Figure 8:   Comparison of transversely, longitudinally
polarized and spin averaged total cross sections.

elastic channel.

Duality hinted at the existence of exotic resonances, mainly coupled to baryons, in 1968, this state
at 2.2 GeV (width ~ 200 MeV) probably is the experimental
answer. Its appearance in the elastic channel does not
make it subject to the ambiguities of interpretation of
its possible brother, the Z* in the K^+p channel.

Finally a word of credit to the MIT bag[17] spectroscopists for whom all this should be considered as
encouraging. Baryonium states mainly coupling to $N\bar{N}$
appear in their model as $3q3\bar{q}$ states and are the analogue
of the $qq\bar{q}\bar{q}$ states associated with the scalar mesons in
the bag model[18], as opposed to $\ell = 1$ $q\bar{q}$ excitations in
the orthodox quark model. Moreover, the same gluon exchange forces which make the N lighter than the Δ bind
6 quark states and predict a stable ΛΛ dihyperon at
2150 MeV[19]. The observed 2.2 GeV dibaryon resonance
could possibly be associated with a NN resonant partner
of the stable dihyperon, although in the specific MIT
bag model these states would not be resonant as it is
not bound by any baryon-baryon channel.

(b1) HIGH ENERGY SCATTERING A LA REGGE[20]

We now return to polarization experiments as a
probe of strong interaction dynamics. I will assume
from the start that we are basically interested in the
very high energy behavior of hadron collisions and will
discuss the art of hunting for its features in medium
energy (2 - 12 GeV/c) experiments. I will limit the
discussion to the case of elastic pp scattering, extensively probed by the ANL polarized beam facility in
the recent past. Both Regge and diffraction models
agree on the dominantly absorptive and helicity

independent nature of very high energy pp collisions

$$\frac{d\sigma}{dt} \simeq [Im\ T_{++}]^2 \equiv [Im\ \mathbb{P}_{++}]^2 \ . \tag{8}$$

The diffractive amplitude T is dominated by the Pomeron
and at high energy when all secondary Regge exchanges
R have become unimportant all polarization effects are
produced by interference of $T_{++} = \mathbb{P}_{++}$ with a helicity
flip component of $\mathbb{P}$, i.e.

$$\mathbb{P} \propto Im\ \mathbb{P}_{++}\ Re\mathbb{P}_{+-} \ . \tag{9}$$

We know the structure of $Im\ \mathbb{P}_{++}$ from measurements of
$d\sigma/dt$, therefore the problem is to calculate $Re\mathbb{P}_{+-}$.
This calculation proceeds in two steps:

(i) we can calculate $Re\mathbb{P}_{++}(t)$ by fixed t dispersion
relations from the energy dependence of $Im\ \mathbb{P}_{++}(s)$ i.e.
by the optical theorem from the energy dependence of
σ_{tot} at high energy (say > 200 GeV/c). The result is
well known[21]

$$Re\ \mathbb{P}_{++}(t) \propto J_0(R\sqrt{-t}) \ , \tag{10}$$

where $R \simeq 1$ fermi.

(ii) next we argue that spin effects are mostly at
the edge of the proton, i.e. the helicity flip amplitude
is peripheral. In analogy to an empirical relation discovered for ρ-exchange in πN scattering[22] we could try[23]

$$\mathbb{P}_{+-}(b) \propto b\mathbb{P}_{++}(b) \ . \tag{11}$$

Here b is the impact parameter. Eq. (11) is schematically shown in Fig. 9 and contrasts the peripherality of

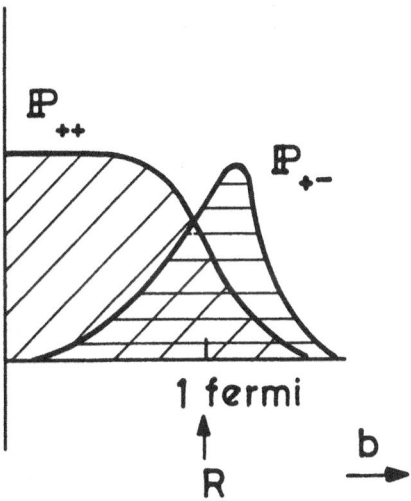

Figure 9: Trends of the impact parameter profiles of
Pomeron helicity flip and non-flip amplitudes.

the helicity flip amplitude with the central character
of the helicity independent diffractive amplitude.
Transforming Eq. (11) to t-space and combining it with
Eq. 10, we obtain our result

$$\text{Re} \mathbb{P}_{+-}(t) \propto \frac{\partial}{\partial \sqrt{-t}} \text{Re} \mathbb{P}_{++}(t) \propto J_1(R\sqrt{-t}) \ . \qquad (12)$$

Eqs. (9) and (12) predict the zero structure of the
pp polarization at high energies

$$P \propto \text{Im } \mathbb{P}_{++} * J_1(R\sqrt{-t}) \qquad . \qquad (13)$$

P is expected to vanish at t values of - .6, - 1.4,
- 2., ... The zeroes at - .6 and - 2. are the zeroes
of the Bessel function in Eq. 13, the zero at - 1.4 is
the diffraction zero of Im $\mathbb{P}_{++}$ corresponding to the dip
structure in dσ/dt at the same value of t. The expected
trend is sketched in Fig. 10a. Before confronting

Eq. 13 with data we have to worry about the contribution of secondary Regge poles to the helicity flip amplitude. We will try to get a handle on this by beating pp and np data against one another. Indeed the I = 1 (ρ,A$_2$) and I = 0 (ω,f) Regge pole contributions add and subtract in the pp and np case respectively. Supposing they are comparable in magnitude we obtain the situation depicted in Fig. 10b. Their contribution to the np polarization is expected to be small. They will add a smooth background, decreasing with energy, to the pp polarization. Therefore two conclusions:

(i) the np polarization at intermediate energies gives us a glimpse into the structure of the pp polarization at high energy. The data[24] in Fig. 11 a,b confirm the expected similarity of the np polarization at 6 GeV/c and pp polarization at 45 GeV/c. Moreover both show the expected zeroes (see Fig. 10a) at t $\simeq$ - .6 and t $\simeq$ - 1.4 (with only marginal evidence for the second).

(ii) The Regge contribution to the pp polarization should transform the high energy zeroes at - .6 and - 1.4 into a valley around t $\simeq$ - 1, changing into a double zero and eventually into two zeroes when the R contribution disappears with increasing energy, as can be seen by essentially adding Fig. 10a and 10b. This trend is clearly exhibited by the data (Fig. 11c and 12, lines in Fig. 11 are drawn to guide the eye). Especially the 12 GeV/c data in Fig. 11c show the asymptotic structure (Fig. 10a) sitting on top of a smooth background (Fig. 10b).

Regge calculations[20] in the literature are mostly complex versions of the above qualitative discussion. Only Kane's model[20] has, however, been able to make this type of approach quantitative to the point that it can be

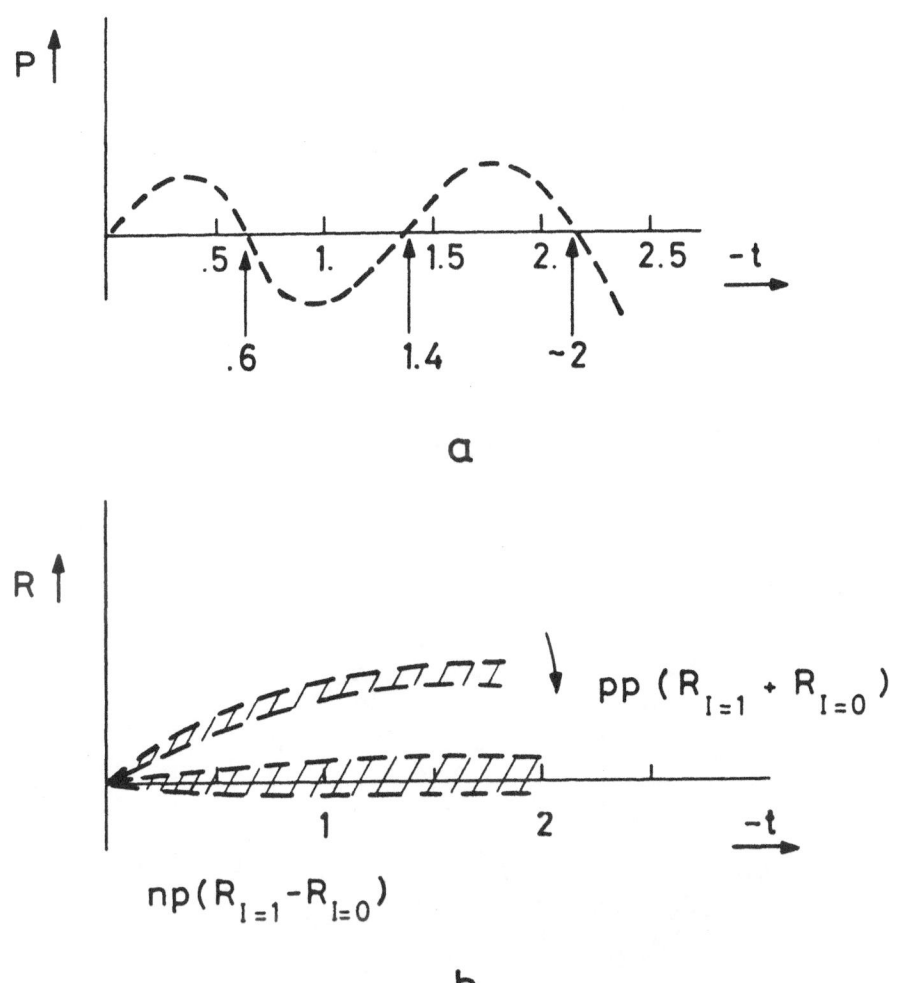

Figure 10 a,b: Expected trends of Pomeron and Regge
contributions to the polarization in pp and np scattering.

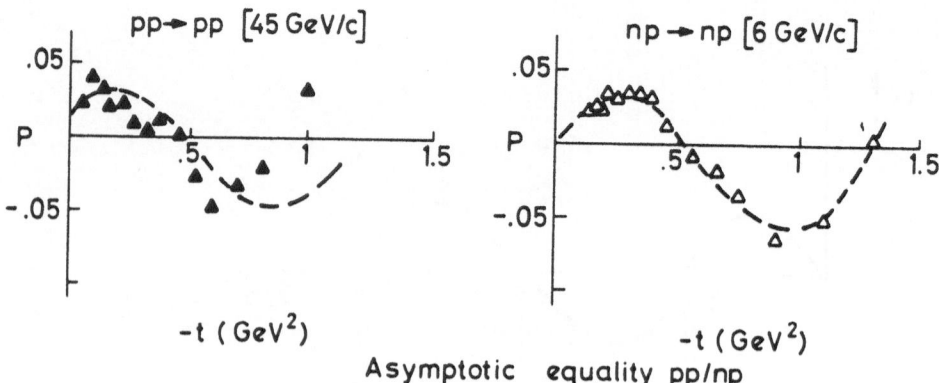

Asymptotic equality pp/np

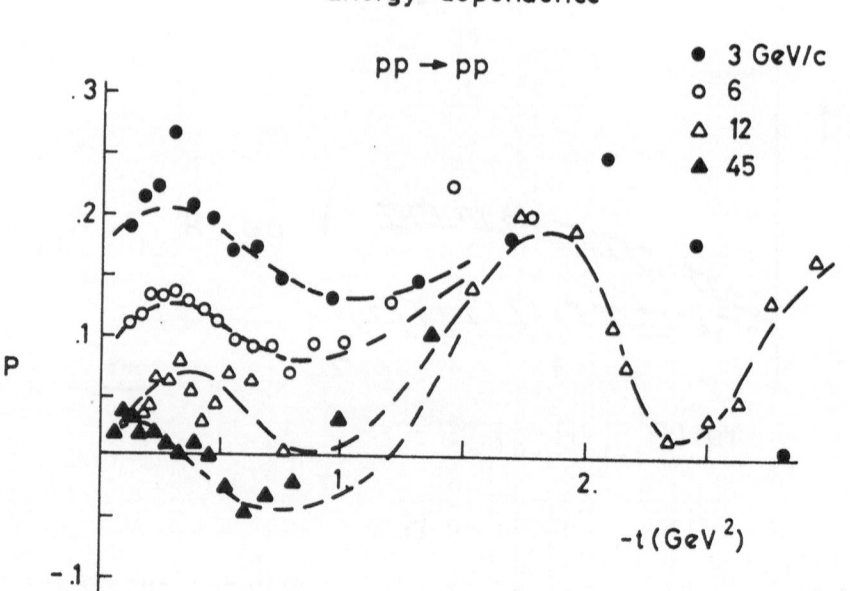

Figure 11:   The trends of Figure 10 compared to data.

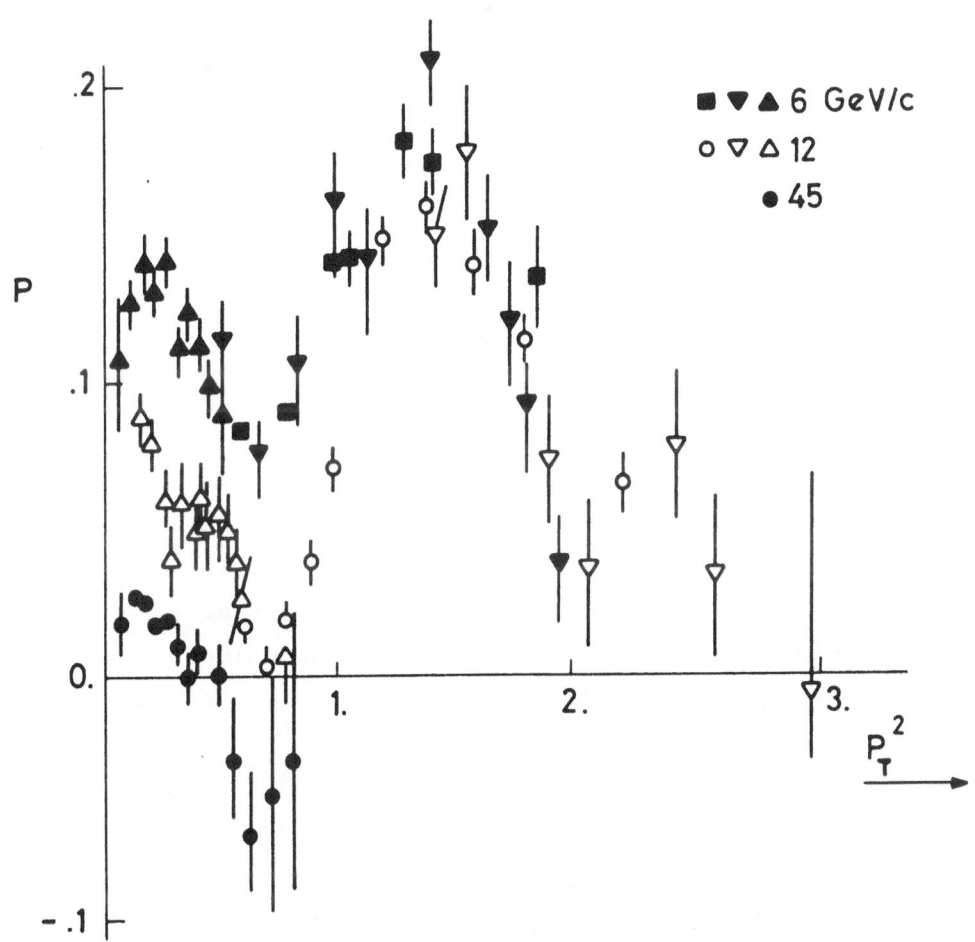

Figure 12:   The trends of Figure 10 compared to data.

(successfully) applied to other measured polarization
parameters. We should especially warn the reader that
in our starting expression of Eq. 9 supplementary contributions have been neglected, which might be important
especially at large (- t) values. The most important
conclusion of this exercise is probably the realization
that the trends of NN polarization parameters (unlike
πN or KN polarizations) reveal the asymptotic structure
of the Pomeron. This has encouraged the application of
diffractive ideas to these data as we will discuss in
the following section.

(b2) DIFFRACTION MODELS OF pp SCATTERING

After the discovery of rising total cross sections
Chou and Yang suggested that spin effects in diffractive
collisions of hadrons might be associated with a physical
rotation of the matter distribution inside the hadrons[25].
If the target proton in Fig. 13 is rotating in the direction shown in the figure the beam will collide with the
dashed left half of the target at a higher energy than
it will collide with the right half. If the cross-

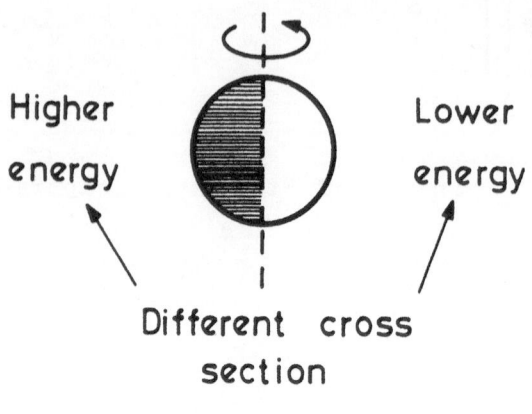

Figure 13

section is changing with energy this will lead to a
left-right asymmetry in the scattering and therefore
lead to polarization effects. As this model has never
been quantified by the authors and no polarization
measurements exist in the energy range where σ_{tot} increases, we mainly mention it to contrast it with the
following models of Low[26] and Durand et al.[27] They
associate spin effects with relative orbital rotations
of the two colliding hadrons. Spin-orbit type forces
are certainly to be expected in a theory of spin $\frac{1}{2}$
fermions interacting by the exchange of vector bosons.

Low and Nussinov attack this problem in the quark-
gluon model. We will follow Low's description using the
bag model[17]. Two hadron bags scatter by exchanging a
color gluon whenever they overlap within the impact parameter. This leads to a picture (Fig. 14) of two color octet bags racing away from each other. In order to confine
color the bags have to break up by the exchange of at

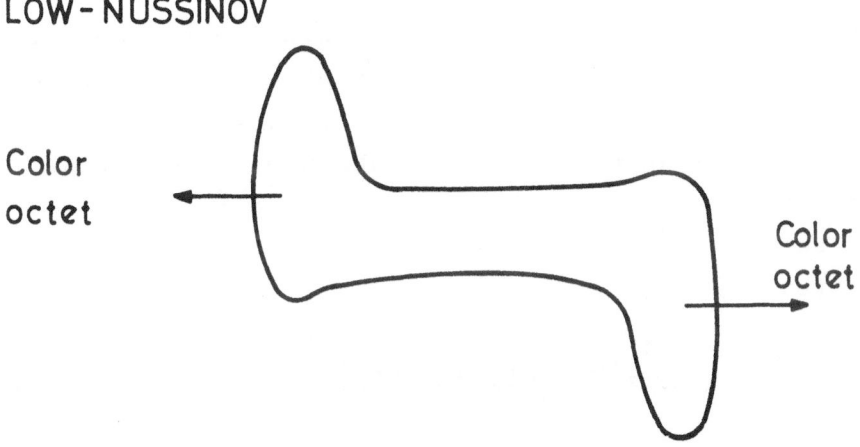

LOW-NUSSINOV

Color
octet

Color
octet

Figure 14: Scattering of quark bags.

least one more colour gluon. The calculation is
described in detail in ref. 26, the result can be
approximately written as

$$T(\sqrt{-t}) \cong i \; \text{Im} \; T(0) \; \exp[i\frac{\sqrt{-t}}{m} \; \vec{\sigma}.\hat{n}] \quad . \tag{14}$$

Eq. 14 exhibits the fact that in the collision the
initial spin is rotated around the normal to the interaction plane. The model predicts that the dominant
polarization parameter is R^{28} ($\equiv D_{ss} \equiv (o,s; o,s)$), all
other polarization effects will disappear at very high
energy,

$$R = - \cos \theta_R + \frac{2\sqrt{-t}}{m} \quad . \tag{15}$$

A positive deviation from the regular ($- \cos \theta_R$) behaviour in the recoil angle θ_R, should be observed. The
predicted pattern of amplitudes is shown in Table 1.

Durand et al.[27] argue that polarization effects in
diffraction scattering are predominantly due to a spin-
orbit force, we will sketch a specific and simple semiclassical picture of their model. It allows back-of-
the-envelope estimates of any polarization and correlation parameter. For the more general and more orthodox
approach to the spin-orbit force model, we refer the
reader to reference 27. In the absence of any spin-
dependent interaction, the scattering of two protons
would be pure shadow scattering, described by an amplitude $M(\theta)$, in this limit

$$\frac{d\sigma_o}{d\theta} \equiv \sigma_o(\theta) = |M(\theta)|^2 \quad . \tag{16}$$

Let us consider first the case of a polarized beam

TABLE 1: Pattern of high energy pp amplitudes in the
diffraction models of refs. 26 and 27 respectively.
The natural (N) and unnatural (U) spin amplitudes are
defined as in ref. 28.   [Units $d\sigma/dt = 1$]

	Low-Nussinov	Durand-Halzen
N_0	i	i
N_1	$-i\,\dfrac{\sqrt{-t}}{m}$	$-\chi\Delta$
N_2	$i\left(\dfrac{\sqrt{-t}}{m}\right)^2$	$-i\chi^2\Delta'$
U_0	0	0
U_1	0	$i\chi^2\dfrac{\Delta}{\sqrt{-t}}$

$\chi<0, \Delta<0, \Delta'>0$
[Δ, Δ' are determined by the
differential cross section,
χ is a parameter]

incident on a spinless target. This situation is
schematically represented in Fig. 15. For incident
spin up relative to the scattering plane, an attractive
spin-orbit interaction leads to an advance in the part
of the incident wave which passes through the upper
half of the target (see Fig. 15d), and to a delay in
the part of the wave which passes through the lower
half. The effect may also be characterized conveniently by regarding the polarized target as a rotating
body which exerts a drag on the incident wave (Fig. 15
a,b). The main effect of the spin-orbit force on the
scattering can be represented as a rotation of the
complete scattered wave (or the shadow) through an
angle δθ about the direction of the spin (we will assume
for simplicity δθ to be small and independent of θ).

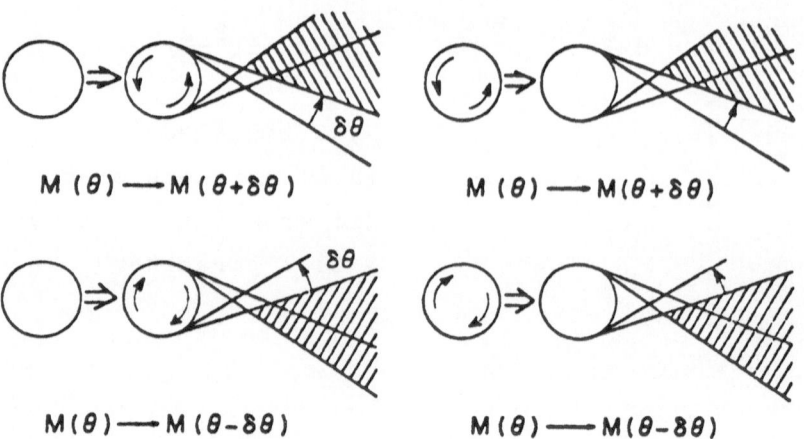

Figure 15: Schematic representation of the rotation of
the scattered wave associated with a weak attractive
spin-orbit interaction. Left: spinless particle incident
on a target with spin up (top) and spin down (bottom).
Right: particles with spin up (top) and spin down
(bottom) incident on a spinless target.

With this assumption, the amplitude for the scattering of a polarized proton with spin up from a spinless target is the same at an angle θ as the spin independent scattering amplitude at the angle $\theta - \delta\theta$,

$$M_\uparrow(\theta) \simeq M(\theta - \delta\theta) \simeq M(\theta) - \delta\theta\frac{\partial M(\theta)}{\partial\theta} \quad . \tag{17}$$

The argument can be repeated for spin down and yields

$$M_\downarrow(\theta) \simeq M(\theta + \delta\theta) \simeq M(\theta) + \delta\theta\frac{\partial M(\theta)}{\partial\theta} \quad . \tag{18}$$

The asymmetry P

$$P \equiv \frac{\sigma_\uparrow - \sigma_\downarrow}{\sigma_\uparrow + \sigma_\downarrow} \simeq \frac{1}{\sigma}(\sigma_\uparrow - \sigma_\downarrow) \tag{19}$$

can now be calculated from Eqs. (17), (18) and (19)

$$P\sigma \simeq \delta\theta \; . \; 2M\frac{\partial M}{\partial\theta} \quad . \tag{20}$$

By combining Eqs. (16) and (20)

$$P\sigma \simeq \delta\theta \; \frac{\partial\sigma}{\partial\theta} \quad . \tag{21}$$

It is easy to generalize the preceding argument to the case of two spins by observing that if both initial protons are polarized in the same direction they will produce a cooperative shift $2\delta\theta$ of the shadow (as illustrated in Fig. 16). The effect of the two spins will however cancel if they are polarized in opposite directions,

$$M_{\uparrow\uparrow}(\theta) = M(\theta - 2\delta\theta) \; , \tag{22}$$

$$M_{\downarrow\downarrow}(\theta) = M(\theta + 2\delta\theta) \quad , \tag{23}$$

$$M_{\uparrow\downarrow}(\theta) = M_{\downarrow\uparrow}(\theta) = M(\theta) . \tag{24}$$

We can therefore calculate the correlation parameter

$$C_{nn} \equiv \frac{1}{\sigma} \left[(\sigma_{\uparrow\uparrow} + \sigma_{\downarrow\downarrow}) - (\sigma_{\uparrow\downarrow} + \sigma_{\downarrow\uparrow}) \right] . \tag{25}$$

The result is

$$C_{nn}\, \sigma \simeq \delta\theta^2 \frac{\partial^2\sigma}{\partial\theta^2} . \tag{26}$$

Although this approach can be pursued to calculate any polarization parameter, it is important to point out that one is simply calculating the effect of an attractive spin orbit interaction corresponding to an eikonal function

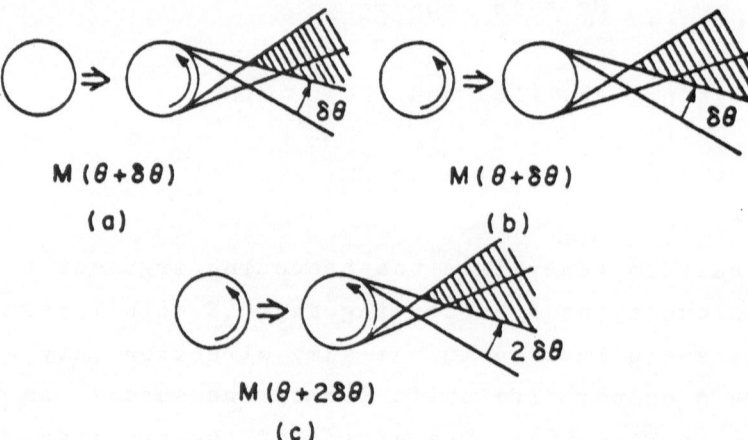

M (θ+δθ)   M (θ+δθ)

(a)   (b)

M (θ+2δθ)

(c)

Figure 16: Schematic representation of the rotation of the scattered wave associated with weak, additive spin-orbit interactions. (a) and (b) represent the rotations associated separately with target and projectile spins; (c) gives the resultant.

$$\chi(\vec{b}) = \chi_c(b) - i\chi_{LS}(b) \; \vec{\sigma} \cdot \vec{L} \quad , \qquad (27)$$

with
$$\vec{L} = \vec{b} \times \hat{\ell} \quad . \qquad (28)$$

It can be shown that the simple model presented above
(with $\delta\theta$ independent of θ) corresponds to χ_{LS} slowly
varying with b or equivalently to a concentration of the
spin dependent amplitudes at the periphery of the
proton.[29] The underlying dynamical assumption is in
fact the same as in the Regge model (Eq. 11).

We will be brave and try to compare this purely
diffractive calculations to data as low as 2 GeV/c.
There are two different types of predictions:

(i) at a given energy the magnitude and t dependence
of all polarization parameters is determined by the
structure of the differential cross section and a single
parameter $\delta\theta$. For a complete discussion of tests of
this model we refer the reader to reference 27, we refer
experimentalists to Table 1 listing the high energy behavior of the amplitudes. Let us just list a sample of
predictions below; they can be presently compared to
data:

$$P \propto \frac{\partial\sigma}{\partial\theta} \quad , \qquad (29a)$$

$$C_{nn} \propto \frac{\partial^2\sigma}{\partial\theta^2} \quad , \qquad (29b)$$

$$C_{ss} \propto - \frac{P}{\sqrt{-t}} \quad , \qquad (29c)$$

$$D_{nn} \simeq 1 \quad , \qquad (29d)$$

$$K_{nn} \simeq C_{nn} \quad . \qquad (29e)$$

The predictions of Eqs. 29a,b are compared to the
6 GeV/c data[30] in Fig. 17. Qualitatively the model
predicts P, C_{nn}, K_{nn} and C_{ss} all to have structure at t
values corresponding to structure in the differential
cross section. This is obviously correct for P, but
questionable for C_{nn} (see Fig. 17) and C_{ss} (Fig. 18).
At 12 GeV/c there is no doubt however that C_{nn} develops
structure similar to P[31]. We expect the C_{ss} data[32],
which were correctly predicted to be negative (Eq. 29c)
to develop similar structure at higher energy. We conclude that the idea of a spin-orbit interaction naturally connects the similar structure of P, C_{nn} and (hopefully) C_{ss} to the break in the differential cross section
around t $\simeq$ - .9. Eqs. 29 d,e are qualitatively in
agreement with observations[30].

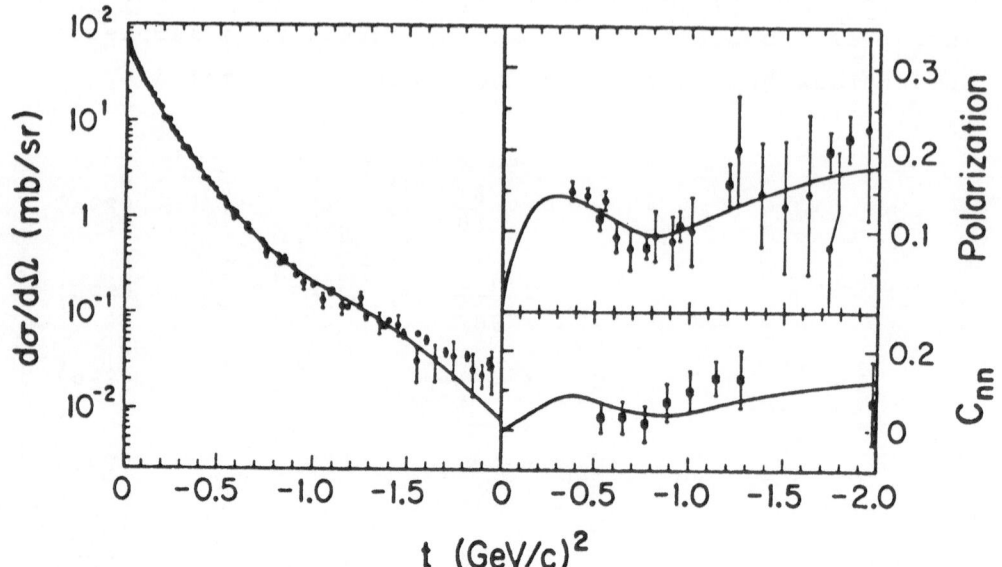

Figure 17: Comparison of Eq. 29 a,b to data at 6 GeV/c.

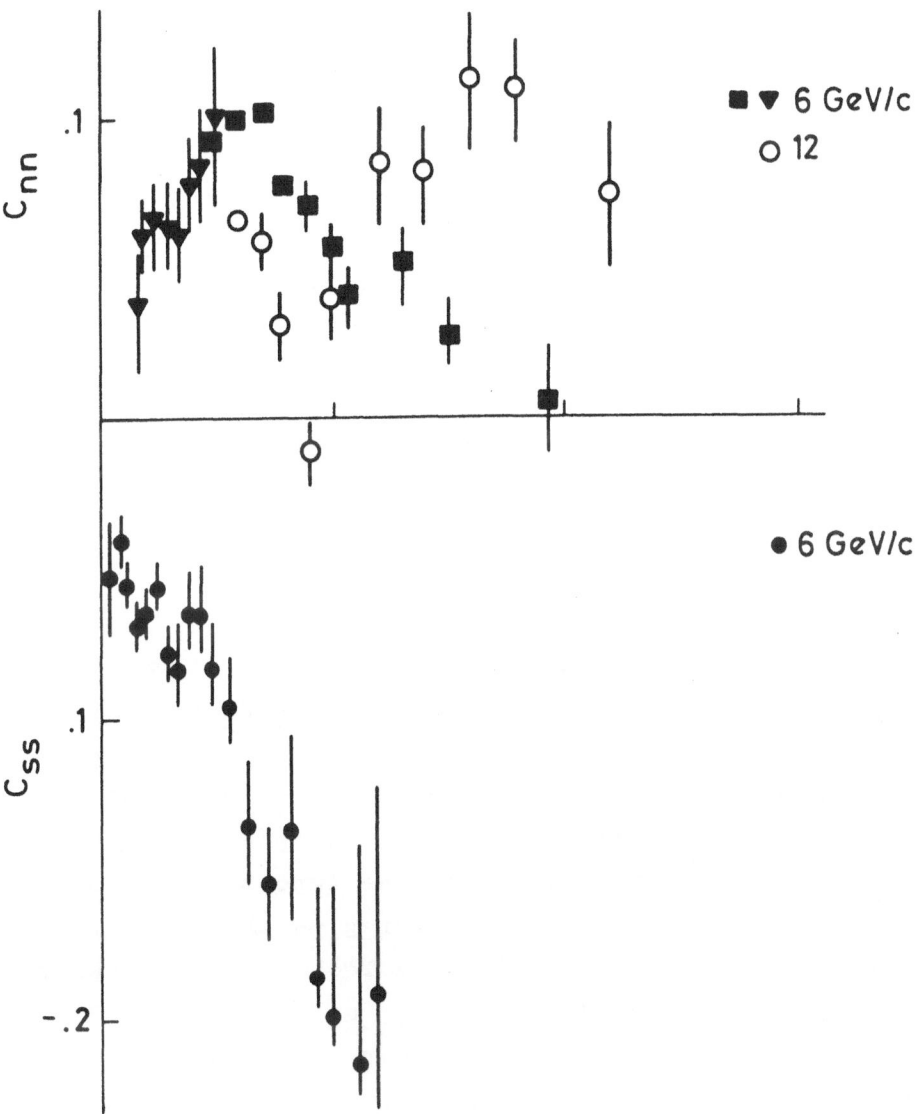

Figure 18: C_{nn} (6 and 12 GeV/c) and C_{ss} (6 GeV/c) data.

(ii) a second type of predictions concerns the
energy dependence of polarization parameters. In this
model all variations in energy are connected with the
energy dependence of $\delta\theta$. As P and C_{nn} depend respectively on $\delta\theta$ and $\delta\theta^2$ we predict the quantity

$$\frac{C_{nn}}{P^2} (\not{s}) \tag{30}$$

to be energy independent. This scaling behavior is experimentally observed over the complete range of energy
where measurements have been made, as can be seen from
Fig. 19.

This model, as the Regge model, predicts the
equality of pp and np polarization at high energy, but,
unlike the Regge model, it is beyond its scope to
account for their difference at lower energy.

As a conclusion we re-emphasize the fact that both
Regge and diffraction approaches agree that medium
energy data give us a hint of the features of very high
energy hadron collisions. Unless we decide some day
that strong interactions of hadrons is a problem we do
not have to solve (is fixed t behavior in QCD a hint at
this?), these measurements will be important. Above
discussion also puts into focus the importance of a
complete set of measurements, giving us the structure
of the amplitudes and hopefully a look at their energy
dependence (note the change in C_{nn} between 6 and 12
GeV/c). Even if none of the patterns in Table 1 emerge,
the hope[33] is that their simplicity will give us the
solution (most likely not mentioned in this review).

ACKNOWLEDGEMENTS

This review has been prepared with the active

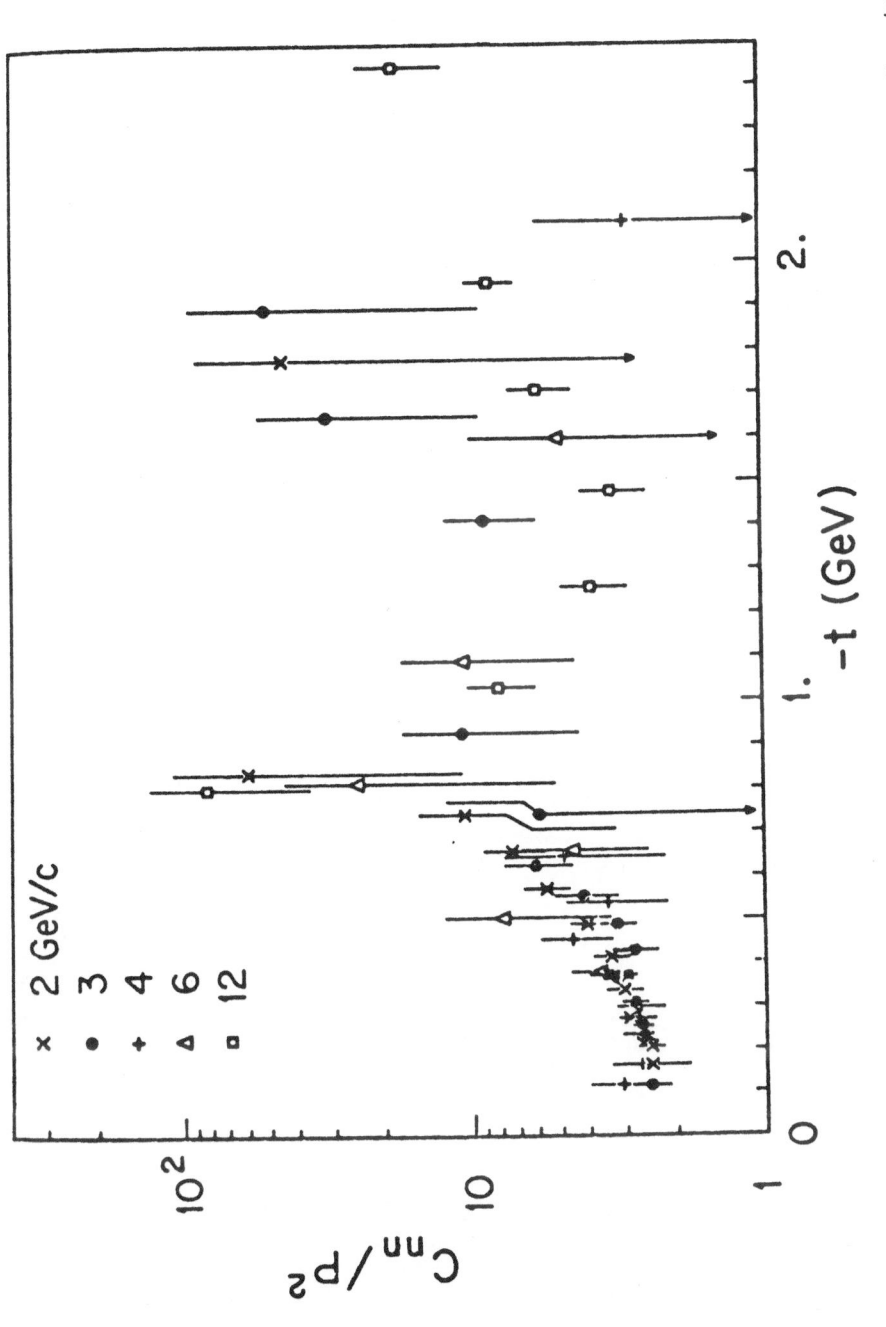

Figure 19: Experimental test of the energy independence of the quantity C_{nn}/P^2 (Eq. 30).

collaboration of H. Miettinen and G. Thomas. We thank
them and also my colleagues at the Rutherford Laboratory
for their advice and suggestions.

REFERENCES

1. For a review of the experimental situation, see
   A.D. Krisch, University of Michigan Preprint UM
   HE 76-38 (1976).

2. L.G. Ratner et al., University of Michigan Preprint
   UM HE 76-15 (1976).

3. See E.A. Peterson, these proceedings.

4. R.D. Field, Proceedings of the Summer Studies on
   High Energy Physics with Polarized Beams, ANL/HEP
   75-02 (1975); F. Paige and D. Sidhu, Phys. Rev.
   (to be published) (1976).

5. G. Bunce et al., Phys. Rev. Letters 36, 1113 (1976).

6. J. Cronin, Proceedings of the VIIth International
   Colloquium on Multiparticle Reactions, Tutzing,
   Germany (1976); H. Frisch, Invited talk at the
   APS Meeting, BNL, November (1976).

7. J. Rosner, Phys. Rev. Letters 21, 950 (1968).

8. F. Halzen, M.G. Olsson and A. Yokosawa, Nuclear
   Physics B113, 269 (1976).

9. A.W. Hendry, Proceedings of the Symposium on High
   Energy Physics with Polarized Beams and Targets,
   ANL (1976).

10. C. Rosenzweig, Phys. Rev. Letters 36, 697 (1976).

11. G.F. Chew and C. Rosenzweig, Nucl. Physics B104,
    290 (1976). G.F. Chew, LBL Preprint LBL-5391
    (1976).

12. J. Dash and H. Navelet, Phys. Rev. D13, 1940 (1976).

13. W. de Boer et al., Phys. Rev. Letters 34, 558
    (1975); J.B. Roberts, private communication.
    A. Yokosawa, private communication.

14. E.L. Berger, P. Pirila and G.H. Thomas, ANL Preprints
    (1976).

15. G.L. Kane and G.H. Thomas, Phys. Rev. D13, 2944.

(1976) and references therein.

16.   A. Yokosawa, these proceedings.

17.   A. Chodos, R.L. Jaffe, K. Johnson, C.B. Thorn and
       V.F. Weisskopf, Phys. Rev. D9, 3471 (1974).

18.   T. De Grand, R.L. Jaffe, K. Johnson and J. Kiskis,
       Phys. Rev. D12, 2060 (1975).

19.   R.L. Jaffe, Phys. Rev. Letters 38, 195 (1976).

20.   For recent papers, see C. Bourrely, J. Soffer and
       D. Wray, Nucl. Phys. B77, 386 (1974); B85, 32
       (1975); and B91, 33 (1975). R.D. Field and P.
       Stevens, Caltech Technical Report CALT-68-534
       (1976). A.C. Irving, Nucl. Physics B101, 263
       (1975). G.L. Kane and A. Seidl, Rev. Mod. Phys.
       48, 309 (1976).

21.   V. Barger, Proceedings of the 17th International
       Conference on High Energy Physics, London (1974).

22.   H. Høgaason, Physica Norvegica 5, 219 (1971).

23.   See e.g. B. Schrempp and F. Schrempp, Nucl. Phys.
       B54, 525 (1973).

24.   For a complete list of references, see A.D. Krisch
       (reference 1) and R. Diebold (these proceedings).

25.   T.T. Chou and C.N. Yang, Stony Brook Preprint ITP
       75-61 (1975).

26.   F.E. Low, Phys. Rev. D12, 163 (1975); S. Nussinov,
       Phys. Rev. Letters 34, 1286 (1975).

27.   L. Durand and F. Halzen, Nucl. Phys. B104, 317
       (1976) and Phys. Rev. D. (to be published).

28.   For the conventional notation and definitions of
       amplitudes and measurements, see F. Halzen and G.
       H. Thomas, Phys. Rev. D10, 344 (1974).

29.   F. Halzen, Proceedings of the Summer Studies on
       High Energy Physics with Polarized beams and Targets,
       ANL HEP 75-02 (1975).

30. For a complete list of references, see A.D. Krisch ref. 1; see also ref. 31, 32.

31. K. Abe et al., Phys. Letters B63, 239 (1976).

32. I.P. Auer et al., Phys. Rev. Letters 37, 1727 (1976).

33. For a discussion, see P.W. Johnson, R.C. Miller and G.H. Thomas, ANL-HEP-PR-76-61 (1976).

pp SCATTERING-AMPLITUDE MEASUREMENTS AND A POSSIBLE DIRECT-CHANNEL RESONANCE IN pp SYSTEM*

A. Yokosawa

Argonne National Laboratory

Argonne, Illinois

INTRODUCTION

In this talk I would like to discuss our results of pp elastic scattering with various spin combinations of polarized beams and polarized targets. Since experimental details have been described in our published articles,[1,2,3] I will minimize my discussion.

Elastic scattering is an old topic in hadronic physics and is the simplest reaction among various hadronic reactions. In spite of numerous elastic-scattering data available at various energies, our understanding of the reaction mechanism is generally poor. This is due to the fact that scattering amplitudes have never been well determined. Most of the data available up to now are differential cross section and polarization, and we need to measure other observables in order to determine scattering amplitudes. Crucial discoveries are yet to come in the field of elastic scattering.

---

*Work supported by the U.S. Energy Research and Development Administration.

Among various elastic scattering processes nucleon-nucleon elastic scattering is relatively much less understood because of the spin-spin interaction. Recent p-p elastic-scattering experiments at ISR (also Fermilab) and Serpukhov revealed remarkable energy dependence in cross sections and polarization respectively. A dip observed in the differential cross section data at $|t| \approx 1.3$ $(GeV/c)^2$ gives a significant clue to the understanding of the reaction mechanism, i.e., the nature of Pomeron exchange. We note that such a dip has never been observed in πp and Kp elastic scattering at Fermi energies. The sign change in polarization (positive at low energies to negative at Serpukhov energies) is equally significant.

The rising total cross section at high energies is also an extremely interesting phenomenon. Recently C.N. Yang suggested that understanding of such a phenomenon has come from the measurement involving a spin effect; interaction between the incident particle and hadronic current in the proton.

Our goal is to establish the energy dependence of pp amplitude from Argonne ZGS energies to Fermilab energies. Eventually, we hope to clarify interesting features, such as dip and secondary bump, observed at high energies. Measurements of total cross-section differences may help to provide an understanding of the rise in the total cross section.

As you know, the ZGS is capable of accelerating a polarized beam. At Fermilab, the production cross section of Λ and $\bar{\Lambda}$ is big enough that one can produce useful polarized proton and antiproton beams.[4] We note that protons decaying from Λ particles are longitudinally polarized ($\approx 60\%$), and necessary spin rotations to the vertical or horizontal direction can be done with standard magnets.[5]

DEFINITION OF SCATTERING AMPLITUDES

We describe scattering amplitudes in three different ways; old notations[6] are not included here.

i) Transversity Amplitudes

Cross section: $|T_1|^2 + |T_2|^2 + 2|T_3|^2 + 2|T_4|^2 + 2|T_5|^2$

ii) S-channel helicity amplitudes

$$\left.\begin{array}{l} <++|++> = \phi_1 \\ <--|++> = \phi_2 \\ <+-|+-> = \phi_3 \end{array}\right\} \quad \text{net helicity nonflip amplitude}$$

$<+-|-+> = \phi_4$ double flip

$<++|+-> = \phi_5$ single flip

iii) t-channel exchange amplitudes[7]

Natural-parity exchange: N_0, N_1, N_2

$N_0 = \frac{1}{2}(\phi_1 + \phi_3)$, $N_1 = \phi_5$, $N_2 = \frac{1}{2}(\phi_4 - \phi_2)$

Unnatural-parity exchange: U_0, U_2

$U_0 = \frac{1}{2}(\phi_1 - \phi_3)$, A_1 exchange

$U_2 = \frac{1}{2}(\phi_2 + \phi_4)$, π exchange

Cross section: $|N_0|^2 + 2|N_1|^2 + |N_2|^2 + |U_0|^2 + |U_2|^2$

(The subscript gives the amount of t-channel helicity flip.)

EXPERIMENTAL OBSERVABLES

If all the measurements are to be made on the horizontal scattering plane, then the spin directions N, L, and S of the polarized beam, the polarized target, and the recoil particles are defined as shown in Fig. 1.

Possible experiments obtaining spin direction of particles by means of polarized beam, polarized target, or spin analysis of final state are listed below. We adopt the notation (Beam, Target; Scattered, Recoil) to express observables;* indicates that spin direction is known; 0 means that spin direction is not known. Assume that spin direction of scattered particles cannot be obtained.

Observables	Description	Symbol
(0,0;0,0)	Cross section	σ
(*,0;0,0) or (0,*;0,0)	Polarization	P
(*,*;0,0)	Correlation tensor	C_{jk}
(*,0;0,*)	Polarization transfer tensor	K_{jk}
(0,*;0,*)	Depolarization tensor	D_{jk}
(*,*;0,*)	High-rank spin tensor	H_{ijk}

These observables are listed in terms of $|t|$-channel exchange amplitudes in Table I. We note that D_{NN}, D_{SS}, and D_{LS} are commonly called the "D parameter," "R parameter," and "A parameter," respectively. In three-spin measurements, we can determine three or four parameters simultaneously for one measurement.

DETERMINATION OF SCATTERING AMPLITUDES

The strategy of this work has been studied by various people, the most recent one being given in ref. 8. In Table II, we illustrate the determination of scattering amplitudes vs. observables at small $|t|$ assuming $|N_0|$ is much larger than $|N_1|$, $|N_2|$, $|U_0|$, and $|U_2|$.

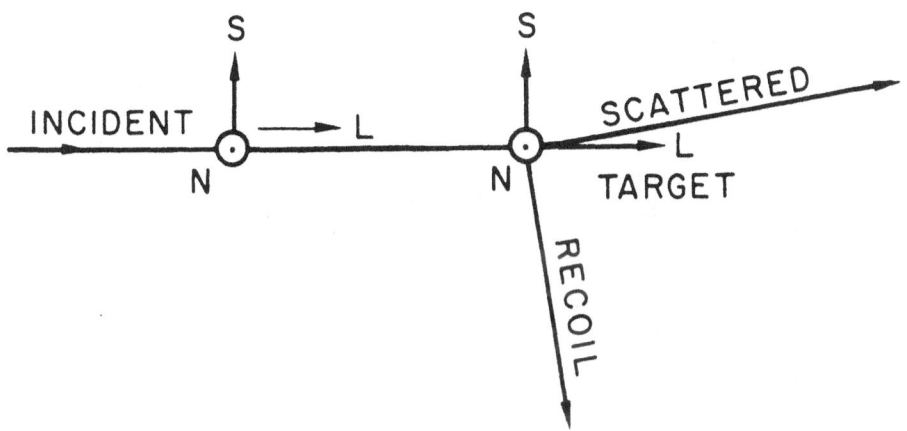

N: NORMAL TO THE SCATTERING PLANE
L: LONGITUDINAL DIRECTION
S = N x L IN THE SCATTERING PLANE

Fig. 1 Unit vectors N, L, and S.

TABLE I

Laboratory Observables (θ_R is the Laboratory Recoil Angle)

Observables $(B,T;S,R)$	Exchange Amplitudes
(Single Scattering)	
$\sigma^{Tot} = (0,0;0,0)$	$\mathrm{Im}N_0$
$\Delta\sigma_T^{Tot} = (N,N;0,0)$	$\mathrm{Im}U_2$
$\Delta\sigma_L^{Tot} = (L,L;0,0)$	$\mathrm{Im}U_0$
$\sigma = (0,0;0,0)$	$|N_0|^2 + 2|N_1|^2 + |N_2|^2+|U_0|^2+|U_2|^2$
$P = (0,N;0,0)$	$-2\mathrm{Im}(N_0-N_2)N_1^*/\sigma$ (Also $(N,0;0,0)$)
$C_{NN} = (N,N;0,0)$	$2\mathrm{Re}(U_0 U_2^*-N_0 N_2^*+|N_1|^2)/\sigma$
$C_{SS} = (S,S;0,0)$	$2\mathrm{Re}(N_0 U_2^*-N_2 U_0^*)/\sigma$
$C_{SL} = (S,L;0,0)$	$2\mathrm{Re}(U_0+U_2)N_1^*/\sigma$
$C_{LL} = (L,L;0,0)$	$-2\mathrm{Re}(N_0 U_0^*-N_2 U_2^*)/\sigma$

TABLE I (continued)

Observables (B,T;S,R)	Exchange Amplitudes
(Double Scattering)	
1) K_{jk} Measurement	
$K_{NN} = (N,0;0,N)$	$-2Re(U_0U_2^* + N_0N_2^* - |N_1|^2)/\sigma$
$K_{SS} = (S,0;0,S)$	$[-2Re(U_2-U_0)N_1^*\ \sin\theta_R - 2Re(N_0U_2^* + N_2U_0^*)\cos\theta_R]/\sigma$
$K_{LS} = (L,0;0,S)$	$[-2Re(N_0U_0^* + N_2U_2^*)\sin\theta_R - 2Re(N_1^*(U_2 \times U_0))\cos\theta_R]/\sigma$
2) D_{jk} Measurement	
$D_{NN} = (0,N;0,N)$	$[|N_0|^2 + 2|N_1|^2 + |N_2|^2 - |U_0|^2 - |U_2|^2]/\sigma$
$D_{SS} = (0,S;0,S)$	$[-2Re(N_0+N_2)N_1^*\ \sin\theta_R - (|N_0|^2 - |N_2|^2 + |U_2|^2 - |U_0|^2)\cos\theta_R]/\sigma$
$D_{LS} = (0,L;0,S)$	$[(|N_0|^2 - |N_2|^2 - |U_2|^2 + |U_0|^2)\ \sin\theta_R - 2Re(N_0 - N_2)N_1^*\cos\theta_R]/\sigma$

TABLE I (continued)

Observables $(B,T;S,R)$	Exchange Amplitudes	Simultaneous Observables
3) Three Spin Measurement		
$H_{SNS}=(S,N;0,S)$	$[2Im(N_0U_2{}^*+N_2U_0{}^*)sin\theta_R+2Im(U_2-U_0)N_1{}^*cos\theta_R]/\sigma$	$(0,N;0,N),(S,0;0,S)$
$H_{HSS}=(N,S;0,S)$	$[-2Im(U_0U_2{}^*-N_0N_2{}^*)sin\theta_R+2Im(N_0+N_2)N_1{}^*cos\theta_R]/\sigma$	$(0,S;0,S),(N,0;0,N)$
$H_{SSN}=(S,S;0,N)$	$-2Im(U_2+U_0)N_1{}^*/\sigma$	$(S,S;0,0),(0,S;0,S),(S,0;0,S)$
$H_{LSN}=(L,S;0,N)$	$2Im(U_0N_0{}^*-U_2N_2{}^*)/\sigma$	$(0,S;0,S),(L,0;0,S)$
$H_{NLS}=(N,L;0,S)$	$[-2Im(N_0+N_2)N_1{}^*sin\theta_R+2Im(U_0U_2{}^*+N_0N_2{}^*)cos\theta_R]/\sigma$	$(0,L;0,S),(N,0;0,N)$
$H_{SLN}=(S,L;0,N)$	$-2Im(N_0U_2{}^*-N_2U_0{}^*)/\sigma$	$(S,L;0,0),(0,L;0,S),(S,0;0,S)$

TABLE II

Determination of Amplitudes vs. Lab Observables

Amplitude	Lab Observable	Sign determined by experiments
$\mathrm{Re}N_1$	$P = (0,N;0,0)$, $H_{NLS} = (N,L;0,S)$	−
$\mathrm{Im}N_1$	$D_{SS} = (0,S;0,S)$	−
$\mathrm{Re}N_2$	$H_{NSS} = (N,S;0,S)$	(data being analyzed)
$\mathrm{Im}N_2$	$C_{NN} = (N,N;0,0)$, $K_{NN} = (N,0;0,N)$	−
$\mathrm{Re}U_0$	$H_{LSN} = (L,S;0,N)$	(to be measured)
$\mathrm{Im}U_0$	$C_{LL} = (L,L;0,0)$, $K_{LS} = (L,0;0,S)$	−
$\mathrm{Re}U_2$	$H_{SNS} = (S,N;0,S)$, $H_{SLN} = (S,L;0,N)$	+
$\mathrm{Im}U_2$	$C_{SS} = (S,S;0,0)$	−

Double-scattering measurements are required to determine
the real part of N_2, U_0, and U_2 amplitudes.

Some of the parameters, e.g., $C_{SL} = (S,L;0,0)$, do
not contain the N_0 term and they are expected to be
approximately zero.

FACILITIES PROVIDING SPIN DIRECTIONS

The spin of polarized protons emerging from the
ZGS is in the N direction. A superconducting solenoid
with a field of 12.0 T·m at 6 GeV/c is used to rotate
the spin of the incident beam from the N to the S direction. The longitudinally polarized beam is produced by
a bending magnet to precess the proton spins until their
polarization is parallel to the beam momentum. This
scheme of operation does not require the vertical ad-
justment of polarized targets. The sign of beam polarization is flipped on alternate pulses; this is essential
to reduce systematic errors.

The polarized proton target is 2 x 2 x 8-cm ethylene
glycol doped with $K_2Cr_2O_7$ and maintained at ∿0.4 K. For
free protons in the target, the polarization is 0.8 to
0.9. Superconducting magnets are used to provide three
directions--N, S, and L-- of spins.

The spin analyses of recoil protons are done by a
carbon polarimeter together with proportional wire-chamber
detectors.

Results

So far we have measured the following parameters
(measurements at $|t| = 0$ are listed later): P, C_{NN},
C_{SS}, C_{SL}, C_{LL}, K_{NN}, K_{SS}, D_{NN}, D_{SS}, D_{LS}, H_{SNS}, and H_{NSS}.
In Figs. 2 and 3, C_{NN} and C_{SS} data are shown.[1,2] The
energy dependence of C_{NN} is remarkable, and the Regge
prediction shown in Fig. 2 is quite inadequte. The

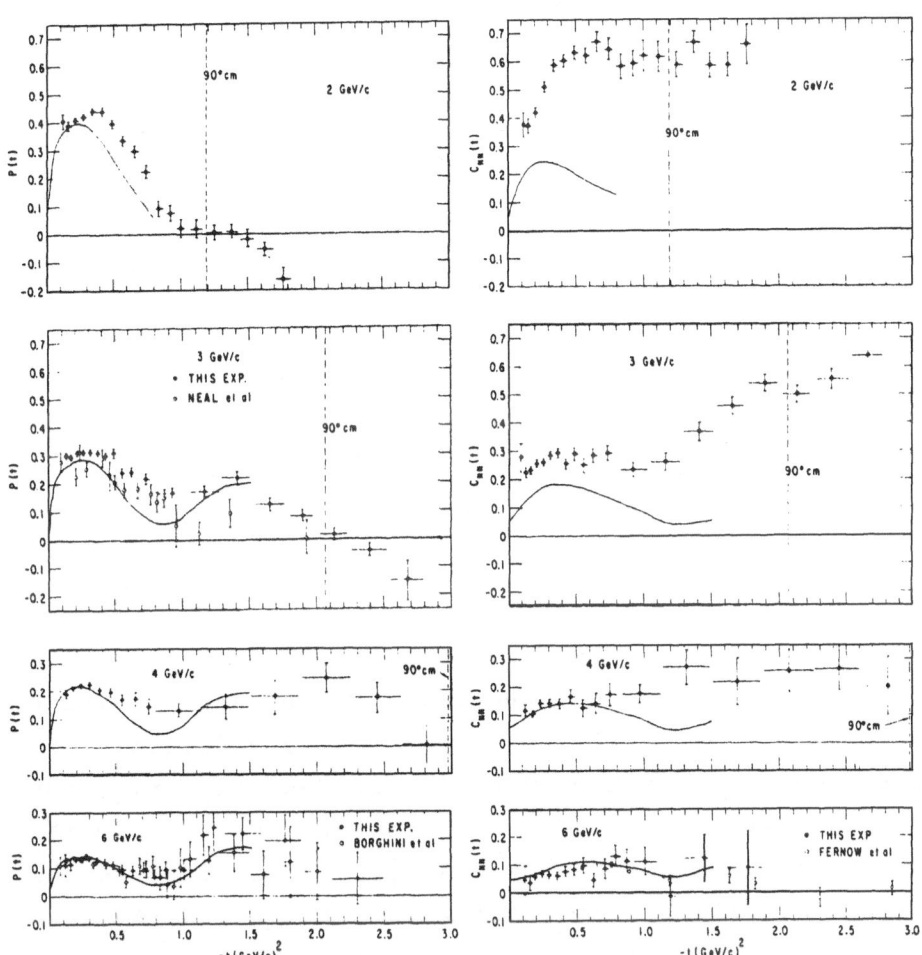

Fig. 2 P and C_{NN} plotted as functions of t at 2, 3, 4, and 6 GeV/c. The smooth curves are predictions based on a Regge-pole model by R. Field and P. Stevens.

previous speaker (F. Halzen) made an interesting explanation of these data. As shown in Table II, the
data shown in Fig. 3 imply that the imaginary part of
the U_2 term (corresponding to π exchange) is negative.
For comparison we show two existing attempts by Field
and Stevens to describe the pp elastic-scattering process.[9] The dashed curve calculated by using the super
Regge Model involves a large number of poles (P, f,
ω, ρ, A_2, π, B) and corrections due to absorption
(Regge cuts). The solid curve calculated by using the
Kane model involves the same Regge poles and absorption
corrections calculated according to the Sopkovich pre-
scriptions,[10] but does not require exchange degeneracy.[11]
In addition, inelastic intermediate states play an important role in the Kane model. The difference of the
two is primarily in the treatment of absorption correction $\pi_c(B_c)$. The data clearly favor the Kane model.

Our preliminary data of the parameter $C_{SL} = (S,L;0,0)$
are nonzero and $(-0.2 \pm 0.1)\%$ for the region of $|t|$
<1.0; this is rather unexpected. Preliminary results of
the parameter $C_{LL} = (L,L;0,0)$ are $(-0.15 \pm 0.10)\%$, for
the region of $|t| < 1.0$. The latter results imply that
the phase between N_0 and A_1 is close to $90°$. (We will
show the evidence of $\mathrm{Im}A_1 \neq 0$ at $|t| = 0$ in the section
below.) It will be interesting to estimate the A_1 trajectory after the phase is established. Preliminary
results of $H_{SNS} = (20 \pm 10)\%$ and $K_{SS} = (3 \pm 10)\%$ in the
region of $0.2 < |t| < 0.6$ are obtained. The real part of
the U_2 term (corresponding to π exchange) is positive.

I am not quite ready to discuss the conclusion of
amplitude measurements, but our hope is to reduce considerably the shaded area shown in Fig. 4 in which the
boundaries of t-channel exchange amplitudes are

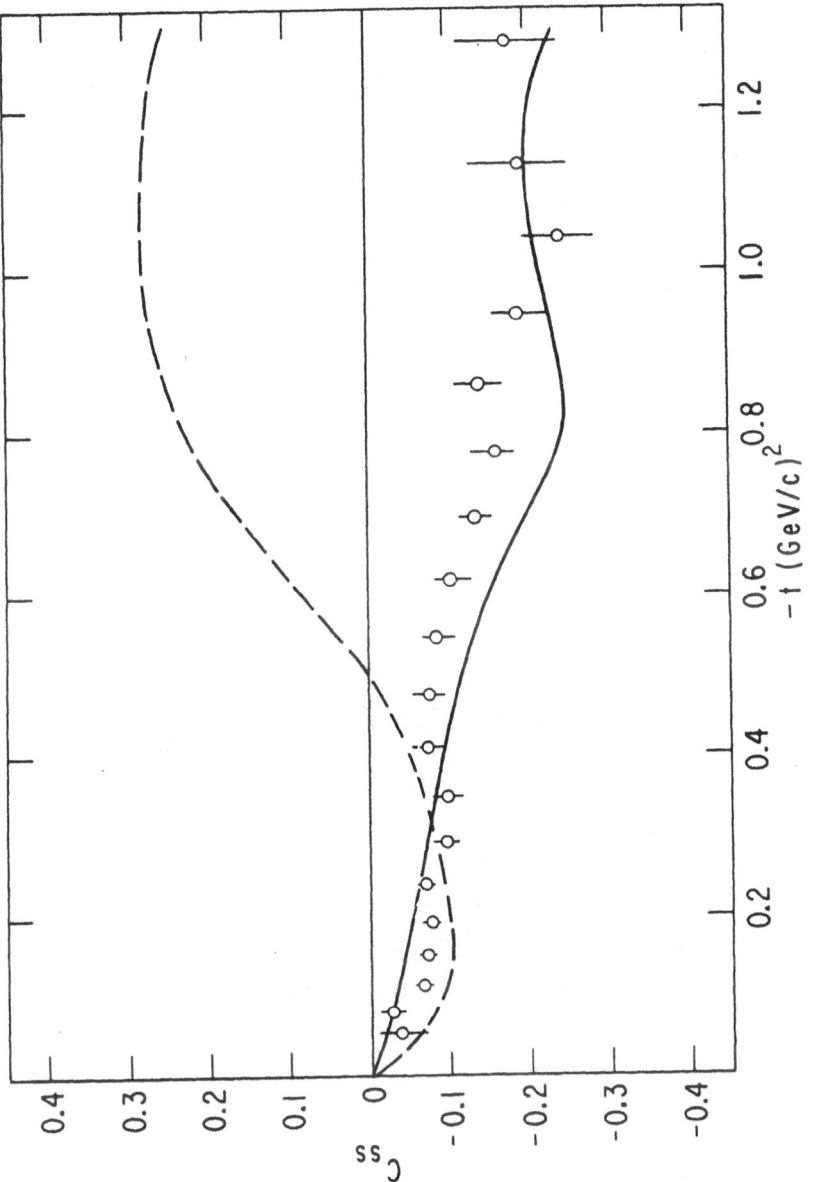

Fig. 3 C_{SS} at 6 GeV/c. The solid and dashed curves are Regge fits as described in the text.

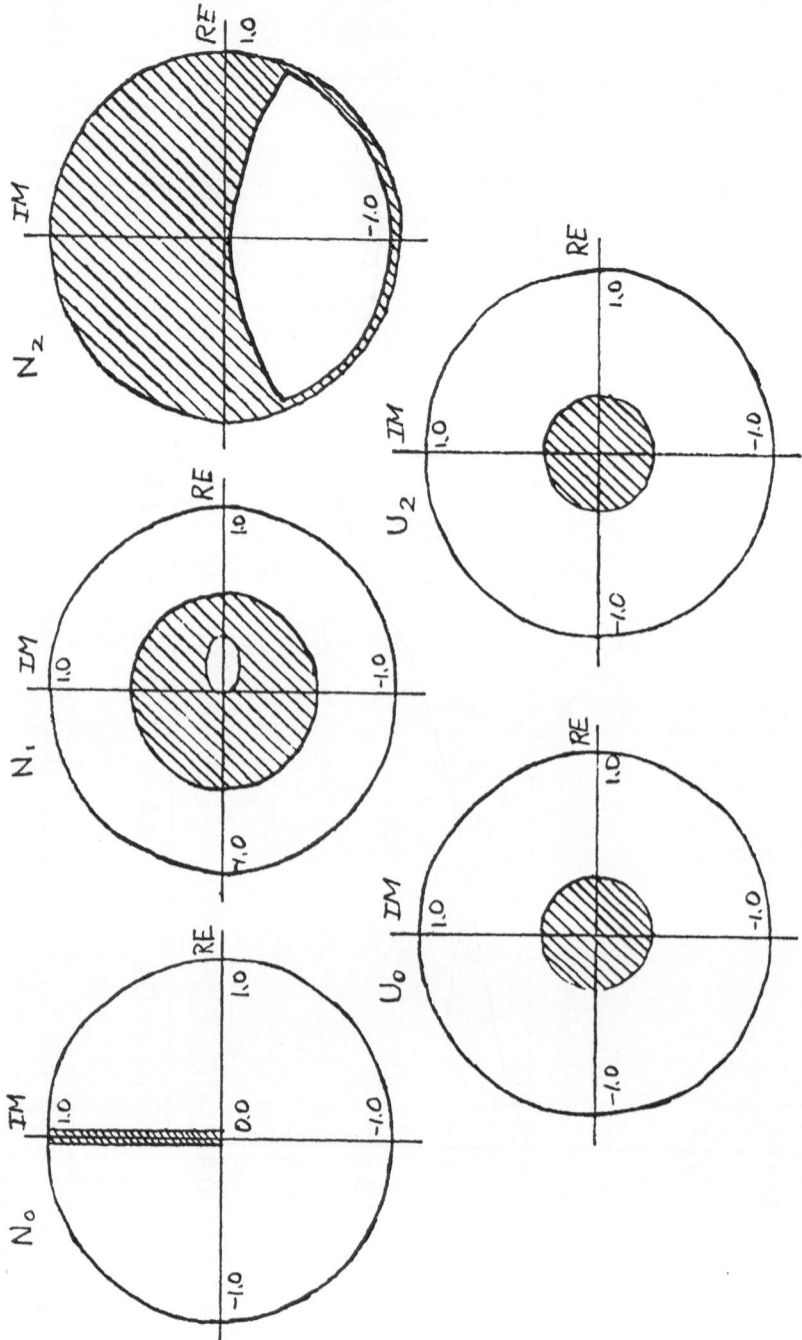

Fig. 4 The boundaries of t-channel Exchange Amplitudes (constrained by the five measurements).

constrained by the five N-type measurements (σ, P, D_{NN}, C_{NN}, and K_{NN}).

We show a typical experimental layout (C_{SS} measurement) in Fig. 5.

TOTAL CROSS-SECTION DIFFERENCE FOR pp SCATTERING

As shown in Table I, measurements of σ^{Tot}, $\Delta\sigma_T^{Tot}$, and $\Delta\sigma_L^{Tot}$ are to be made for the unknowns $Im\phi_1(0)$, $Im\phi_2(0)$, and $Im\phi_3(0)$. Our previous speaker (J. Roberts) discussed $\Delta\sigma_T^{Tot}$ measurements. I would like to discuss $\Delta\sigma_L^{Tot}$ measurements[3] from 1.17 to 2.50 GeV/c and preliminary results at 3.0 and 6.0 GeV/c. Our attempt was also motivated by interesting theoretical predictions by Berger, Pirila, and Thomas. This measurement was done in a standard transmission experiment. A striking energy dependence is observed with a maximum difference of -17 mb at P_{lab} = 1.47 GeV/c as shown in Fig. 6. In Fig. 7, $\sigma^{Tot}(\rightleftarrows)$ and $\sigma^{Tot}(\rightleftharpoons)$ are shown. The imaginary parts of $\phi_1(0)$ and $\phi_3(0)$ are shown in Fig. 8; $Im\phi_1(0)$ seems to vary linearly with log k over our energy range including 3.0 and 6.0 GeV/c. Our preliminary result of $\Delta\sigma_L^{Tot}$ at 6 GeV/c is 1.0 ± 0.1 mb and this implies that there exists an A_1 exchange term in pp scattering.

Several attempts to find dibaryon states have been unsuccessfully made but mostly in the reaction p + p $\rightarrow \pi^{\pm}$ + $X^{+,+++}$. However, a candidate for X^{++} has been the s-wave N-N* resonance which is fed from the pp initial state of $^3D_2$.

The question is whether the structure in $\sigma^{Tot}(\rightleftarrows)$ as shown in Fig. 6 is due to the NΔ threshold effect or a direct-channel resonance in the pp system. We note here that $\sigma^{Tot}(\rightleftarrows)$ consists only of partial waves in the triplet state and thus has nothing to do with an s-wave NΔ

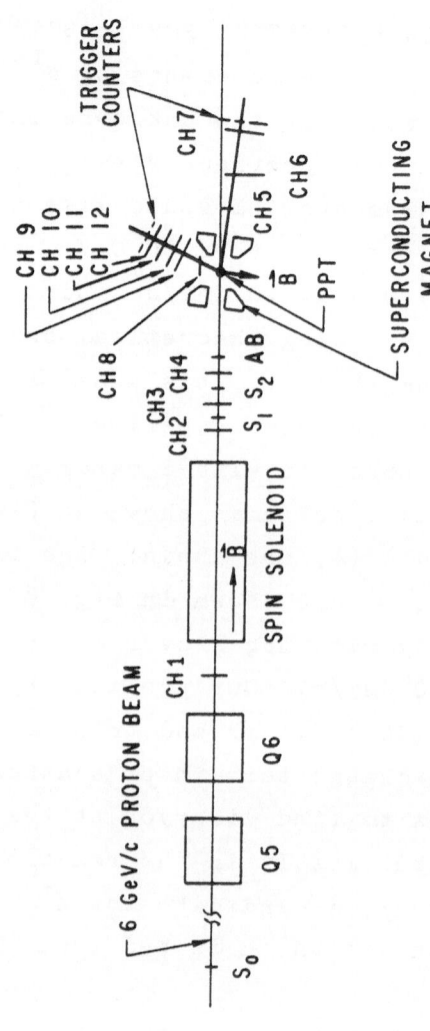

Fig. 5 Experimental apparatus. CH1 to CH12 are multi-wire proportional chambers. The S_0, S_1, S_2 and AB are scintillation counters. The drawing is illustrative only and not to scale.

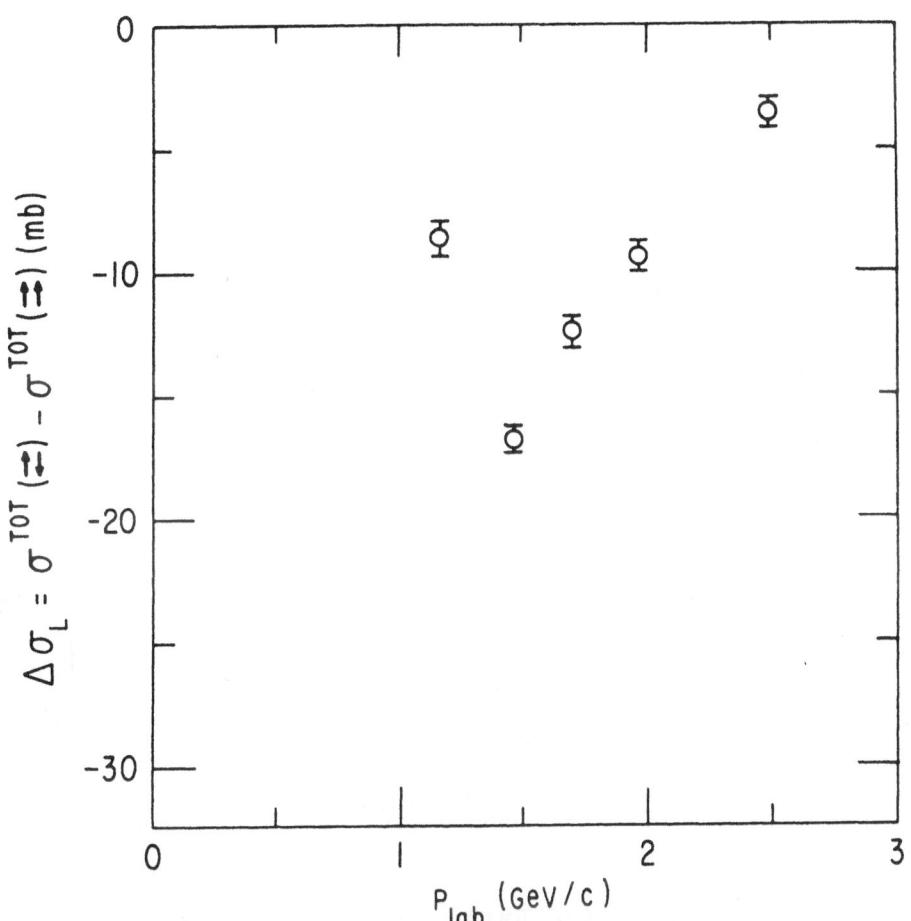

Fig. 6 Total cross-section difference $\Delta\sigma_L = \sigma^{Tot}(\overrightarrow{\leftarrow})$
$\sigma^{Tot}(\overrightarrow{\rightarrow})$. The error bars include statistical errors only.

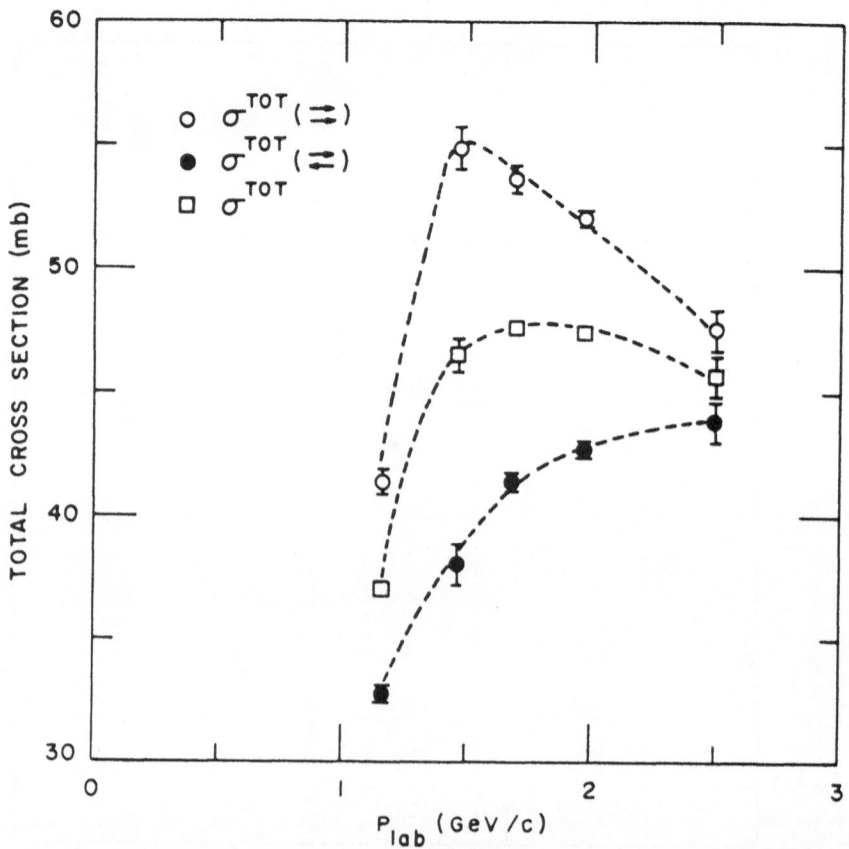

Fig. 7.

Total cross-sections for pure initial spin states. These
cross sections are calculated from $\Delta\sigma_L$ and previous
measurements of the spin-averaged total cross section.
The dotted curves are only to guide the eye.

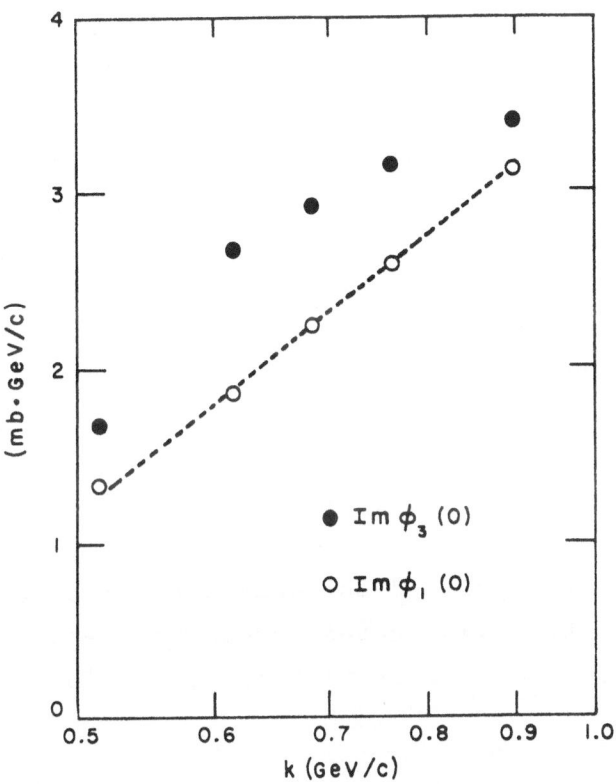

Fig. 8 The imaginary parts of the s-channel helicity
amplitudes ϕ_1 and ϕ_3 at zero angle. The dotted straight
line is to guide the eye.

threshold effect. One may test for the possibility of
a resonance effect by examining the behavior of the
imaginary and real parts of scattering amplitudes at
$|t| = 0$ and comparing them to the Breit-Wigner formula.
Existing data show that $\text{Re}(\phi_1 + \phi_3)$ at $|t| = 0$ varies
from positive to negative as the energy increases,
passing zero at ~1.7 GeV/c.[12]

In Fig. 9, the total cross section is shown together
with the elastic total cross section, $\Delta\sigma_{el}^{Tot}$, and in Fig.
10, σ_{el}^{Tot} is shown together with $\frac{1}{2}\sigma^{Tot}(\vec{\rightleftarrows})$. We observe a
similar bump structure in both curves. This elastic
bump occurs in the vicinity of the $N\Delta$ threshold; consequently, the bump has been hidden. One would expect
a similar structure in polarization results. In Fig.
11, maximum polarization at small $\theta_{c.m.}$ is plotted with
respect to incident momenta. The position of the bump
is around 1.3 GeV/c. A similar energy dependence is
observed in average polarization between $|t| = 0.1$ and
0.2 $(\text{GeV/c})^2$.

It has been pointed out that the effect of resonances can be investigated through the energy dependence
of the product of differential cross section and polarization.[13]

$$\sigma \cdot P = -2\text{Im}(N_0 - N_2)N_1^*, \text{ where}$$

$$N_0 = \tfrac{1}{2}(\phi_1 + \phi_3), \quad N_2 = \tfrac{1}{2}(\phi_4 - \phi_2), \text{ and } N_1 = \phi_5.$$

The structure in $\sigma^{Tot}(\vec{\rightleftarrows})$ is due to the ϕ_3 term whose
characteristics can be extracted from the experimental
data as

$$\sigma \cdot P = [\text{Im}\phi_3 \cdot \text{Re}\phi_5 - \text{Re}\phi_3 \cdot \text{Im}\phi_5] + \text{B.G.}$$

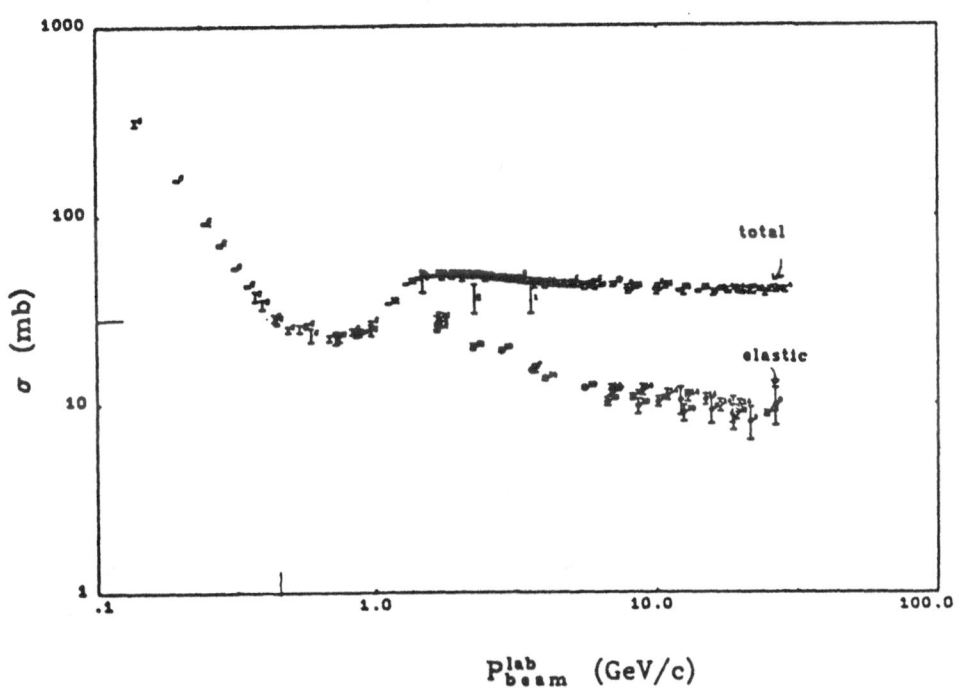

Fig. 9 Total cross section and elastic total cross section.

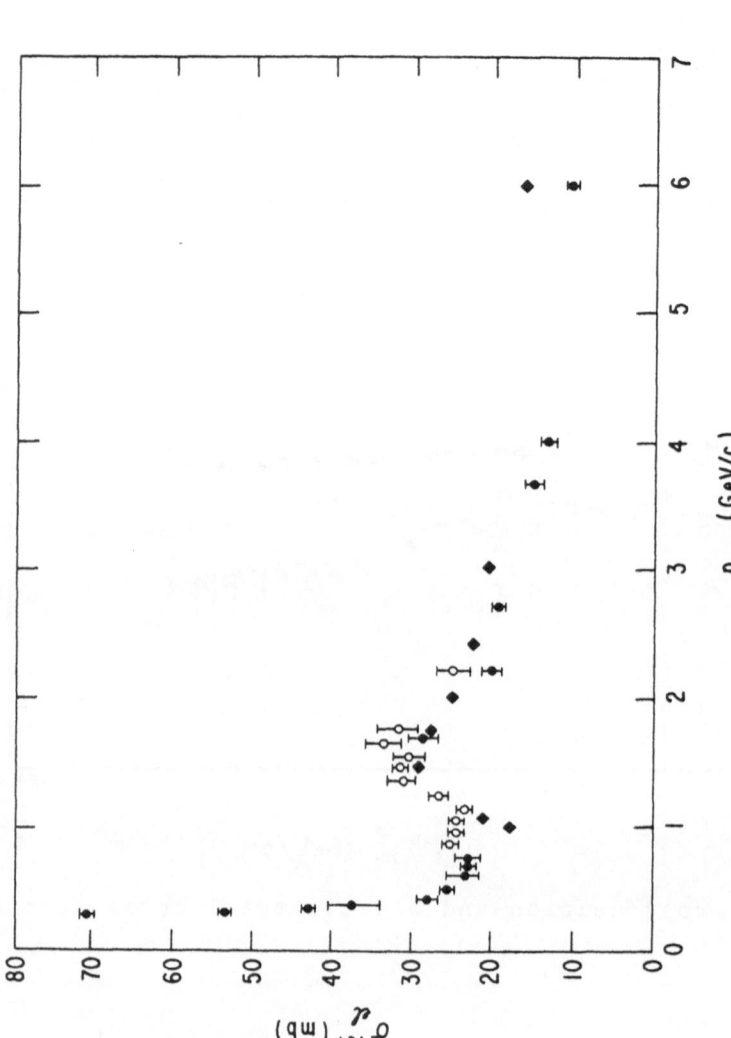

Fig. 10 Elastic total cross section by circles (white circles are calculated by integrating the differential cross sections) together with ½σ$^{Tot}(\vec{+})$ shown by diamonds (a point at 1 GeV/c was from a phase-shift solution).

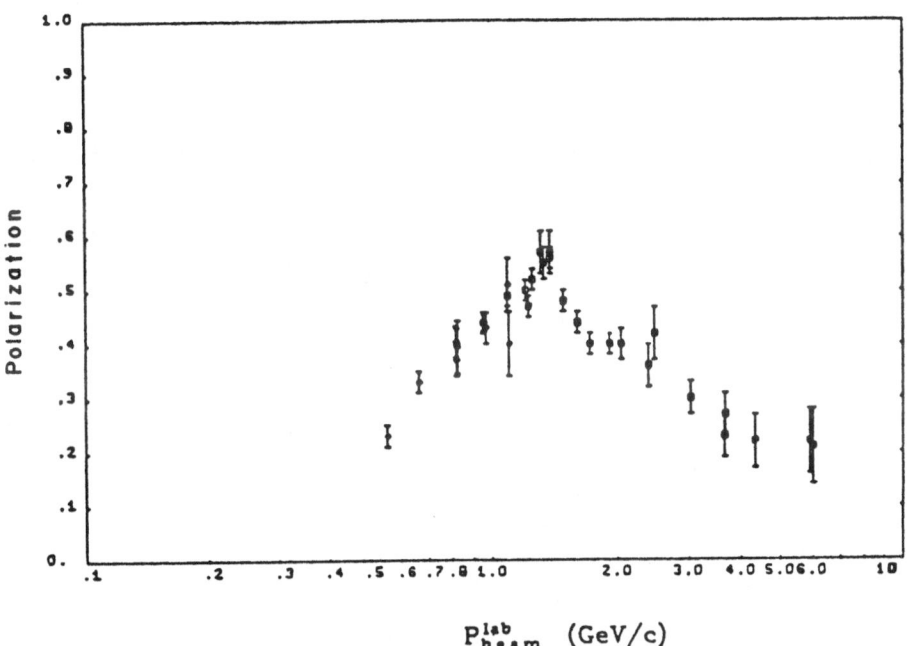

Fig. 11 Energy dependence of polarization

We are currently working on this subject.

PROPERTIES OF A DIBARYON RESONANCE

Results discussed in the previous section strongly suggest an existence of a dibaryon resonance, and we will list the properties of such a resonance.

Mass:	2250 MeV
Width:	$\sim$250 MeV
Charge:	2
Quantum number:	RJJ triple state (for I = 1 amplitudes, only for odd J); S=1, L=1, and J=1. An assignment of J=1 comes from the structureless behavior in cross sections and polarization.

In order to firmly establish this resonance, further measurements may be needed.

REFERENCES

1. D. Miller, C. Wilson, R. Giese, D. Hill, K. Nield,
    P. Rynes, B. Sandler and A. Yokosawa, Phys. Rev.
    Lett. 36, 763 (1976).

2. I. P. Auer, D. Hill, R. C. Miller, K. Nield,
    B. Sandler, Y. Watanabe, A. Yokosawa, A. Beretvas,
    D. Miller and C. Wilson, Phys. Rev. Lett. 37,
    1727 (1976).

3. I. P. Auer, E. Colton, D. Hill, K. Nield, B. Sandler,
    H. Spinka, Y. Watanabe, A. Yokosawa and A. Beretvas,
    to be published in Phys. Lett.

4. There exists a plan to construct an enriched anti-
    proton beam produced from $\bar{\Lambda}$ (private communication
    with B. Cox).

5. We are actually doing these to depolarize our pre-
    sent beam for FNAL #61 experiment (Argonne-Berkeley-
    FNAL-Harvard-Yale collaboration).

6. For instance, see M.J. Moravesik, The Two-Nucleon
    Interaction, Clarendon Press, Oxford (1963).

7. For details, see Halzen & Thomas, Phys. Rev. D10,
    344 (1974).

8. P. Johnson, G. Thomas and R.C. Miller, ANL-HEP-PR-
    76-61 (1976).

9. R. Field and P. Stevens, ANL/HEP CP 75-73, p. 28
    (1975).

10. N. J. Sopkovich, Nuovo Cimento 26, 186 (1962).

11. G. L. Kane and A. Seidl, Rev. Mod. Phys. 48,309
    (1976). (The whole amplitude rather than the pole
    term is approximately degenerate.)

12. A. Diddens, Proceedings of the XVII International
    Conference, London, July 1974, p. I-41.

13. A. Yokosawa, Phys. Rev. Lett. 16, 1019 (1966).

LARGE $P_\perp^2$ SPIN DEPENDENCE OF p-p ELASTIC SCATTERING[*]

K. Abe

Randall Laboratory of Physics

The University of Michigan

Ann Arbor, Michigan 48109

I would like to talk about our experiment which measured the pp elastic scattering differential cross sections using the Argonne ZGS polarized beam and our polarized target. The measurement was done at 11.75 GeV/c, which is presently the highest possible energy for this type of measurement. The data I am going to present here are based on a one month run[1] in February, 1976, when we made measurements up to $P_\perp^2 = 2.2$ (GeV/c)^2 and a two month run in November and December 1976, when we took measurements up to $P_\perp^2 = 3.6$ (GeV/c)^2. We are thus exploring the very high $P_\perp$ region where the inner structure of the proton should manifest itself in our measurement. At 11.75 GeV/c 90° scattering in the CM system corresponds to $P_\perp^2 = 5.2$ (GeV/c)^2.

---

[*]Work supported by U.S. Energy Research and Development Administration.

The cross section in this region is extremely small, it is about 100 nb at $P_\perp^2 = 3.6 \ (GeV/c)^2$. Since we must measure difference between the cross sections in different spin states at about the 1% level, we must accumulate about 20,000 events at each point to achieve the precision necessary to extract any meaningful physics. This was made possible by obtaining a very high intensity polarized beam and by developing a reliable method to anneal the polarized target after it lost polarization due to radiation damage.

The people who participated in this experiment are listed below.

MICHIGAN   K. Abe

                                           R.C. Fernow

                                           A.D. Krisch

                                           T.A. Mulera

                                           A.J. Salthouse

                                           B. Sandler

                                           K.M. Terwilliger

ARGONNE   J.R. O'Fallon

                                           L.G. Ratner

                                           P.F. Schultz

NORDITA/CERN   H.E. Miettinen

MAX PLANCK INST.   W. DeBoer

I will briefly discuss pp scattering amplitudes in the transversity representation,[2] which is what we directly measure. In our experiment the spins of both the beam and target protons are polarized in the so-called $\vec{n}$ direction which is perpendicular to the scattering plane. There are four combinations of initial spin states (beam, target = ↑↑ , ↑↓ , ↓↑ , ↓↓). There are also four combinations in the final states (scattered,

recoil $= \uparrow\uparrow, \uparrow\downarrow, \downarrow\uparrow, \downarrow\downarrow$). Therefore we have sixteen amplitudes. Of these the eight single flip amplitudes are
zero by parity conservation. Time reversal invariance
requires that $\sigma(\uparrow\uparrow \rightarrow \downarrow\downarrow)$ must be equal to $\sigma(\downarrow\downarrow \rightarrow \uparrow\uparrow)$.
Rotational invariance of space requires that for
identical particles $\sigma(\uparrow\downarrow \rightarrow \uparrow\downarrow) = \sigma(\downarrow\uparrow \rightarrow \downarrow\uparrow)$ and
$\sigma(\uparrow\downarrow \rightarrow \downarrow\uparrow) = \sigma(\downarrow\uparrow \rightarrow \uparrow\downarrow)$. Hence we are left with five
complex amplitudes labeled a,b,c,d,e:

$$
\begin{bmatrix} \uparrow\uparrow \\ \uparrow\downarrow \\ \downarrow\uparrow \\ \downarrow\downarrow \end{bmatrix}_{\text{final}}
=
\begin{bmatrix} a & o & o & b \\ o & c & d & o \\ o & d & c & o \\ b & o & o & e \end{bmatrix}
\begin{bmatrix} \uparrow\uparrow \\ \uparrow\downarrow \\ \downarrow\uparrow \\ \downarrow\downarrow \end{bmatrix}_{\text{initial}}
$$

These 5 quantities are the fundamental amplitudes for
p-p elastic scattering. These amplitudes are the ones
that can be calculated directly from the theory of
strong interactions that we hope to someday find. The
spin averaged cross sections which are measured with
conventional beams and targets are a complicated average
of these 5 quantities. Notice that if we measure the
spin of one of the final states in addition to the spin
of both initial states, we directly determine the magnitude of each amplitude[3]. In the present experiment we
specify the spins of both initial states but we know
nothing about the final spin states. Thus we are
measuring the sums of the squares of the amplitudes:

$$d\sigma/dt(\uparrow\uparrow \rightarrow 00) = 1/4\ [|a|^2 + |b|^2] = \langle d\sigma/dt\rangle\ (1 + 2A + C_{nn})$$

$$d\sigma/dt(\downarrow\downarrow \rightarrow 00) = 1/4\ [|b|^2 + |e|^2] = \langle d\sigma/dt\rangle\ (1 - 2A + C_{nn})$$

$$d\sigma/dt(\uparrow\downarrow\rightarrow 00) = 1/4 \; [|c|^2 + |d|^2] = <d\sigma/dt> \; (1 - C_{nn})$$

$$= d\sigma/dt \; (\downarrow\uparrow\rightarrow 00)$$

where 0 means not measured. We can express these two-
spin cross sections in terms of the spin averaged cross
section $<d\sigma/dt>$, the asymmetry or polarization parameter
A, and the spin-spin correlation parameter C_{nn}. The
parameter A measures the difference in cross section
depending on whether the initial spins are parallel or
antiparallel to the orbital angular momentum. Therefore
it measures the spin-orbit force. The spin-spin correlation parameter measures the difference in cross
section depending on whether the spins of the two
protons are parallel or antiparallel to each other.
Thus it measures the spin-spin force.

The experimental set up is shown in Fig. 1. The
11.75 GeV/c polarized protons first elastically scatters
in the liquid hydrogen target. The elastic events were
detected by two identical double arm spectrometers L
and R. From the observed left-right asymmetry and the
A parameter known from other experiments we determined
the beam polarization, which was typically about 50%.

$$P_B = 1/A \; (\frac{L-R}{L+R})$$

The beam continued downstream and then scatters in
the polarized proton target (PPT). The elastic events
from the PPT were detected by the FB double arm
spectrometer for each of the four initial spin states
($\uparrow\uparrow$, $\uparrow\downarrow$, $\downarrow\uparrow$ and $\downarrow\downarrow$).

The beam intensity during the last month was
typically 7×10^9 protons per pulse. This high

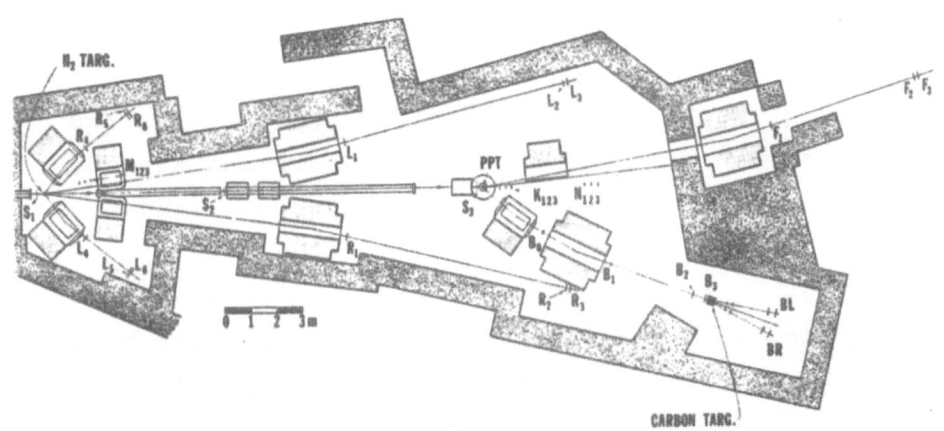

Fig. 1

intensity and the large fraction of non-free protons in
the polarized proton target made the counting rate of
each detector extremely high. However, by tightly con-
straining the angle and momentum of both the scattered
and recoil particles, we obtained a clean elastic
signal.

The target polarization was typically 65% and its
direction was reversed every several hours. The beam
polarization was reversed from pulse to pulse. Thus,
for each of the four different initial spin states we
measured the elastic event rate from which we obtained
the various $d\sigma/dt(ij \to 00)$ and A and C_{nn}. They are listed
in Table 1.

Fig. 2 shows A and C_{nn} plotted against $P_{\perp}^2$. The
black dots are from our February measurement and the
white squares are from the November-December runs. Note
that A has a sharp zero at $P_{\perp}^2 = 0.7$ $(GeV/c)^2$ and then a
maximum at $P_{\perp}^2 = 1.4$ $(GeV/c)^2$. It then falls smoothly
toward zero in the larger $P_{\perp}^2$ region. There has been
speculation for a few years that there might be a second
minimum at around $P_{\perp}^2 = 2.0$ $(GeV/c)^2$. In fact an Eikonal
model[4] tends to favor such structure. However it appears
to be ruled out by this data.

The structure of C_{nn} is quite interesting. It has
a very sharp minimum at $P_{\perp}^2 = 0.9$ $(GeV/c)^2$ and develops
into a broad maximum just like A. It then falls toward
zero at about $P_{\perp}^2 = 3(GeV/c)^2$. Although the present
statistical precision is limited it appears that C_{nn}
may rise again at $P_{\perp}^2 = 3.6$ $(GeV/c)^2$.

Both A and C_{nn} increase near the region where the
spin averaged cross section starts to break from the
steep diffractive peak.[5] Since the cross section has
another break at around $P_{\perp}^2 = 3.8(GeV/c)^2$, it is

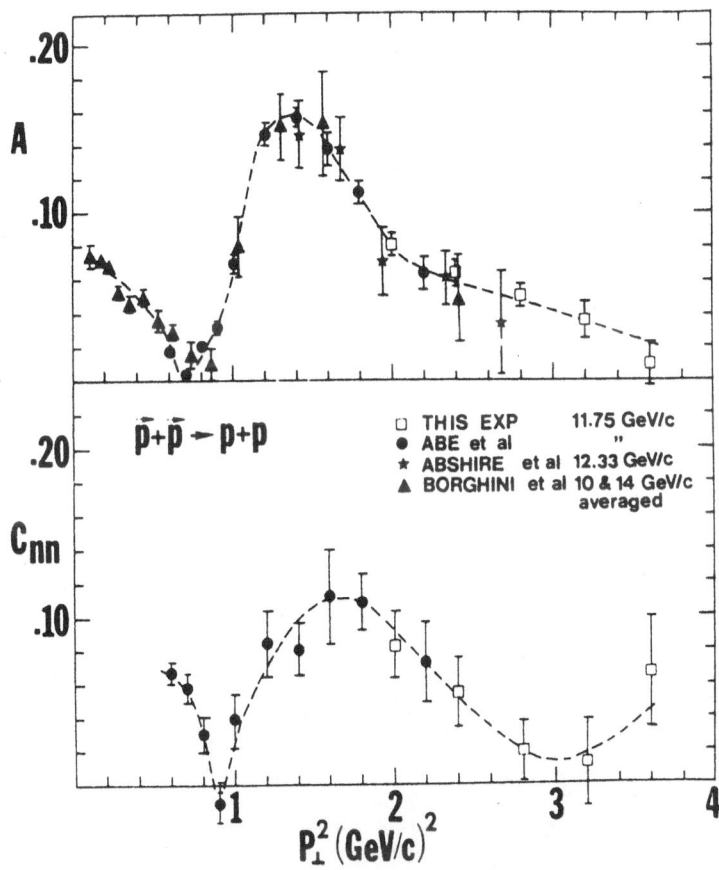

Fig. 2

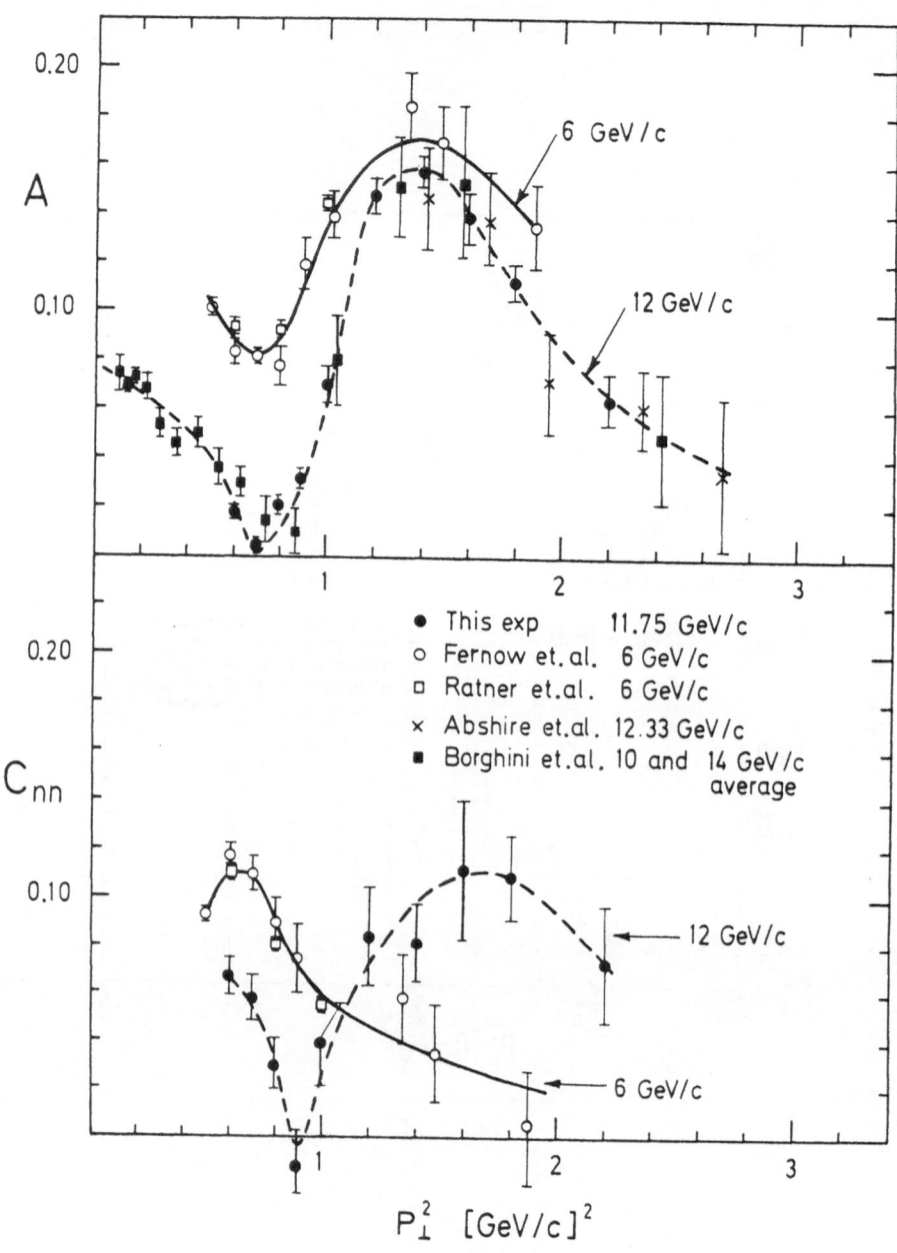

Fig. 3

particularly interesting to find out if this rise of C_{nn} is a real effect or not. We intend to explore this very large $P_\perp^2$ region this coming spring.

Fig. 3 shows our earlier data plotted together with the 6 GeV/c data to see the energy dependence of A and C_{nn}. We can see that they depend on the incident energy quite differently in the small $P_\perp^2$ diffraction peak and in the large $P_\perp^2$ region. In the diffractive peak both A and C_{nn} decrease continuously with energy, indicating that the diffractive interaction becomes more dominant as the energy increases. However, at large $P_\perp^2$ A is roughly independent of energy and C_{nn} is growing with incident energy. This seems to be telling us that spin plays an important role in large $P_\perp^2$ interactions at high energy.

Fig. 4 shows our data plotted in terms of two-spin cross sections. This gives an intuitive picture as to what role spin is playing in pp scattering. In the forward peak, we can hardly see the difference between the three two-spin cross sections. However, the spin dependence becomes very large just after the break. At $P_\perp^2 = 1.4$ $(GeV/c)^2$ $d\sigma/dt(\uparrow\uparrow)$ is twice as large as $d\sigma/dt(\downarrow\downarrow)$. This is the $P_\perp^2$ region in which the spin averaged cross section develops a dip at momenta above 200 GeV/c.[6]

The data so far indicate that the first break in the cross section is somehow associated with a large spin dependence which stays large even toward higher energies where people thought the spin dependence would be negligible. We are very eager to see what the spin dependence looks like at the next break in the cross section which occurs near $P_\perp^2 = 3.8(GeV/c)^2$. We hope to get some data in this very large $P_\perp^2$ region this spring.

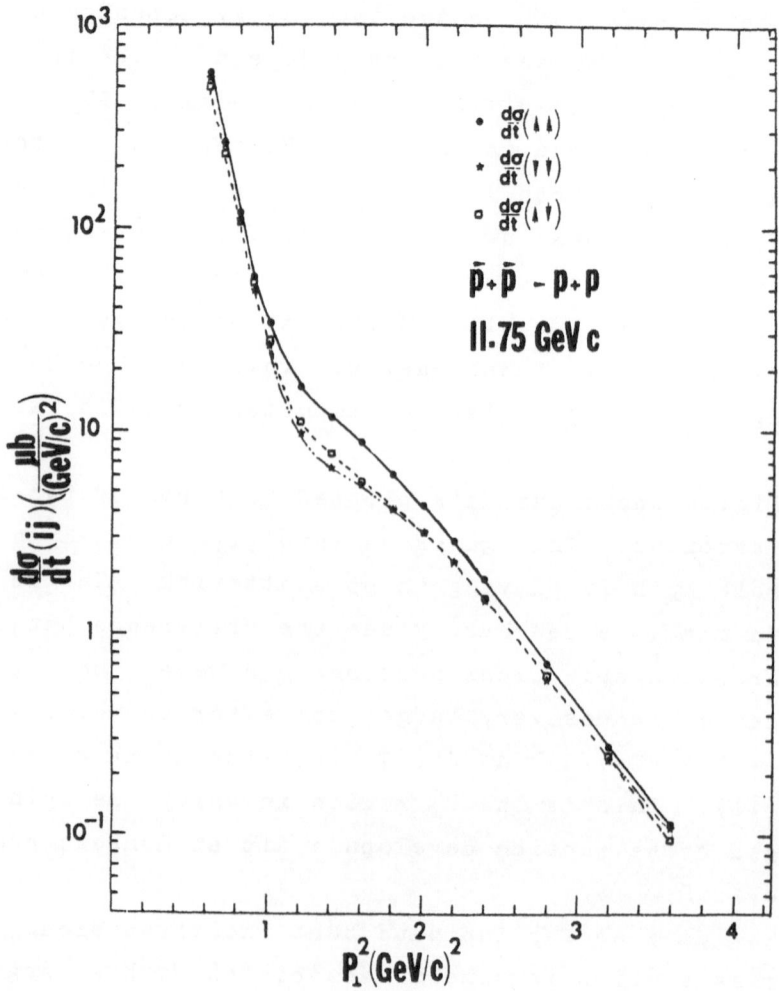

Fig. 4

Maybe this will somehow tell us if these different regions are due to spinning layers in the proton or due to the proton containing constituent quarks with spin.

TABLE I

$P_\perp^2$ $(GeV/c)^2$	A (%)	C_{nn} (%)
2.0	8.1± .7	8.3±2.0
2.4	6.4± .8	5.5±2.1
2.8	5.1± .7	2.0±1.8
3.2	3.6±1.1	1.3±2.6
3.6	1.0±1.3	6.7±3.3

REFERENCES

1. K. Abe et al., Phys. Lett. 63B, 239 (1976).

2. R.C. Fernow, American Journal of Physics Vol. 44, No. 6, June 1976.

3. L.G. Ratner et al., Phys. Rev. D15, (1977).

4. A.W. Hendry et al., Phys. Rev. D10, 3662 (1974).

5. C.W. Akerlof et al., Phys. Rev. 5, 1138 (1967).

6. C.W. Akerlof et al., Phys. Rev. D14, 2864 (1976).

ELASTIC AND INELASTIC POLARIZATION EFFECTS OBSERVED WITH THE ARGONNE EFFECTIVE MASS SPECTROMETER*

D.S. Ayres, D. Cohen, R. Diebold, S.L. Kramer,

A.J. Pawlicki and A.B. Wicklund

(Presented by R. Diebold)

Argonne National Laboratory, Argonne, Ill.60439

ABSTRACT

The Argonne polarized proton beam has been used together with the Effective Mass Spectrometer to study several different aspects of pp and pn interactions. The polarization asymmetry for pn elastic scattering was measured for the first time above cyclotron energies. The results show that both $I = 0$ and $I = 1$ exchanges have important contributions to the t-channel single-flip amplitudes. While the $I = 1$ amplitude has the energy dependence expected, the $I = 0$ contribution falls much faster with energy. For $p_\uparrow p \to \Delta^{++} n$ we find a left-right production asymmetry of about 40% for $-t \geq 0.5$ GeV^2, independent of energy. Absorbed one-pion-exchange together with quark-model calculations give a qualitative explanation of the Δ^{++} decay distributions and their

---

*Work supported by the U.S. Energy Research and Development Administration.

dependence on the beam polarization. In the diffrac-
tive process $p_\uparrow p \to \Delta^{++} \pi^- p$ we observe asymmetries in the
low mass $\Delta\pi$ system that can be described by interference
between an S-wave Deck background and a P-wave N^*_{1470}
resonance contribution.

ELASTIC SCATTERING ASYMMETRY

The polarization asymmetry for pp elastic scattering
has been measured extensively over the years and has
been found to fall more rapidly with energy than expected by Regge-exchange models.[1] The asymmetry parameter for pn elastic scattering is closely related to
that for pp, and together these parameters can be used
to separate the I = 0 and I = 1 t-channel exchange
contributions to the single spin-flip amplitude. Pure
I = 0 exchange, as might be expected in optical models,
would result in equal asymmetries for the two reactions.
A single spin-flip amplitude with pure I = 1, on the
other hand, would give mirror symmetry, A(pn) = -A(pp),
similar to that for $\pi^\pm p$ elastic scattering.

Both A(pp) and A(pn) were measured at 2,3,4 and 6
GeV/c with typically 400 000 events at each energy.[2]
More recently we have taken over 10^6 events at 12 GeV/c.
The spectrometer[3] is shown in Fig. 1. The polarized
beam was scattered in a 20-inch liquid deuterium target;
the angle and momentum of the fast scattered proton were
measured in the spectrometer, magnetostrictive wire
spark chambers placed about an SCM-105 magnet (66-cm
high gap by 2 m wide with 11.4 kGm bend). The resolution
on the beam and spectrometer momentum was sufficient to
reject inelastic events, and the recoil counters were
used to distinguish pp from pn scatters. Various small
corrections were made for cross talk between the pp and

pn samples and for coherent elastic scattering off the deuteron as a whole.

The 2 to 6 GeV/c results are shown in Fig. 2. Both the pp and pn asymmetries show a broad positive maximum near -t = 0.3 GeV^2. As shown by previous experiments above 1.5 GeV/c, A(pp) = 0.75/p at -t = 0.3 GeV^2, where p is the beam momentum in GeV/c. The pn

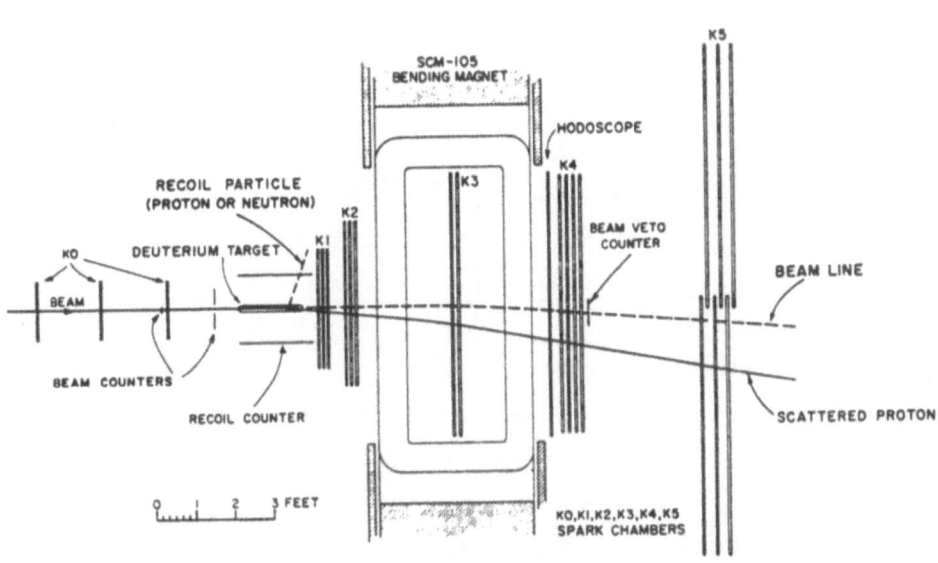

Fig. 1   Sketch of the Argonne Effective Mass Spectrometer as used to study the polarization asymmetry of pp and pn elastic scattering.

asymmetry falls considerably faster with energy; the
ratio A(pn)/A(pp) at $-t$ = 0.3 GeV^2 falls from 0.78 ±
0.02 at 2 GeV/c to 0.22 ± 0.03 at 6 GeV/c.

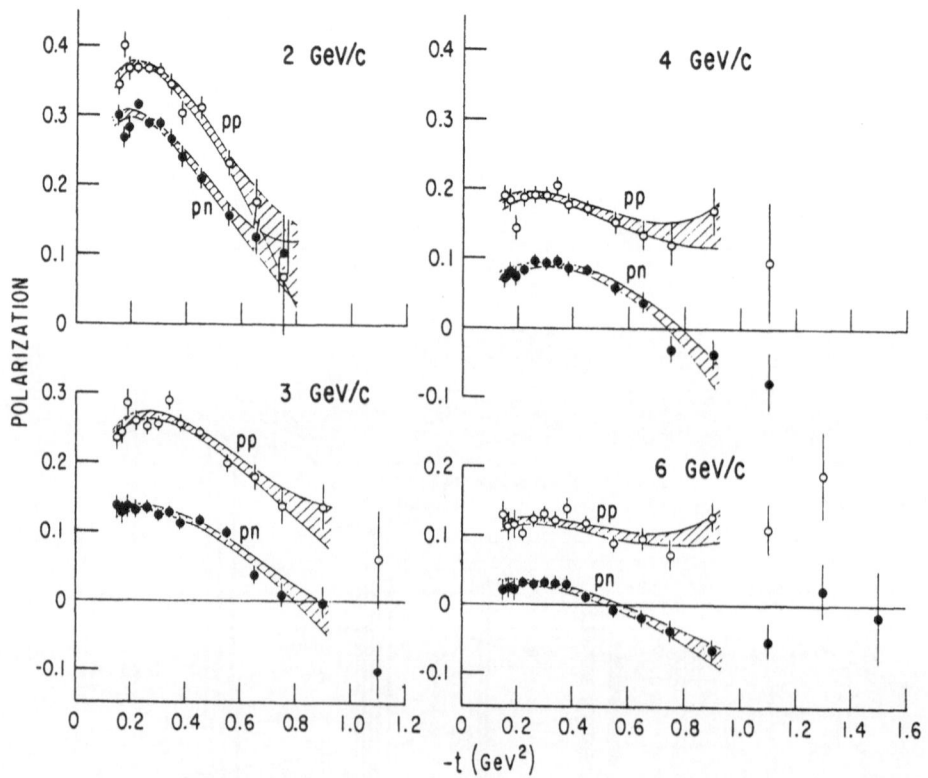

Fig. 2 Polarization asymmetries for pp and pn elastic
scattering at incident momenta of 2 to 6 GeV/c (Ref. 2).
The errors are statistical only and do not include the
±6% scale uncertainty from the beam polarization. Fits
to the data from 0.15 to 1.0 GeV^2 using the form
$P = \sqrt{-t}\ (a+bt+ct^2)$ are shown as bands (±1 standard
deviation).

We can combine the pp and pn results to separately determine the energy dependence of the I = 0 and
I = 1 exchange contributions to the single flip amplitude. Using the notation of Halzen and Thomas,[4]

$$\frac{d\sigma}{dt} = |N_0|^2 + 2|N_1|^2 + |N_2|^2 + |\pi|^2 + |A|^2 , \qquad (1)$$

$$A \frac{d\sigma}{dt} = 2 \text{ Im } (N_0-N_2)^* N_1 \qquad (2)$$

Near the forward direction the diffractive (I = 0) nonflip amplitude N_0 dominates, and the I = 0 and I = 1
contributions to the component of the single spin-flip
amplitude orthogonal to N_0 in the complex plane are
given by

$$4N_{1\perp}(I=0) \approx (A\sqrt{d\sigma/dt})_{pp} + (A\sqrt{d\sigma/dt})_{pn} , \qquad (3)$$

$$4N_{1\perp}(I=1) \sim (A\sqrt{d\sigma/dt})_{pp} - (A\sqrt{d\sigma/dt})_{pn} .$$

These quantities were fit to the form $N_1 \propto p^{\alpha_{eff}-1}$ over
the range 3 to 6 GeV/c, and the results for the effec-
tive trajectories are shown in Fig. 3. The I = 1 values
for α_{eff} are consistent with the trajectory expected for
ρ and A_2 Regge exchanges, but the I = 0 amplitude has
a much steeper energy dependence, with a typical value
of $\alpha_{eff} = -0.6$ at $-t = 0.3$ GeV^2.

The t dependence of the I = 0 and 1 flip amplitudes
for pp and pn elastic scattering[2] are compared in Fig. 4
with those for $\pi^{\pm}p$ scattering.[5] The double zero observed for the I = 1 exchange in πp scattering, usually
explained in terms of $\alpha_\rho = 0$ near $-t = 0.6$ GeV^2, is
presumably filled in by the A_2-exchange contribution to
pp and pn scattering.

The ratio A(pn)/A(pp) at 6 GeV/c is shown in
Fig. 5. Since the pp and pn differential cross
sections are the same to within measurement errors,
optical models would naturally predict unity for this
ratio. The spin-orbit model[6] discussed at this
conference by Halzen has an adjustable coupling strength
and would predict the ratio to be constant, although
not necessarily +1. The ratio at 6 GeV/c does not
follow this expectation, however, going from +0.25 at
$-t = 0.3$ GeV^2 to -0.5 near 1 GeV^2.

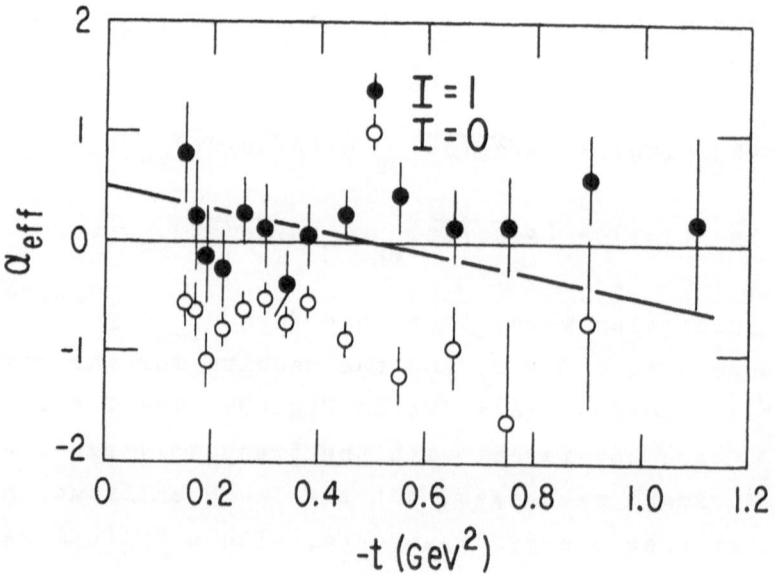

Fig. 3 Effective Regge trajectories derived from fits
to the form $N_1 \propto p^{\alpha_{eff}-1}$ from 3 to 6 GeV/c, with
N_1 for I=0, 1 given by Eqs. (3 and 4). In addition
to the statistical errors shown, there is a systematic
uncertainty in α_{eff} of ± 0.12. The straight line shows
the energy dependence expected from Regge-pole exchange
with $\alpha_{eff} = 0.5 + t$.

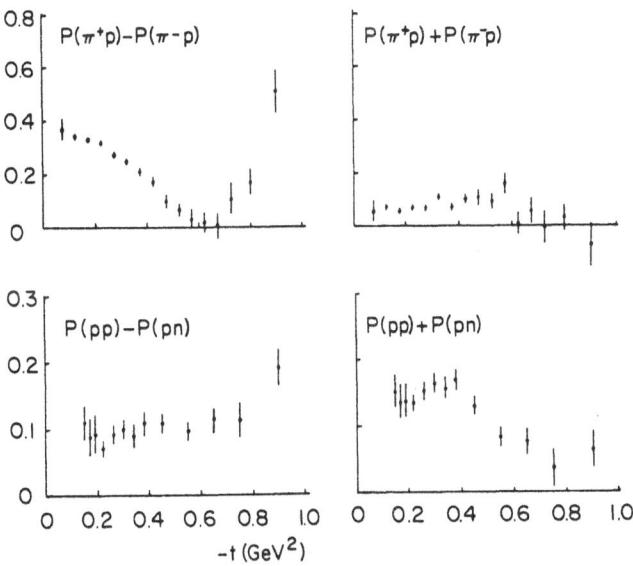

Fig. 4 Sums and differences of the polarization asymmetries for $\pi^{\pm}p$ (Ref. 5) and pp, pn (Ref. 2) elastic
scattering at 6 GeV/c. The sums (differences) indicate
the form of the I=0 (I=1) spin-flip amplitudes for the
two reactions.

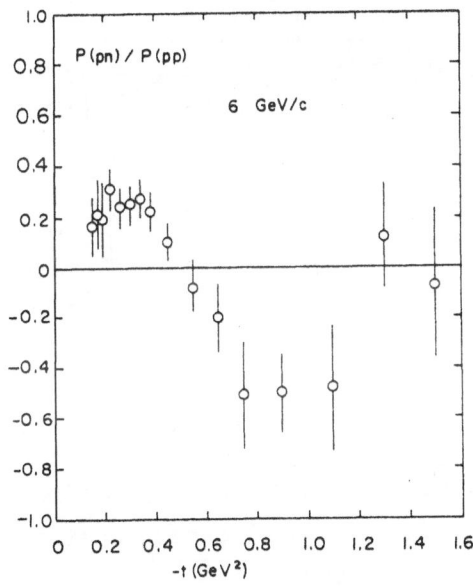

Fig. 5 The ratio of np to pp polarization asymmetry at
6 GeV/c.

The 2 to 6 GeV/c results inspired several theoretical models which made the predictions shown in
Fig. 6 for A(pn) at 12 GeV/c. Preliminary experimental
results are in good agreement with the models of Field-
Stevens[7] and Dash-Navelet[8] which explicitly put in
low-lying I=0 trajectories. The models of Irving[9] and
of Bourrely et al.,[10] which include a Pomeron flip
amplitude (out of phase with the Pomeron nonflip), do
not seem to do as well.

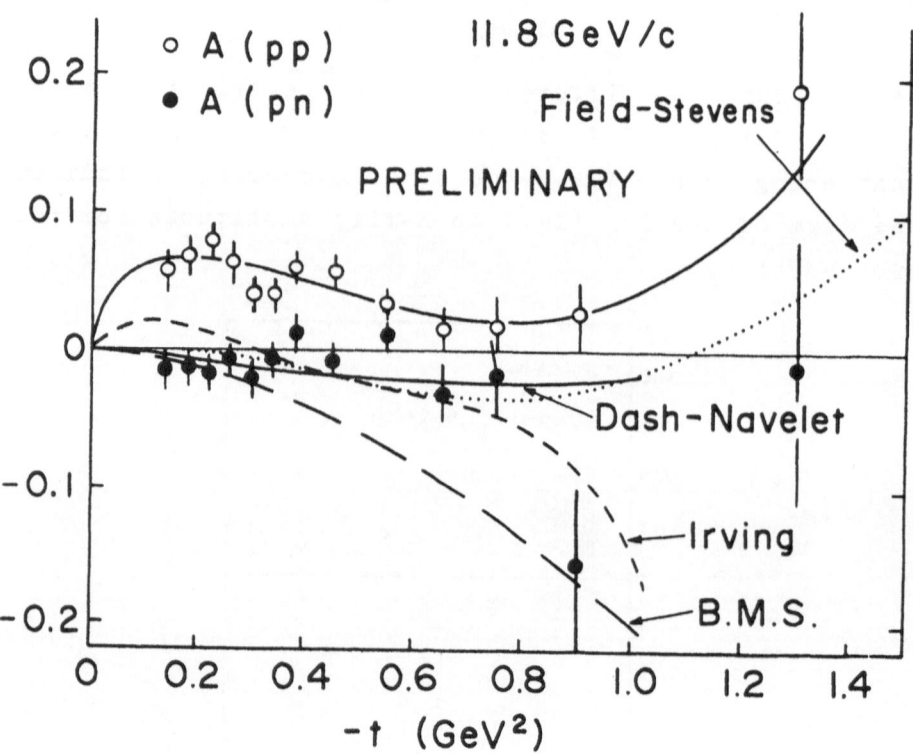

Fig. 6 Preliminary results on pn and pp polarization
asymmetries at 11.8 GeV/c, compared with the predictions of Refs. (7-10).

INELASTIC POLARIZATION EFFECTS

We have accumulated considerable data on various exclusive inelastic processes. Some of the early results have been discussed at previous symposia.[11] For these studies we have used a large aperture threshold gas Cerenkov counter behind the spectrometer to help distinguish pions and protons. The 40-counter hodoscope was used to require two or three particles through the spectrometer.

The spin dependence of inelastic reactions of the type $p_\uparrow p \rightarrow N^* N$ reflects both the spectroscopy and the production dynamics of the N^* resonances. In the production of a pure state as in $p_\uparrow p \rightarrow \Delta^{++} n$, polarization effects arise from interference of different exchange mechanisms such as π, B, ρ, and A_2. In the case of diffractive reactions, specifically $p_\uparrow p \rightarrow (p\pi^+\pi^-)p$, the physics interest lies more in the spectroscopy than in the production mechanisms. While the partial wave interference effects cannot be predicted reliably, as they can for π-exchange processes, the spin dependence can be used to probe the partial wave content.

The mass distributions for $p\pi^\pm$ at 6 GeV/c are shown in Fig. 7, corrected for acceptance. The figure represents about 140 000 $p_\uparrow n \rightarrow p\pi^- p$ events and 800 000 $p_\uparrow p \rightarrow p\pi^+ n$ events. Note that we are observing the excitation of the polarized beam proton into a $p\pi^\pm$ system. While the $\Delta^{++}(1236)$ dominates the $p\pi^+$ spectrum, there is also a substantial enhancement in the 1900-MeV region. The acceptance is dropping rapidly, but there are still ~ 1000 events/5 MeV in this region. The first, second and third resonances are all visible in the $p\pi^-$ system.

These processes are dominated at low t by scattering off the virtual pion cloud of the target, as indicated

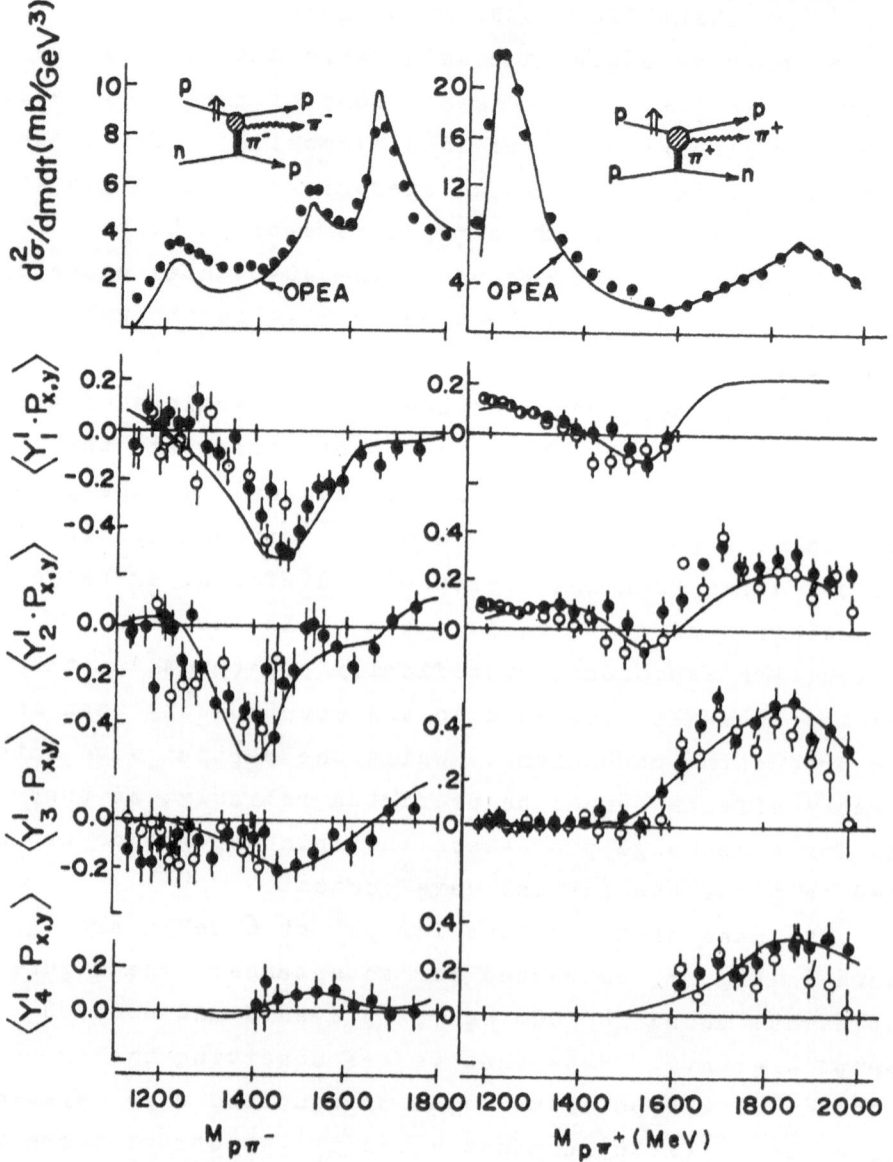

Fig. 7 Preliminary results on the cross sections and
m=1 moments for pn → pπ⁻p and pp → pπ⁺n at 6 GeV/c and
-t ≤ 0.2 GeV², compared with the OPE predictions. The
open (solid) points show the correlation of the m=1
moments with beam polarization along (orthogonal to)
the production normal.

by the Feynman diagrams in Fig. 7. The mass distributions are given qualitatively by the OPEA curves, while
the polarization effects in the m=1 moments, $\langle Y_L^1 P_{x,y} \rangle$,
follow the polarization effects seen in formation ex-
periments with pion beams on a polarized target.

The overall production asymmetry for $p_\uparrow p \to \Delta^{++} n$ was
measured with about 500 000 events each at 3, 4, and
6 GeV/c. These asymmetries are large except in the
small -t region where π exchange dominates (Fig. 8),
and they are consistent with an energy-independent
smooth curve. Since we measure spin dependence only at
the $p_\uparrow \to \Delta^{++}$ vertex, and not at the recoil $p \to n$ vertex,
A_1-like exchange amplitudes contribute only in qua-
drature and are presumably negligible. Thus the
polarization which we observe arises mainly from inter-
ferences such as $\pi.B$, $\rho.A_2$, etc.

To understand the mechanics responsible for the
large production asymmetry, we must examine all the
correlations between the proton spin and Δ^{++} decay
angles (after correcting for nonresonant background
contributions) in terms of the underlying production
amplitudes. We make the usual assumption that s-channel
helicity nonflip terms are negligible at the nucleon
vertex (e.g., A_1 or nonflip ρ, A_2 exchange). The
physics then depends on eight production amplitudes including four natural and four unnatural-parity exchange
terms (N.P. and U.P. for short). We denote these by
N^{2M} (for N.P.) and U^{2M} (for U.P.), which describe the
four possible transitions from a helicity $+\frac{1}{2}$ proton to
a helicity $M(\pm 3/2, \pm\frac{1}{2})$ Δ^{++}. The N.P. and U.P. terms do
not interfere, but their contributions cannot be sepa-
rated without a model.

In a preliminary analysis,[12] it was shown that the

quark model, which gives a recipe for separating N.P.
and U.P. observables, successfully relates the unpolarized density matrix elements in Δ^{++} production to
those in the exotic channel $K^+n \rightarrow K^{*0}p$. In the latter
reaction, N.P. exchange is suppressed compared with the
nonexotic channel $K^-p \rightarrow \bar{K}^{*0}n$; this suppression occurs
because the exchange degenerate (EXD) $\rho + A_2$ contribution, which is mainly real, is cancelled by the $\pi+B$ cut
contributions.[13] If the quark model is at least qualitatively valid, then this suppression, which depends on

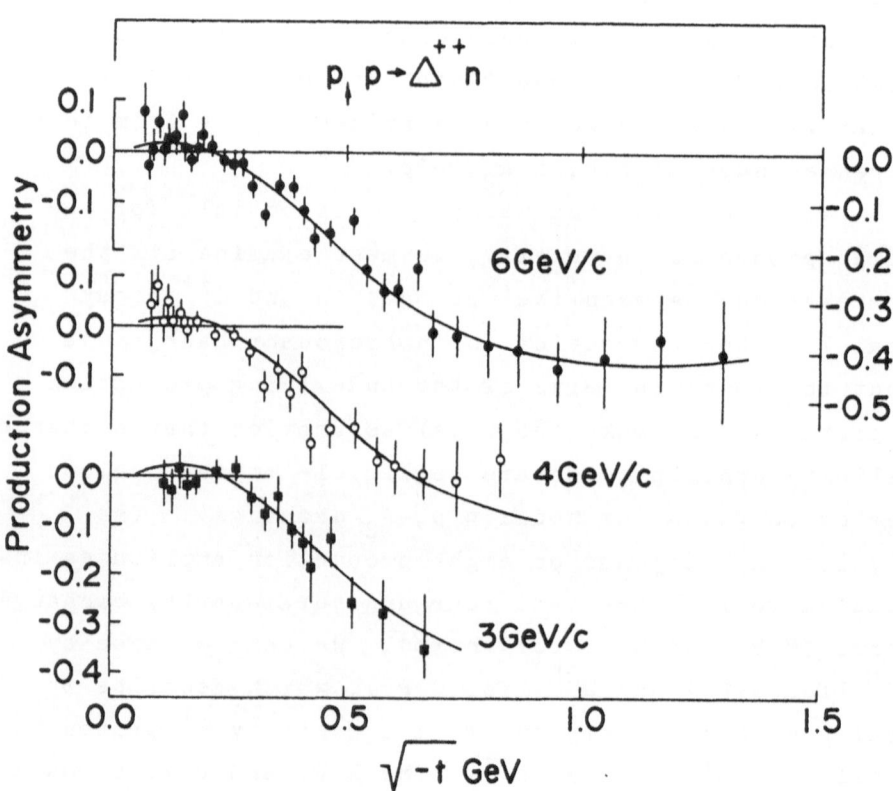

Fig. 8 Overall left-right asymmetries in $p_\uparrow p \rightarrow \Delta^{++}n$ at
3, 4, and 6 GeV/c. The curve is an eyeball interpola-
tion of the 6 GeV/c data.

the signs of the coupling constants, should also
characterize pp → Δ^{++}n. In that case the N.P. Δ^{++}
production cross section is quite small (less than 20%
of the cross section for -t <0.5 GeV^2) and unless the
phases of the N.P. amplitudes violate EXD maximally,
the N.P. contribution to polarization effects should be
negligible.

The 10 observables are shown in Fig. 9. The curves
show the results from a model having the following in-
gredients: the N.P. terms are given by the quark model
as stated above; the U.P. amplitudes U^3, U^1, U^{-1}, and
U^{-3} are specified by "Poor Man's Absorption Model" π exchange;[14] the U^1 amplitude also has a B-exchange contri-
bution defined by $U^1(B)/U^1(\pi) = -i\pi/2|t|$. The phase of
the B-exchange contribution is related by the quark
model and SU(3) to measured ρ-ω interference phases.[15]
This model breaks EXD, as it must to give nonzero spin
correlations, by constructing the amplitude U^1 to be
out of phase with U^3, U^{-1} and U^{-3}. The reason for
assuming that B couples mainly to U^1 (helicity $\frac{1}{2}\Delta$'s) is
that a similar coupling pattern is observed in the quark-
model-SU(3) related processes $\pi^-p \to \rho^0 n$ and $\pi^-p \to \omega n$.

The model then predicts that A_{33}, $A_{31} - I_{31}$, and
$A_{3-1} + I_{3-1}$ vanish, in rough agreement with Fig. 9. On
the other hand, observables that involve U^1 do not vanish
and their signs are predicted unambiguously, namely
A_{11}, $A_{31} + I_{31}$, and $A_{3-1} - I_{3-1}$. Although there are
discrepancies (especially in $A_{3-1} - I_{3-1}$), the model
shows how the physics of pp → Δ^{++}n can be related to
that in analogous vector meson production processes, and
provides a qualitative explanation for the EXD breaking
manifested in the polarization effects.

We now turn to the diffractive reaction $p_\uparrow p \to \Delta^{++}\pi p$.

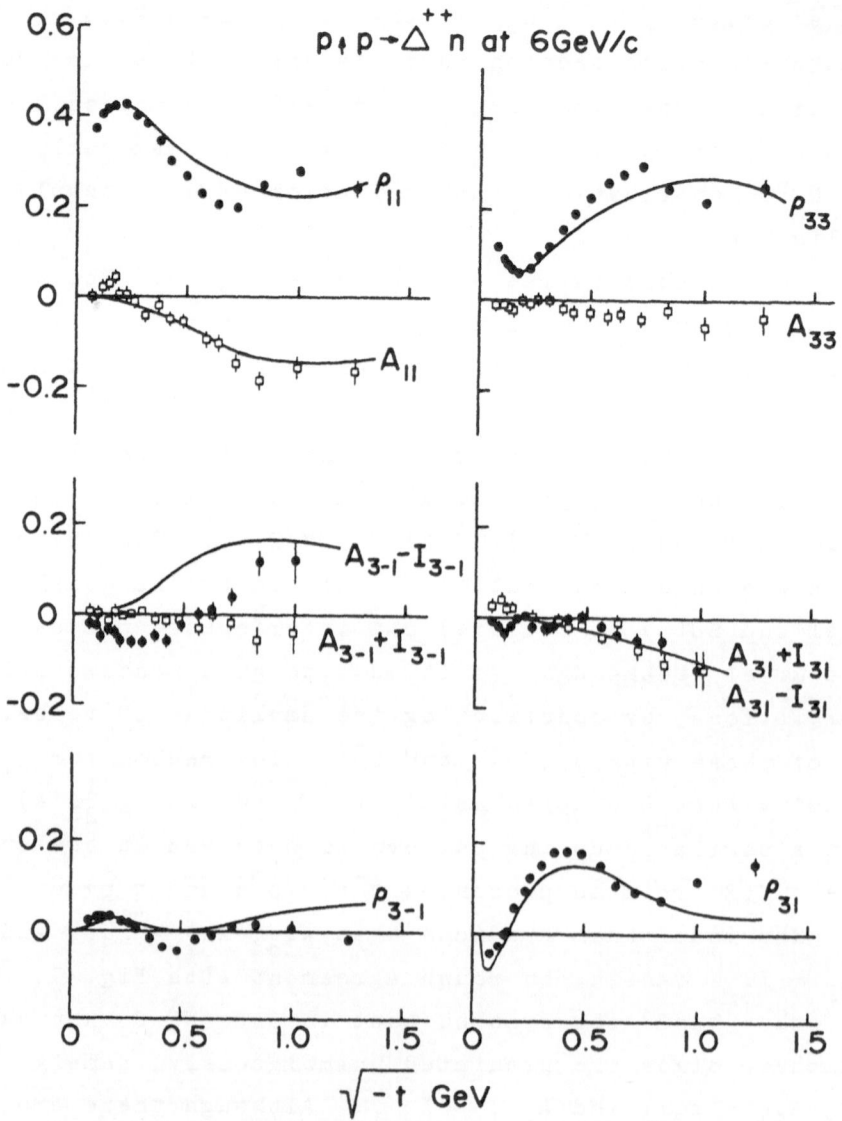

Fig. 9 Density matrix elements and spin dependent observables in $p_\uparrow p \to \Delta^{++} n$ at 6 GeV/c, corrected for nonresonant background contributions. The curves are from
a model described in the text. The quantities A_{33},
$A_{3-1} + I_{3-1}$ and $A_{31} - I_{31}$ (open points) are predicted
to vanish.

Figure 10 shows uncorrected mass spectra at 6 GeV/c
for two regions of momentum transfer. These spectra
show structure common to proton diffraction dissociation
at other energies, namely enhancements around 1425,
1520, and 1660 MeV. The purpose of this experiment is
to examine the spin-parity content of the $p\pi^+\pi^-$ system
using high statistics (500 000 events) and the unique
spin correlations available with the polarized beam.

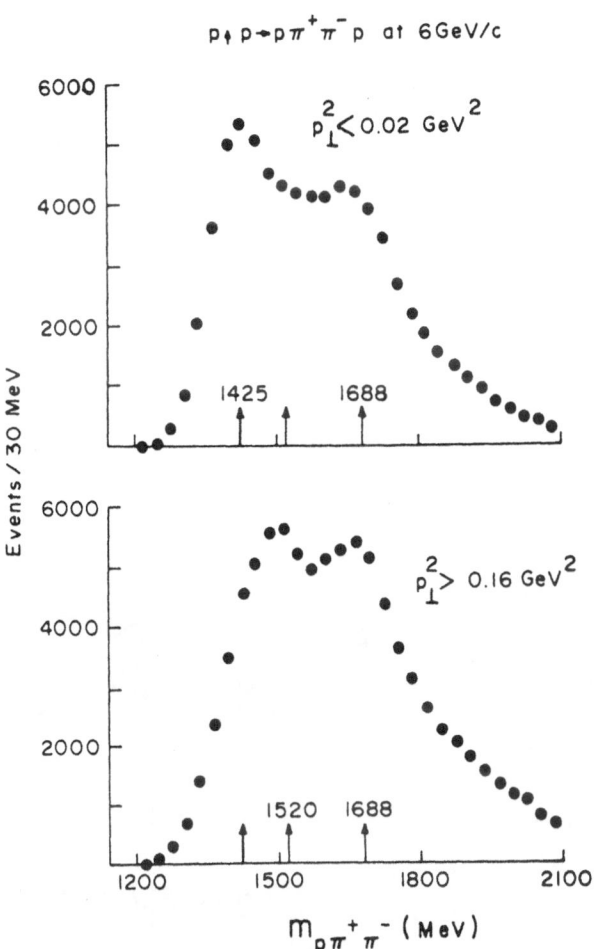

Fig. 10   Uncorrected mass spectra for pp → $(p\pi^+\pi^-)$p at
6 GeV/c for different $p_\perp^2$ regions.

In principle this process is complicated and involves many $p\pi^+\pi^-$ partial waves and many production
amplitudes. In order to see the main outlines of the
physics, we make the following assumptions and restrictions:

(1) We select the $\Delta^{++}\pi^-$ final state with a mass
cut 1160 < $m_{p\pi}+$ <1260 MeV and ignore interference
effects from other final states such as $\Delta^\circ\pi^+$, which
would be included in a complete partial wave analysis.
Empirically the spin dependence that we see is significantly weaker outside of this Δ^{++} mass cut.

(2) We assume that the dominant production amplitude is Pomeron exchange (Fig. 11), which we take to be

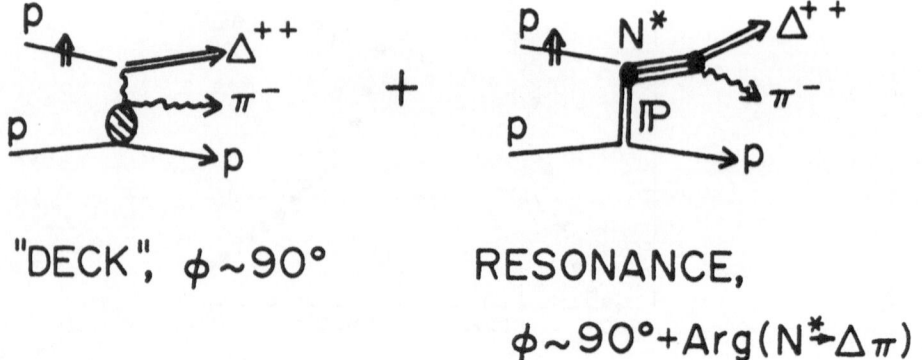

Fig. 11 Diagrams illustrating diffractive proton dissociation by the Deck mechanism and by direct resonance
production.

natural-parity exchange and helicity nonflip at both beam and target vertices. To make this assumption reasonable, we cut on $p^2 < 0.12$ GeV^2 to the recoil proton.

(3) The polarization which we measure is the correlation between the spin vector and the decay normal for the $\Delta^{++}\pi^-$ system. This is equivalent to looking at the asymmetry in the virtual process $p_\uparrow P \rightarrow \Delta^{++}\pi^-$, where the Pomeron, P, behaves like a spinless (hence helicity nonflip) particle. This polarization arises from interference between different $\Delta^{++}\pi^-$ waves. In particular, as illustrated in Fig. 11, the interference between Deck amplitude and diffractively produced N* resonances can lead to polarization because of the Breit-Wigner phases of the N*'s.

The N*'s that are known to couple strongly to $\Delta\pi$ are the N^*_{1470} ($J^P = \frac{1}{2}^+$), N^*_{1520} ($3/2^-$), N^*_{1660} ($5/2^-$) and N^*_{1688} ($5/2^+$).[16] We assume that the Deck amplitude is mainly S-wave with $J^P = 3/2^-$. Consequently in the low mass region only the N^*_{1470} can give polarization by interfering with Deck; the N^*_{1520} has the same J^P as Deck and their interference cannot give polarization. Since the N^*_{1470} can decay only into helicity $\frac{1}{2}$ Δ^{++}'s, polarization in the low mass region should be confined to helicity $\frac{1}{2}$ Δ's and should vanish for helicity $3/2$. Finally, the N^*_{1470} is a very broad state ($\Gamma \approx 200$ MeV) and cannot be directly responsible for the enhancement at 1425 MeV in Fig. 10. We expect the N^*_{1470}-Deck interference to change only slowly with mass.

These expectations are borne out rather well by the data (Fig. 12). Below 1600 MeV the asymmetry is small for helicity $-3/2$ Δ's, but as large as -50% for helicity $-\frac{1}{2}$, depending on the decay cosine. The curves in Fig. 12 are shapes predicted by assuming a $90°$ phase difference

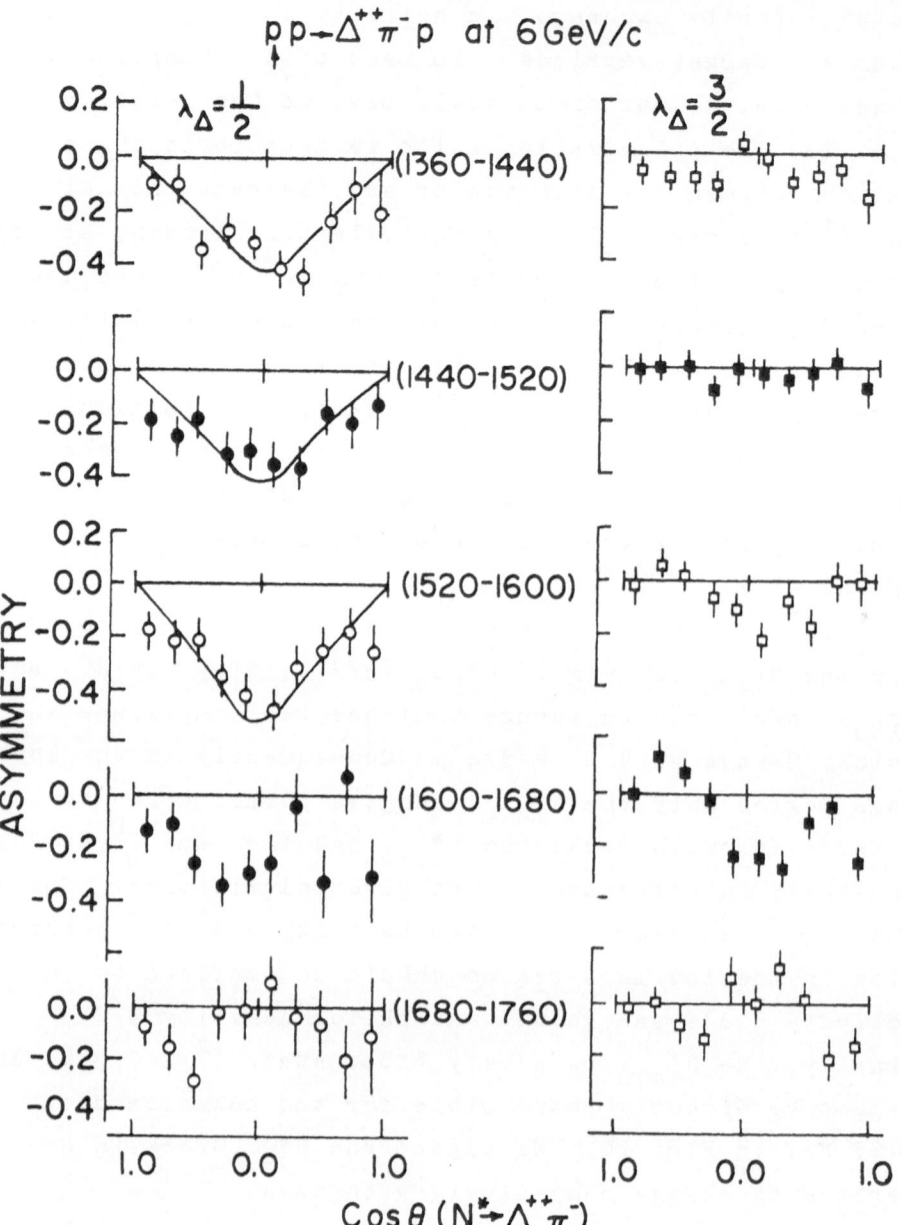

Fig. 12   Polarization correlation in $p_\uparrow p \rightarrow (\Delta^{++}\pi^-)p$ versus s-channel Δ^{++} decay cosine, comparing helicity $\frac{1}{2}$ and $3/2$ Δ^{++} production; different $\Delta^{++}\pi^-$ mass intervals are denoted by parentheses.

between the $\frac{1}{2}^+$ and $3/2^-$ amplitudes. They imply an amplitude ratio $N^*_{1470}/Deck \approx 0.2$ from 1400 to 1600 MeV; the sign of the polarization implies that the N^*_{1470} phase is 90° ahead of the Deck amplitude. Of course, the $\frac{1}{2}^+$ wave could conceivably be explained purely as a Deck effect rather than as a resonance, in which case the $\frac{1}{2}^+$ phase would require a more sophisticated Deck model than suggested by Fig. 11.

FUTURE WORK

This past fall we recorded a sample of approximately 1.5×10^6 events at 12 GeV/c of the reaction $p_\uparrow p \rightarrow p \pi^+ n$, which will be used to better determine the energy dependence of Δ^{++} production. We also took a short run on $p_\uparrow p \rightarrow p \pi^+ \pi^- p$ at 12 GeV/c (400 000 events) as a preview to a longer run a year from now. We also plan to use a V° trigger to obtain $\sim$ 400 000 events of the specific final state $p_\uparrow p \rightarrow \Lambda K^+ p$. The Λ decay can, of course, be used to measure the Λ polarization. The ΛK^+ are expected to come mainly from diffractively produced N*'s in the 1600 to 1800 MeV region.

Three more density matrix elements could be obtained for $pp \rightarrow p\pi^+n$ in the Δ^{++} region with a longitudinally polarized beam. The additional information would allow a partial separation of natural and unnatural parity exchanges.

A set of chambers to measure the direction of the recoil protons is being built. These chambers will not only help to clean up the $pp \rightarrow \Lambda K^+ p$ sample, but will also allow a separation of the pK^+K^-p final states.

REFERENCES

1. A. Gaidot et al., Phys. Letters 61B, 103 (1976).

2. R. Diebold et al., Phys. Rev. Letters 35, 632 (1975).

3. For more details on the spectrometer, see, for example, L. Ambats et al., Phys. Rev. D9, 1179 (1974).

4. F. Halzen and G. Thomas, Phys. Rev. D10, 344 (1974).

5. M. Borghini et al., Phys. Letters 31B, 405 (1970).

6. L. Durand and F. Halzen, Nucl. Phys. B104, 317 (1976).

7. R.D. Field and P.R. Stevens, ANL-HEP-CP-75-73, p 28.

8. J. Dash and H. Navelet, Phys. Rev. D13, 1940 (1976).

9. A.C. Irving, Nucl. Phys. B101, 163 (1975).

10. C. Bourrely, A. Martin, and J. Soffer, preprint (1976).

11. A.B. Wicklund, Section XV of ANL/HEP 7440, Section III of ANL/HEP 75-02, and Proc. of Symp. on High Energy Physics with Polarized Beams and Targets, AIP Conf. Proc. 35, p 198 (1976); R. Diebold, ibid, p 92.

12. R.D. Field, Some Aspects of Two Body Phenomenology (XVII Int. Conf. on High Energy Physics, London 1974) p I-185.

13. G.C. Fox and C. Quigg, Production Mechanisms of Two-to-Two Scattering Processes at Intermediate Energies (Ann. Rev. of Nucl. Sci., Vol. 23, 1973) p 219.

14. P.K. Williams, Phys. Rev. D1, 1312 (1970).

15. S.L. Kramer et al., Phys. Rev. Lett. 33, 505 (1974).

16. D.J. Herndon et al., Phys. Rev. D11, 3183 (1975).

ORBIS SCIENTIAE 1977 OFFICIAL "HOSTESSES"
(from left) Linda Scott, Helga Billings, Elva Brady, and Yvonne Leber

INCLUSIVE ASYMMETRIES

Earl A. Peterson

University of Minnesota

Minneapolis, Minnesota 55455

The advent of the polarized proton beam at the Argonne zero-gradient synchrotron (ZGS) has permitted considerable improvement in technique for experiments which study spin dependence in inelastic strong inter-actions. Although these experiments are possible with a polarized proton target, the backgrounds due to carbon and oxygen in these targets make inclusive measurements difficult. In a recent experiment, we have used the polarized proton beam to measure left-right asymmetries for the reactions $p_{pol} + p \rightarrow p$, $K^{\pm}$, $\pi^{\pm}$ + anything by placing a liquid hydrogen (LH_2) target in the extracted polarized proton beam (p_{pol} denotes polarized proton). The beam polarization is vertical and normal to the plane defined by the incident and scattered momenta. A portion of the experiment was run with a target fill of liquid deuterium, so that some information on the reaction $p_{pol} + n \rightarrow p$, $K^{\pm}$, $\pi^{\pm}$ + anything was obtained. The data reported here are preliminary in nature and were accumulated at an incident proton momentum of 11.8 GeV/c.

We have used ZGS Beam 5, shown schematically in
Figure 1, as a single-arm spectrometer to detect the
scattered particle. The spectrometer has an angular
acceptance of order 10^{-4} sr, depending somewhat on the
kinematic setting, a ± 5% momentum acceptance, and two
ethylene-filled, threshold Cerenkov counters, each with
two optically independent sections, for particle identification. Charged particles produced in a 10-cm-long,
3.8-cm-diam target are restored to the axis of quadrupoles
X5Q1-3 by steering magnets X5B1 and X5SB1. For some
points on the edges of the kinematic range, dipole magnets
X5B2 and X5B3 are also used in the steering process. The
angular range of the spectrometer depends on the momentum
and polarity of the scattered pion; the data reported here
include laboratory angles between 0^0 and 17^0 and momenta
between 2 and 9 GeV/c.

The direction, size, and position of the incident
proton beam are determined by two sets of x-y proportional
chambers read out in an integrated mode. The relative
intensity (typically 5×10^8 protons per 500-msec pulse)
and polarization (typically 55% at 11.8 GeV/c) of the
incident beam are monitored by four scintillation-counter
telescopes (L, R, U, and D). L and R view a thin poly-
ethylene target and act as a polarimeter with an analyzing
power of 0.020 ± 0.001 at p_0 = 11.8 GeV/c. This polarimeter has been calibrated at both incident momenta against
an absolute elastic-scattering polarimeter located in
another experimental area. Telescopes U and D, located
in the vertical plane, monitor the proton intensity on
the LH_2 target.

The proton target asymmetries are obtained from the
equation

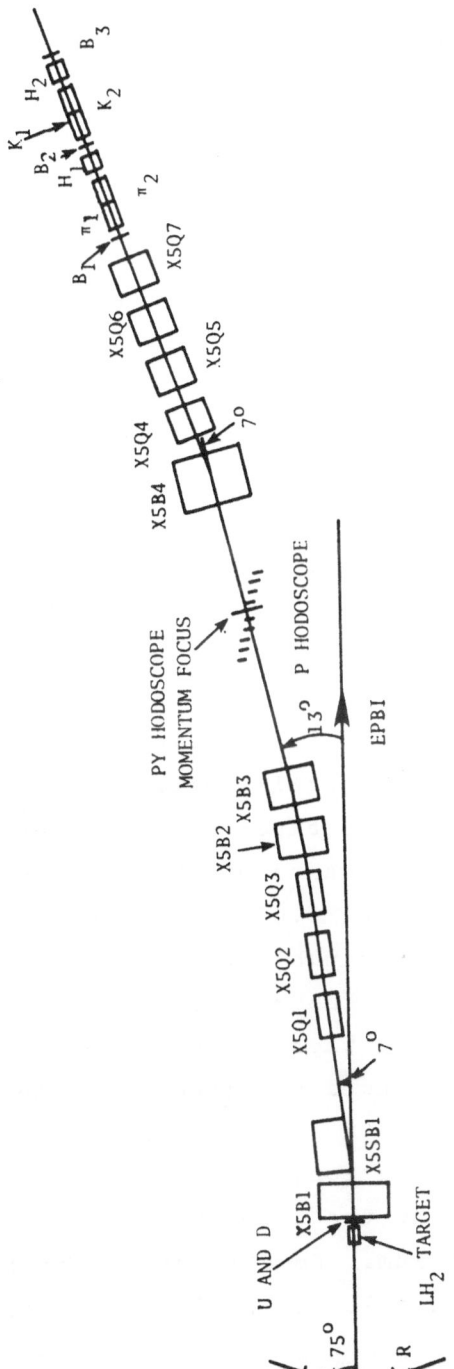

Figure 1. Experimental arrangement

$$A_x = \frac{1}{P_B} \frac{N(\uparrow) - N(\downarrow)}{N(\uparrow) + N(\downarrow)}, \qquad\qquad (1)$$

where $N(\uparrow)$, $[N(\downarrow)]$ are the number of x-type coincidences $(B_1 B_2 B_3 C_1 C_2)$ recorded for incident beam polarization up (down) normalized to the intensity monitors and P_B is the beam polarization. The symbols C_1 and C_2 denote the Cerenkov tags appropriate to the detection of particle x in the final state (e.g., both π_1 and π_2 for $\pi^{\pm}$, no signal for p,...). The asymmetry defined here is positive when more particles are produced to the left in the horizontal plane looking in the direction of the incident beam.[1] Corrections due to multiple scattering, nuclear absorption, decays, and uncertainty in the spectrometer acceptance have no effect on the asymmetry. The sign of the beam polarization is reversed on every accelerator pulse to minimize systematic errors. Target-empty background runs have been taken for every data point; the target-empty rate is typically between 10 and 25% of the target-full rate. As an experimental check, we have made single-arm measurements of the pp elastic polarization at p_0 = 6 GeV/c and $0.07 < t < 0.3$ $(GeV/c)^2$; the results are consistent with the published data.[2]

The neutron target asymmetries are calculated as in equation (1) above, where N now refers to the normalized number of coincidences expected from the neutron content of the deuterium target. The normalization includes a target empty subtraction, equalization of incoming beam polarization and the differing densities of liquid hydrogen and deuterium. No correction has been made for multiple scattering or other nuclear effects, and the data presented here are preliminary in that respect.

Our measurements of the asymmetry for pion production

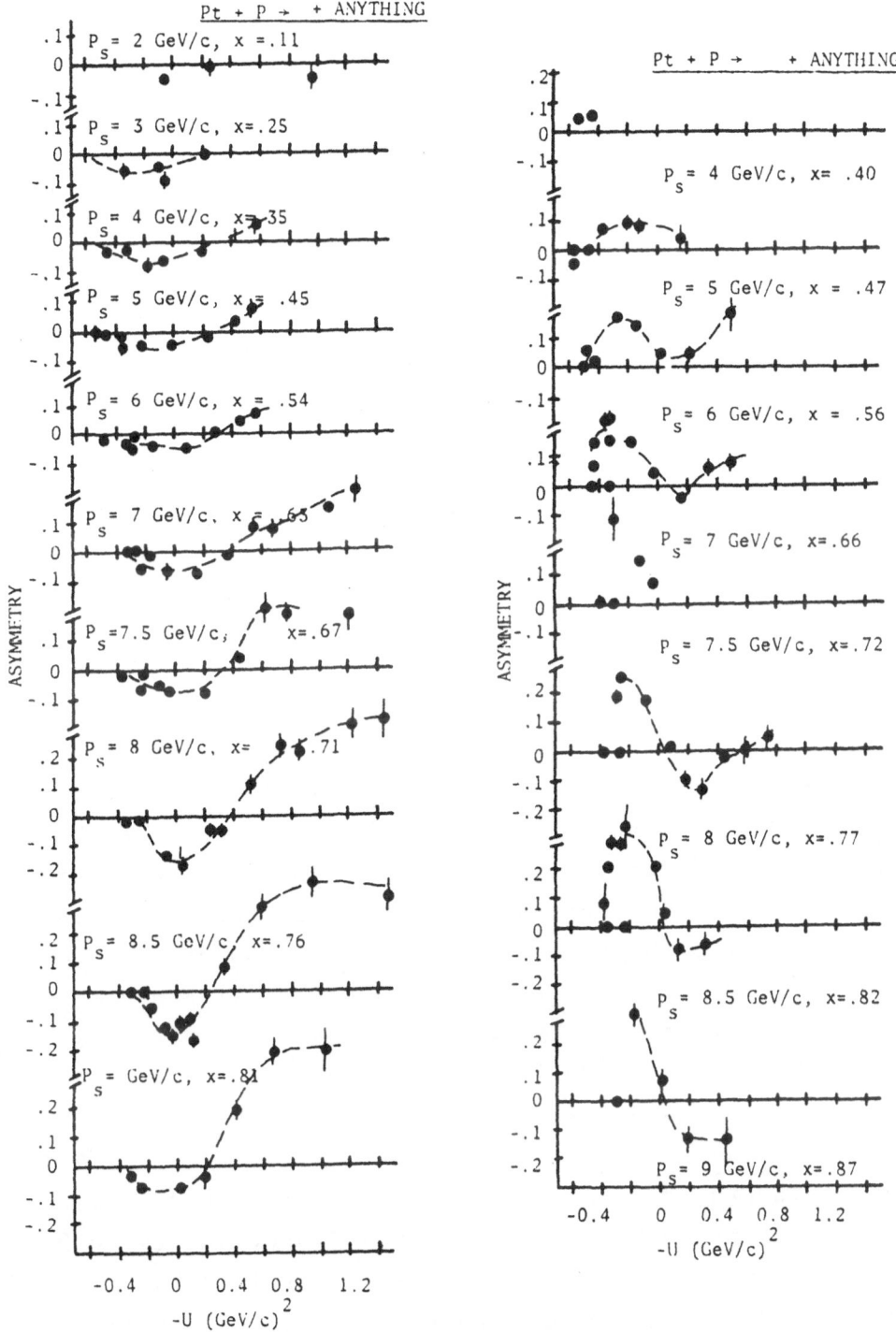

Figure 2.   Inclusive pion asymmetries in pp

are shown in Figure 2, which also includes data taken
in an earlier run at 6 GeV/c (the few points plotted as
open circles at equivalent values of the Feynman x parameter).[3] We have chosen to represent the data as a
function of u, the square of the four-momentum transfer
from the incident proton to the outgoing pion, in order
to emphasize the similarities between these data and the
polarization in backward π p elastic scattering. The
resolution in u is $\Delta u \leq \pm 0.1$ $(GeV/c)^2$. The data for
the two pion charges are considerably different for any
particular kinematic point, but there are some overall
similarities. (1) The 6- and 11.8 GeV/c asymmetries
are consistent, except where one of the asymmetries is
forced to zero by the requirement of no asymmetry for
zero production angle (open and filled triangles). (2)
The magnitudes of the asymmetries are consistently larger
for larger x but the shape of the dependence on u, in
particular, the location of the maxima and zeros, is
roughly independent of x. (3) The effects of the a-
symmetry zero for 0° production are of limited extent in
u. For example, the π⁻ asymmetry at large x rises to
20% within 0.6° of the forward direction. We have verified this effect by noting a reversal in the asymmetry
for pions from the other side of the proton beam.

The asymmetries for pion production off a neutron
target appear to be very similar to the proton target
data. These are shown in Figures 3 and 4 for secondary
momenta of 8 and 9 GeV/c, respectively. The latter
figure includes the existing data on backward π⁺p elastic
scattering (at 6 GeV/c) as well.[4] Figure 5 presents the
data on π⁻ production off protons and neutrons (at 8.5
GeV/c) as well as the corresponding elastic data.

It is apparent from Figures 4 and 5 that there is

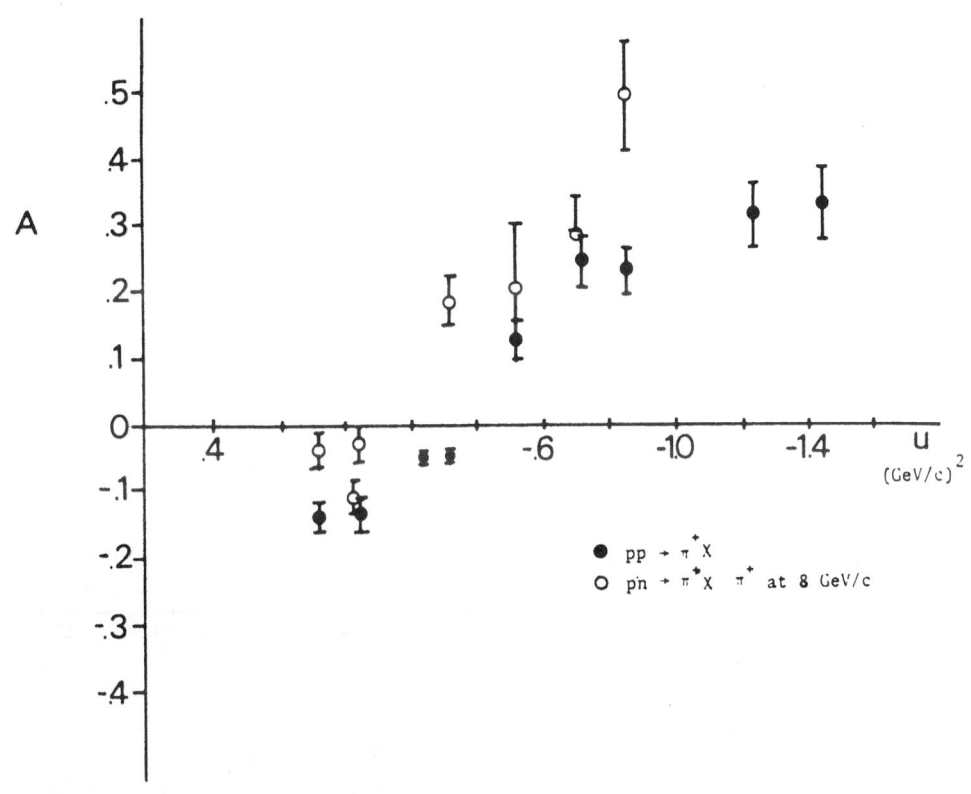

Figure 3.   Inclusive π^+ asymmetries

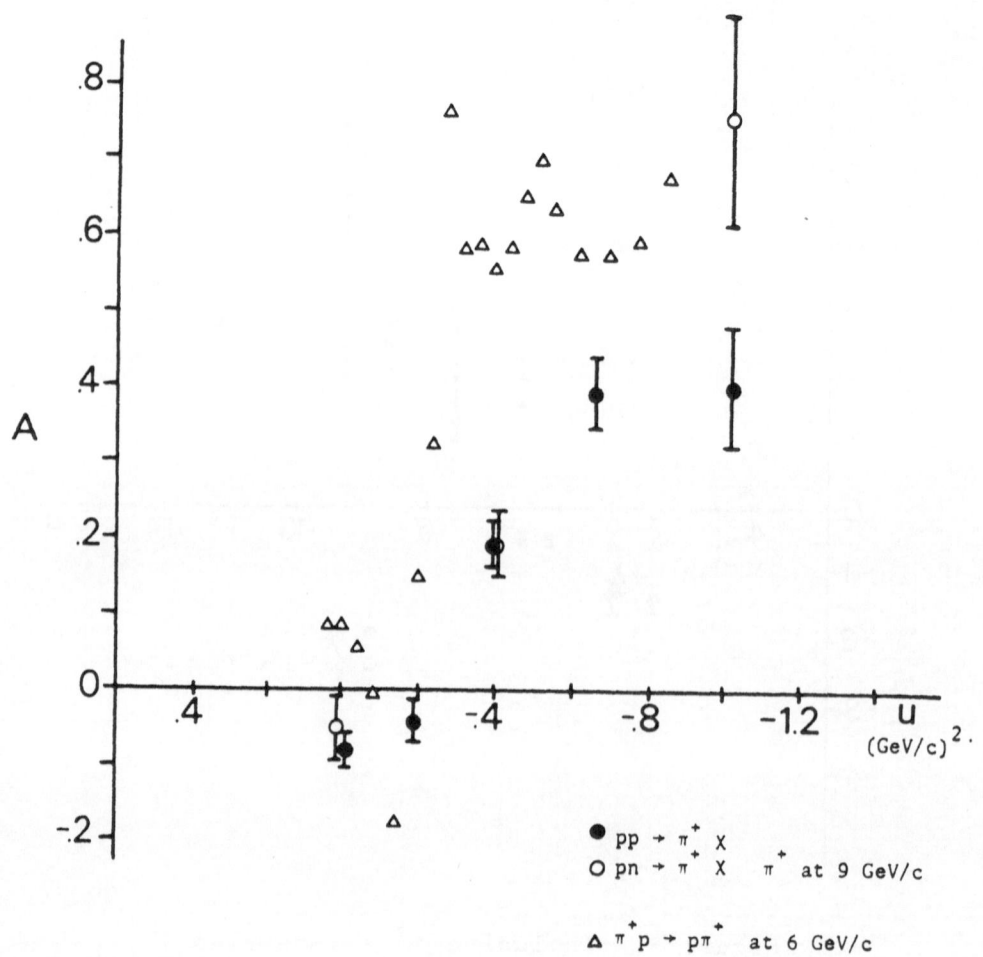

Figure 4.   Inclusive π^+ asymmetries

a strong correlation between the structure in the inclusive and backward pion elastic data, at least in the
region (large pion momentum) where the inclusive process
is likely to be dominated by baryon exchange. At lower
momenta (smaller x), the structure seems to be diluted
by other effects, presumably s-channel in nature. A
crude model for this correlation may be based on the
work of Ader, Meyers and Salin,[3] who showed that for
baryon exchange reactions,

$$A\sigma \propto |N|^2 - |U|^2 , \qquad (2)$$

where N and U are the natural and unnatural-parity exchange amplitudes. The diagrams in Figure 6 exhibit
the correlation[6]: 6a) is a simple exchange model for
backward elastic scattering leading to an asymmetry of
the form

$$A\sigma \propto |\beta_N|^2 - |\beta_u|^4 , \qquad (3)$$

where β_N, β_u are the effective natural and unnatural
residue functions. These same residues appear in the
closure graph, 6b) for the inclusive process, multiplied
by an ersatz Reggeon-nucleon total cross section, leading
to

$$A\sigma \propto |\beta_N|^2 \sigma_T(R_N p) - |\beta_u|^2 \sigma_T(R_u p). \qquad (4)$$

If these total cross sections are approximately equal
for both "N" and "U" scattering from protons or neutrons,
the similarity of all three sets of data can be naively
understood.

     If the arguments presented above can be made more

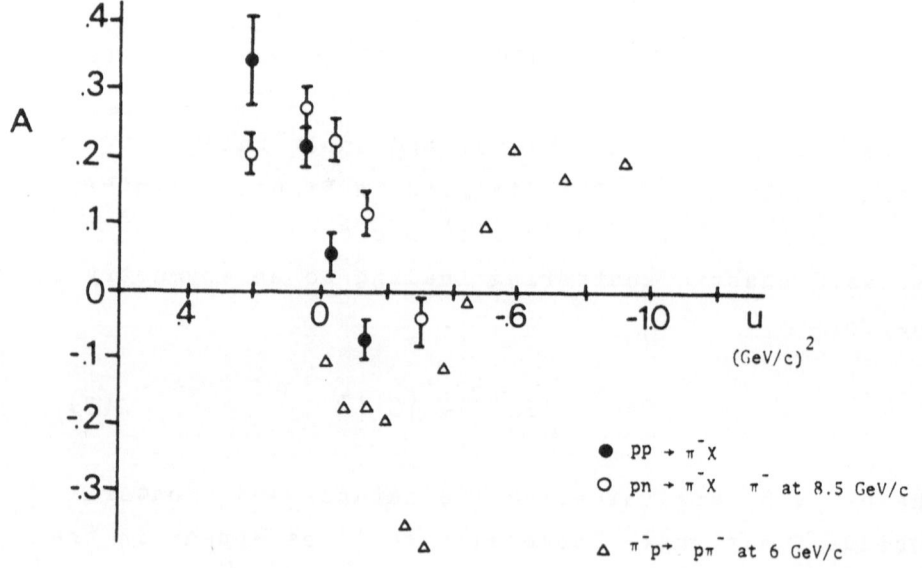

Figure 5.   Inclusive π^- asymmetries

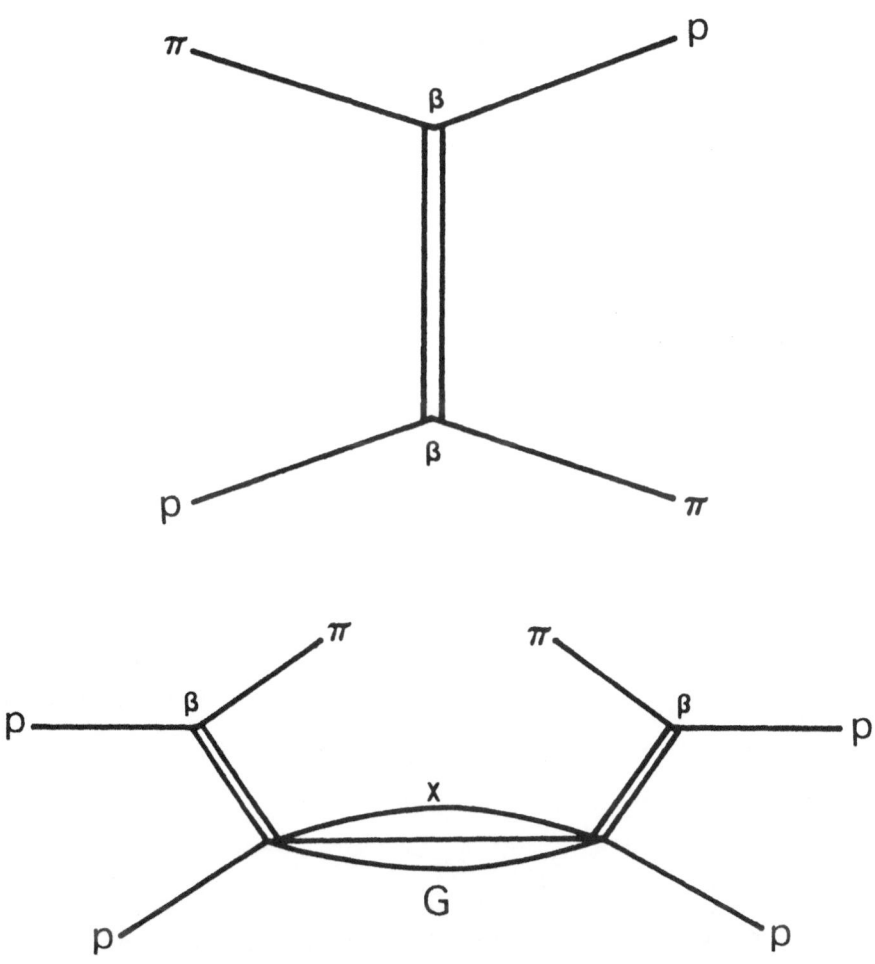

Figure 6. a) Schematic Regge diagram for the backward elastic amplitude. b) Schematic closure diagram for the inclusive cross section.

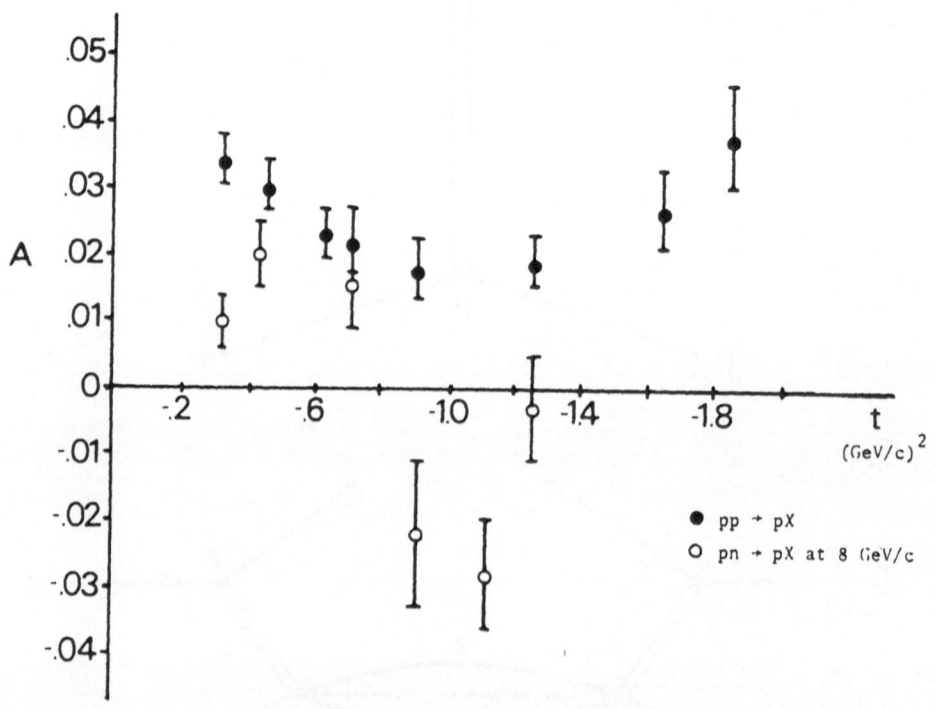

Figure 7.   Inclusive proton asymmetries

quantitative, the inelastic data offer one advantage:
it is relatively easy to explore the positive-u region
with inelastic kinematics. Elastic experiments (from
a polarized target) are difficult in this region, where
the backward scattered pion has a very low energy.

    The inclusive production of baryons shows no such
apparent regularity. Figure 7 exhibits the asymmetry
for secondary protons at 8 GeV/c from both proton and
neutron targets. The asymmetries are small (about half
the size of elastic scattering data) and, as in the
elastic case,[7] strikingly different for proton and neutron
targets. It is apparent that both the I = 1 and I = 0
amplitudes are significant in these processes.

Work supported by the U.S. Energy Research and Development Administration and by the Graduate School of the
University of Minnesota.

REFERENCES

1. This convention yields a positive asymmetry for pp elastic scattering which is the same sign as results from the use of the Basel convention for polarized-target scattering. This consistency is the result of the properties of the identical particles in pp scattering. In general, to compare our results with p backward scattering from a polarized target, an additional minus sign must be introduced into the asymmetries.

2. D. R. Rust et al., Phys. Lett. 58B, 114 (1975).

3. R. D. Klem et al., Phys. Rev. Lett. 36, 929 (1976).

4. L. Dick et al., Nucl Phys. B43, 522 (1972) and B64, 45 (1973).

5. J. P. Ader, C. Meyers and Ph. Salin, Nuclear Physics B58, 621 (1973).

6. This correlation was first noticed by B. Wicklund.

7. R. Diebold, et al., Phys. Rev. Lett. 35, 632 (1975).

PRELIMINARY POLARIZATION RESULTS AT FERMILAB ENERGIES*

M.D. Corcoran, S.E. Ems, F. Fredericksen,

S.W. Gray, B. Martin, H.A. Neal, H.O. Ogren,

D.R. Rust, J.R. Sauer, and P. Smith

Presented by H.O. Ogren for H.A. Neal

Indiana University, Bloomington, In. 47401

ABSTRACT

This paper describes the jet target, the superconducting spectrometer and the polarimeter used to
measure the recoil proton polarization in an Indiana
University experiment at the Fermi National Accelerator
Laboratory. Analysis procedures and data checks are
explained and preliminary results are presented and
compared with model predictions. The future plans for
the experiment are also discussed.

INTRODUCTION

Recently there have been experimental and theoretical indications that the polarization in pp elastic
scattering may be surprisingly large at high energies

---

*Work supported by U.S. Energy Research and Development
Administration under Contract E(11-1)2009, Task A.

at certain t values. The ISR measurements of the dip
in the differential cross section (Ref. 1) suggest the
possibility of large polarizations. The Pumplin-Kane
(Ref. 2) model predicts that polarization effects will
persist to very high energies. Polarization measurements at Serpukhov suggest large polarization at
$|t| \geq .7$ at 45 GeV/c (Ref. 3).

Our group has undertaken a program to measure the
polarization of protons from both elastic and inclusive
interactions in the Internal Target Area (ITA) at Fermilab. Together with the ITA staff and another group of
experimenters (E198A, a University of Rochester, Rutgers
University, and Imperial College of London collaboration)
we constructed a spectrometer to detect pp elastic
scattering. To this we appended a polarimeter designed
to measure recoil proton polarization with a minimum of
systematic bias.

JET TARGET

The target utilized was a hydrogen gas jet pumped
by large diffusion pumps (see Figure 1). Hydrogen gas
at ∼ 10 atmospheres pressure is ejected from a 3 mil
nozzle through the circulating Fermilab beam. Most of
the gas is caught by a special mylar cone and directed
into a 1 m^3 buffer volume. This large volume reduces
the pressure to a level which the two ten inch diffusion
pumps can handle. Two more ten inch pumps are on the
main vacuum chamber. Other pumps up and downstream
minimize the spread of the pressure bump. To further
limit the amount of gas introduced into the main ring
the jet itself is pumped from behind the nozzle after
the pulse is over.

The jet has a density of ∼ $.5 \times 10^{-7}$ g/cm^3 and a

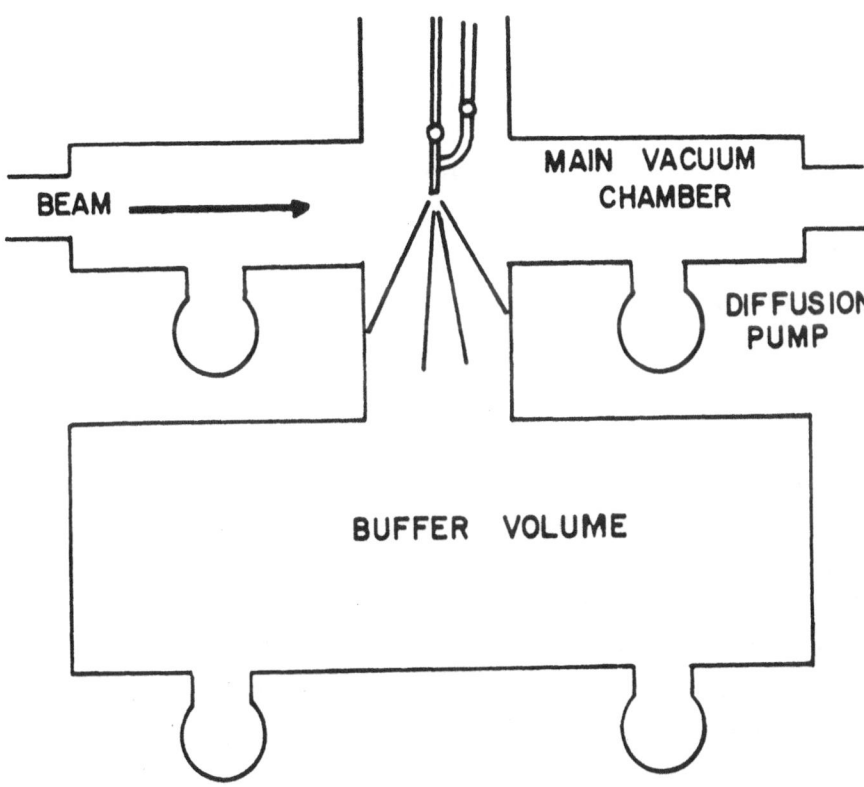

Fig. 1   Illustration of hydrogen gas jet.

width of ~ 6mm FWHM. Typically we used three 100 ms jets
during each ramp. The limit in integrated luminosity
was set by the maximum permissible beam loss in the Internal Target Area.

SPECTROMETER

The spectrometer uses a superconducting quadrupole
doublet and a superconducting dipole to identify elastic
protons produced in beam-jet interactions (see Figure 2).
The quadrupoles are tuned to act as a field lens and
to focus protons with the same production angle to a
point; this allows the measurement of the angle with a
single position measurement. The chamber unit used to
measure the production angle θ has two x planes with
1.3 mm spacing (staggered by one half wire) and a single
y plane. A similar chamber assembly about 1 meter
downstream permits a momentum dependent correction to
the angle measurement and serves as the first point in
the momentum determination. Two more modules are
mounted on each side of the dipole. The final two
chambers are 15° tilted u,v chambers with 2 mm wire
spacing. Two trigger counters and 3 hodoscopes are
placed between the first 2 chamber modules. Another
trigger counter and another hodoscope are at the end of
the spectrometer.

Two experiments can read the chambers asynchronous-
ly. Our amplifiers look at the output of the spectrometer amplifiers and treat the signals similarly to the
signals from our own polarimeter chambers. The spectrometer acceptance is ~ 10 mr horizontally by ~ 40 mr
vertically. The momentum bite is ~ ±5%. The momentum
resolution is presently approximately 1%. The missing
mass resolution is 100 MeV (FWHM) at 100 GeV/c. At

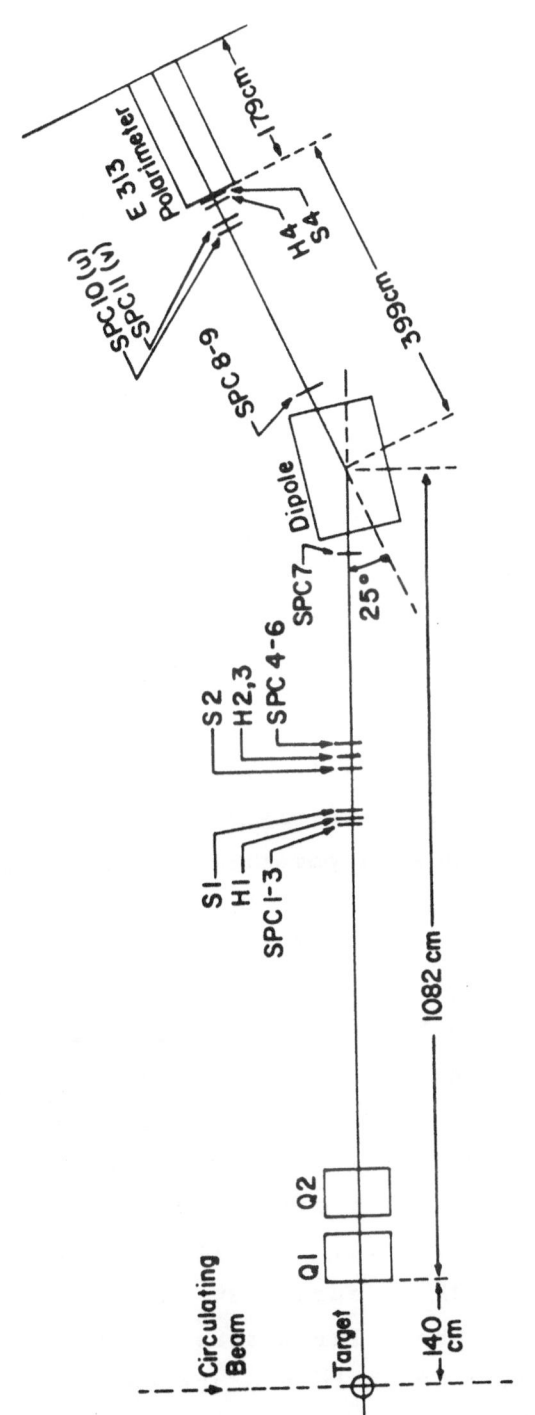

Fig. 2 Layout of the internal target superconduction spectrometer. Q_1 and Q_2 are quadrupole magnets. SPC-11 are multiwire proportional chambers. H1-4 are hodoscopes and S1-4 are trigger counters.

$t = -.3 \ (\text{GeV/c})^2$ and $p_{LAB} < 100$ GeV/c the spectrometer
produces a beam of elastically scattered recoil protons
with essentially negligible inelastic contamination.
A typical missing mass distribution for $t = -.8 \ (\text{GeV/c})^2$
at $p_{LAB} = 40$ GeV/c is shown in Figure 3.

POLARIMETER

The polarimeter consists of proportional chamber
telescopes (both x and y) on each side of a carbon re-
scattering target (shown in Figure 4). The carbon
analyzer has an effective analyzing power between .2
and .3 for accepted projected scattering angles between
6° and 22°. The analyzer was calibrated in the Argonne
ZGS polarized proton beam. The x and y hodoscopes are
used to resolve track ambiguities. Range counters with
variable absorbers can be used to "enrich" the elasticity
of the double scatters and improve the analyzing power.
A key feature is the polarimeter's ability to rotate
about its axis allowing left and right to be interchanged;
all first order instrumental asymmetries then average to
zero.

A second important feature of the polarimeter is a
hardwired computer. Since only a few percent of the
elastic scatters detected by the spectrometer actually
interact in the carbon target, most fast logic triggers
are uninteresting. The polarimeter computer uses information from the proportional chambers before they are
read out to enrich the fraction of recorded usable double
scatters by a factor of 20 - 80. It first checks that
there is an incident track in each view (x and y)
approximately normal to the polarimeter. Next it
searches for a companion track in the chambers after
the carbon target which could signal a simple

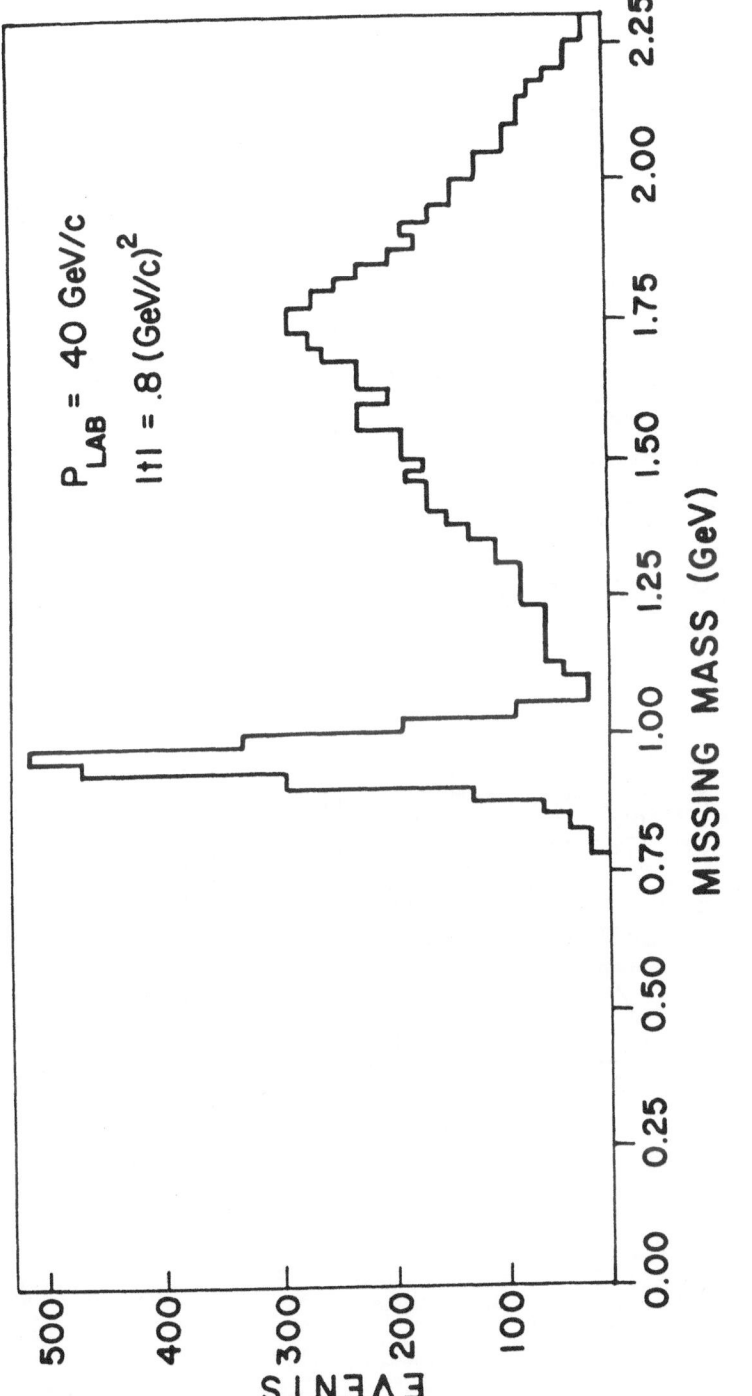

Fig. 3 Missing mass distribution at t = -.8 (GeV/c)^2.

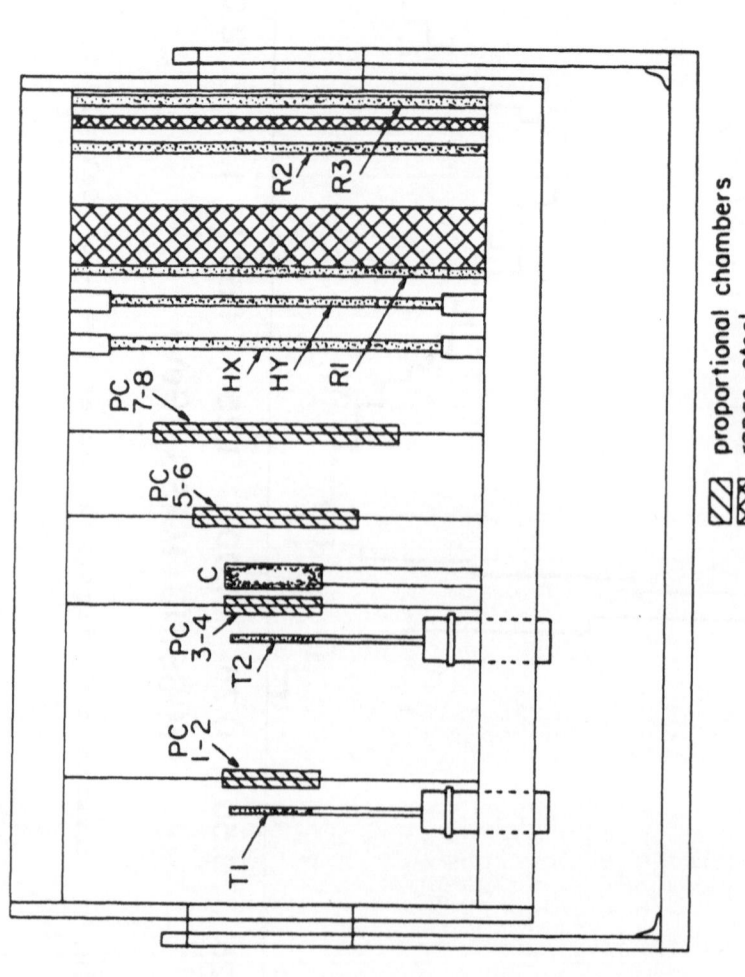

scale: 8:1

Fig. 4 Layout of polarimeter. T_1 and T_2 are trigger counters. PC1-4 are multiwire proportional chambers. HX and HY are hodoscopes. R1-R3 form a range telescope.

straight-through. It also searches for a companion
track in the rear chambers which looks like a true
scatter. The choice, tolerances, and specifications
of the tests appropriate for each recoil momentum and
each thickness of carbon target used can be remotely
selected from the electronics trailer. An illustration
of the effectiveness of the polarimeter computer in
enhancing the recorded number of useful double scatters
is shown in Figure 5.

ANALYSIS PROCEDURE

In the polarization analysis only those events
were retained which reconstructed in both views and
which had consistent chamber and hodoscope trajectory
information. Cuts were placed on the re-scattering
vertex to ensure that all events utilized in the polarization determination did indeed scatter in the carbon
target. A typical vertex distribution in the coordinate
along the polarimeter axis is shown in Figure 6. Only
double scatters within the projected angle range 6° -
22° were utilized. This was done to maximize the net
analyzing power and analyzing efficiency and to correspond to the experimental conditions in the calibration
measurements.

To minimize the effects of instrumental asymmetries
several precautions were taken. The most important
protection against biases came from the averaging out
of first order instrumental asymmetries by rotating the
entire polarimeter by 180° about its axis. This flipping
of the polarimeter about the nominal recoil beam
centroid permitted "left" and "right" to be effectively
interchanged so that effects due to asymmetric chamber
or counter inefficiencies are cancelled when the results

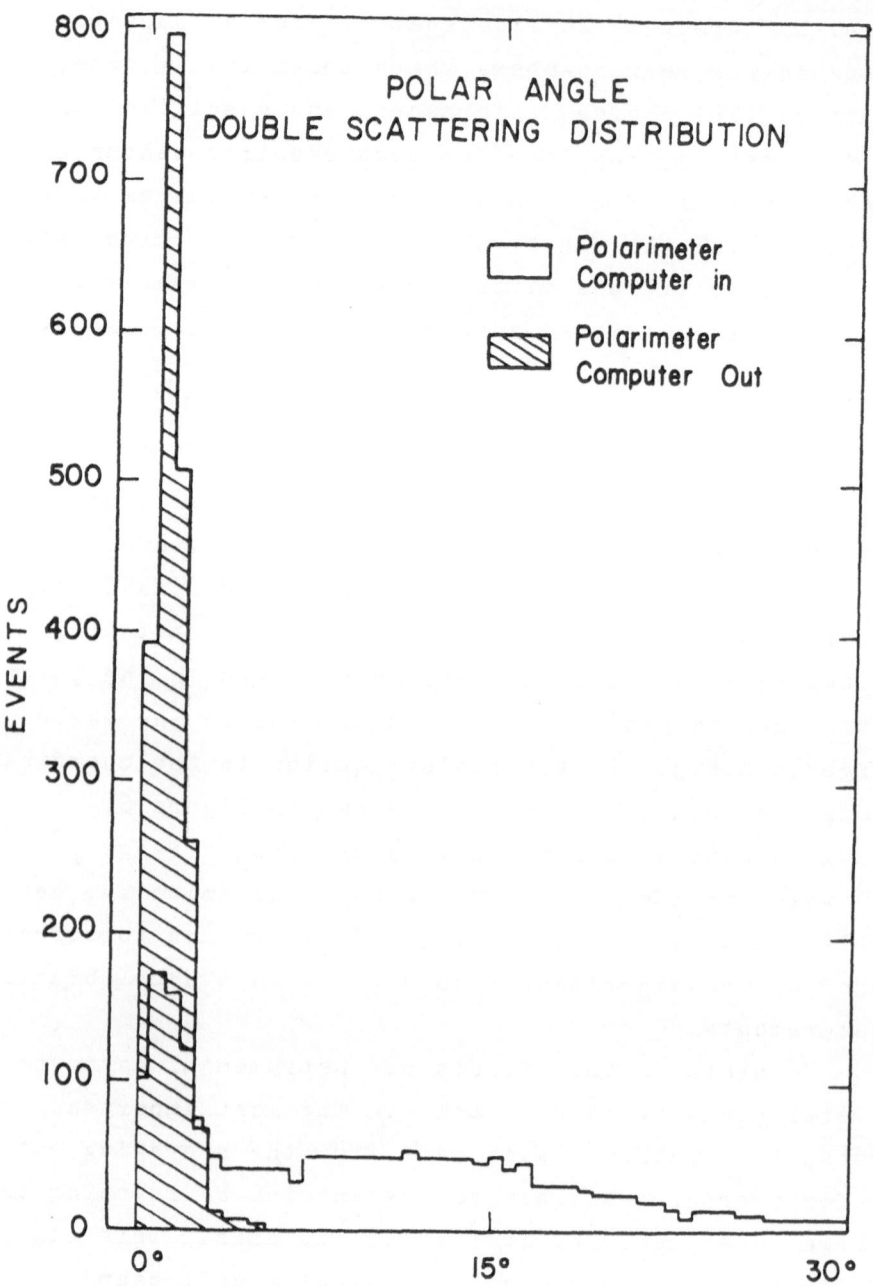

Fig. 5 Illustration of the effectiveness of the polarimeter computer in suppressing small angle double scatters.

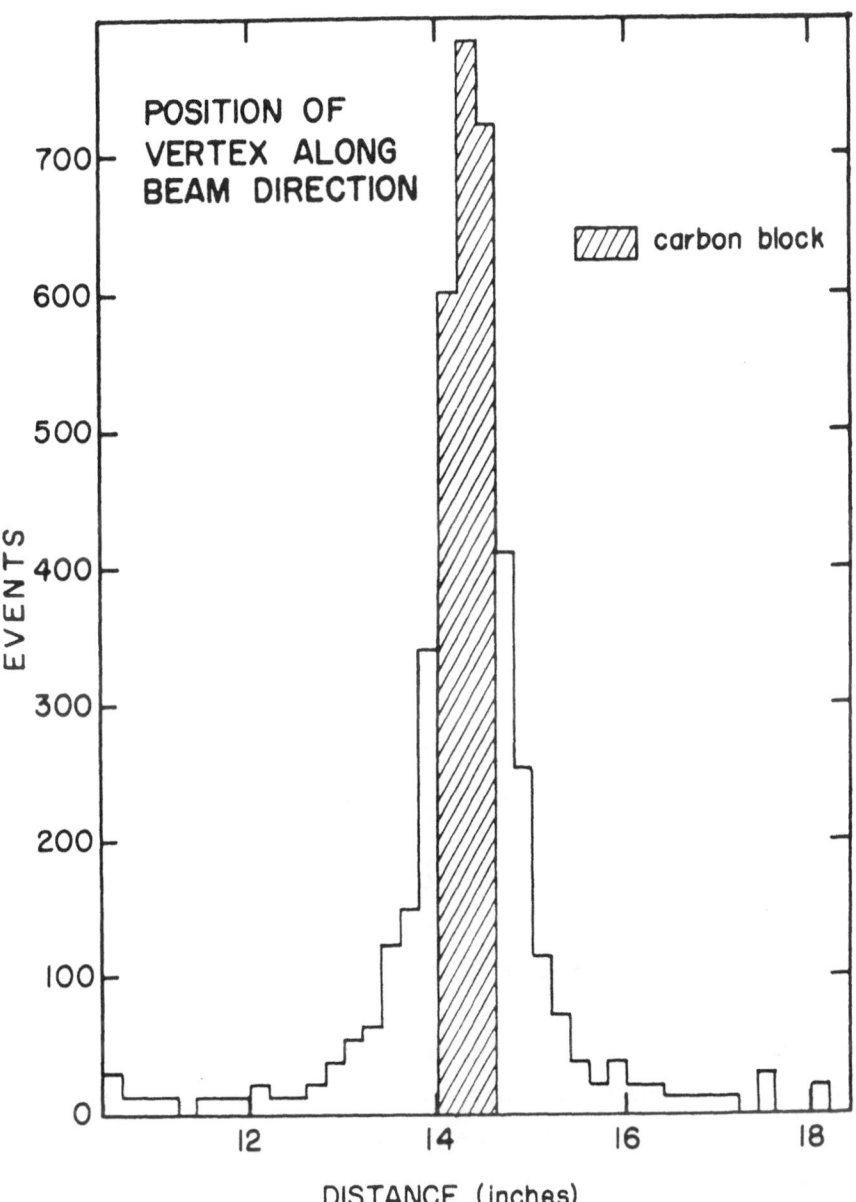

Fig. 6   Double scattering vertex distribution along
         polarimeter axis.

from the two supplementary orientations are averaged.
The relative polarimeter chamber alignment was monitored at frequent intervals by analyzing straight-through
events. The efficiencies of all chambers and hodoscope were continually monitored as a function of position by using several independent schemes.

Other checks involved a measurement of the up-down
instrumental asymmetry with the polarimeter at its $0°$,
$90°$, and $180°$ azimuthal orientations and the measurement of the asymmetry resulting from the double scattering
of pions. In all cases the results of these tests were
consistent with the polarimeter having a left-right bias
less than $\Delta\varepsilon$ = .01. The effect of this bias on results
obtained by averaging data from the $0°$ and $180°$
orientations is found to be consistent with zero
($\Delta\varepsilon$ < .001). Furthermore, the polarimeter was used to
measure the polarization at t = -.3 $(\text{GeV/c})^2$ near
17 GeV/c where the polarization is well-known from
previous measurements. This check was successful and
illustrates the polarimeter's capability to detect finite
proton polarizations.

The polarization P is determined from the relations

$$\varepsilon \equiv \frac{L - R}{L + R}, \quad P = \frac{1}{A}\varepsilon,$$

where L and R are the number of double scatters in the
left and right angular regions corresponding to horizontally projected angles between $6°$ and $22°$. A is the
analyzing power which was determined in a calibration
experiment in the ZGS polarized beam for a large range
of proton energies (Ref. 4).

RESULTS

Preliminary results from this experiment at $t = -.3$ and $t = -.8$ $(GeV/c)^2$ at several s values are given in Figures 7, 8 along with other data at these t values. At $t = -.3$ $(GeV/c)^2$ the polarization is seen to drop from a very large value of $\sim 35\%$ at $s = 6$ GeV^2 to $\sim 1\%$ at $s = 100$ GeV^2. At $t = -.8$ $(GeV/c)^2$, however, a significant negative polarization appears to develop near $s = 100$ GeV^2. This sizeable negative polarization which was also suggested in earlier Serpukhov data at $p_{LAB} = 45$ GeV/c, is in contrast to the small positive value of the polarization at this t value at all $p_{LAB} < 20$ GeV/c. Theories must confront the question of why such spin effects exist at high energies. It is of interest to note that the small positive polarization at small $|t|$ $(|t| < .4$ $(GeV/c)^2)$ and the rather large negative polarizations for $|t|$ in the vicinity of $.7 - 1.0$ $(GeV/c)^2$ is a prediction of the Pumplin-Kane model.[2] The agreement of a prediction of this model with the fixed t data is illustrated in Figures 7,8. The model's predictions arise from an exploration of the consequences of the Pomeron having quantum numbers different from the vacuum. The pursuit of the implications of this fundamental idea seems to be absolutely essential.

FUTURE PLANS

Our present plans are to explore the development in energy of polarization effects near $t = -.8(GeV/c)^2$. The experiment is particularly well-adapted to fixed t studies since the analyzing power and acceptance are nearly independent of s. Later we plan to complete s sweeps of the polarization at other t values in the

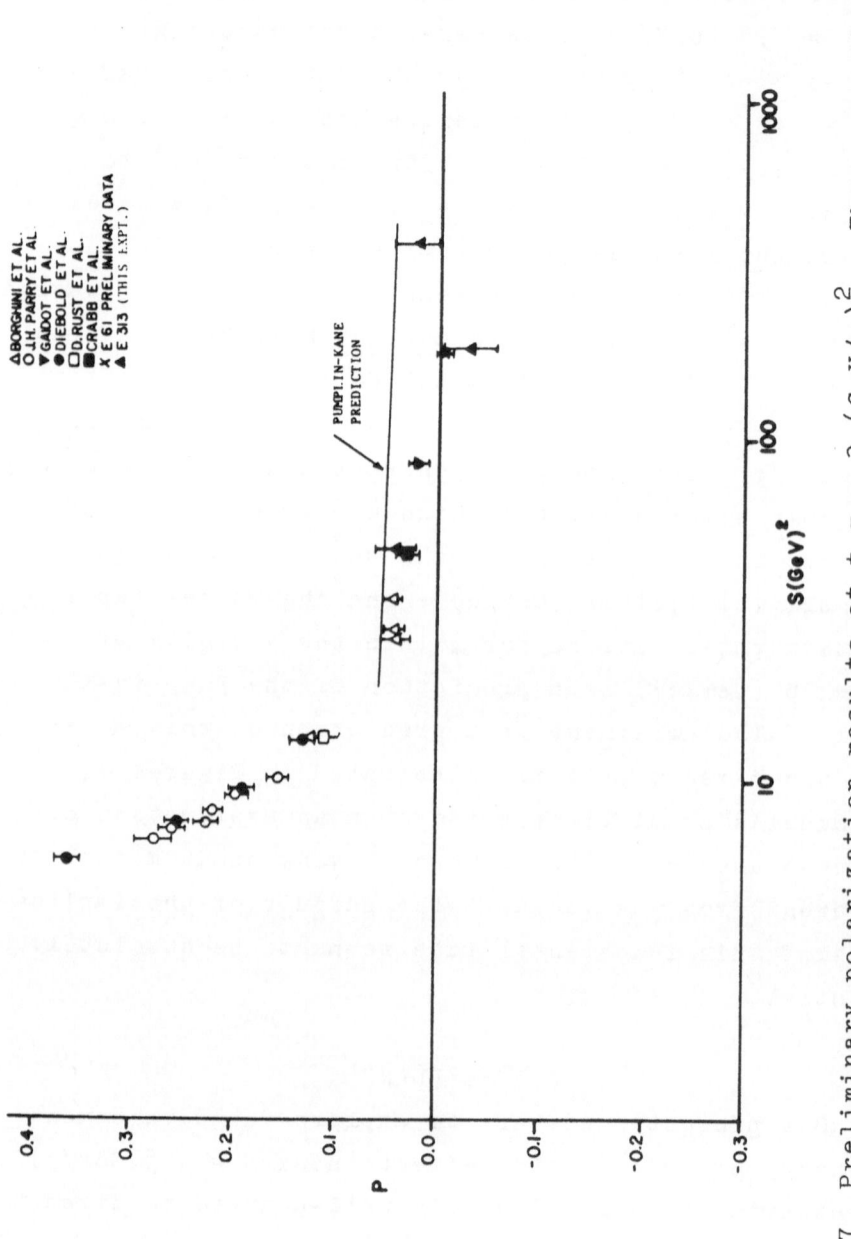

Fig. 7 Preliminary polarization results at t = -.3 (GeV/c)². The curve shown
is a prediction of the model in Ref. 2.

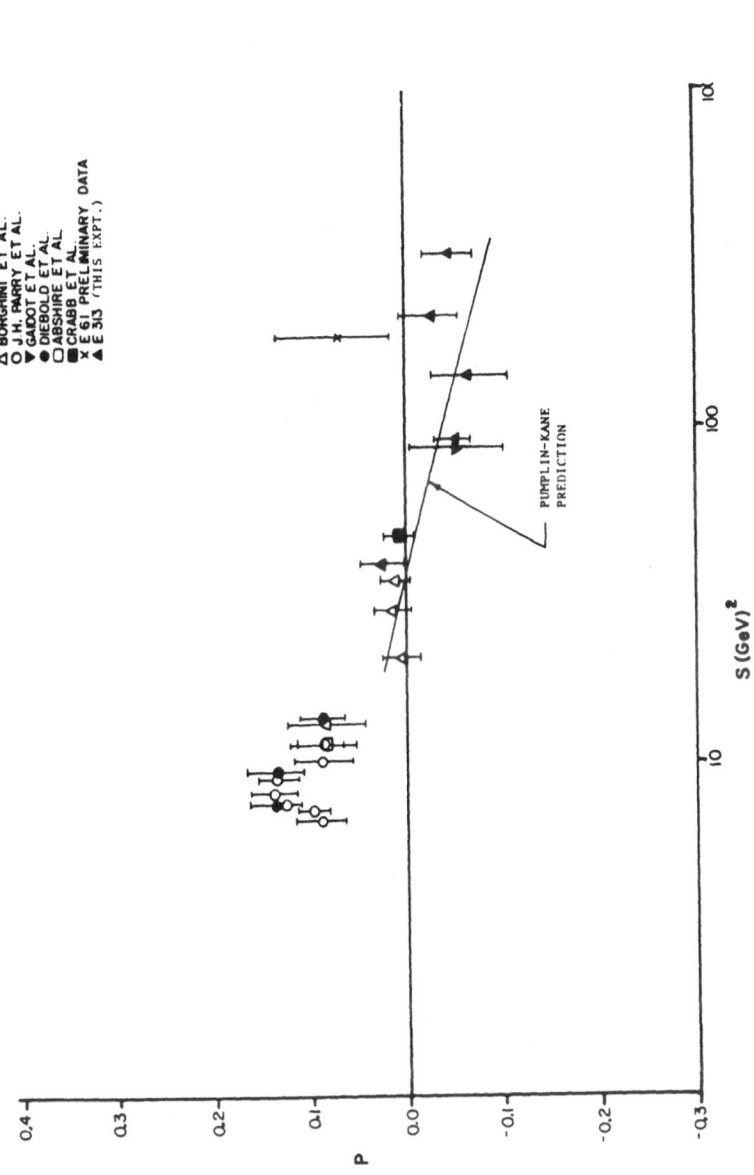

Fig. 8 Preliminary polarization results at $t = -.8$ $(GeV/c)^2$. The curve shown is a prediction of the model in Ref. 2.

region $.3 < |t| < 1.0$ $(GeV/c)^2$. We also plan some less
precise explorations near the dip in the elastic differential cross section where effects at the 30 - 50% level
are predicted by some models. In addition, we plan to
study the polarization of protons produced inclusively
over a wide range of s.

ACKNOWLEDGEMENTS

We wish to thank the Fermilab Internal Target
Group for their role in building the spectrometer room,
the spectrometer itself, and the jet target. We also
wish to acknowledge the E-198 experimenters from the
University of Rochester, Rutgers University, and the
Imperial College of London for their contributions to
the spectrometer effort and for the use of some of
their equipment.

REFERENCES

1. A. Bohm et al., Phys. Letters 49B, 491 (1974).

2. J. Pumplin and G.L. Kane, Phys. Rev. D11, 1183 (1975).

3. A. Gaidot et al., Phys. Letters 61B, 103 (1976).

4. G.W. Bryant, H.A. Neal, D.R. Rust, "Proton-Carbon Analyzing Power Measurements for Proton Kinetic Energies between .150 GeV and .440 GeV", Indiana University Internal Report COO-2009-102.

POLARIZED TARGET EXPERIMENT AT FERMILAB

Owen Chamberlain

Lawrence Berkeley Laboratory

University of California, Berkeley, CA. 94720

I am reporting today on Experiment 61 at Fermilab,
which is a large collaboration aimed at measuring the
polarization in π^+p, π^-p, and p-p elastic scattering.
We have preliminary results from our first run at 100 GeV.

The experimenters are: From Harvard, Walter Johnson
(actually Suffolk University), Bob Kline, Margaret Law,
and Frank Pipkin. From Yale, Jim Snyder and Mike Zeller.
From the Argonne, Paul Auer, Dan Hill, Bernie Sandler,
and Aki Yokosawa. From Fermilab, Alan Jonckheere and
Peter Koehler. From Berkeley, Walter Brückner,
Owen Chamberlain, Gil Shapiro, and Herb Steiner.

The apparatus is a double-arm spectrometer, as
shown in plan view in Fig. 1. The scattering target is
a polarized proton target (PPT) 8 cm in length as
measured along the beam direction. The target material
is ethylene glycol. Its hydrogen can be polarized to
80% at about 0.4 K.

Each spectrometer arm involves magnetic analysis.
In all there are 16 planes of proportional wire chambers
(PWC) to determine particle trajectories. No Cherenkov

counters are placed in the beam, as the beam is thought
to be too intense to allow them to be practicable. Two
threshold Cherenkov counters are located in the forward
arm to identify the forward (scattered) particle. One
counter should count pions, the other both pions and
kaons. Ninety-seven percent of pions are identified as
such by the first Cherenkov counter.

Identification of elastic scattering of pions or
protons on protons is relatively straightforward.
Discarding, for present purposes, the momentum measurement
on the fast forward (scattered) particle (on the basis
of its being relatively inaccurate) we have effectively
a 3-constraint fit to elastic scattering on free protons.
One expression of this fit is the calculation of 3
components of excess momentum. A plot of numbers of
events versus one component of this excess momentum is
shown in Fig. 2, which shows a free hydrogen peak of
width about 20 MeV/c standing on a broader background
consisting mainly of quasi-elastic scattering events
(approximately elastic scattering from bound protons
exhibiting Fermi motion).

Alternatively we may calculate a χ^2 value for each
event expressing its discrepancy from an apparent elastic
kinematics in the plane of the scattering as defined by
the common plane of beam center line and recoiling particle
and then separately calculate the degree of noncoplanarity,
expressed as $\Delta\phi/\epsilon_\phi$ (the descrepancy in azimuthal angle ϕ
divided by the expected error in ϕ for that event).
Figure 3 shows the χ^2 distribution for 2 classes of events
based on values of $\Delta\phi/\epsilon_\phi$. The peak at low χ^2 shows the
elastic scatterings on free protons. Figure 4 shows the
distribution in $\Delta\phi/\epsilon_\phi$ for 2 classes of events based on
χ^2 value. The peak at zero shows the predominant

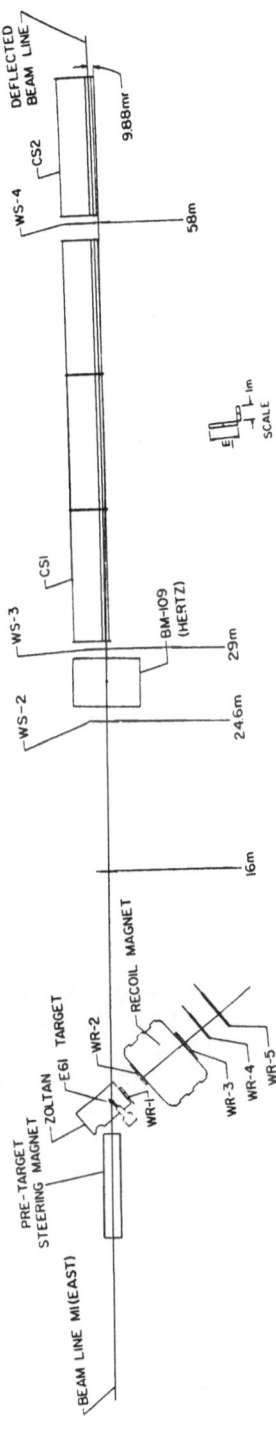

Fig. 1. Plan view of the apparatus. The magnet for the polarized proton target is called Zoltan. Wire proportional chambers are designated W. The Recoil Magnet and the magnet Hertz are used for momentum analysis. The Cherenkov counters are denoted by C. Dimensions are given for the configuration used at 100 GeV, but notice that the drawing shows a considerably foreshortened version of the forward arm of the system.

Table I.   Available beam monitors and our primary
            objections to relying on each.

Monitor Type	Principal Objections
1. Counter telescope at 6°	Affected by beam steering
2. Total counts in recoil arm	Is possibly polarization dependent
3. Ionization chamber in beam	Affected by beam steering
4. Coincidences between pole-tip counters	Affected by beam halo
5. Counts in quasi-elastic background	Affected by level of liquid ^{3}He in the polarized target, therefore potentially different for different polarizations

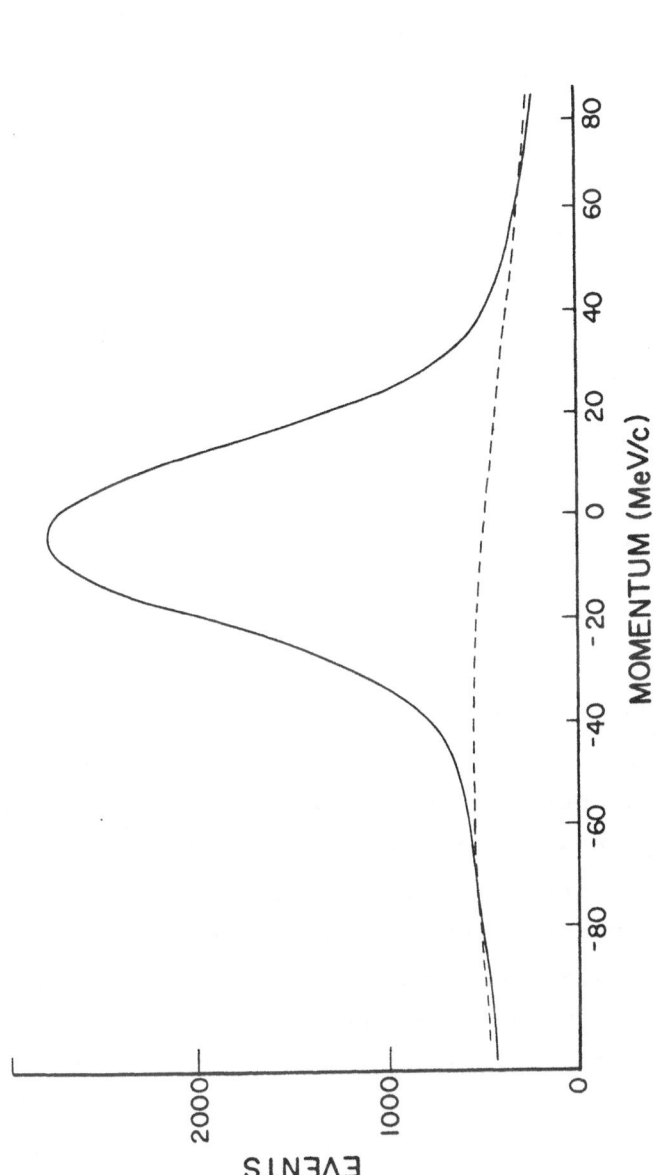

Fig. 2. Distribution of events in one component of missing momentum. The width of the free-hydrogen peak (about 20 MeV/c) is explained by angular variations among incident protons and such effects as multiple Coulomb scattering. The broader background is interpreted as quasi-elastic scattering on bound protons having Fermi motion.

coplanarity of the low-χ^2 events and confirms that there
are two criteria giving satisfactory agreement as to
which events are elastic scattering on free protons.

A more controversial question concerns what beam
monitor is to be used to compare the amount of effective
beam on target for runs with positive and negative
(upward and downward) target polarization. Table I lists
the five available monitors in this experiment and the
objections that may most easily be raised against relying
on each. Noise levels in each monitor (as judged using
the other monitors) are being investigated at present.
Correlations with target polarization are also being
investigated.

Fig. 5 shows our preliminary results for polarization
in elastic π^--p scattering at 100 GeV. Strictly speaking,
it is the asymmetry in the scattering off polarized protons
that is measured. We rely on time-reversal invariance
when we term it the polarization. Clearly these data are
consistent with the polarization being everywhere very
small at this energy.

For comparison there is shown in Fig. 6 the expected
polarization based on scaling down the polarization
results from lower energies in accordance with Regge theory
and the accepted parameters of pomeron and ρ trajectories.
Comparison of Figs. 5 and 6 indicates that our preliminary
results are in accord with Regge theory.

Fig. 7 shows our preliminary results for π^+-p elastic
scattering. Again the values of polarization tend to be
quite small. A comparison with results from lower
energies (not shown) again indicates no disagreement with
Regge theory.

Our results are consistent with the mirror symmetry
observed at lower energies--the positive-pion polarization

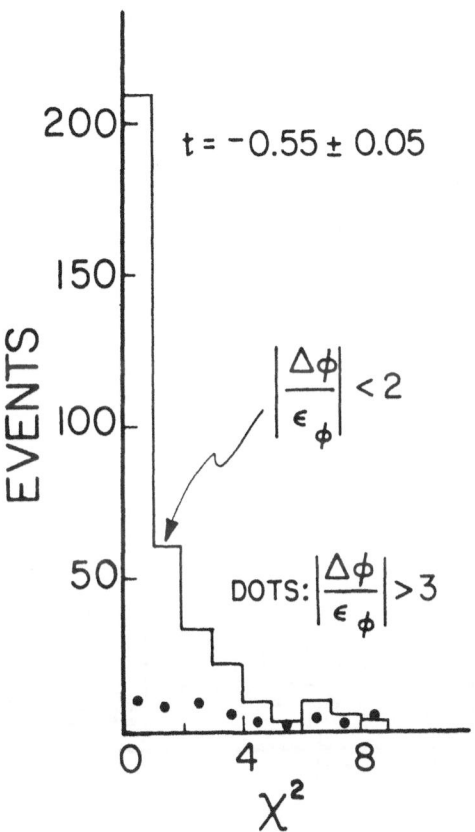

Fig. 3. χ^2 distributions for coplanar ($\Delta\phi/\varepsilon_\phi$ small) and non-coplanar ($\Delta\phi/\varepsilon_\phi$ large) events. The χ^2 values reflect only the characteristics of each event as projected onto the scattering plane.

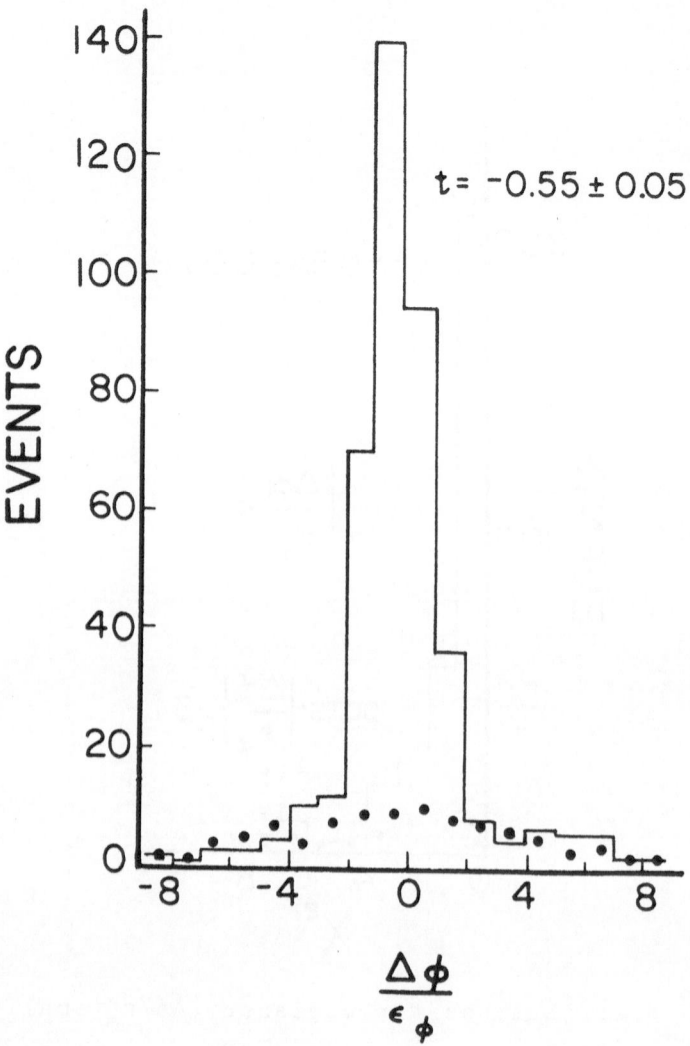

Fig. 4. Distributions in $\Delta\phi/\epsilon_\phi$, the measure of deviation from coplanarity, for 2 classes of event based on χ^2. Low χ^2 means good fit to elastic scattering on a free proton for the event projected onto the scattering plane.

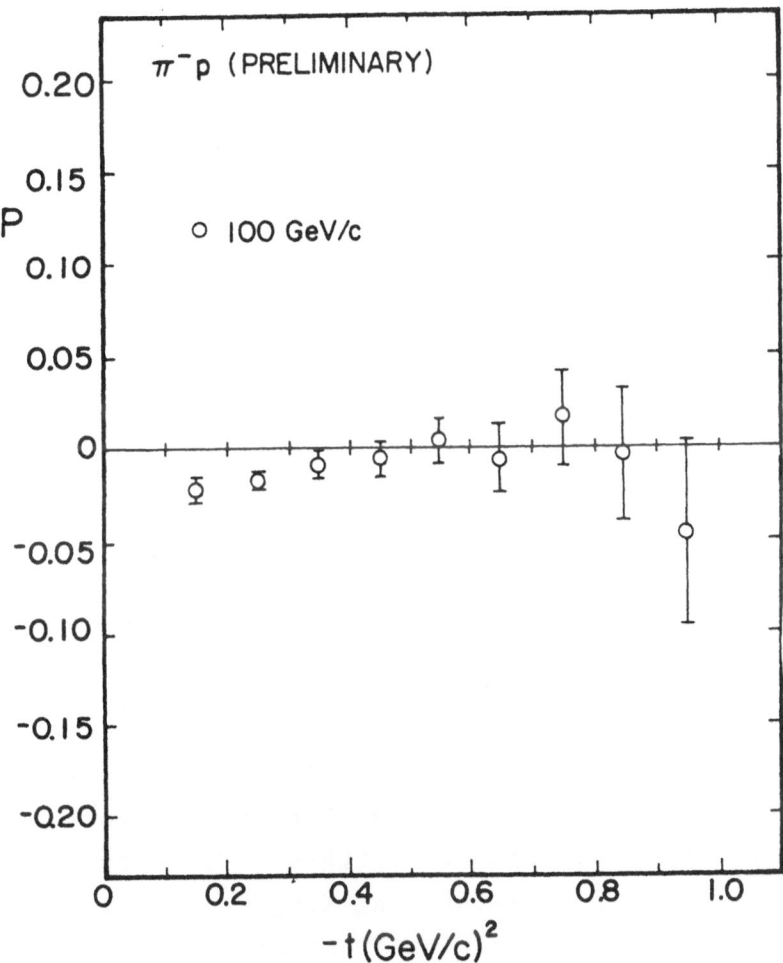

Fig. 5. Our preliminary results for the polarization in π^--p elastic scattering at 100 GeV.

being positive to about the same extent the negative-
pion polarization is negative. This is telling us that
the ρ trajectory and the pomeron trajectory couple in
similar ways.

I had hoped to present comparable polarization
results for p-p scattering. However, it has been
pointed out to us that the results of Bunce, Handler,
March, Martin, Pondrom, Sheaff, Heller, Overseth,
Skubic, Devlin, Edelman, Edwards, Norem, Schachinger,
and Yamin,[1] showing that Λ hyperons produced at high
energies may be highly polarized, suggest that our
proton beam might be somewhat polarized. If there should
be a component of beam polarization normal to our
(horizontal) scattering plane, then our measurements of
polarization P would be contaminated with some (unknown)
contribution from the correlation coefficient C_{nn}. For
that reason no p-p results are presented here.

In the future we will be trying to extend the
polarization measurements to larger values of -t.
Large polarization values are expected to be found
near the cross-section dip at $-t = 1.4$ $(GeV)^2$ when the
lab energy is about 300 GeV. It is amusing to attempt
to predict whether or not it will be easier to get
polarizations significantly different from zero near
the dip. Crudely speaking, the larger polarization
should be a help and the low cross section should be
a hinderance. The following argument gives a guide as
to what one may expect.

We assume there is one large amplitude called
$A(\propto \phi 1 + \phi 3)$, one small amplitude called $B(\propto \phi 5)$, and
3 other amplitudes small enough to be neglected. Then,
assuming for the purposes of this argument that A is
purely imaginary, we have

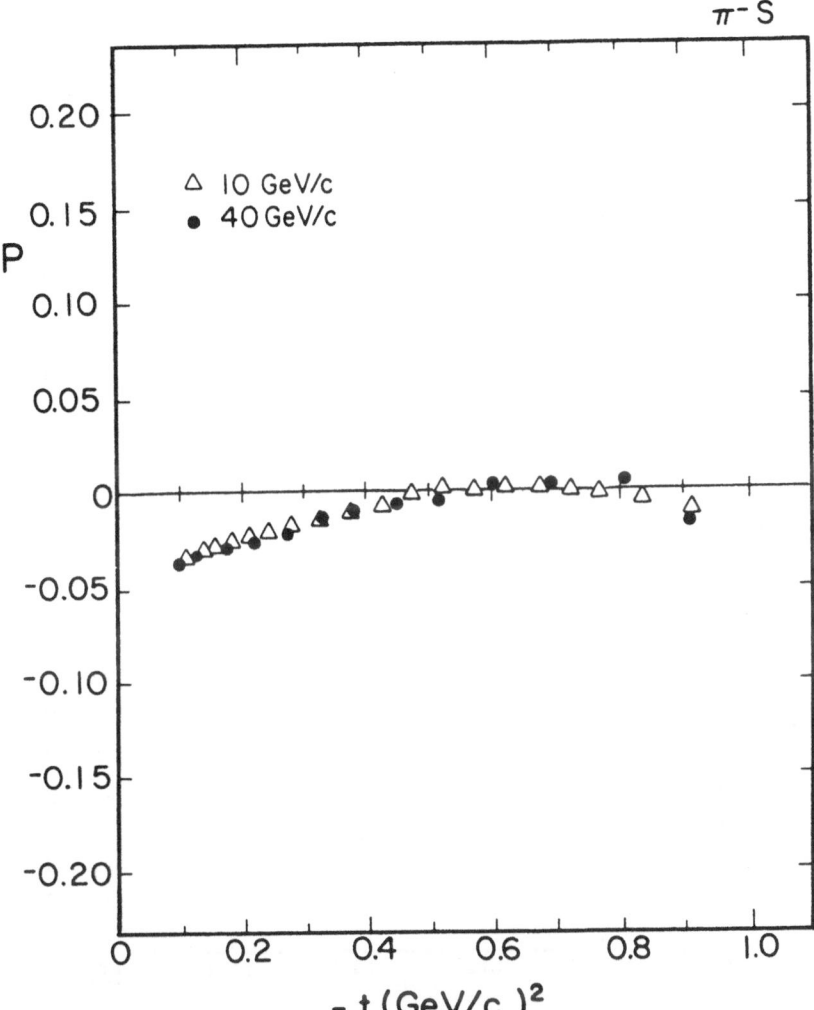

Fig. 6. Expected results for the π^- -p polarization at 100 GeV, based on lower-energy results as adjusted downward according to Regge theory. Note the similarity to the results shown in Fig. 5.

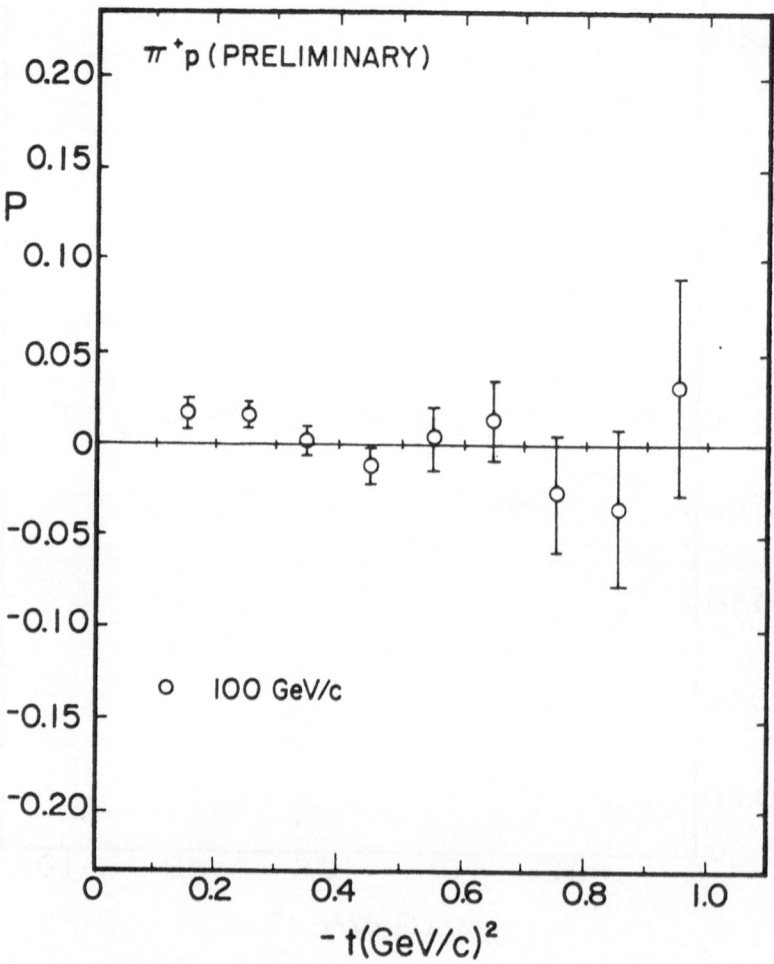

Fig. 7. Our preliminary results for the polarization in
elastic π[+] -p scattering at 100 GeV.

$$\frac{d\sigma}{dt} \simeq |A|^2 + |B|^2 \simeq (Im\ A)^2,$$

$$P\ \frac{d\sigma}{dt} \simeq 2\ Im\ [A*B] \simeq -2\ (Im\ A)\ (Re\ B),$$

$$P \simeq \frac{-2\ Re\ B}{Im\ A}\ .$$

This polarization is to be compared with the uncertainty in the polarization, here assumed to be dominated by statistical uncertainties. For a given beam intensity and length of run the uncertainty in polarization P may be taken as

$$\Delta P = (C)^{-1/2}(\frac{d\sigma}{dt})^{-1/2} \simeq (C)^{-1/2}\ (Im\ A)^{-1}\ ,$$

where C is propotional to the beam intensity times the length of run. Then

$$\Delta P/|P| \simeq 1/\{2C^{1/2}|Re\ B|\}\ .$$

This suggests that to get values of $\Delta P/P$ smaller than 1/3, so polarization is at least 3 standard deviations from zero, one should seek out regions in which Re B is suitably large in magnitude. It does not help, in first approximation, to have a small value of Im A.

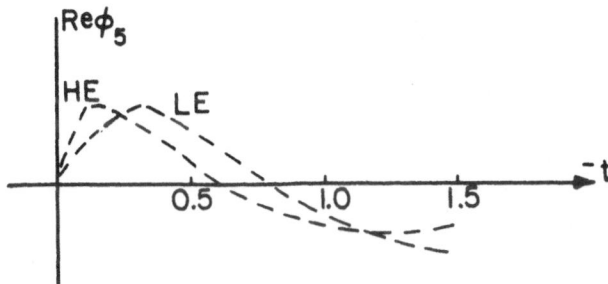

Fig. 8. Estimate given by Kane of the t dependence of the real part of the amplitude ϕ_5, suggesting it may have a magnitude near -t=1.3 nearly as large as at small values of -t.

However, there is hope in the estimates given by Gordon Kane[2], shown in Fig. 8, that $B(\propto \phi_5)$ may be almost as large in magnitude at $-t=1.3$ as at 0.3 $(\text{GeV}/c)^2$.

In conclusion, our preliminary results appear to show no significant deviations from the predictions based on Regge theory and polarization measurements at lower beam energies. In the future we hope to press to as large values of $-t$ as possible, in the hope of finding surprises. We will also be trying to increase our beam intensity with the aim of getting improved accuracy.

REFERENCES

1. G. Bunce et al., Phys. Rev. Lett. <u>36</u>, 1113 (1976).

2. G. L. Kane, Proc. Conf. High Energy Physics with
    Polarized Beams and Polarized Targets (M. L. Marshak,
    Ed., 1976) Amer. Inst. Phys., New York, 1976, p.43.

LOCAL SUPERSYMMETRY AND INTERACTIONS

R. Arnowitt[*]

Northeastern University

Boston, Massachusetts 02115

ABSTRACT

An examination of some of the results of gauge supersymmetry is given. Topics discussed include unification of gravitation and electromagnetism and appearance of minimal electromagnetic couplings, the absence of the cosmological constant, possible origin of internal symmetries, and models involving color and flavor.

1. INTRODUCTION

The interesting aspects of theories based on supersymmetry center around the feature that in such models bosons and fermions are placed in the same multiplet. Thus supersymmetry affords a unification of bosons and fermions not found in more conventional group symmetries. At present it is the only theoretical framework where such a phenomena occurs, and thus represents an interesting framework for constructing

---
[*]Research supported in part by the National Science Foundation.

179

unified gauge theories of interactions.

The conventional supersymmetry of Volkov and
Akulov, Wess and Zumino, and Salam and Strathdee[1] is
a global symmetry. The first attempt to construct a
local supersymmetry was made by Wess and Zumino in a
two dimensional model[1], and the first four dimensional
local supersymmetry (which we will call gauge supersymmetry) was developed by P. Nath and myself[2] and
examined subsequently[3]. An alternate approach, supergravity, was developed by Freedman, van Nieuwenhuizen
and Ferrara, and Deser and Zumino[4] and also studied
extensively[5].

In this talk I will discuss some of the basic results obtained in the gauge supersymmetry formalism,
and perhaps in this way bring out the similarities and
differences between the two approaches. In particular
we will examine the following topics:

1. Unification of gravitation and electromagnetism
and appearance of minimal electromagnetic couplings.

2. The absence of the cosmological constant.

3. Possible origin of internal symmetries.

4. Preliminary models with color and flavor.

To summarize briefly each of these - it appears
that gauge supersymmetry achieves unification of
Einstein-Maxwell interactions and Maxwell-Dirac (minimal)
interactions by the presence of a set of intermediary
auxilliary tensor and vector fields of heavy mass. [Thus
the conventional gravitational and electromagnetic interactions can be thought of as being "threshold theories"
of the more general gauge supersymmetry interactions,
much as the Fermi weak interaction theory is the threshold limit of the Weinberg-Salam model[6].] The existence
of these auxilliary fields implies that the multiplets

needed to achieve such unifications must be quite large
(and it might be difficult to do things this way with
simple multiplets).

The mass growth arising in the auxilliary fields
(and elsewhere) is achieved by a spontaneous breakdown
of gauge supersymmetry which appears to imply a _zero_
cosmological constant at the tree level (and perhaps
also at the one loop level). Since the gauge invariance
of gauge supersymmetry almost completely determines all
the field interactions, the spontaneous symmetry breaking
also will determine what internal symmetries remain unbroken. Thus there is the rather unique result that the
basic gauge invariance can determine at least part of the
structure of the internal symmetry group, a phenomenon
which does not occur in conventional gauge theories.

For simplicity, we will discuss items (1) - (3)
mainly for the N = 2 space with a doublet of fermi coordinates (though a more general analysis can be given[3]).
The analysis of item (4) is still in a very preliminary
phase, and we present here only the simplest (unsatisfactory!) model involving color and flavor. Models with
color and flavor seem to require large N.

2. GLOBAL SUPERSYMMETRY AND GAUGE SUPERSYMMETRY

The simplest type of global supersymmetry can be
characterized by introducing a supersymmetry space
$z^A \equiv (x^\mu, \theta^\alpha)$ with one anti-commuting fermi coordinate[7]
θ^α. The supersymmetry transformation is then generated
by the _linear_ coordinate transformation

$$z^A = z^{A'} + \xi^A(z) \quad , \tag{2.1}$$

where

$$\xi^\mu = i\beta\bar\lambda\gamma^\mu\theta \quad ; \quad [\beta] = \text{mass}$$

$$\xi^\alpha = \lambda^\alpha \quad , \tag{2.2}$$

λ^α is an infinitesimal anticommuting constant and β is a constant. If one introduces a metric $g_{AB}^{(o)}(z)$ in the supersymmetry space

$$ds^2 = dz^A g_{AB}^{(o)}(z) dz^B \quad . \tag{2.3}$$

then the general form of $g_{AB}^{(o)}$ which is preserved by Eq. (2.2) and the Poincare algebra is[7]

$$g_{\mu\nu}^{(o)}(z) = \eta_{\mu\nu} \quad ; \quad g_{\mu\alpha}^{(o)}(z) = -i\beta(\bar\theta\gamma^a)_\alpha \eta_{a\mu}$$

$$g_{\alpha\beta}^{(o)}(z) = k\eta_{\alpha\beta} + \beta^2(\bar\theta\gamma^a)_\alpha(\bar\theta\gamma^b)_\beta \eta_{ab}. \tag{2.4}$$

Here k is an arbitrary constant. If k is not zero, one can always scale it to unity, so in effect there are only two independent possibilities, k = 0, 1. The difference between the two cases is that for $k \neq 0$ an inverse metric exists

$$g_{(o)}^{AB} g_{BC}^{(o)} = \delta^A_C \quad , \tag{2.5}$$

where

$$g_{(o)}^{\mu\nu}(z) = \eta^{\mu\nu}(1 - \frac{\beta^2}{k^2}\bar\theta\theta) \tag{2.6}$$

$$g_{(o)}^{\mu\alpha}(z) = -i(\frac{\beta}{k})(\gamma_a\theta)^\alpha\eta^{a\mu}; \quad g_{(o)}^{\alpha\beta} = \frac{1}{k}\eta^{\alpha\beta} \quad .$$

Suppose now one wished to extend this flat space metric to include gravitation. This would involve replacing $\eta_{\mu\nu}$ by the Einstein field $g_{\mu\nu}(x)$ and $\eta_{a\mu}$ by the vierbein field $e_{a\mu}(x)$ where $g_{\mu\nu}(x) \equiv e_{a\mu}e_{b\nu}\eta^{ab}$:

$$g_{\mu\nu}(z) = g_{\mu\nu}(x) \quad ; \quad g_{\mu\alpha}(z) = -i\beta(\bar{\theta}\gamma^a)_\alpha \, e_{a\mu}(x)$$

$$\text{(2.7)}$$

$$g_{\alpha\beta}(z) = k\eta_{\alpha\beta} + \beta^2(\bar{\theta}\gamma^a)_\alpha(\bar{\theta}\gamma^b)_\beta\eta_{ab}.$$

It is interesting to see how easily supersymmetry can accommodate gravitation, Eq. (2.7) now being covariant with respect to the general coordinate transformations of Einstein:

$$\xi^\mu(z) = \xi^\mu(x) \quad ; \quad \xi^\alpha(z) = 0 . \qquad \text{(2.8)}$$

[One can easily extend Eq. (2.7) to also make it local vierbein invariant.] However, it is no longer invariant with respect to the simple linear transformation with constant coefficients of global supersymmetry, Eq. (2.2). To obtain an invariance algebra, gauge supersymmetry generalizes Eq. (2.2) (in a fashion similar to Einstein theory in bose space) by requiring that the metric tensor $g_{AB}(z)$,

$$ds^2 = dz^A g_{AB}(z)dz^B , \qquad \text{(2.9)}$$

be the fundamental gauge (super) field of the theory, with the gauge algebra generated by <u>arbitrary</u> coordinate transformations in supersymmetry space:

$$z^A = z^{A'} + \xi^A(z) \ . \tag{2.10}$$

Thus g_{AB} transforms as[8]

$$g'_{AB}(z') = \frac{\partial_L z^C}{\partial z^{A'}} \, g_{CD}(z) \, \frac{\partial_R z^D}{\partial z^{B'}} \ , \tag{2.11}$$

and hence the infinitesimal gauge transformation, $\delta g_{AB} \equiv g'_{AB}(z) - g_{AB}(z)$ is given by

$$\delta g_{AB}(z) = g_{AC}\left(\frac{\partial_R \xi^C}{\partial z^B}\right) + \left(\frac{\partial_L \xi^C}{\partial z^A}\right)g_{CB} + \left(\frac{\partial_R g_{AB}(z)}{\partial z^C}\right)\xi^C. \tag{2.12}$$

There still remains the question of whether g_{AB} possesses an inverse or not (corresponding to whether $k \neq 0$ or $k = 0$ in the empty space metric of Eq. (2.4)). Gauge supersymmetry now further assumes that an inverse does exist ($k \neq 0$) and hence that supersymmetry space is Riemannian. This condition now fixes the theory in an almost unique way. As in Einstein theory, the only possible second order field equations are

$$R_{AB}[g_{CD}] = \lambda g_{AB} \ ; \quad [\lambda] = (\text{mass})^2 \ , \tag{2.13}$$

where R_{AB} is the contracted curvature tensor of super-symmetry space[2,3] and $\lambda = $ const. Thus the gauge in-variance determines the dynamics to within only the ambiguity of whether $\lambda = 0$ or $\lambda \neq 0$. As shown by Nath, Zumino and myself, these field equations are derivable from a local action principle[3]. One has then a theory where all fields are gauge fields, and the gauge in-variance determines the dynamics. The gauge invariance also leads to a set of Bianchi identities

$$(-1)^b [R^{BA} - \frac{1}{2} g^{BA} R]; _B = 0 . \qquad (2.14)$$

One may expect that these Bianchi identities will help
guarantee the consistency of the field equations, as
they do in supergravity[4,5].

The above discussion has been carried out for the
case of a single Majorana coordinate, but one may easily
generalize to an arbitrary N-tuplet:

$$z^A = (x^\mu, \theta^{\alpha a}) \quad ; \quad a = 1, \ldots N . \qquad (2.15)$$

The index a then carries the internal symmetry label
and this allows internal symmetries to be included in
the gauge group. In the following we will always assume
that this generalization has been made, but will suppress
the internal symmetry index when no ambiguity arises.

Almost all the special properties of gauge supergravity that we will describe below arise from the assumption of the existence of an inverse metric i.e.
$k \neq 0$. It is this that leads to a unique set of field
equations Eq. (2.13), rather than an infinite number of
dynamical possibilities. One may still examine the singular case, however, as a limiting process $k \to 0$. To
see the significance of this, consider a few of the terms
that arise in the superfield expansion of $g_{AB}(z)$ for the
case $N = 2$ of Eq. (2.15):

$$g_{\mu\nu}(z) = g_{\mu\nu}(x) + \bar{\psi}_{\mu\nu}(x)\theta + \bar{\theta}\varepsilon_a\theta p^a_{\mu\nu}(x) + \ldots$$

$$g_{\mu\alpha}(z) = \frac{k}{2}\,\bar{\psi}_{\mu\alpha}(x) - i\beta(\bar{\theta}\gamma^a)_\alpha e_{a\mu}(x) + (\bar{\theta}\varepsilon F)_\alpha A_\mu(x)$$

$$+ (\bar{\theta}\varepsilon_a F)_\alpha B^a_\mu(x) + (\bar{\theta}\varepsilon F)_\alpha \bar{\psi}(x) i\gamma_\mu\theta + \ldots$$

$$\tag{2.16}$$

$$g_{\alpha\beta}(z) = k(\eta F(x))_{\alpha\beta} + \beta^2(\bar{\theta}\gamma^a)_\alpha(\bar{\theta}\gamma_a)_\beta + \ldots,$$

where $\varepsilon_0 = 1$, $\varepsilon_{1,2,3}$ are Pauli matrices, in the internal symmetry space ($a = 0,1,3$), and $\varepsilon \equiv i\varepsilon_2$. We have seen that $g_{\mu\nu}(x)$ and $e_{a\mu}(x)$ are gravitational field variables. A_μ is a vector field with $J^{PC} = 1^{--}$, $\psi_{\alpha a}(x)$ is a Dirac field ($\psi_{\alpha \text{Dirac}} \equiv \psi_{\alpha 1} - i\psi_{\alpha 2}$), $\bar{\psi}_{\mu\alpha}(x)$ is a Rarita-Schwinger field etc. Eq. (2.13) implies that A_μ couples minimally to the Dirac field and so represents the Maxwell field. One may now show[9] that the N = 1 supergravity coupling[4] of one Majorana component of $\psi_{\mu\alpha}(x)$ to $g_{\mu\nu}(x)$ arises in the singular limit k → 0. (These two fields form the simplest supergravity multiplet.) Thus this singular limit of gauge supersymmetry is closely related to super-gravity[10]. In order to deduce the dynamics of Eq. (2.13) (and other results below) however, it is necessary to deal with a non-singular metric, since only then does the theory possess sufficient internal constraints.

3. SPONTANEOUS BREAKDOWN AND THE COSMOLOGICAL CONSTANT

In gauge supersymmetry, all fields are gauge fields and hence a priori massless. A spontaneous symmetry breakdown allowing mass growth is thus essential for the theory to have physical content. This spontaneous break-down is also related to the role of global supersymmetry

in the theory. Thus the global transformation Eq. (2.2)
has been submerged in the general gauge algebra Eq.
(2.10). At the tree level, at least it re-emerges as
a symmetry of the vacuum state and hence as a symmetry
of the S-matrix. Thus the spontaneous breakdown does
a number of things simultaneously: it grows mass, re-
establishes the role of global supersymmetry, and as we
will see below eliminates the cosmological constant,
determines what internal symmetries are preserved, and
allows for the Einstein-Maxwell-Dirac unification.

   To see how this goes, one proceeds in the standard
way, and expands $g_{AB}(z)$ around its vacuum expectation
value $g_{AB}^{vac} \equiv <0|g_{AB}(z)|0>$:

$$g_{AB}(z) = g_{AB}^{vac} + h_{AB}(z) \quad , \qquad (3.1)$$

$h_{AB}(z)$ thus represents the dynamical modes. At the
tree level g_{AB}^{vac} is a solution of the classical field
equations,

$$R_{AB}(g_{CD}^{vac}) = \lambda g_{AB}^{vac} \quad . \qquad (3.2)$$

Since the invariances of the S-matrix must be invariances
of g_{AB}^{vac}, we have looked for solutions of Eq. (3.2) which
have the form of a generalized global supersymmetry
metric:

$$g_{\mu\nu}^{vac}(z) = \eta_{\mu\nu} \quad ; \quad g_{\mu\alpha}^{vac}(z) = -i(\bar{\theta}\Gamma_\mu)_\alpha$$

$$g_{\alpha\beta}^{vac}(z) = k\eta_{\alpha\beta} + (\bar{\theta}\Gamma^\mu)_\alpha(\bar{\theta}\Gamma_\mu)_\beta \quad . \qquad (3.3)$$

Here $\eta\Gamma_\mu$ is a general symmetric matrix in Dirac and
internal symmetry space consistent with proper Lorentz
invariance:

$$\Gamma^\mu = \gamma^\mu M_{(s)} + i\gamma^\mu\gamma^5 M_{(a)} \quad . \qquad (3.4)$$

$M_{(s,a)}$ are symmetric (anti-symmetric) matrices in the
internal symmetry space. The vacuum expectation g_{AB}^{vac}
is invariant under the generalized supersymmetry transformation of Eq. (2.2) with $\beta\gamma^\mu$ replaced by Γ^μ, and
is the most general supersymmetry S-matrix invariance
according to the theorem of Haag et al[11]. Inserting
Eq. (3.3) into Eq. (3.2) gives then the following constraints on Γ_μ:

$$-k^{-2}\mathrm{tr}\Gamma_\mu\Gamma_\nu = \lambda\eta_{\mu\nu} \quad , \qquad (3.5a)$$

$$-2k^{-2}\Gamma_\mu\Gamma^\mu = \lambda \quad . \qquad (3.5b)$$

Note that these equations depend upon k being non-zero.
    Equations (3.5) are the necessary conditions at
the tree level that a spontaneous breakdown exist to a
vacuum state with generalized global supersymmetry.
The analysis has not yet been extended to the loop level,
and this unfortunately means that one cannot check
whether any of the solutions of Eqs. (3.5) are an absolute minimum. We will assume in the following that a
minimum does exist for solutions that are close to the
tree results, though this clearly requires further analysis.
    One of the immediate consequences of Eqs. (3.5) is
the vanishing of the cosmological constant. This follows

from the fact that the l.h.s. of Eq. (3.5a) is just
$R_{\mu\nu}(g_{CD}^{vac})$. From Eq. (3.4) one has $R_{\mu\nu}(g_{CD}^{vac}) = -k^{-2}c\eta_{\mu\nu}$
(where C = const.) and so Eq. (3.5a) reads $\lambda = -k^{-2}c$.
Now the Einstein field $g_{\mu\nu}(x)$ enters in the $(\theta^{\alpha})^{o}$ part
(i.e. the θ^{α} independent part) of $g_{\mu\nu}(z)$ in Eq. (2.16)
and so the Einstein equations come from the $(\theta^{\alpha})^{o}$ terms
in $R_{\mu\nu}(g_{CD}(z))$. The cosmological term thus comes from
those $(\theta^{\alpha})^{o}$ terms in $R_{\mu\nu}(g_{CD}(z)) - \lambda g_{\mu\nu}(z)$ which are
proportional to $g_{\mu\nu}(x)$:

$$[R_{\mu\nu}(g_{CD}(z)) - \lambda g_{\mu\nu}(z)]_{cosmol.} = \Lambda g_{\mu\nu}(x) \quad , \quad (3.6)$$

where Λ is the true cosmological constant of gravitational theory. Taking the vacuum expectation value of
Eq. (3.6) gives

$$[-k^{-2}c\eta_{\mu\nu} - \lambda\eta_{\mu\nu}] = \Lambda\eta_{\mu\nu} \quad , \quad\quad\quad (3.7)$$

and so $\Lambda = 0$, as a consequence of the spontaneous
symmetry breaking condition.

It is clear that the above argument does not depend
upon the particular choice Eq. (3.3) for the vacuum metric
since the condition that $R_{\mu\nu}(g_{CD}^{vac})$ be proportional to
$\eta_{\mu\nu}$ just represents the Lorentz covariance of the vacuum
state. The result may also hold at the higher loop level
as well (though here the infinities of the theory require
careful discussion).

4. SPONTANEOUS BREAKDOWN AND INTERNAL SYMMETRIES

In conventional gauge theories, the structure of
the internal symmetries is controlled by the choice of
Higgs potential, which determines which symmetries

spontaneously break, and which are maintained. Thus
the gauge invariance, while limiting the form of the
Higgs potential, leaves undetermined most properties
of the internal symmetry. To a large extent, one has
freedom to arrange which symmetries are broken by appropriate choice of the Higgs potential. In gauge
supersymmetry, a new phenomenon arises due to the fact
that here the gauge invariance completely determines
the dynamics (aside from whether λ of Eq. (2.13) vanishes
or not). Thus one no longer has the freedom to add
additional Higgs potentials. The gauge invariance itself determines which symmetries break and which are
preserved, essentially by fixing the Higgs potential.

Internal symmetry gauge transformations appear in
gauge supersymmetry as linear transformations on the
internal index of Eq. (2.15). A general gauge transformations with both even and odd parity parts is given
by

$$\xi^{\alpha a}(z) = [\lambda^A(x) + i\gamma^5\lambda_5^A(x)]\theta^{\alpha b}(M_A)_{ba} . \qquad (4.1)$$

Here $\lambda^A(x)$, $\lambda_5^A(x)$ are gauge functions and M_A, $A=1....N^2$
are a complete set of real $N \times N$ matrices in the in-
ternal symmetry space. The unbroken elements of this
gauge group must, of course, preserve the vacuum state
for the correspoinding global transformations, and hence
leave the vacuum metric Eq. (3.3) invariant. The condition that δg_{AB}^{vac} vanish, then leads to the following
theorem:

After spontaneous breakdown of gauge supersymmetry,
the unbroken internal symmetry algebra is the subalgebra
of the orthogonal algebra constructed from those linear
combinations of generators M_ℓ,

$$M_\ell = c_\ell{}^A M_A + c_{5\ell}{}^A (i\gamma^5) M_A \quad ; \quad \tilde{M}_A = -M_A, \quad (4.2a)$$

which commute with Γ_μ i.e.

$$[\Gamma_\mu, M_\ell] = 0 . \qquad (4.2b)$$

That the unbroken symmetries reside in the compact $O(N)$ group was independently discovered by Freund[12], and the general theorem stated above is due to Nath and myself[13].

To find the unbroken algebra, then, one must first solve the spontaneous breaking equations Eq. (3.5) for Γ_μ and then determine which linear combinations M_ℓ commute with Γ_μ[(Eqs. (4.2)]. We illustrate this process now explicitly for the simple case of $\lambda \neq 0$.

We first note [by taking the trace of Eq. (3.5b)] that Eqs. (3.5) are consistent only if $N = 2$. Thus the spontaneous symmetry breaking condition determines the dimensionality of the internal symmetry space for $\lambda \neq 0$. Equations (3.5) do not have a unique solution for Γ_μ. If, however, one requires that the vacuum be parity invariant, the solution is indeed unique and turns out to be

$$\Gamma_\mu = \beta\gamma_\mu \quad ; \quad \frac{\beta^2}{k^2} = \frac{\lambda}{8} . \qquad (4.3)$$

We see that for $\lambda \neq 0$, the spontaneous symmetry breaking equations have determined Γ_μ so that the vacuum metric Eq. (3.3) has precisely the $N = 2$ global supersymmetry form Eq. (2.4). Also β has been evaluated in terms of λ.

In arriving at Eq. (4.3), we have looked for solutions of the tree equations Eq. (3.2) that have the

specific form of Eq. (3.3). It would be of interest
to see if there are any other parity conserving spontaneous breaking solutions of Eq. (3.2). If there are
not, this would imply the <u>deduction</u> of global supersymmetry from gauge supersymmetry. [This question is
currently under investigation by Nath and myself.] We
next examine Eqs. (4.2) to see which gauges are unbroken.
Since $N = 2$, the M_A are the set of Pauli matrices:

$$M_A: \quad \varepsilon_0 \equiv 1, \; \varepsilon_1, \; \varepsilon_3 \quad ; \quad \varepsilon \equiv i\varepsilon_2. \qquad (4.4)$$

There are 8 different gauges, and Eq. (4.2b) determines
that only <u>one</u>, the even parity gauge $\xi^\alpha = \lambda(x)(\varepsilon\theta)^\alpha$ is
unbroken. This is precisely the $O(2)[\equiv U(1)]$ electro-
magnetic gauge of the Maxwell field $A_\mu(x)$ of Eq. (2.16).
Thus for $\lambda \neq 0$, the gauge invariance of gauge supersymmetry has determined the entire structure of the internal symmetry space with only one additional assumption
of parity conservation.

As in usual gauge theories, the vector mesons associated with the broken gauges grow masses by absorbing
fictitious Goldstone bosons. The general theory of this
has been worked out[13]. We note here that the $B_\mu^a(x)$
mesons of Eq. (2.16) are in fact the gauge mesons for the
broken vector gauges generated by $M_A = \varepsilon_0, \; \varepsilon_1, \; \varepsilon_3$, and
the associated fictitious Goldstone fields are the scalar
fields $f^a(x)$ where $F(x) = k + f^a(x)\varepsilon_a$. One can of course
extend the discussion to other more complicated super
gauge transformations and discuss the mass growth of
other fields[14]. For example, the $p_{\mu\nu}^a(x)$ of Eq. (2.16)
grow masses by absorbing the $B_\mu^a(x)$ and $f^a(x)$. It is
this type of mass growth that produces the unification
of the gravitational field with the other interactions

of the theory.

5.   UNIFICATION OF INTERACTIONS

The Einstein gauge transformations, $\xi^\mu(z) = \xi^\mu(x)$, $\xi^\alpha = 0$ and the internal symmetry gauges transformations $\xi^\mu = 0$, $\xi^\alpha(z) = \lambda^A(x)(\theta M_A)^\alpha$ are special cases of the general transformations $\xi^A(z)$ of gauge supersymmetry. Thus in a group theoretical sense, gauge supersymmetry unifies space-time invariances with internal symmetries. It is necessary to verify, however, that this unification proceeds correctly at a dynamical level from Eqs. (2.13). We summarize briefly here the results for the Einstein-Maxwell-Dirac sectors for the $\lambda \neq 0$ model. Basically, it is the spontaneous symmetry breaking that produces a valid dynamical unification by producing mass growth in the appropriate fields.

We start by examining the Maxwell-Dirac system. The part of the metric directly related to the coupling between these two fields for the N = 2 case reads

$$g_{\mu\nu}(z) = g_{\mu\nu}(x) + p_{\mu\nu\alpha\beta}(x)\theta^\alpha\theta^\beta + \ldots\ldots$$

$$g_{\mu\alpha}(z) = -i\beta(\bar{\theta}\gamma_\mu)_\alpha + (\bar{\theta}\epsilon)_\alpha[eA_\mu(x) + \bar{\psi}(x)i\gamma^\mu\theta] + (\bar{\theta}\epsilon)_\alpha B_{\mu\beta\gamma}\theta^\beta\theta^\gamma + \ldots$$

$$g_{\alpha\beta}(z) = k\eta_{\alpha\beta} + \beta^2(\bar{\theta}\gamma^a)_\alpha(\bar{\theta}\gamma_a)_\beta + \ldots\ , \quad (5.1)$$

where $\epsilon = i\epsilon_2$ is the charge matrix in the O(2) Maxwell internal symmetry space. In Eq. (5.1) we have factored out a parameter e from the photon field $A_\mu(x)$ which we will see is the elementary charge of the Dirac field $\psi(x)$. The field $B_{\mu\alpha\beta}(x)$ is the same "sector" as the

photon field but two orders in θ^α higher in the superfield expansion. Similarly $p_{\mu\nu\alpha\beta}$ is two orders higher
than the gravitational field $g_{\mu\nu}$.

The wave equation for the Dirac field occurs in the
$(\bar{\theta}\epsilon)_\alpha$ sector of $R_{\alpha\beta}$:

$$R_{\alpha\beta} - \lambda g_{\alpha\beta} = (\bar{\theta}\epsilon)_{[\alpha} D_{\beta]} + \ldots \ . \qquad (5.2)$$

By direct calculation, one finds that the Maxwell-Dirac
part of D_α reads

$$[\bar{\psi}\gamma^\mu(\overleftarrow{\partial}_\mu + e\epsilon A_\mu)]_\alpha \ , \qquad (5.3)$$

(where we have neglected gravitational effects for
simplicity). This is precisely the minimal electromagnetic coupling (in Majorana notation) with elementary
charge e. Thus minimal coupling arises naturally in the
theory. The Maxwell equation for $A_\mu(x)$ occurs in the
$(\bar{\theta}\epsilon)_\alpha$ sector of $R_{\alpha\mu}$:

$$R_{\alpha\mu} - \lambda g_{\alpha\mu} = (\bar{\theta}\epsilon)_\alpha D_\mu + \ldots \ . \qquad (5.4)$$

Here the Maxwell-Dirac part of D^μ reads

$$e\partial_\lambda F^{\mu\lambda} + \frac{1}{k} B^\mu \ , \qquad (5.5)$$

where $F_{\mu\nu} = \partial_\mu A_\nu - \partial_\nu A_\mu$ and $B_\mu \equiv B_{\mu\alpha\beta}\eta^{\beta\alpha}$. The Dirac
field's current does not appear directly in the Maxwell
equation, but rather B^μ plays the role of the source.
The equation for B_μ comes from the $(\theta^\alpha)^3$ sector of
$R_{\alpha\mu} - \lambda g_{\alpha\mu} = 0$ and reads

$$\partial_\lambda B^{\mu\lambda} + \frac{\beta^2}{k^2} B^\mu = J^\mu \ , \qquad (5.6)$$

where $B_{\mu\lambda} \equiv \partial_\lambda B_\mu - \partial_\mu B_\lambda$. The mass term $(\beta^2/k^2)B^\mu$ arises in Eq. (5.6) as a consequence of the spontaneous break-down. A term proportional to the Dirac current, $\bar{\psi}\epsilon\gamma^\mu\psi$, enters in the source J^μ of the B^μ field. For "low" momentum phenomena, $p^2 << \beta^2$, one may neglect the gradients in Eq. (5.6) and write $\beta^\mu \cong (k^2/\beta^2)J^\mu$. Inserting this into Eq. (5.5) results in a Dirac current appearing correctly as the source of the Maxwell field. Thus the conventional Maxwell-Dirac theory is a "threshold" approximation in gauge supersymmetry. B_μ plays the role of the W meson of weak interactions. The currents of other charged fields arise as sources of the Maxwell field in a similar fashion by means of the couplings to massive auxilliary fields.

A similar result holds in the gravitational sector. Here the Einstein equations come from the $(\theta^\alpha)^0$ sector of $R_{\mu\nu}(z) - \lambda g_{\mu\nu} = 0$. In terms of the fields in Eq. (5.1), one finds

$$G^E_{\mu\nu} = - \frac{1}{k} \left(p_{\mu\nu} - \frac{1}{2} g_{\mu\nu}(x) \, p^\lambda_{\ \lambda} \right) , \qquad (5.7)$$

where $G^E_{\mu\nu}$ is the usual l.h.s. of the Einstein equations, and $p_{\mu\nu} \equiv p_{\mu\nu\alpha\beta} \eta^{\beta\alpha}$. Again we see it is $p_{\mu\nu}$ that appears as the source, rather than a direct coupling arising e.g. to the Maxwell stress tensor $T_{\mu\nu}^{\ \ \text{Max}}$. The spontaneous breaking cause $p_{\mu\nu}$ to obey a wave equation with mass, and the $p_{\mu\nu}$ source contains a term proportional to $e^2 T_{\mu\nu}^{\ \ \text{Max}}$. Thus in the threshold limit, one may eliminate $p_{\mu\nu}$ in favor of its source, leading to a term in Eq. (5.7) proportional to $e^2(k^2/\beta^2)T_{\mu\nu}{}^{Max}$. Thus there is Maxwell-Einstein unification with a gravitational coupling constant

$$G_E \sim \frac{e^2 k^2}{\beta^2} , \qquad (5.8)$$

or $G_E \sim e^2/\lambda$ by Eq. (4.3).

Eq. (5.8) shows that $\sqrt{\lambda}$ is the super-heavy mass ($\sim 10^{18}$ GeV) and so for practical applications one could always eliminate the superheavy fields as we have done in the two examples above. It is then the resultant Applequist-Carrazone system[15] (after such an elimination) that is to be compared with the physics at realizable energies. In particular, the existence of more than one spin 2 meson produces no a priori difficulty provided only one remains massless.

6.   MODELS WITH FLAVOR AND COLOR

The previous discussion of the $\lambda \neq 0$ model illustrates how electromagnetism and gravitation enters in gauge supersymmetry. However, a realistic model must include the other interactions of nature, and this appears feasible only for larger N. For the $\lambda = 0$ case, the spontaneous symmetry breaking equations at the tree level, Eqs. (3.5), no longer restrict N, and thus this case allows one to consider more complicated fermi spaces. More important is the remarkable result[3] that the only non-vanishing solutions of Eq. (3.5) for $\lambda = 0$ require parity breakdown. Thus the $\lambda = 0$ possibility seems to be the relevant one if one wishes to include weak (and hopefully strong) interactions into the model, since the

basic result of P and C violation arises automatically
in the theory.

Current phenomenology requires that there exist
at least four flavors and three colors of quarks as
well as at least four leptons. In the previous discussion, e.g. Eq. (5.1), we have seen that the fields
observable at low (i.e. presently available) energy
appear in the low θ^α sectors of the metric. (Many of
the higher θ^α sector fields e.g. B_μ and $p_{\mu\nu}$ become
superheavy and so can be eliminated as discussed above.)
It is natural then to see if one can accomodate all the
quarks and leptons in the low θ^α sectors.

The simplest way, perhaps, to do this is to construct a Pati-Salam[16] type model of gauge supersymmetry.
This can be done by giving to the fermi coordinates a
flavor and color index (in addition to the charge index
of the N = 2, $\lambda \neq 0$ model). We write

$$\theta^{\alpha q c f} \quad ; \quad q = 1, 2 \quad ; \quad c, f = 1, \ldots 4. \quad (6.1)$$

Here q is the charge index (spanned as before by the
Pauli matrices ε_o, ε_1, ε_3, $\varepsilon = i\varepsilon_2$) and f and c are the
flavor and color indices. We have assumed four flavors
and four colors. [The fourth color represents the leptons
and any reasonable theory must then explain why the
leptons are so different from the quarks.] The color
and flavor spaces are spanned by a set of U(4) matrices
C_A and F_A, A = 1,..16 respectively (we chose C_A, F_A to
be real anti-symmetric matrices[17]). The Dirac fields,
which appear in the quadratic sector of $g_{\mu\alpha}$ (z) then
carry the same quantum numbers as the fermi coordinates,
$\bar{\psi}_{\alpha q c f}(x)$, and represent the array of quarks and leptons.

To analyze which of the gauges remain unbroken, one follows the procedure of Sec. 4: One first solves the spontaneous breakdown equations, Eqs. (3.5), for Γ_μ. These now read

$$-k^{-2} \text{tr} \Gamma_\mu \Gamma_\nu = \lambda \eta_{\mu\nu} = 0, \qquad (6.2a)$$

$$-2k^{-2} \Gamma_\mu \Gamma^\mu = \lambda = 0. \qquad (6.2b)$$

One then determines which gauges remain unbroken from Eqs. (4.2). Unfortunately, unlike the $\lambda \neq 0$ case, Eqs. (6.2) possess a large number of allowed solutions for the higher N fermi spaces. (Hopefully, when loop corrections are included, this ambiguity will be reduced.) We have thus attempted a more modest program of seeing whether there exists a solution of Eqs. (6.2) such that Eqs. (4.2) leads to the current phenomenological group structure. Thus we require, for any acceptable Γ_μ, the following conditions to hold:

(i)   The $U(1)_\gamma$ of electromagnetism and $SU(3)^{\tilde{c}}$ of color be perfectly conserved.

(ii)   In flavor space, the gauges leading to weak interaction theory be least broken.

(iii)   There are no other perfectly conserved gauges.

Concerning condition (2), by "least broken" we mean have the smallest mass. Thus mass growth can arise from violating Eq. (4.2b) (and hence not preserving $g_{\mu\alpha}^{vac}$) and/or violating the orthogonal algebra condition ($\tilde{M}_A = -M_A$) of Eq. (4.2a) (and hence not preserving $g_{\alpha\beta}^{vac}$). The weak interaction gauges [e.g., of $SU(2) \times U(1)$ or of

$SU(2)_L \times SU(2)_R \times U(1)$] are assumed to preserve Eq. (4.2b) but not Eq. (4.2a) and hence grow only moderately heavy masses, i.e., < 100 GeV. All other flavor gauges must violate both of Eqs. (4.2) and hence grow larger masses and perhaps become very heavy (if not superheavy). In this way the theory picks out from the large number of gauges present, the ones that dominate electromagnetic, weak and strong interactions in the "low energy: (i.e., $\lesssim 10^3$ GeV) domain.

Since all solutions of Eqs. (6.2) violate parity, it is convenient to express Γ_μ in terms of its chiral parts

$$\Gamma_\mu = [\Gamma_+ P_+ + \Gamma_- P_-] \gamma_\mu , \qquad (6.3)$$

where $P_\pm$ are the chiral projection operators

$$P_\pm = \tfrac{1}{2}[1 \pm i\varepsilon\gamma^5] , \qquad (6.4)$$

and $\Gamma_\pm$ are matrices in internal space. ($P_\pm$ maximally violate P and C.) Eqs. (6.2) then require

$$\Gamma_+\Gamma_- = 0 = \Gamma_-\Gamma_+ \qquad (6.5)$$

showing the breakdown of parity.

A model based on the four flavors and four colors of Eq. (6.1) leads to theory with a weak group of the $SU(2)_L \times U(1)_y$ type. There exists a Γ_μ obeying Eq. (6.5) (with $\Gamma_+ = 0$) satisfying condition (i) above, and partially satisfying (ii). One can satisfy both (i) and (ii) if one doubles the number of quarks and leptons by adding an additional number ("heaviness") to the fermi

space: $\theta^{\alpha qcfh}$, $h = 1, 2$. The state $h = 1$ corresponds
to the "light" quarks (with masses $\leqslant 2$ GeV) and the
usual leptons. The state $h = 2$ corresponds to heavy
quarks (with masses $\geqslant 5$ GeV) and heavy leptons. We
let the s_a be Pauli matrices in the h-space and let
$P_{(L,Q)}$ be the (lepton,quark) projection operator in
$U(4)^c$ space, i.e.,

$$P_L = \frac{1 - \sqrt{6} \, C_{15}}{4} \; , \qquad P_Q = \frac{3 + \sqrt{6} \, C_{15}}{4}, \qquad (6.6)$$

where C_{15} is the 15th SU(4) color matrix. Then a possible choice for $\Gamma_\pm$ is

$$\Gamma_\pm = \frac{1 \mp s_3}{3} \left[\beta_\pm P_L + (\gamma_\pm^{(1)} F_{(1)} + \gamma_\pm^{(2)} \varepsilon F_{(2)}) P_Q \right] ,$$

$$(6.7)$$

where $\beta_\pm$, $\gamma_\pm$ are constants and $F_{(1,2)}$ are matrices in
flavor space:

$$F_{(1)} = \begin{pmatrix} 0 & \vdots & \tau_1 \\ \cdots & \vdots & \cdots \\ \tau_1 & \vdots & 0 \end{pmatrix} \quad ; \quad F_{(2)} = \begin{pmatrix} 0 & \vdots & \tau_1 \\ \cdots & \vdots & \cdots \\ -\tau_1 & \vdots & 0 \end{pmatrix} .$$

$$(6.8)$$

To see how Eq. (6.7) implies (i) and (ii), we first
note that $\Gamma_\pm$ obey Eq. (6.5) due to the "heaviness" projection operators $\frac{1}{2}(1 \pm s_3)$. The off diagonal $F_{(1,2)}$
flavor matrices commute with the charge generator and
so the $U(1)_\gamma$ gauge is rigorously conserved. The C_{15}
matrix of P_L and P_Q commutes only with the subalgebra
SU(3)^c generators, and so the other elements of SU(4)^c
are badly broken. Thus only the hadrons possess the
strong, conserved, SU(3) interaction (even though the

theory initially started with an $SU(4)^C$ symmetry). In
this way the vacuum metric (6.7) distinguishes leptons
from hadrons. Of the 30 $SU(4)$ chiral flavor gauges,
only the following generators

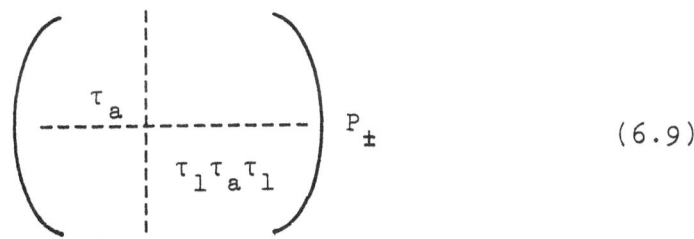

$$\left(\begin{array}{c|c} \tau_a & \\ \hline & \tau_1 \tau_a \tau_1 \end{array}\right) P_\pm \qquad\qquad (6.9)$$

commutes with $F_{(1,2)}$ and hence with $\Gamma_\pm$. Thus the flavor
gauges which are least broken are $SU(2)_L \times SU(2)_R \times$
$U(1)$ with μ-e universality. This establishes condition
(ii). The model does not satisfy condition (iii), however, at least at the tree level. There are additional
conserved $U(1)$ gauges (e.g., representing fermion number
conservation). More serious is the additional conserved
non-abelian gauges. For example, the presence of conserved charge and color generators leads also to a conserved gauge generated by the product of the charge and
color generators. Whether closed loops will eliminate
these unwanted conserved gauges requires detailed examination.

While the above model is still unsatisfactory, it
is interesting to see that a large amount of present
phenomenology can be fitted within the framework of
gauge supersymmetry. It remains to be seen whether the
theory will produce a unique satisfactory model. At
present we are cautiously optimistic.

REFERENCES

1. D. V. Volkov and V. P. Akulov, Phys. Lett. 65B,
    73 (1973); J. Wess and B. Zumino, Nuclear Phys.
    B76, 477 (1974); A. Salam and J. Strathdee,
    Nuclear Phys. B76, 477 (1974). For a recent review
    see P. Fayet and S. Ferrara, Physics Reports (to
    be pub.).

2. P. Nath and R. ARnowitt, Phys. Lett. 56B, 171 (1975).

3. R. Arnowitt, P. Nath and B. Zumino, Phys. Lett. 56B,
    81 (1975); R. Arnowitt and P. Nath, Gen. Rel. Grav.
    7, 89 (1976); P. Nath and R. Arnowitt, J. de Physique
    37, C2 (1976); P. Nath, Proc. of Conf. on Gauge
    Theories and Modern Field Theory, Boston 1975 (MIT
    Press, Cambridge, Mass. 1976); R. Arnowitt and P.
    Nath, Phys. Rev. Lett. 36, 1526 (1976); S.S. Chang,
    Phys. Rev. 14D, 447 (1976); R. Arnowitt and P. Nath.
    "Origin of Internal Symmetry", to be pub. Phys. Rev.
    Feb. 15 issue (NUB #2300), P. Nath and R. Arnowitt,
    "Structure of Spontaneously Broken Local Gauge
    Supersymmetry", sub. to Nuclear Phys. (NUB #2302).

4. D. Z. Freedman, P. van Nieuwenhuizen and S. Ferrar,
    Phys. Rev. D13, 3214 (1976); S. Deser, and B. Zumino,
    Physics Letters 62B, 335 (1976).

5. S. Ferrar, J. Scherk and P. van Nieuwenhuizen Phys.
    Rev. Lett. 37, 1035 (1976); D. Z. Freedman and J. H.
    Schwarz, Stony Brook preprint ITP-SB-76-41 (1976);
    S. Deser, B. Zumino, P. di Vecchia and P. Howe,
    CERN preprint Th 2208; S. Ferrar, F. Gliozzi and
    J. Scherk, Ecole Normale Sup. preprint P.T.E.N.S.
    76/19; S. Ferrara, D. Z. Freedman and P. van
    Nieuwenhuizen, Stony Brook Preprint ITP-SB-76-46;
    B. de Wit, University of Leiden Preprint; S. Ferrara
    and P. van Nieuwenhuizen, Phys. Rev. Lett. 37, 1669

(1976); S. Ferrara, J. Scherk and B. Zumino, Ecole Normale Sup. preprint P.T.E.N.S. 76/23; D. Z. Freedman, Stony Brook preprint ITP-SB-76-58, D. Z. Freedman and A. Das, Stony Brook preprint ITP-SB-76-64.

6. S. Weinberg, Phys. Rev. Lett. $\underline{19}$, 1264 (1967); A. Salam, Proc. of Eighth Nobel Symposium (John Wiley, N.Y., 1968).

7. The θ^α obey $\{\theta^\alpha, \theta^\beta\} = 0$ where $\alpha, \beta = 1 \ldots 4$ is the Dirac index. The Majorana condition is $\theta_\alpha = \theta^\beta \eta_{\beta\alpha}$ where $\eta = -C^{-1}$ (C = charge conjugation matrix) and $\theta_\alpha = (\theta^+ \gamma^0)^\alpha$. Our Dirac matrices obey $\{\gamma^\mu, \gamma^\nu\} = -2\eta^{\mu\nu}$ with the Lorentz metric obeying $\eta^{oo} = -1$. We use the convention where θ^α and x^μ have the same dimension so that β of Eq. (2.2) has dimensions of mass.

8. The symbols ∂_L and ∂_R stand for left and right derivatives respectively.

9. P. Nath and R. Arnowitt, Physics Lett. $\underline{65B}$, 73 (1976).

10. Non-Riemannian geometries have been studies by B. Zumino, Proc. of Conf. on Gauge Theories and Modern Field Theory, Boston 1975 (MIT Press, Cambridge, Mass. 1976); and V. P. Akulov, D. V. Volkov and V. A. Sorkoa, JETP Letters $\underline{22}$, 396 (1975). These are related to the singular geometry arising in the $k \to 0$ limit.

11. R. Haag, J. T. Lopuszanski and M. Sohnius, Nucl. Phys. $\underline{B88}$, 257 (1975).

12. P. G. O. Freund, Jour. Math. Phys. $\underline{17}$, 424 (1976).

13. For a detailed discussion of this result, see the seventh paper of Ref. (3).

14. See the last paper of Ref. (3).

15. T. Appelquist and P. Carrazone, Phys. Rev. $\underline{D11}$,

2856 (1975).

16. J. C. Pati and A. Salam, Phys. Rev. $\underline{D8}$, 1240 (1973).

17. A representation of C_A is the following: $C_A = \{i\lambda_A^{(a)}, \varepsilon\lambda_A^{(s)}\}$ where $\lambda_A^{(s,a)}$ are the symmetric (antisymmetric) $U(4)$ matrices.

SUPERGRAVITY FIELD THEORIES AND THE ART OF CONSTRUCTING THEM

Daniel Z. Freedman

Institute for Theoretical Physics

State University of New York at Stony Brook

Stony Brook, New York 11794

During the past year Lagrangian field theories in four dimensional space-time have been developed which have a new gauge principle, namely, local supersymmetry. The gauge field of supersymmetry transformation is the spin 3/2 Rarita-Schwinger field and it necessarily occurs in these theories together with the vierbein field which describes gravitation. A variety of constructions of supergravity field theories have now been given which include vector, spinor, and scalar fields. Although none of them seems to apply directly to experiment, I take the attitude that they are all interesting because they are the basic constructions associated with a gauge principle of considerable mathematical elegance. In qualitative terms, what has been achieved is an extension of general relativity in which gravitation is closely linked with the fundamental concept of anti-commuting fermion fields. Unification of lower spin fields with the graviton occurs in some models, and supergravity practitioners hope that a corresponding unification of

gravitation with other particle interactions can be
achieved. Since there are as yet no signals from
experiment that nature is aware of our efforts, we look
for theoretical signals. The improved renormalizability
situation in supergravity, which will be discussed by
Peter van Nieuwenhuizen at this conference, may be one
such signal.

In this review of supergravity field theories,
I will not give the specific attribution of results
discussed in the text. Instead, I will give an annotated
bibliography at the end which I hope will be a useful
guide to the literature of this field.

GLOBAL SUPERSYMMETRY

Since supergravity is the gauge theory of
ordinary or global supersymmetry, I start with a brief
non-historical discussion of that subject. A rich
theoretical development may be summarized in the
statement that global supersymmetry is the study of
quantum field theories with conserved Majorana spinor
charges Q_α which satisfy the (graded) commutation rules

$$\{Q_\alpha, \bar{Q}_\beta\} = (\gamma^a)_{\alpha\beta} P_a \quad , \tag{1}$$

$$[M^{ab}, Q_\alpha] = -i(\sigma^{ab})_{\alpha\beta} Q_\beta \quad , \tag{2}$$

$$[P^a, Q_\alpha] = 0 \quad , \tag{3}$$

involving the generators P^a and M^{ab} of translations and
Lorentz transformations. The irreducible representations
of this system tell us what sets of particles span a
"supermultiplet" and therefore what kinds of field theories
can admit a supersymmetry. For massless particles, an

irreducible representation has the spin (more properly, helicity) content $(j, j - 1/2)$ for $j = 1/2, 1, 3/2 \ldots$ and therefore consists of a real boson and Majorana fermion of adjacent spins. Representations for massive particles can be regarded as combinations of massless representations and we will not discuss them here.

Globally supersymmetric field theories for the $(1/2, 0)$ and $(1, 1/2)$ multiplets have been studied extensively, and the latter case is perhaps the most interesting. The vectors and spinors are assigned to the adjoint representation of an arbitrary internal symmetry group and the corresponding minimally coupled Yang-Mills Lagrangian

$$L = - \frac{1}{4} F_{ab}{}^i F^{abi} + \frac{1}{2} i \bar{\chi}^i \gamma^a (D_a \chi)^i \qquad (4)$$

is supersymmetric. It changes by a total derivative under the infinitesimal field variations

$$\delta A_a^i = \frac{i}{\sqrt{2}} \bar{\epsilon} \gamma_a \chi^i \quad , \qquad (5)$$

$$\delta \chi^i = \frac{1}{\sqrt{2}} \sigma^{bc} F_{bc}{}^i \epsilon \quad , \qquad (6)$$

where the supersymmetry parameter ϵ is a space-time independent anti-commuting Majorana spinor. The proof of invariance is straightforward. It involves a) detailed Dirac algebra, namely, the relations $[\sigma^{bc}, \gamma^a] = \gamma^b \eta^{ac} - \gamma^c \eta^{ab}$ and $[\sigma^{bc}, \gamma^a] = - i \epsilon^{abcd} \gamma_5 \gamma_d$; b) recognition of the gauge field Bianchi identity $\epsilon^{abcd} (D_a F_{bc})^i = 0$; and c) use of a Fierz rearrangement to show that $f^{ijk} (\bar{\chi}^{ia} \chi^j) (\bar{\epsilon} \gamma_a \chi^k) = 0$. Analogous manipulations are also necessary in supergravity constructions to which we now turn.

SUPERGRAVITY

From the basic algebra (1-3), we see that a local theory of supersymmetry necessarily involves the gauging of Poincaré generators along with the Q_α. Thus supersymmetry and gravitation will be unified. The (2,3/2) multiplet of the representation theorem is the natural candidate for the particle content and the corresponding interacting field theory of vierbein ($V_{a\mu}$) and Rarita-Schwinger (ψ_μ) fields has been constructed. It is invariant under supersymmetry transformations involving an arbitrary anti-commuting Majorana spinor function $\varepsilon(x)$.

In the second order relatively formalism of most textbooks, the Lagrangian is

$$
L = - \frac{1}{4\kappa^2} V\, V^{a\mu} V^{b\mu}\, R_{\mu\nu ab} - \frac{1}{2}\, \varepsilon^{\lambda\rho\mu\nu}\, \bar{\psi}_\lambda\, \gamma_5\, \gamma_\mu\, D_\nu\, \psi_\rho
$$

$$
- \frac{1}{16} V\kappa^2 [\, (\bar{\psi}^\lambda \gamma^\mu \psi^\rho)\, (\bar{\psi}_\lambda\, \gamma_\mu \psi_\rho + 2\, \bar{\psi}_\mu\, \gamma_\lambda\, \psi_\rho) \qquad (7)
$$

$$
- 4\, (\psi \cdot \gamma \psi^\rho)\, (\bar{\psi} \cdot \gamma \psi_\rho)\,]\quad,
$$

with curvature tensor

$$
R_{\mu\nu ab} = \partial_\mu\, \omega_{\nu ab} - \partial_\nu\, \omega_{\mu ab} + \omega_{\mu a}{}^c \omega_{\nu cb} - \omega_{\nu a}{}^c\, \omega_{\mu cb} \qquad (8)
$$

and covariant derivative

$$
D_\nu\, \psi_\rho = \partial_\nu \psi_\rho - \Gamma_{\nu\rho}{}^\sigma \psi_\sigma + \frac{1}{2}\, \omega_{\nu ab}\, \sigma^{ab}\, \psi_\rho\quad. \qquad (9)
$$

The vierbein connection $\omega_{\mu ab}$ and Christoffel symbol $\Gamma_{\mu\nu}{}^\rho$ are determined in terms of $V_{a\mu}$ by

$$\omega_{\mu ab} = \omega^o_{\mu ab}$$

$$= \frac{1}{2} [V_a^\nu (\partial_\mu V_{b\nu} - \partial_\nu V_{b\mu}) + V_a^\rho V_b^\sigma (\partial_\sigma V_{c\rho}) V_\mu^c] - [a \to b] ,$$

$$(10)$$

$$\Gamma_{\mu\nu}^{\ \rho} = V^{a\rho} [\partial_\mu V_{a\nu} + \omega_{\mu a}^{\ \ b} V_{b\nu}] . \tag{11}$$

The Lagrangian changes by a total derivative under the supersymmetry variations

$$(12)$$

$$\delta V_{a\mu} = -i \kappa \bar{\epsilon} \gamma_a \psi_\mu ,$$

$$\delta \psi_\rho = \kappa^{-1} [D_\rho \epsilon - \frac{1}{4} \kappa^2 (2 \bar{\psi}_\rho \gamma_a \psi_b + \bar{\psi}_a \gamma_\rho \psi_b) \sigma^{ab} \epsilon] . \tag{13}$$

The theory was constructed by starting from the minimally coupled spin 2 and spin 3/2 actions and the variations $\delta V_{a\mu}$ of (12), which was suggested by analogy with the variation (5) of supersymmetric Yang-Mills theory, and $\delta \psi_\rho \sim D_\rho \epsilon$, which is the covariant generalization of a gauge-like invariance of the free massless spin 3/2 action essentially noted in 1941 by Rarita and Schwinger. A beautiful cancellation of the terms in δL which are linear in ψ_ρ was then noted. It requires an intimate mix of Dirac algebra and use of the Ricci and cyclicity identities

$$[D_\mu, D_\nu] \epsilon = \frac{1}{2} R_{\mu\nu ab} \sigma^{ab} \epsilon , \tag{14}$$

$$\epsilon^{\lambda\rho\mu\nu} R_{\lambda\rho\mu\sigma} = 0 . \tag{15}$$

Full invariance required the addition of the contact term in (7) and the additional bilinear term in (13). The procedure was arduous in the first construction,

but even more complicated procedures are now routinely
carried out by supergravity practitioners. It is a
useful lesson that complicated algebraic manipulations
can be mastered!

   An important simplification can be achieved by
introducing the first order or torsion formalism. This
method was first used in relativity theory by Cartan
and Palatini, and the essential feature is that the
connection field $\omega_{\mu ab}$ is treated as an independent
Lagrangian variable. It is a non-propagating or auxiliary
field, and its equation of motion can be solved
algebraically yielding

$$\omega_{\mu ab} = \overset{o}{\omega}_{\mu ab} + K_{\mu ab} \qquad , \qquad (16)$$

$$S_{\mu\nu}{}^{\rho} = -\frac{1}{2}\, [K_{\mu\nu}{}^{\rho} - K_{\nu\mu}{}^{\rho}] \qquad (17)$$

$$= \frac{1}{2}\, [\Gamma_{\mu\nu}{}^{\rho} - \Gamma_{\nu\mu}{}^{\rho}] \qquad ,$$

where S and K are the torsion and contortion tensors and
are usually bilinear expressions in the matter fields to
which gravity is coupled in the particular problem
studied.

   As applied to supergravity, the first order
Lagrangian is

$$L = -\frac{1}{4\kappa^{2}}\, V\, V^{a\mu}\, V^{b\nu}\, R_{\mu\nu\ ab} - \frac{1}{2}\, \epsilon^{\lambda\rho\mu\nu}\, \bar{\psi}_{\lambda}\, \gamma_{5}\, \gamma_{\nu}\, \mathcal{D}_{\nu}\, \psi_{\rho} ,$$
$$(18)$$

with the curvature tensor (8). It is locally supersymmetric provided one takes the derivative

$$D_\nu \psi_\rho = (\partial_\nu + \frac{1}{2} \omega_{\nu ab} \; \sigma^{ab}) \psi_\rho \quad . \tag{19}$$

This is not a true covariant derivative (as can be seen by comparison with Eq. 9), but the curl formed from it is covariant and only the curl enters the Lagrangian. Further, it is only with the derivative (19) that the first order spin 3/2 action is Weyl conformal invariant. The torsion associated with (18) is

$$S_{\mu\nu\rho} = - \frac{1}{2} i \; \kappa^2 (\bar{\psi}_\mu \gamma_\rho \psi_\nu) \quad . \tag{20}$$

This leads exactly to the contact term in (7).

The action formed from (18) is invariant under the field variations

$$\delta V_{a\mu} = -i \; \kappa \bar{\varepsilon} \; \gamma_a \; \psi_\mu \quad ,$$

$$\delta \psi_\rho = \kappa^{-1} D_p \; \varepsilon \quad ,$$

$$\delta \omega_{\mu ab} = B_{\mu ab} + \frac{1}{2} V_{a\mu} B_{\lambda b}{}^\lambda - \frac{1}{2} V_{b\nu} B_{\lambda a}{}^\lambda \quad ,$$

$$B_\mu{}^{\nu\rho} = \kappa V^{-1} \varepsilon^{\nu\rho\sigma\tau} \bar{\varepsilon} \; \gamma_5 \; \gamma_\mu D_\sigma \psi_\tau \quad . \tag{21}$$

Note that the contact term in (7) and the extra bilinear term in (13) are both torsion effects and therefore do not appear explicitly in the first order form of supergravity. The proof of invariance is also simpler. However, the second order form seems to be best for quantization and calculation of Feynman graphs.

Subsequent supergravity constructions have been performed in both the first and second order formalisms. They are always completely equivalent and differ in technical details only. Lagrangians are simpler in first

order although the simplification is more modest than
above, and the variation $\delta\omega_{\mu ab}$ is generally very
complicated.

When one finds a new symmetry of a field theory
it is usually instructive to form commutators of the
transformations in order to determine the full algebra
of invariance. This commutator algebra has been
investigated in supergravity. The commutator of
variations δ_1 (and δ_2) with spinor parameters ϵ_1 (and ϵ_2)
involves the spinor bilinear $\xi^a(x) = i\bar{\epsilon}_1(x)\gamma^a\epsilon_2(x)$ and
one finds the equations

$$[\delta_1, \delta_2]V_{a\mu} = D_\mu \xi_a \quad,$$

$$[\delta_1, \delta_2]\psi_\mu = \xi^\nu(D_\nu\psi_\mu - D_\mu\psi_\nu) \quad,$$

$$[\delta_1, \delta_2]\ \omega_{\mu ab} = \xi^\nu(R_{\nu\mu ab} - \kappa^2 V^{-1}\epsilon_{ab}{}^{\sigma\tau}\ \psi_\nu\gamma_5\gamma_\mu\ D_\sigma\psi_\tau) \quad. \tag{22}$$

These commutators have a uniform interpretation as the
sum of

i)   a general coordinate transformation with
displacement parameter ξ^ν.

ii)   a local Lorentz transformation with field
dependent parameter $\lambda_{ab} = \xi^\nu\omega_{\nu ab}$ which serves to make
the general coordinate transformation a covariant
operation on spinor and local Lorentz tensors.

iii)   and the original supersymmetry transformation
with field dependent parameter $\epsilon' = -\kappa\xi^\nu\psi_\nu$.
Analogous features appear in other supergravity models.
The commutator algebra on fields appears to be more
complicated than the original global supersymmetry
algebra (1), but in fact the action on the Hilbert space
of physical states of spin 2 and spin 3/2 particles is

exactly that of (1). A field-dependent Yang-Mills gauge transformation appears in the commutator algebra of supersymmetric Yang-Mills theory, so that the field dependent transformations in ii) and iii) above are not an entirely new feature.

There are certain other terms in the commutators which vanish as a consequence of the equations of motion which have been omitted in (22) and in the discussion. These are separately invariances of the theory, but are expected to be of little consequence because they vanish on-shell. Analogous terms also occur in global supersymmetry models when auxiliary fields are omitted, as in the form of supersymmetric Yang-Mills theory given previously. This has led to a line of thinking that an improved treatment of supergravity theories will result if one can find a set of auxiliary fields which eliminates the equation of motion terms. Some progress in this direction has been made.

From the mathematical standpoint, the fact that the supergravity commutators contain general coordinate and local Lorentz transformations means that a non-trivial factorization of the general coordinate group has occurred. This is the generalization of the non-trivial factorization of the Poincaré group which appears in global supersymmetry.

Another significant aspect of supergravity theories is that previous consistency difficulties of an inter-acting spin 3/2 field are avoided because of the fermionic gauge invariance.

One way to see this is to study the constraints implicit in the Euler-Lagrange equations

$$\frac{\delta L}{\delta \bar\psi_\lambda} = 0 \ .$$

(23)

In previously considered interacting spin 3/2 field
theories, the application of a derivative to the Euler
variation gives an expression which does not vanish
on-shell and leads to an interaction-dependent constraint
on the initial data of the Cauchy problem for ψ_ρ. This
constraint is mainly responsible for previous consistency
problems. In supergravity one finds that

$$\mathcal{D}_\lambda \ \frac{\delta \ L}{\delta \bar{\psi}_\lambda} \tag{24}$$

vanishes on-shell, and no such constraint is present.
This is a simple consequence of the fermionic gauge
invariance and is true in all supergravity constructions.

One can also show quite directly by using the
Faddeev-Popov procedure for quantization of supergravity
theories that acausal propagation of spin 3/2 waves is
avoided and that the residue of spin 3/2 propagator poles
in S-matrix elements is positive as required by unitarity.
This has been demonstrated in the classical or tree
approximation. Either acausal propagation or ghosts or
both difficulties always occurred in previous interacting
spin 3/2 theories, which were treated in an external
field approximation roughly comparable to the tree
approximation.

If supergravity remains a viable field of research,
a name will have to be given to the spin 3/2 partner of
the graviton. The name super-graviton is natural in this
context but I hope that physicists will have more
imagination. The name hemitrion has been attempted in
one paper, but journal editors were not impressed by its
classic credentials. A potentially important contribution
has been made by two supergravity non-practitioners from
well known eastern institutions who suggest the name

<u>gravitino</u>. I wonder who will be first to try that in the literature.

Up to now we have mainly discussed the pure supergravity theory, i.e., the $(2,3/2)$ gauge multiplet. Subsequently developed theories can be grouped in three categories: A) matter multiplet couplings; B) extended supergravity theories, and C) theories with a minimal vector gauge coupling of the Rarita-Schwinger field.

MATTER MULTIPLET COUPLINGS

If local supersymmetry is a true gauge principle, then it should be possible to take any global supersymmetric field theory and implement local invariance by introducing interactions with the $(2,3/2)$ gauge multiplet. The analogous problem in Yang-Mills theory and ordinary gravitation is solved by the recipe of introducing covariant derivatives and simple tensor analysis. At present there is no such simple recipe in supergravity. Instead, a general strategy has evolved which requires a combination of general principles and hard work in each case. We will discuss the strategy after presenting examples of both matter couplings and extended supergravity.

One interesting example is the locally supersymmetric extension of the $(1,1/2)$ multiplet which was previously discussed at the global level. The Lagrangian (in first order form) is a sum of four terms

$$L = L_{SG} + L_M + L_N + L_C \tag{25}$$

and it should be emphasized that it is only the sum which is invariant. The separate parts are not. The first order gauge Lagrangian L_{SG} is given in (18) while the matter kinetic Lagrangian

$$L_M = -\frac{1}{4} V g^{\mu\nu} g^{\rho\sigma} F_{\mu\rho} F_{\nu\sigma} + \frac{i}{2} V \bar{\chi} \not{D} \chi \qquad (26)$$

is the covariant generalization of (4), and contains a combined first order gravitational and Yang-Mills covariant derivative $D_\mu \chi$. Internal symmetry indices are suppressed.

The term

$$L_N = \frac{-i}{\sqrt{2}} \kappa V \bar{\psi}_\lambda \sigma^{\mu\nu} F_{\mu\nu} \gamma^\lambda \chi \qquad (27)$$

contains the spin 3/2 field coupled to the conserved Noether current of the global supersymmetric theory. Of course, this is to be expected since ψ_λ is the gauge field of supersymmetry transformations. An additional contact term

$$L_C = -\frac{1}{2} V \kappa^2 (\bar{\psi}_\lambda \sigma_{\mu\nu} \gamma^\lambda \chi)(\bar{\psi}_\mu \gamma_\nu \chi) \qquad (28)$$

is necessary for complete invariance. Since contact terms quadratic in the gauge fields are necessary to treat boson fields in Yang-Mills theory, we should not be surprised to see them here.

The field variations under which L is invariant are

$$\delta V_{a\mu} = -i\kappa \bar{\epsilon} \gamma_a \psi_\mu \quad,$$

$$\delta \psi_\rho = \kappa^{-1} D_\rho \epsilon + \frac{i}{4} \kappa (\bar{\chi} \gamma_5 \gamma^\nu \chi) \gamma_\nu \gamma_\rho \gamma_5 \epsilon \quad,$$

$$\delta A_\mu = \frac{i}{\sqrt{2}} \bar{\epsilon} \gamma_\mu \chi \quad,$$

$$\delta \chi = \frac{i}{\sqrt{2}} \bar{F}_{\mu\nu} \sigma^{\mu\nu} \epsilon \quad,$$

$$\bar{F}_{\mu\nu} = F_{\mu\nu} - \frac{i}{\sqrt{2}} \kappa (\bar{\psi}_\mu \gamma_\nu - \bar{\psi}_\nu \gamma_\mu) \chi \quad. \qquad (29)$$

These variations consist of the terms of pure supergravity and the global variations plus additional terms
in which the matter and gauge fields are thoroughly
mixed. We do not give the variation $\delta\omega_{\mu ab}$ which is
very complicated and in the literature. Note that $\bar{F}_{\mu\nu}$
is a supercovariant field strength which is constructed
so that $D_\mu\epsilon$ terms cancel in its variation. This supercovariant derivative is a useful notion in supergravity.

Unfortunately, scalar fields seem to be complicated
in supergravity although they are simple in global
supersymmetry. The local extension of the action for the
$(1/2,0)$ multiplet has been constructed but has qualitatively more complicated features than the case $(1,1/2)$.
The worst of these is that an apparently non-polynomial
Lagrangian results when the mass term of the $(1/2,0)$
multiplet is extended to local invariance.

A very interesting feature appears with the $(3/2,1)$
multiplet. When the global supersymmetric free kinetic
action is extended to local invariance by coupling to
the $(2,3/2)$ gauge multiplet, one finds a global $SO(2)$
symmetry among the two different spin 3/2 fields. The
local supersymmetry of the model can be correspondingly
extended to involve two independent Majorana spinor
parameters $\epsilon^1(x)$ and $\epsilon^2(x)$. One can also formulate this
model in terms of complex Dirac spinors, but the Majorana
basis is most suited for manipulations in supergravity
and for the ensuing development. This model is actually
the simplest example of extended supergravity which we
now discuss from a more general viewpoint.

EXTENDED SUPERGRAVITY

In extended supersymmetry one deals with a system
of N Majorana spinor charges $Q_\alpha{}^i, i = 1, 2, \ldots\ldots N$. The

anti-commutation rule

$$\{Q_\alpha{}^i, \bar{Q}_\beta{}^j\} = \delta^{ij} (\gamma^a)_{\alpha\beta} P_a \qquad (30)$$

replaces (1) while (2) and (3) have an obvious extension. The rule (30) is almost uniquely selected by an axiomatic study of possible supersymmetries of a relativistic S-matrix. The additional allowed possibilities involve central charges which we do not discuss here. Because the basic commutation rules are symmetrical in the indices i which distinguish the spinor charges, we expect a similar symmetry in the corresponding field theories, and this will be discussed later.

We will now obtain the massless irreducible representations of extended supersymmetry by a very simple argument. It is best to introduce the equivalent Weyl spinor description where one has two-component complex spinors $Q_\alpha{}^i$ and their adjoints $Q_\alpha{}^{i\dagger}$ which satisfy

$$\{Q_\alpha{}^i, Q_\beta{}^{j\dagger}\} = \delta^{ij} (\sigma^a)_{\alpha\beta} P_a \quad ,$$

$$\{Q_\alpha{}^i, Q_\beta{}^j\} = 0 \qquad , \qquad (31)$$

$$\{Q_\alpha{}^{i\dagger}, Q_\beta{}^{j\dagger}\} = 0 \qquad ,$$

where the 2x2 Weyl matrices ($\sigma^a = 1, \underline{\sigma}$) appear. We shall also need the commutation relations

$$[J_3, Q_\alpha{}^i] = -\frac{1}{2} (\sigma^3)_{\alpha\beta} Q_\beta{}^i \quad ,$$

$$[J_3, Q_\alpha{}^{i\dagger}] = +\frac{1}{2} (\sigma^3)_{\alpha\beta} Q_\beta{}^{i\dagger} \quad , \qquad (32)$$

with the z-component of angular momentum which follows
from (2).

    We then pick a basis of one-particle helicity states
$|\bar{P}, \lambda>$ with momentum $\bar{P}^a = (\omega,0,0,\omega)$ in the z-direction.
On this basis (31) reduces to

$$\{Q_2^i , Q_2^{j\dagger}\}= 2\omega\delta^{ij} , \qquad (33)$$

with all other anti-commutators vanishing. Further (32)
tells us that Q_2^i raises and $Q_2^{i\dagger}$ lowers helicity by
1/2 unit. Therefore extended supersymmetry reduces to
the anti-commutation rules of N independent fermion
creators and annihilators, and it is duck soup to find
the irreducible representations!

    The maximum helicity state $|\bar{P},\bar{\lambda} >$ of a basis of
the representation space may be chosen to have the
properties

$$Q_2^i |\bar{P}, \bar{\lambda} > = Q_1^i |\bar{P}, \bar{\lambda} > = Q_1^{i\dagger} |\bar{P}, \bar{\lambda} > = 0 \qquad (34)$$

and is a singlet of the internal symmetry. By applying
the creator $Q_2^{i\dagger}$ one finds the sequence of states

$$|\bar{P}, \bar{\lambda}-\tfrac{1}{2},i> = Q_2^{i\dagger} |P^-,\bar{\lambda} > ,$$

$$|\bar{P}, \bar{\lambda}-1,[ij]> = Q_2^{j\dagger} Q_2^{i\dagger}| \bar{P}, \bar{\lambda} >, \text{etc.}$$

$$(35)$$

States of helicity $\bar{\lambda} - \tfrac{1}{2}m$ have multiplicity $\tfrac{1}{m!} N(N-1)...$
$(N-m+1)$ and the sequence stops when the singlet state
of helicity $\bar{\lambda} - 1/2N$ is reached. To get a representation
compatible with local field theory, we must add the CPT
conjugate states of reversed helicity which can also be
obtained by starting from the singlet $|\bar{P}, - \bar{\lambda} >$ and

applying the helicity-raising operator $Q_2{}^i$ which is the CPT conjugate of $Q_2{}^{i\dagger}$. The CPT conjugate states are already present in the initial sequence if $N = 4\bar{\lambda}$. Therefore field theories with $N = 4\bar{\lambda}$ and $N = 4\bar{\lambda} - 1$ have the same number of relatisvistic fields. They are likely to be identical because supersymmetric constructions are highly unique. All irreducible representations of extended supersymmetry can be viewed as reducible superpositions of $(j, j-1/2)$ multiplets of ordinary supersymmetry.

Extended supergravity, which is the gauge theory of an extended supersymmetry, is based on irreducible representations with $\bar{\lambda} = 2$, i.e. representations which contain a singlet graviton, N spin 3/2 gravitinos, $\frac{1}{2} N(N-1)$ vectors, and (for $N \geqslant 3$) spinors, and scalars with appropriate multiplicities. All these particles are unified with the graviton, a unification which is particularly remarkable for spinors and scalars which are not associated with gauge fields.

It is generally thought that field theories corresponding to the extended supergravity representations can be constructed for $N \leqslant 8$. Complete constructions have been given for $N = 2$ and 3 and substantial but partial results are known for $N = 4$. For $N \geqslant 9$, the helicity $\lambda = 2$ states occur with multiplicity greater than one, which does not agree with usual ideas about gravitation. An additional difficulty is that particles of spin 5/2 or greater occur in the representations and it is folklore that their interactions are inconsistent. The folklore may well be correct, but since consistency problems for spin 3/2 are avoided in supergravity, perhaps one should keep an open mind about higher spin fields (especially in particular combinations).

The correlation between spin content and internal symmetry is one of the most intriguing features of extended supergravity. Both $SO(N)$ and $SU(N)$ internal symmetries can be defined in the Hilbert space of the representations. An $SO(N)$ symmetry is natural in the corresponding field theories, because the number of vector fields is $\frac{1}{2} N(N-1)$. In principle the internal symmetry can be gauged, and this has been done for $N = 2$ and 3. A trivial reflection can also be included in these models so that one can speak of an $O(N)$ symmetry. It has also been shown that the global $SO(N)$ symmetry can be extended to an $SU(N)$ symmetry by defining global, chiral and dual transformations of the spinor and vector fields of the $N = 2$ and $N = 3$ models. This extension is apparently incompatible with local internal symmetry, and it remains for future developments to show which concept is more useful.

Extended supergravity theories are conceptually elegant and so far have favorable renormalizability properties. Can they be realistic? The case $N = 8$ is particularly attractive because it is CPT self-conjugate and contains enough spin 1/2 states, namely 56 to accomodate the ever growing number of quarks and leptons. Although this model may have a global $SU(8)$ internal symmetry group, the $SO(8)$ invariance appears to be the largest symmetry which can be gauged, and $SO(8)$ does not contain the subgroup $SU(3) \times SU(2) \times U(1)$ which is the minimal symmetry in the conventional pictures of strong, electromagnetic and weak interactions. There is more freedom in extended supergravity coupled to matter multiplets, although unfavorable renormalizability properties are anticipated. Futher, the restriction $N \leqslant 4$ is probably necessary because matter multiplets

of maximum spin one or less exist only up to this value.
Inconsistencies are expected if spin 3/2 fields are
present in the matter multiplet, rather than the
gravitational multiplet.

It is time to examine specific extended supergravity theories. We present the SO(3) model
Lagrangian, in which the SO(2) model appears as a
special case. Historically SO(2) came first. The
SO(3) theory has the spin content (2, 3/2, 3/2, 3/2, 1
1, 1, 1/2), and the fields $V_{a\mu}$, $\psi_\rho{}^i$, $A_\mu{}^i$, χ where i is
the SO(3) group index. In first order formalism the
Lagrangian is

$$L = L_K + L_N + L_C \quad , \tag{36}$$

with general covariant kinetic term

$$L_K = -\frac{1}{4\kappa^2} VR - \frac{1}{2} \epsilon^{\lambda\rho\mu\nu} \bar\psi_\lambda{}^i \gamma_5 \gamma_\mu D_\nu \psi_\rho{}^i$$

$$- \frac{1}{4} Vg^{\mu\nu}g^{\rho\sigma} F_{\mu\nu}{}^i F_{\rho\sigma}{}^i + \frac{i}{2} V\bar\chi \not{D}\chi \qquad . \tag{37}$$

The second term contains SO(3) invariant structures
suggested by the Noether currents of the component
multiplets (ϵ^{ijk} is the SO(3) structure constant)

$$L_N = -\frac{1}{2} \kappa\epsilon^{ijk}\bar\psi_\mu{}^i (VF^{\mu\nu k} - \frac{1}{2} i \gamma_5 \tilde{F}^{\mu\nu k})\psi_\nu{}^j$$

$$- \frac{i}{\sqrt{2}} \kappa V \bar\psi_\lambda{}^i \sigma^{\mu\nu}F_{\mu\nu}{}^i \gamma^\lambda\chi \tag{38}$$

and contact terms

$$L_C = -\frac{1}{4}\kappa^2\varepsilon^{ijk}\varepsilon^{k\ell m} V (\bar{\Psi}_\mu^{\ i}\ \Psi_\nu^{\ j}) \times$$

$$\times [(\bar{\Psi}^{\mu\ell}\Psi^{\nu m}) - \frac{i}{2} V^{-1}\varepsilon^{\mu\nu\rho\sigma} (\bar{\Psi}_\rho^{\ \ell}\ \gamma_5\Psi_\sigma^{\ m})]$$

$$-\frac{i}{\sqrt{2}}\kappa^2\ \varepsilon^{ijk} V(\bar{\Psi}_\mu^{\ i}\ \Psi_\nu^{\ j})\ (\bar{\Psi}_\lambda^{\ k}\ \sigma^{\mu\nu}\gamma^\lambda\chi)$$

$$-\frac{1}{2}\kappa^2 V(\bar{\Psi}_\lambda^{\ i}\ \sigma^{\mu\nu}\gamma^\lambda\chi)\ (\bar{\Psi}_\mu^{\ i}\ \gamma_\nu\chi)\ . \tag{39}$$

We refer readers to the literature for the supersymmetry variations which involve a triplet of Majorana spinors $\varepsilon^i(x)$. The SO(2) model Lagrangian is obtained by setting $\Psi_\rho^3 = A_\mu^1 = A_\mu^2 = \chi = 0$, a truncation which is compatible with the field variations, so that local supersymmetry is maintained (for $\varepsilon^3(x) = 0$).

The only coupling constant in this Lagrangian is the gravitational constant κ. There is not as yet any minimal gauge interaction of the vector fields, and the SO(3) internal symmetry is global only. A global SU(3) symmetry is also present which consists of the SO(3) transformations plus added transformations which involve γ_5 rotations of χ and Ψ_ρ^i and dual transformations of the field strengths $F_{\mu\nu}^i$.

In the SO(4) extended supergravity model there are the singlet graviton, quartets of spin 3/2 and spin 1/2 fields, a sextet of vectors, and singlet scalar and pseudoscalar fields A and B. A partial derivation of this theory through terms of order κ^2 in the variation δL has been very recently obtained which includes some terms of order κ^3 in L. A great deal of hard work was involved in finding even this partial solution, and

unfortunately there are complicated features compared
with the previous case, namely, a strongly indicated
non-polynomial structure in the scalar fields A and B
and terms such as $AF_{\mu\nu}^{2}$ and $BF_{\mu\nu}F^{\mu\nu}$. These complications
are also likely to be present in extended supergravity
theories for $N > 4$.

MINIMAL GAUGE COUPLING FOR SPIN 3/2 FIELDS

One of the motivations for the construction of
supergravity theories was the prejudice that local
invariance is more elegant and powerful than global
invariance. For the same reason, it is desirable to
gauge the internal symmetry in extended supergravity,
and this has been done for the $SO(2)$ and $SO(3)$ models.
The resulting theories are quite interesting but have
problems of interpretation as we shall discuss.

The additional terms necessary to extend the
Lagrangian of (36-39) so that local $SO(3)$ invariance
is achieved are quite simple. One first replaces the
derivative $D_{\nu}\psi_{\rho}^{i}$ in (37) by the combined gravitational
and Yang-Mills derivative

$$(\hat{D}_{\nu}\psi_{\rho})^{i} = (\partial_{\nu} + \frac{1}{2}\omega_{\nu ab}\sigma^{ab})\psi_{\rho}^{i} + e\,\varepsilon^{ijk}A_{\nu}^{j}\psi_{\rho}^{k} \quad , \tag{40}$$

where e is the $SO(3)$ gauge charge. To maintain local
supersymmetry it is then necessary to add

$$L' = -e\kappa^{-1}\,V\bar{\psi}_{\lambda}^{i}\,\sigma^{\lambda\rho}\,\psi_{\rho}^{i} + \frac{3}{2}e^{2}\kappa^{-4}V \quad , \tag{41}$$

which are a spin 3/2 mass (with $m = e\,\kappa^{-1}$) and
cosmological term. The $SO(2)$ result is entirely analogous.
The uniqueness of these extensions has not been shown

but I suspect that they are unique because local supersymmetry is a very restrictive requirement.

These theories exhibit an unusual correlation
among several concepts:

1) minimal gauge coupling of a spin 3/2 field,

2) spin 3/2 mass (of combined electromagnetic
   and gravitational origin),

3) spontaneous breakdown of global supersymmetry
   (because $m_\psi \neq 0$ but $m_A = m_\chi = 0$),

4) cosmological term.

It is the cosmological term which causes
difficulties. Direct comparison with the experimental
limit from astronomical measurements gives the terribly
stringent upper limit $e^2 < 10^{-120}$. This naive comparison
with experiment may not be correct (because of
renormalization effects, for example) but the result
is still discouraging. Further difficulty arises
because the appropriate background geometry for
quantization of field theories with cosmological term
is a de Sitter space. (Minkowski space is not a solution
of the background field equations). Although some work
has been done in this area, the physical interpretation
of the resulting quantum field theory is unconventional
and not, I believe, fully understood.

The cosmological term also occurred in a previous
supergravity model where the spin 3/2 field has a minimal
axial vector coupling. It has been suggested that this
term always appears when supersymmetry is spontaneously
broken, and some form of breakdown is needed in any
realistic application of the theories. We therefore
must learn to deal with the cosmological term. It is
not clear how and I note here only that, because this
term appears with both positive and negative signs in

different models, it may be possible to achieve
cancellation by combining the mechanisms.

There are interesting special cases of the model
discussed in this section. They involve the restrictions

i) $\psi_\rho^2 = \psi_\rho^3 = A_\mu^2 = A_\mu^3 = 0$, and

ii) $A_\mu^1 = \chi = 0$.

If both i) and ii) are imposed one finds the pure
supergravity (2,3/2) Lagrangian plus fermion mass and
cosmological term. If one removes the restriction ii)
the vector multiplet is added. These restrictions are
compatible with the field variations (for $\varepsilon^1 = \varepsilon^2 = 0$)
so that local supersymmetry is maintained. The minimal
vector coupling is eliminated in these special cases but
they may be useful to explore questions associated with
the mass and cosmological terms.

GENERAL STRATEGY OF SUPERGRAVITY CONSTRUCTIONS

We now list the essential elements common to the
construction of all the supergravity theories we have
discussed:

1) The irreducible representation theorems give
information as to what fields can be coupled. This
is always a superposition of (j,j-1/2) multiplets of
ordinary supergravity.

2) One forms a starting Ansatz for the Lagrangian
consisting of covariant kinetic terms for the individual
fields plus Noether current terms for the component
(j, j-1/2) multiplets. In extended supergravity these
terms must be SO(N) invariant.

3) A similar starting Ansatz for the transformation
rules is based on the known pure supergravity variations

and global supersymmetry variations.

4)   Independent terms in the variation δL of
increasing order in κ are then carefully studied.   The
requirement that they vanish determines the parameters
of the starting Ansatz and correction terms to both the
starting Lagrangian and transformation rules.   Auxiliary
techniques based on the consistency condition (24) can
help to manage the algebraically complex situation which
occurs at some level of the calculation.   An accurate
Fierz matrix is an absolute requirement as is a steady
hand at Dirac algebra.   There is also room for inspired
algebraic intuition which makes some of the hard work
worthwhile.

5)   There is an important "building up principle"
in these constructions.   Namely, identities used and the
specific form of terms found in one model can often be
used in the construction of more complicated models.
One example of this is visible in Eqs. (28) and (39)
which show that the same $\psi^2\chi^2$ contact term occurs in the
(1,1/2) matter coupling problem and in SO(3) extended
supergravity.

As a result of remarkable progress during the last
nine months, the basis features of supergravity field
theories have been delineated.   There are still unsolved
problems in several areas, including field theories not
yet constructed, a geometrical interpretation which may
give additional insight, and further study on the
important question of regularization and renormalizability.
Finally, there is the question of where the real physics
is.   Supergravity practitioners hope that it lies somewhere among the ideas discussed.   Only future work will
tell where!

Notation

We use latin indices a,b,c etc. and the metric
tensor $\eta_{ab}=(+,-,-,-)$ for flat Minkowski space and local
Lorentz coordinate frames in curved space. Dirac
matrices satisfy $\{\gamma^a,\gamma^b\} = 2\eta^{ab}$, $\gamma_5 = i\gamma^0\gamma^1\gamma^2\gamma^3$ while
$\varepsilon^{0\,1\,2\,3} = 1$. Greek indices μ,ν,ρ etc. are used for
Riemannian coordinates. Note that $\varepsilon^{\mu\nu\rho\sigma}$ transforms
as a tensor density but we lower indices as in the
example $\varepsilon_{ab}{}^{\rho\sigma} = V_{a\mu}V_{b\nu}\varepsilon^{\mu\nu\rho\sigma}$. The notation V denotes
$\det(V_{a\mu})$.

BIBLIOGRAPHY

Global Supersymmetry:

Early Papers:

Y. A. Golfand and E. P. Liktman, JETP Lett. 13, 452 (1971)

D. V. Volkov and V. P. Akulov, JETP Lett. 16, 438 (1972) and Phys. Lett. 16B, 109 (1973)

J. Wess and B. Zumino, Nucl. Phys. B70, 39 (1974)

Global (1,1/2) Multiplet:

J. Wess and B. Zumino, Nucl. Phys. B78, 1 (1974)

A. Salam and J. Strathdee, Phys. Lett. 51B, 353 (1974)

S. Ferrara and B. Zumino, Nucl. Phys. B79, 413 (1974)

B. deWitt and D. Z. Freedman, Phys. Rev. D12, 2286 (1975)

Reviews:

B. Zumino, in Proceedings of the XVII International Conference on High Energy Physics, Landen 1974, edited by J. R. Smith (Rutherford Laboratory, Chilton Didcot, Berkshire, England 1974)

A. Salam and J. Strathdee, Phys. Rev. D11, 1521 (1975)

S. Ferrara, Rivista del Nuovo Cimento 6, 105 (1976)

Spin (2,3/2) Gravitational or Gauge Multiplet:

D. Z. Freedman, P. van Nieuwenhuizen and S. Ferrara, Phys. Rev. D13, 3214 (1976). (Construction of the Gauge Action in Second Order Formalism)

S. Deser and B. Zumino, Phys. Lett. <u>B62</u>, 335
(1976). (First Order Treatment of the Gauge
Action and Consistency Viewpoint)

D. Z. Freedman and P. van Nieuwenhuizen, Phys.
Rev. <u>D14</u>, 912 (1976). (Commutator Algebra;
$[\delta_1\delta_2]\omega_{\mu ab}$ has recently been calculated by
van Nieuwenhuizen, unpublished)

A. Das and D. Z. Freedman, Nucl. Phys. <u>B114</u>,
271 (1976). (Path Integral Quantization and
Causality and Ghost-free Properties)

M. T. Grisaru, H. Pendleton and P. van
Nieuwenhuizen, Phys. Rev. D. to be published,
(S-matrix approach to supergravity)

D. Z. Freedman, Stony Brook preprint ITP-SB-76-39
(Conformal invariant extension of the gauge
action)

P. K. Townsend, Stony Brook preprint ITP-SB-77-2
(Supergravity with spin 3/2 mass and cosmological
term)

Matter Multiplet Couplings:

S. Ferrara, J. Scherk and P. van Nieuwenhuizen,
Phys. Rev. Lett. <u>37</u>,1035 (1976) (Abelian
(1,1/2) Multiplet)

S. Ferrara, F. Gliozzi, J. Scherk, and
P. van Nieuwenhuizen, Nucl. Phys., to be
published (non-Abelian (1,1/2) multiplet
and (1/2,0) multiplet)

D. Z. Freedman and J. Schwarz, Phys. Rev. <u>D15</u>,
to be published (non-Abelian (1,1/2) multiplet
and its commutator algebra)

S. Ferrara, D. Z. Freedman, P. van Nieuwenhuizen,
P. Breitenlohner, F. Gliozzi and J. Scherk,
Phys. Rev. <u>D15</u>, to be published ((1/2,0) multiplet

including its mass term and commutator
algebra)

B. deWitt, Phys. Lett., to be published (Local
extension of super quantum electrodynamics)

D. Z. Freedman, Phys. Rev. D15, to be published
(Axial gauge coupling of the spin 3/2 field)

Extended Supergravity:

R. Haag, J. T. Lopusanski and M. Sohnius, Nucl.
Phys. B88, 257 (1975) (Axiomatic study of
supersymmetry algebras)

A. Salam and J. Strathdee, Nucl. Phys. B76, 477
(1974) (Techniques for finding irreducible
representations).

M. Gell-Mann and Y. Ne'eman, unpublished
(Extended supergravity representations)

S. Ferrara and P. van Nieuwenhuizen, Phys. Rev.
Lett. 37,1662 (1976) (SO(2) field theory)

D. Z. Freedman, Phys. Rev. Lett. 38, 105 (1977)
(SO(3) field theory)

S. Ferrara, J. Scherk and B. Zumino, Phys. Lett.,
to be published (SO(3) field theory and its
commutator algebra)

D. Z. Freedman and A. Das, Nucl. Phys., to be
published (Gauge internal symmetry in SO(2)
and SO(3) field theories)

S. Ferrara, J. Scherk and B. Zumino, Nucl. Phys.,
to be published (Global SU(2) × U(1) and SU(3)
internal symmetry)

A. Das, Stony Brook preprint, ITP-SB-77-4
(SO(4) field theory)

Renormalization Properties of Supergravity:

For references see the presentation of
P. van Nieuwenhuizen, this proceedings.

Geometrical Approaches to Supergravity:

        D. V. Volkov and V. P. Akulov, JETP Lett. 18, 312 (1973) (An early superspace approach)

        R. Arnowitt and P. Nath. Riemmannian superspace approach--For references see the presentation of R. Arnowitt, this proceedings.

        J. Wess and B. Zumino, Affine Superspace Approach-- For references see the presentation of B. Zumino, this proceedings.

        F. Mansouri, Yale University preprint, 1976 (Fiber bundle approach)

        N. Chamseddine and P. C. West, Imperial College preprint, 1976 (Fiber bundle approach).

        P. Breitenlohner, Phys. Lett., to be published (Auxiliary fields which "close" the commutator algebra).

RENORMALIZABILITY OF SUPERGRAVITY

P. van Nieuwenhuizen

Institute for Theoretical Physics

State University of New York at Stony Brook

Stony Brook, L.I., New York 11794
(presented by P. van Nieuwenhuizen) and

M.T. Grisaru
Brandeis University, Waltham, Mass. 02154

ABSTRACT

Supergravity[1,2,3] has better renormalizability properties than Einstein gravitation. Explicit calculations and theoretical proofs indicate that the so-called SO(n) models are one- and two-loop finite. Nothing is yet known about three-loop properties.

I.   INTRODUCTION

The unification of quantum theory and relativity is one of the fundamental problems of theoretical physics. It used to be called quantization of gravitation, but since the theory of quantization of nonabelian gauge theories (and their application to gravitation[4]) we know now how to quantize gravitation: as the nonabelian gauge theory of massless spin 2 bosons. The question then is

whether this quantized theory is renormalizable, and
here an important difference with other gauge theories
arises.   Because the gravitational coupling constant
has the dimension of a mass, counter terms have a
differential functional form from the original action.
For example, to the Einstein action $\kappa^{-2}R$ belong one-
loop counter terms proportional to $R_{\mu\nu}^2$ and R^2.   It
follows that one cannot use the usual renormalization
program to rescale the physical parameters in the theory
in order to absorb ΔL back into L.   Therefore gravitational theories are either finite, or nonrenormalizable,
but the case in between (true renormalizability) can
never occur.   One must thus hope for finiteness, and as
it turns out, Green's functions are never finite, but
S-matrices are sometimes finite.

Einstein gravitation has been investigated concerning its finiteness properties and three sets of
results have been obtained:

(i) pure Einstein gravitation (without matter constituents) has a one-loop finite S-matrix.   Nothing
is (yet) known about its two-loop finiteness
(paradoxically, in supergravity this question is already solved);

(ii) coupling to individual matter fields (scalars[4],
fermions[5], photons[6] or Yang-Mills bosons[7]) always
leads to a nonrenormalizable theory;

(iii) coupling the two most elegant and best tested
field theories, quantum electrodynamics and Einstein
gravitation, leads again to a nonrenormalizable
theory.[8]

Hence, coupling three things together (photons, electrons
and gravitons) does not improve matters and one might
begin to despair of the possibility that some magical

combination of matter fields would lead to a finite
quantum theory. As we will see, supergravity provides
such a magical combination, and it does lead to finite
S-matrix elements.

One-and-a-half years ago, faced with the series of
negative results in Einstein quantum gravity, it seemed
that at least three alternative attacks might be tried:[9]

(i) Add extra symmetries to a theory of gravitation
plus matter in addition to the gravitational gauge
invariance. The two prime candidates mentioned at
the time were conformal symmetry and supersymmetry;

(ii) nonperturbative methods (always easier to suggest
than to carry out) seem, if anywhere applicable, most
appropriate for gravitation. According to the van
Dam-Veltman[10] theorem, there does not exist a nearby
theory of massive gravitational fields which reduces
in the $m \rightarrow 0$ limit to Einstein gravitation (unlike,
of course, the case of electromagnetism.) If it is
then really so essential that gravitons be massless,
one should consider around any massive particle a
cloud of soft gravitons, hence nonperturbative methods
might be called for.[11]

(iii) Give up the criteria of renormalizability proper,
but substitute other criteria for obtaining finite
answers from an essentially nonrenormalizable theory;
or perhaps change the quantization rules once more
(as they were changed in the transition from abelian
to nonabelian quantization techniques.)

In these proceedings we will consider what happens
if one adds supersymmetry to Einstein gravitation. Why
should supergravity, as this marriage is called, hold
out any hopes for the renormalization of gravitation?
First of all, it is known that certain globally

supersymmetric field theories are in fact much more
convergent in their quantum corrections than one would
expect from naive power counting alone.[12] One might
hope that similar cancellations occur also in supergravity and are in fact so strong as to leave no divergences at all. Also, it is known that the zero
point infinities of bosons and fermions precisely cancel
each other in certain globally supersymmetric models.[13]
On the other hand, there exists a model of global supersymmetry with a dimensional coupling constant (as in
gravitation) which, though somewhat less divergent than
could have been the case, is nevertheless nonrenormalizable.[14] Therefore, let us investigate supergravity to
see to which side it falls.

II.   WHAT IS SUPERGRAVITY?

Supergravity[1,2,3] is the combination of <u>local</u>
supersymmetry and gravitation. It is impossible to
combine global supersymmetry with gravitation (or local
supersymmetry with flat spacetime) as one might expect
from the commutator of two global supersymmetry transformations with parameters ϵ_1 and ϵ_2.

$$[\epsilon_1 Q, \; \bar{\epsilon}_2 Q] = 2i(\bar{\epsilon}_2 \gamma^\alpha \epsilon_1) \; P_\alpha \quad . \tag{1}$$

For local supersymmetry, $\epsilon = \epsilon(x)$ and the right hand
side describes a spacetime dependent translation, i.e.
a general coordinate transformation which requires curved
spacetime. And conversely, if spacetime is curved (when
there is gravitation) then an initially constant ϵ picks
up a spacetime dependence after a gravitational transformation; hence gravitation requires local supersymmetry.
Therefore, in quantizing supergravity, one has to apply

the covariant quantization rules not only to the gravitational gauge invariance, but also to the local supersymmetry gauge invariance.

The interesting property of supergravity is that
it combines three fundamental concepts of theoretical
physics: spin-statistics (because it describes a symmetry between bosons and fermions) and the structure of
spacetime (as the equivalence principle tells us any
gravitational theory must do) in the context of a very
elegant gauge theory. Supergravity is an irreducible
and nontrivial extension of Einstein theory [eq. (1)
promises that supergravity is the square root of general
relativity] and the first example of a consistent field
theory for spin 3/2 fields. It is not different from
Einstein theory in its classical predictions because the
exchange of fermions, spin 3/2 here, is always short
range. But it does differ at the quantum level as we
now discuss.

III. RENORMALIZATION OF SO_2 SUPERGRAVITY

Any nonabelian gauge theory is quantized as follows:

(i)   Choose a gauge invariant action with gauge
      parameters $\eta_1, \ldots \eta_N$.

(ii) Add a gauge fixing term $L_B = -\frac{1}{2} \sum_{i=1}^{N} (A_i)^2$ to
      obtain propagators.

(iii) Add a ghost term $L_G = \bar{C}^i (\partial A_i / \partial \eta_j) \, C^j$ to restore
      unitarity.

(iv) Since propagator and vertices are now known, con-
      struct Feynman diagrams.

(v)   Regulate the divergent amplitudes associated with
      these diagrams by means of a regularization
      scheme that preferably maintains the symmetry
      of the theory (though this is not necessary.)

(vi) The question then is whether the theory is renormalizable (which is understood, in the case
of gravitation, to mean: do the infinities in the
S-matrix cancel?)

Let us illustrate in an example how one performs
these steps in supergravity. The model we consider
(SO_2 theory) is the first extended supergravity model;
it was constructed by Ferrara and one of us[15], and its
quantum corrections (plus a theoretical proof of their
finiteness) were explicitly evaluated by Vermaseren and
us[16]. It contains one graviton, one spin 1 particle and
two spin 3/2 particles (which occur symmetrically and
hence lead to two local supersymmetries as we will see).
The action is given by

$$L = L(\text{kinetic}) + L(\text{Noether}) + L(\text{torsion}) + L(\text{supercovariant}),$$

$$L(\text{kinetic}) = -(2\kappa^2)^{-1}\, eR - \sum_{i=1}^{2}\, \tfrac{1}{2}\, \bar{\psi}_\mu^i\, \varepsilon^{\mu\nu\rho\sigma}\, \gamma_5\gamma_\nu D_\rho \psi_\sigma$$

$$\tfrac{1}{4}\, eF_{\mu\nu}F^{\mu\nu}\ ,$$

$$L(\text{Noether}) = \frac{\kappa}{2\sqrt{2}}\, \varepsilon^{ij}\, \bar{\psi}_\mu^i(eF^{\mu\nu} + \tfrac{1}{2}\, \gamma_5\, \tilde{F}^{\mu\nu})\psi_\nu^j\ ,$$

$$\tilde{F}^{\mu\nu} = \varepsilon^{\mu\nu\rho\sigma}F_{\rho\sigma}\ ,$$

$$L(\text{torsion}) = -\frac{e\kappa^2}{32}\, [\bar{\psi}^{\mu,i}\gamma^\nu\psi^{\rho,i}][\bar{\psi}_\mu^j\gamma_\nu\psi_\rho^j + 2\bar{\psi}_\nu^j\gamma_\mu\psi_\rho^j$$

$$-\, 4g_{\mu\rho}\bar{\psi}_\nu^j\gamma^\lambda\psi_\lambda^j]\ ,$$

$$L(\text{supercov.}) = (-\kappa^2/8)(\bar{\psi}_\mu^i\varepsilon^{ij}\psi_\nu^j)(e\bar{\psi}_\mu^k\psi_\nu^\ell + \tfrac{1}{2}\, \varepsilon^{\mu\nu\rho\sigma}\bar{\psi}_\rho^k\, \gamma_5\psi_\sigma^\ell)\varepsilon^{k\ell}.$$

$$(2)$$

In fact, one can omit the $L(\text{torsion})$ and $L(\text{supercovar.})$ if one uses first order (torsion Palatini) formalism and introduces supercovariant derivatives in the $L(\text{Noether})$, but for our purposes an explicit representation as given above is more suitable.

The action has an impressive number of local and global symmetries. The local symmetries with their parameters (and the number of them) and associated gauge fixing terms are[16,18]

gen. coord. transf.; ξ^α (4); $-\frac{1}{2}[\partial_\mu(e_{\mu\nu}+e_{\nu\mu}-\delta_{\mu\nu}\,e_{\alpha\beta}\,\delta^{\alpha\beta})]^2$,

local Lorentz transf.; ω_{ab}(6); $-\frac{\alpha}{2}(e_{\mu\nu}-e_{\nu\mu})^2$ with $\alpha \to \infty$,

two supersymm. transf.; ε^1, ε^2 (8); $-\frac{1}{2}\bar{\psi}^i_\mu\,\delta^\mu_a\gamma^a\,\not{\partial}\gamma^b\,\delta^\nu_b\,\psi_\nu$,

electromag. gauge inv.; Λ(1); $-\frac{1}{2}(\partial_\mu A_\mu)^2$.   (3)

In addition, the theory has a global U_2 symmetry group,[17] consisting of SU_2 chiral rotations

$$\delta\begin{pmatrix}\psi^1_\mu\,,L\\\psi^2_\mu\,,L\end{pmatrix}=e^{i\vec{\omega}\cdot\vec{\sigma}}\begin{pmatrix}\psi^1_\mu\,,L\\\psi^2_\mu\,,L\end{pmatrix},\delta\begin{pmatrix}\psi^1_\mu\,,R\\\psi^2_\mu\,,R\end{pmatrix}=\left(e^{i\vec{\omega}\cdot\vec{\sigma}}\right)^*\begin{pmatrix}\psi^1_\mu\,,R\\\psi^2_\mu\,,R\end{pmatrix},$$

(4)

where $\psi_\mu,L = \frac{1}{2}(1+\gamma_5)\psi_\mu$, and a U_1 global combined duality-chirality invariance

$$\delta\hat{F}_{\mu\nu} = e\,\varepsilon_{\mu\nu\rho\sigma}\,\hat{F}^{\rho\sigma}\,,\quad \delta\psi^i_\mu = -\gamma_5\psi^i_\mu\,.\qquad (5)$$

A partial invariance of this kind (eq. (5) without hats) was first found in ref. (18) and was sufficient for the proof of one-loop finiteness of SO_2 theory, but the extension to the full group U_2 and to exact invariance was performed in ref. (17) and might be useful for higher-loop finiteness proofs.

The ghost terms consist of three types. The boson-boson sector is the ghost matrix of Maxwell-Einstein theory.[5] The fermion-fermion sector consists of two commuting spin 1/2 ghosts which are decoupled thanks to our choice of supersymmetry gauge fixing term. Finally, there are cross terms between boson and fermion ghosts, but their vertices contain also spin 3/2 fields and will not contribute to the processes we consider below.

The propagators one obtains are the familiar ones:

$$\text{photon} = -ik^{-2}, \quad \text{graviton} = -2i\,(\eta_{\mu\rho}\eta_{\nu\sigma}+\eta_{\mu\sigma}\eta_{\nu\rho}-\eta_{\mu\nu}\eta_{\rho\sigma})k^{-2},$$

$$\text{spin } 1/2 = -\,\slashed{k}\,k^{-2}, \quad \text{spin } 3/2 = \tfrac{1}{2}\,(\gamma_\nu\,\slashed{k}\,\gamma_\mu)\,k^{-2}\,, \tag{6}$$

except for the simple form of the spin 3/2 propagator. This simple form[18] is again due to our choice of gauge fixing term; for example, for a massive spin 3/2 field the propagator is much more complicated.

Let us now choose a physical process with as few and as simple one-loop diagrams as possible. The choice we came up with is photon-photon elastic scattering because here neither L(torsion) nor L(supercov.) contribute at the one-loop level (they would at the two-loop level). The set of diagrams is depicted in figure one. The first diagram with ME in it stands for the one-loop divergences of photon-photon scattering in ordinary Maxwell-Einstein theory. Only the sum of all

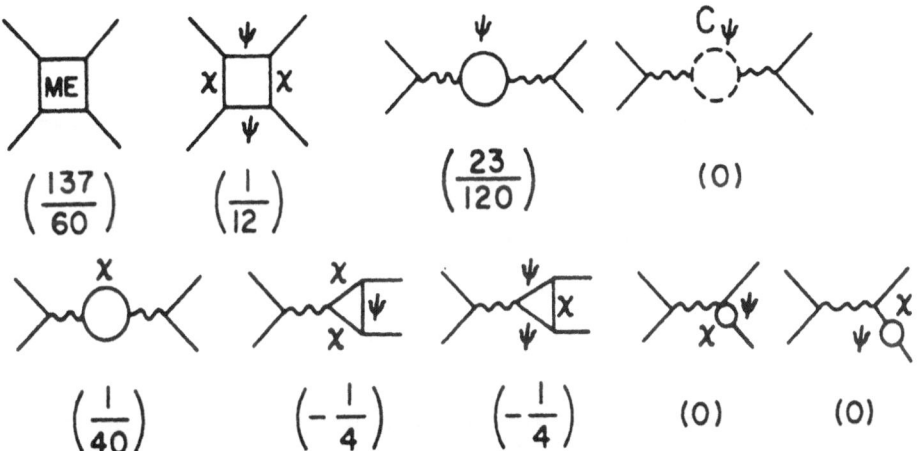

Fig. 1   The divergences of photon-photon scattering in the Maxwell-Einstein theory with spin content $(2,3/2,1, 1/2)$. The sum of all divergences equals $(25/12)$ $\kappa^4 (n-4)^{-1} [T_{\mu\nu}(\gamma)]^2$.

divergences of this system had previously been evaluated in the background field formalism,[3] but since the S-matrices (but not the Green's functions) of the background field method and normal field theory are equal and since clear recipes exist in the literature on how to extract from a background field calculation the divergences of a particular S-matrix,[19] the contribution of these diagrams was easily obtained,

$$S(ME, \text{ div}) = \frac{\kappa^4}{n-4} \frac{137}{60} [T_{\mu\nu}(\text{photon})]^2 . \qquad (7)$$

(The number 137 in the ME system caused, at the time, some hilarity). The other diagrams are what supersymmetry adds to this process. The last two diagrams vanish, as they should, because of gauge invariance, and the moment of truth came when all numbers in figure one were added. Amazingly, their sum was zero. In fact, shortly afterwards a general proof of the finiteness of all 4-particle processes in this model was given, which we will discuss later.[16]

IV.   OTHER EXPLICIT CALCULATIONS IN SUPERGRAVITY

Several other models of supergravity have been considered and for several processes the one-loop divergences of their S-matrices have been determined. The first supergravity calculation was actually not the one in the last section, but photon-photon scattering in the supersymmetric Maxwell-Einstein system[18] obtained by coupling the spin (2,3/2) gauge multiplet to the spin (1,1/2) matter multiplet. The action is given by[20]

$$L = L(\text{gauge}) + L(\text{kinetic matter}) + L(\text{Noether})$$
$$+ L(\text{four-fermion}),$$
$$L(\text{Noether}) = \frac{1}{4} e \kappa (\bar{\psi}_\mu \gamma^\alpha \gamma^\beta \gamma^\mu \lambda) F_{\alpha\beta} , \tag{8}$$

and again the four-fermion seagulls do not contribute.
The set of diagrams is depicted in figure 2. In addition to the Maxwell-Einstein diagrams, we now have box
and triangle graphs, each with both spin 3/2 and spin
1/2 fields and constructed from the Noether vertex, as
well as various self-energy graphs and the same two
vanishing graphs we encountered before. All divergences
are again proportional to the photon energy tensor
squared and the sum of all coefficients does not vanish.
(Since all these results are needed in the next calculation where everything cancels, we know that these numbers
are correct). The spin $(2,3/2,1,1/2)$ supergravity system
is therefore nonrenormalizable.

One difference between the SO_2 system with $(2,3/2,$
$3/2,1)$ content and this $(2,3/2,1,1/2)$ system is that
only in the former case all particles are connected by
(local or global) symmetries, as indicated below

$$2 = 3/2 \qquad\qquad 2 = 3/2 \qquad\qquad \varepsilon = \text{supersymmetry}$$

$$3/2 = 1 \qquad\qquad 1 = 1/2 \quad \xi = SO_2 \text{ invariance}$$
$$\tag{9}$$

(The SO_2 symmetry $\delta\psi_\mu^1 = \psi_\mu^2$ and $\delta\psi_\mu^2 = -\psi_\mu^1$
follows from the above-mentioned U_2 global symmetry by
taking $\omega_x \neq 0$, $\omega_y = \omega_z = 0$). It therefore seems interesting to investigate whether the $(2,3/2,1,1/2)$ system
becomes renormalizable by adding enough matter multiplets
to fill the gap between 3/2 and 1. This is possible and

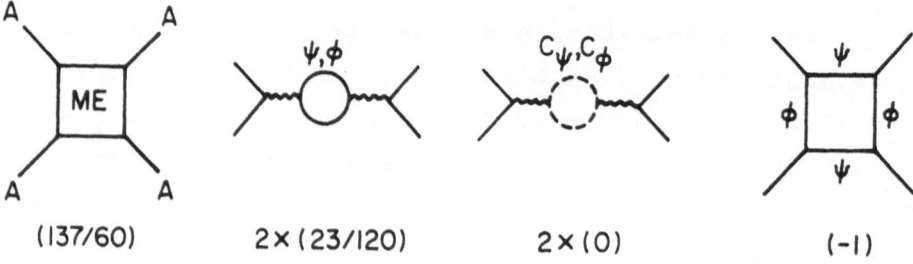

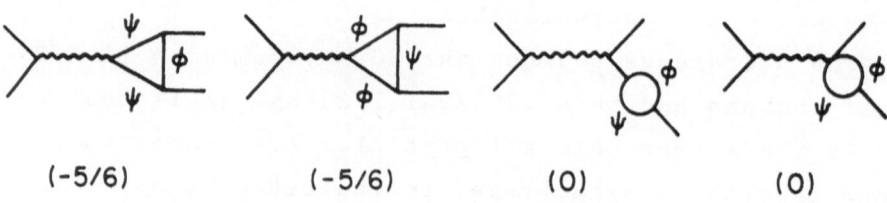

Fig. 2   The divergences of photon-photon scattering in the SO_2 supergravity theory. The numbers below the graphs denote the coefficients of $(n-4)^{-1} \kappa^4 [T_{\mu\nu}(\gamma)]^2$ where $T_{\mu\nu}(\gamma)$ is the photon energy tensor, while ME stands for all graphs present in ordinary Maxwell-Einstein theory.

leads to SO_3 theory. The multiplets involved and the corresponding action are[22]

$$2 = 3/2 \qquad\qquad L = L(\text{gauge}) + L(\text{kinetic})$$

$$3/2 = 1 \qquad\qquad L(\text{Noether}) + L(\text{four-fermion})$$

$$3/2 = 1$$

$$1 = 1/2$$

$$L(\text{Noether}) = \frac{\kappa}{2\sqrt{2}} \, \varepsilon^{ijk} \overline{\psi}_\mu^{\; i} (eF^{\mu\nu,j} + \frac{1}{2} \gamma_5 \, \widetilde{F}^{\mu\nu,j}) \psi_\nu^{\; k}$$

$$+ \frac{1}{4} \, e \, \kappa (\overline{\psi}_\mu^{\; i} \, \gamma^\alpha \gamma^\beta \gamma^\mu \lambda^\varepsilon) \, F_{\alpha\beta}^{\; i} \quad . \tag{10}$$

Again $L(\text{four-fermion})$ does not contribute to photon-photon scattering, and all diagrams, presented in figure 3, are equal to those of figure 1 and 2 times integers. For example, we need twice the photon self-energy because it is already counted once in the ME diagrams. Adding all coefficients of the divergences, one finds that this model is one-loop renormalizable as far as this process is concerned. Again, a general proof is now available[21]; it uses the fact that this model has an SO_3 global invariance which rotates the three ψ fields and the three photons simultaneously into each other. (According to general theories, all these models with n ψ fields must have a U_n global invariance except SO_4 which has a SU_4 invariance. However, only subgroups are known for some of the Lagrangian models already constructed.) Finally, let us discuss SO_4 supergravity.[23] The multiplet structure is

$$(2,3/2) + \text{three } (3/2,1) + \text{three } (1,1/2) + (\tfrac{1}{2},0,0) \tag{11}$$

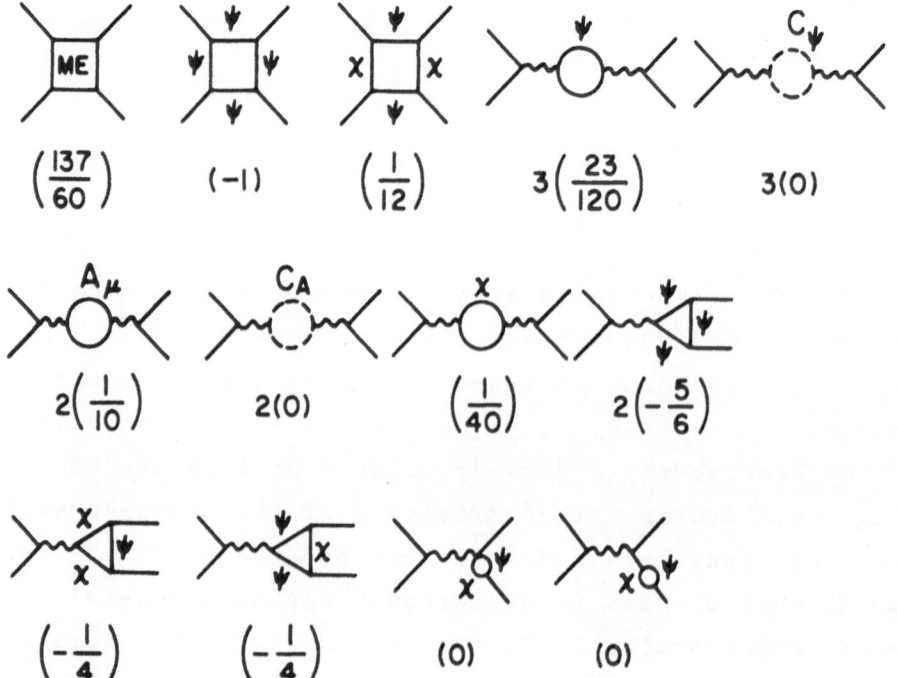

Fig. 3 The divergences of photon-photon scattering
with identical colors in the SO_3 supergravity theory
with spin content $(2,(3/2)^3, (1)^3,1/2)$. The sum of all
divergences vanishes. The same holds for unequal colors.

and the action is now more complicated

$$L = L(gauge) + L(kinetic) + L(Noether) + L(scalar-photon)$$
$$+ L(seagulls\ with\ fermions),$$

$$L(Noether) = \frac{\kappa}{2\sqrt{2}}\ \bar{\psi}^i_\mu\ (eF^{ij}_{\mu\nu} + \frac{1}{2}\ \gamma_5\tilde{F}^{ij}_{\mu\nu})\psi^j_\nu$$

$$+ \frac{1}{4}\ e\kappa(\bar{\psi}^i_\mu\ \gamma^\alpha\gamma^\beta\gamma^\mu\ \lambda^j)\ F^{*k\ell}_{\alpha\beta}$$

$$+ \frac{\kappa}{2}\ e\ \bar{\psi}^i_\mu\ (\not{A} + i\gamma_5\ \not{B})\ \lambda^i\ ,$$

$$L(scalar-photon) = -\ \frac{e\kappa}{2\sqrt{2}}\ (AF^{*ij}_{\mu\nu} - iBF^{*ij}_{\mu\nu})\ F^{\mu\nu,ij}$$

$$-\ \frac{e\kappa^2}{4}\ (AF^{*ij}_{\mu\nu} - iBF^{*ij}_{\mu\nu})^2\ ,\tag{12}$$

where $\tilde{F}_{\mu\nu} = \varepsilon_{\mu\nu\rho\sigma}\ F^{\rho\sigma}$, $F^{*ij} = \frac{1}{2}\ \varepsilon^{ijk\ell}\ F^{k\ell}$ and the
photon-tensor indices ij are antisymmetric. The seagulls
with fermions contain terms like $A(\bar{\psi}\lambda)F$ and (fortunately)
cannot contribute to photon-photon scattering when all
four photons have the same color (say i,j = 3,4) because
there is only an AFF* coupling for the disintegration
of the scalar A into two photons. The set of all diagrams is given in figure 4 and again the (expected)
miracle happens: all divergences cancel. Hence, SU_4
supergravity is also one-loop finite (a general proof
exists also).[21]

V. PROOFS OF RENORMALIZABILITY IN SO(n) SUPERGRAVITY
     The proof that all four particle amplitudes are
one-loop finite in the SO(n) models is based on two
facts:

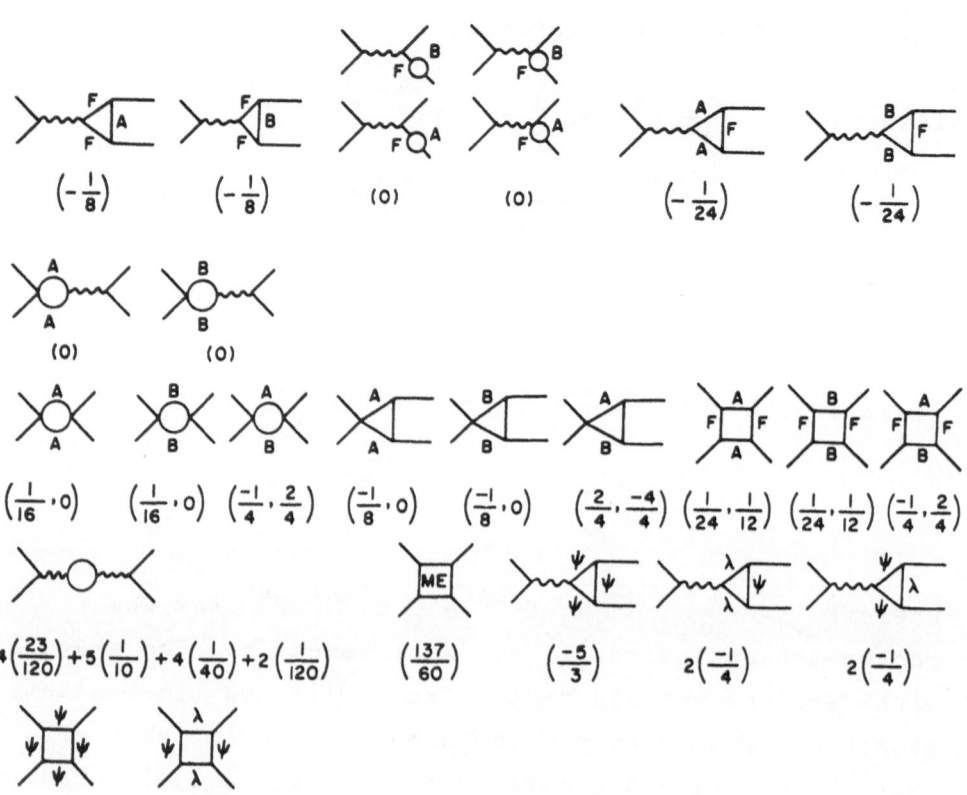

Fig. 4 The divergences of photon-photon scattering with identical colors in SO_4 supergravity theory with spin content $(2, (3/2)^4, (1)^6, (1/2)^4, 0, 0)$. The sum of all divergences cancels. The pairs of numbers are the co-efficients of (F^4) and $(F^2)^2$, while the simple numbers are the coefficients of $(T_{\mu\nu})^2$.

a) Graviton-graviton scattering is always one-loop finite whether or not matter is present. If matter is present, graviton-scalar and graviton-spin 1/2 fermion scattering is finite as well, as is graviton-photon scattering in a theory invariant under photon duality transformations.[16,21]

b) In a theory having global supersymmetry, four-particle S-matrix elements are related to each other[24] in such a way that from the finiteness of the amplitudes listed above one can conclude that all amplitudes are one-loop finite.

In the background field method the one-loop counter Lagrangian has the form

$$\Delta L^{(1\ \text{loop})} = \alpha R_{\mu\nu}^2 + \beta R^2 + \gamma R_{\mu\nu} M^{\mu\nu}(\phi) + \delta R_{\mu\nu\lambda\rho} L^{\mu\nu\lambda\rho}(\phi)$$
$$+ \varepsilon N(\phi) , \qquad (13)$$

where M, L, N are tensors constructed out of the matter fields ϕ and the gravitational field and are at least quadratic in the matter fields. The divergences of the S-matrix are obtained by inserting into the above expression solutions of the classical field equations (with "in-field" boundary conditions). This means that one can also use the classical field equations to simplify the counter Lagrangian. In particular $R_{\mu\nu}$, R can be replaced by expressions involving $T_{\mu\nu}$(matter) and containing at least two matter fields, so that the α,β,γ terms are replaced by terms containing at least four matter fields.

On dimensional and tensorial grounds L cannot contain terms quadratic in scalar or spin 1/2 fields while the only spin 1 possibility, $F^{\mu\nu} F^{\lambda\rho}$ is not invariant

under duality transformations and is therefore absent in
a theory possessing such invariance. Finally, while N
can depend a priori on only two matter fields (and a
sufficient number of derivatives), use of the field
equations eliminates such terms in favor of terms with
more matter fields. If we now examine again the
divergencies of the S-matrix we discover that, since
there are no purely gravitational terms or terms with
only two matter fields left (except possibly for spin
3/2 fields), all purely gravitational processes and
processes with two spin 0, 1/2 or 1 particles and an
arbitrary number of gravitons are finite.

We now show how S-matrix elements are related by
supersymmetry. From local supersymmetry we derive the
action of the supersymmetry charges on the "in-field"
annihilation and creation operators. We illustrate the
method in the SO_2 case.

The theory is invariant under the local transformations $(i, j = 1, 2)$

$$\delta e_{a\mu} = \kappa \, \bar{\eta}^i \, \gamma_a \, \psi_\mu^i \, ,$$

$$\delta \psi_\mu^i = \frac{2}{\kappa} D_\mu \eta^i + \frac{1}{\sqrt{2}} \, \epsilon^{ij} \, \sigma_{\rho\sigma} \gamma_\mu \, \eta^j \, \hat{F}^{\rho\sigma} \, , \qquad (14)$$

$$\delta A_{\bar{\mu}} = \sqrt{2} \, \epsilon^{ij} \, \bar{\eta}^i \, \psi_\mu^j \, ,$$

from which we deduce[24], for space-time independent η

$$[Q^i(\eta), a_2(p)] = \Gamma(\eta, p) \, a_{3/2}^i(p) \, ,$$

$$[Q^i(\eta), a_{3/2}^j(p)] = \delta^{ij} \Gamma^*(\eta, p) a_2(p) + \frac{1}{\sqrt{2}} \, \epsilon^{ij} \Gamma(\eta, p) a_1(p),$$

$$\qquad (15)$$

$$[Q^i(\eta), a_1(p)] = \sqrt{2}\ \varepsilon^{ij}\ \Gamma^*(\eta,p)\ a_{3/2}^j(p) \quad .$$

Here $a_s^i(p)$ is the "in" annihilation operator for a particle of spin s, SO_2 charge i and positive helicity (for the negative helicity operators replace Γ by $-\Gamma^*$),

$$\Gamma(\eta,p) = \sqrt{2p_0}\ [\eta_1\ \cos\theta/2\ e^{-i\phi/2} + \eta_2\ \sin\theta/2\ e^{i\phi/2}], \tag{16}$$

$$p = p_0[1,\ \sin\theta\ \cos\phi,\ \sin\theta\ \sin\phi,\ \cos\theta] \ ,$$

and η_1, η_2 are two arbitrary, anticommuting complex numbers in terms of which the Majorana spinor η is parametrized.

Since the $Q^i(\eta)$ commute with the S-matrix and are assumed to annihilate the vacuum, the relation

$$< in|[Q^i(\eta), S]\ |\ in> = 0 \tag{17}$$

allows one to derive equations involving S-matrix elements by commuting Q^i through the "in" operators on the left and right. These equations are linear in the arbitrary numbers η_1, η_2, η_1^*, η_2^* and give many relations between individual S-matrix elements. We find (omitting momentum labels we just indicate helicity and SO_2 charge) relations such as

$$<2,2|\ -\ 2,2> = <2,2|\ -\ 2,\ -2> = 0 \ , \tag{18}$$

$$<3/2^i,2|3/2^j,\ 2> \sim \delta^{ij}\ <2,2|2,2> \ , \tag{19}$$

$$<1,2|\ 3/2^i,\ 3/2^j> \sim \varepsilon^{ij}\ <1,2|1,2> \ , \tag{20}$$

etc. We have omitted certain kinematical factors.

It turns out that various four particle amplitudes
are either zero, proportional to purely gravitational
ones (which are finite) or to photon-graviton amplitudes (which are also finite since SO_2 is invariant
under the duality transformations of Eq. (3)).

Similar methods can be applied to the SO_3 and SO_4
cases (and of course to pure supergravity). In the
SO_4 case one finds that all nonvanishing amplitudes are
related either to graviton-graviton, graviton-scalar,
graviton-spin 1/2 or graviton-photon amplitudes. But
we know that these amplitudes are finite. We therefore
conclude that in $SO(n)$ as well all four-particle amplitudes are one-loop finite.

It is remarkable that in supergravity it is
possible[25] to determine two-loop renormalizability
properties whereas the corresponding properties in
ordinary Einstein gravity remain unsettled. This is because in supergravity a helicity conservation property
holds, as exemplified by Eq. (18). (In Einstein theory
such conservation holds at the tree level but nothing is
known beyond that).

It has been shown[26] that, after using the equations
of motion the only local infinity of the two-loop
S-matrix for gravitational processes has the form

$$\Delta L^{(2 \text{ loop})}(\text{grav}) = \alpha \ R^{\mu\nu}_{\cdot\lambda\rho} \ R^{\lambda\rho}_{\sigma\zeta} \ R^{\sigma\zeta}_{\mu\nu} \ . \quad (21)$$

In supergravity this term is part of a locally supersymmetric set of terms whose actual form is not known.
It was also shown in ref. (26) that if the above counterterm is indeed present it corresponds to infinities of
the helicity nonconserving graviton-graviton amplitudes

<2,2|2,-2>, <2,2|-2,-2>. But in supergravity such amplitudes are identically zero. It follows that supersymmetry requires $\alpha=0$ and all local divergences of the two-loop S-matrix for gravitational processes cancel.

Arguments have been presented by Tomboulis[27] to suggest that all two-loop divergencies are local. If this is indeed the case (his argument assumes some unproven properties of one-loop ghost counterterms) we can conclude that gravitational processes are completely two-loop finite. At least in pure supergravity this has the immediate consequence that all four-particle processes are two-loop finite.

A similar approach fails at the three-loop level. A possible counterterm is $[R^{\mu\nu}{}_{\lambda\rho} R^{\lambda\rho}{}_{\mu\nu}]^2$ which only contributes to helicity conserving processes and therefore cannot a priori be excluded. At the present time there seems to be no easy way to settle the three- and higher-loop question.

To conclude, we wish to mention that while our approach is to a large extent based directly on properties of the S-matrix, a more traditional approach to the divergence problem, based on a direct examination of the counter Lagrangian in supersymmetric theories has been also presented by Deser, Kay and Stelle[28]. These authors agree with our conclusions and claim to have extended them to many particle amplitudes as well.

OUTLOOK

At last there is progress in unifying quantum theory and general relativity. Explicit calculations and theoretical proofs indicate that SO(n) models are one- and two-loop finite, although some of the proofs are as yet incomplete. Nothing is yet known about finiteness of three- and higher loops.

When the SO(n) symmetries are gauged, cosmological terms appear in the action.[29] The quantization of such theories offers altogether new problems, since here one must quantize the fluctuations about a curved background, for example de-Sitter space. It is an important question for future applications to real physics to know whether such theories are finite as well.

Anomalies might spoil some of the theoretical proofs (but not, of course, the explicit calculations). One might expect axial anomalies; whether they would imply supersymmetry anomalies as well is an open question. There is of course a deep connection between the axial current, the source of the supersymmetry gauge field and the energy momentum tensor.[30] Also this question is of great importance.

All higher loop theoretical arguments assume that a proper regularization scheme exists. In global supersymmetry one can use higher derivative kinetic terms in the action[12] or dimension regularization,[31] but the former definitely cannot be used in local supersymmetry. Nor do locally supersymmetric R^2 actions regularize supergravity, as simple power counting shows. Whether dimensional regularization also saves the situation here, is as yet unknown.

Thus many problems remain to be solved. Nevertheless the spectacular cancellations that take place at the one-loop level (and apparently at the two-loop level) are a reason for optimism and encourage one to believe that supergravity theories may be finite to all orders.

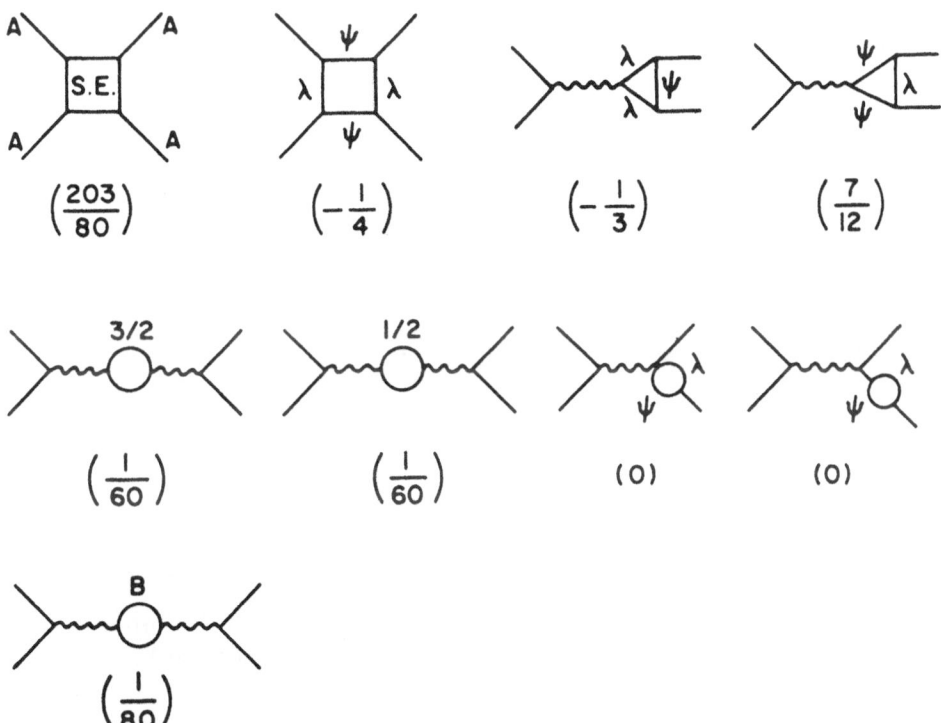

Fig. 5 The divergences of scalar-scalar scattering in the scalar multiplet coupling to supergravity with spin content $(2,3/2,1/2,0,0)$. The sum of all divergencies equals $(\frac{31}{12})\ \kappa^4 (n-4)^{-1}\ (\partial_\mu A)^4$.

REFERENCES

1. D.Z. Freedman, P. van Nieuwenhuizen and S. Ferrara, Phys. Rev. D13, 3214 (1976).

2. S. Deser and B. Zumino, Phys. Lett. 62B, 335 (1976).

3. D.Z. Freedman and P. van Nieuwenhuizen, Phys. Rev. D14, 912 (1976).

4. G. 't Hooft and M. Veltman, Ann. H. Poincaré 20, 69 (1974).

5. S. Deser and P. van Nieuwenhuizen, Phys. Rev. D10, 401 (1974).

6. Eidem, Phys. Rev. D10, 411 (1976).

7. Eidem, with H.-S. Tsao, Phys. Rev. D10, 3337 (1974).

8. M.T. Grisaru, P. van Nieuwenhuizen and C.C. Wu, Phys. Rev. D12, 1813 (1975).

9. P. van Nieuwenhuizen in "Proceedings Marcel Grosman meeting", Trieste 1975, North-Holland.

10. H. van Dam and M. Veltman, Nucl. Phys. B22, 397 (1970).

11. A related possibility is to use the dimensionlessness of $g_{\mu\nu}$ to use as basis $g_{\mu\nu} + \kappa^2(n-4)^{-1}(\alpha R_{\mu\nu}+\beta g_{\mu\nu}R)$ on which the theory might be truly renormalizable ('t Hooft and Veltman).

12. J. Iliopoulos and B. Zumino, Nucl. Phys. B76, 310 (1974).

13. B. Zumino, Nucl. Phys. B89, 535 (1975).

14. W. Lang and J. Wess, Nucl. Phys. B81, 249 (1974).

15. S. Ferrara and P. van Nieuwenhuizen, Phys. Rev. Lett. 37, 1669 (1976).

16. M.T. Grisaru, P. van Nieuwenhuizen and J.A.M. Vermaseren, Phys. Rev. Lett. 37, 1662 (1976).

17. S. Ferrara, J. Scherk and B. Zumino, LPTEN 76/24 (Paris).

18. P. van Nieuwenhuizen and J.A.M. Vermaseren, Phys.

Lett. $\underline{65B}$, 263 (1976).

19. M.T. Grisaru, P. van Nieuwenhuizen and C.C. Wu,
     Phys. Rev. $\underline{D12}$, 3203 (1975).

20. S. Ferrara, J. Scherk and P. van Nieuwenhuizen,
     Phys. Rev. Lett. $\underline{37}$, 1035 (1976); S. Ferrara, J.
     Scherk, F. Gliozzi and P. van Nieuwenhuizen, Nucl.
     Phys. $\underline{B117}$, 333 (1976).

21. M.T. Grisaru and H. Pendleton, (to be published).

22. D.Z. Freedman, Phys. Rev. Lett. $\underline{18}$, 105 (1977).
     S. Ferrara, J. Scherk and B. Zumino, LTPENS 76/23
     (Paris).

23. A. Das, Stony Brook preprint (ITP-SB-77-4).

24. M.T. Grisaru, H. Pendleton and P. van Nieuwenhuizen,
     Phys. Rev. D, to be published.

25. M.T. Grisaru, Phys. Lett. $\underline{66B}$, 75 (1977).

26. P. van Nieuwenhuizen and C.C. Wu, J. Math. Phys. $\underline{18}$,
     182 (1976).

27. E. Tomboulis, Princeton preprint.

28. S. Deser, J. Kay and K. Stelle, Brandeis preprint.

29. See the contribution of D.Z. Freedman in these
     proceedings.

30. S. Ferrara and B. Zumino, Nucl. Phys. $\underline{B87}$, 207
     (1975).

31. R. Delbourgo and M. Ramon Medrano, Imp. College
     preprint ICTP/75/11.

TOPICS IN SUPERGRAVITY AND SUPERSYMMETRY

Bruno Zumino

CERN

Geneva, Switzerland

ABSTRACT

We discuss the supersymmetric Higgs effect which occurs when a supersymmetric matter system, in which supersymmetry is spontaneously broken, is coupled with supergravity. A second topic discussed is the formulation of supergravity as a geometry of superspace.

SUPERSYMMETRIC HIGGS EFFECT

Rigorous supersymmetry implies the existence of supermultiplets made up of fermions and bosons with equal masses. If supersymmetry is to be relevant for the physical world, it must be broken, either softly or spontaneoulsy. Spontaneous breaking of global supersymmetry gives rise to the appearance of a Goldstone fermion. When global supersymmetry is promoted to a local invariance by coupling supersymmetric matter to supergravity, the Goldstone fermion disappears as a consequence of a phenomenon analogous to the Higgs effect of ordinary gauge theories. We describe now in some detail this supersymmetric Higgs effect and consider its possible

application to the construction of realistic models.
We follow some recent work by S. Deser and the author[1],
to appear shortly[*]. Observe that the supersymmetric
Higgs effect gives a possible solution to the problem
of the apparent nonexistence in Nature of the Goldstone
fermion of spontaneously broken supersymmetry. As we
know, this cannot be identified with the electron neutrino, because it would satisfy low energy theorems which
contradict observed properties of the neutrino spectrum[3].

   Spontaneous breaking of global supersymmetry generates a Majorana spin $\frac{1}{2}$ Goldstone fermion[4,5], which we
shall call λ. Irrespective of the particular field
theory in which it arises, its properties can be described, following Volkov and Akulov[4], by means of the
nonlinear realization of global supersymmetry

$$\delta\lambda = \frac{1}{a}\,\alpha + ia\,\bar{\alpha}\,\gamma^n\lambda\,\partial_m\lambda, \qquad (1)$$

where α is the infinitesimal supersymmetry parameter
and a is a constant which measures the strength of the
spontaneous breaking of supersymmetry. The nonlinear
Lagrangian for λ, invariant (up to a divergence) under
(1), is given by

$$L_\lambda = -\frac{1}{2a^2}\,\det\,(\delta_m{}^n + ia^2\bar{\lambda}\gamma^n\partial_m\lambda)$$

$$= -\frac{1}{2a^2} - \frac{i}{2}\,\lambda\,\gamma\cdot\partial\lambda + \ldots \qquad . \qquad (2)$$

---

[*] Volkov and Soroka[2] were the first to point out the
possible occurence of a supersymmetric Higgs effect.
However, in spite of its formal similarity, their point
of view is essentially different from ours.

This description is perfectly analogous to that used
for the pion in nonlinear pion dynamics. However, while
the chiral group $SU(2) \times SU(2)$ is also broken explicitly
by a pion mass term, if we assume that supersymmetry is
broken only spontaneously the above description is expected to be rigorous and to be actually valid for a
suitably defined field λ in any renormalizable model in
which a Goldstone fermion emerges.

Let us now try to promote (1) to a local transformation with parameter $\alpha(x)$ and to make (2) invariant
under it by coupling λ to the supergravity field $e_m{}^a$
and ψ_m. Without describing the complete Lagrangian,
which is rather complicated, one can easily find the
first terms in an expansion in the coupling constants
a and κ (gravitational constant). Under the transformation laws

$$\delta\lambda = \frac{1}{a}\,\alpha(x) + \ldots,$$

$$\delta e_m{}^a = -i\kappa\bar\alpha\gamma^a\psi_m ,$$

$$\delta\psi_m = -\frac{2}{\kappa}\,\partial_m\alpha + \ldots,$$

(3)

the Lagrangian

$$L_\lambda = -\frac{1}{2a^2}\,e - \frac{i}{2}\,\lambda\gamma\cdot\partial\lambda - \frac{i}{2a}\,\bar\lambda\gamma\cdot\psi + \ldots,$$

$$e = \det e_m{}^a ,$$

(4)

changes by a divergence. To (4) one must add the usual
supergravity Lagrangian[6]

$$L = \frac{1}{2\kappa}\,eR - \frac{i}{2}\,\varepsilon^{\ell m n r}\,\bar\psi e\gamma_5\gamma_m D_n\psi_r .$$

(5)

The sum $L + L_\lambda$ is invariant under (3). The transformation low for the field λ shows that it corresponds to a
pure gauge degree of freedom and that it can be transformed to zero by means of a suitably chosen local supersymmetry transformation. In other words, the field λ
can be absorbed into a redefinition of the fields $e_m{}^a$
and ψ_m. The resulting theory is described by the Lagrangian (5) of supergravity plus a cosmological term
(plus possible additional terms from the supersymmetric
matter part which gave rise to spontaneous symmetry
breaking).

This result is puzzling and disappointing. It is
puzzling because the disappearance of the Goldstone
particle (Higgs effect) gave rise to a cosmological term,
instead of generating a mass term for the spin 3/2 gauge
field, as one would have expected. It is disappointing
because the empirical smallness of the cosmological constant $-1/2a^2$ seems to destroy any hope that the spontaneous breaking of supersymmetry will be large enough
to be responsible for the observed mass splitting between bosons and fermions. In the following we discuss
these points.

The above puzzle is immediately resolved if one
observes that, in presence of a cosmological term, one
cannot quantize in a Minkowski background, but one must
take instead as a background space a solution of the
Einstein equations with cosmological term. The simplest
and most natural is the corresponding de Sitter space.
Now, in de Sitter space the concept of mass is rather
delicate.

It has been recently observed[7,8] that one can add
to the supergravity Lagrangian (5) the sum of a cosmological term and of a spin 3/2 mass term

$$3\frac{\mu^2}{\kappa^2} e - \frac{i}{2} \mu \, \varepsilon^{emnr} \, \bar{\psi}_e \gamma_5 \Sigma_{mn} \psi_r,$$

$$\Sigma_{mn} = \frac{1}{4}[\gamma_m, \gamma_n],$$

(6)

without spoiling local supersymmetry. Indeed the sum
of (5) and (6) is invariant under a modified supersymmetry transformation in which the usual transformation law

$$\delta\psi_m = -\frac{2}{\kappa} D_m \alpha$$

(7)

is replaced by

$$\delta\psi_m = -\frac{2}{\kappa} D_m \alpha - \frac{\mu}{\kappa} \gamma_m \alpha$$

(8)

(there is a corresponding change in $\delta\omega_{mab}$). The existence of this local supersymmetry shows that, in spite
of the apparent mass term in (6), the spin 3/2 field has
the number of degrees of freedom appropriate to the massless case. It is also easy to see that the Lagrangian
(5) plus (6) admist a global supersymmetry, obtained by
taking α independent of x. It is the global supersymmetry
of the corresponding de Sitter space of radius μ^{-1}, which
has $0(3,2)$ as the maximal Lie subalgebra[*]. When the
cosmological term and the spin 3/2 mass term are related
in the particular way given in (6) there is local supersymmetry and, with a sensible definition of mass, the
spin 3/2 field is massless. When they are not related

---

[*] Here we disagree with Ref. 7, where the spin 3/2 mass
term is interpreted as giving rise to a breaking of
global supersymmetry.

as in (6), there is no local supersymmetry and the spin
3/2 field is massive (this is true in particular when
there is only a cosmological term, which resolves the
agove mentioned paradox). Then the equation of motion
imply the constraints

$$\gamma^m \psi_m = 0,$$

$$\partial_m \psi^m + \ldots = 0,$$

(9)

where the dots are terms of higher order in κ. These
constraints are exactly of the kind that gives the right
number of degrees of freedom for a massive spin 3/2 field.
The classical equations of motion are still consistent,
even though there is no local supersymmetry, and no
anomalous propagation hypersurfaces occur.

The existence of the invariant (6) can be used to
resolve the second problem mentioned above. We observe
that the sign of the cosmological term in (6) is fixed,
and corresponds to a de Sitter space with $O(3,2)$ invariance. On the other hand, that of the cosmological
term in (4) is also fixed and is the opposite. Adding
(4), (5) and (6), one can adjust the constants so that
the cosmological terms cancel. Before, however, one
must modify (4) so as to make it invariant under the
new transformation law (8). This is not difficult, to
the order considered here, and requires adding to (4) a
term

$$-i\mu \, \bar{\lambda}\lambda + \ldots \qquad .$$

(10)

The meaning of this term can be understood as belonging

to the invariant Lagrangian for a Goldstone spinor in
a de Sitter space of radius μ^{-1}.

Now one can cancel the cosmological terms between
(4) and (6)

$$\frac{1}{2a^2} = \frac{3\mu^2}{\kappa^2} \quad . \tag{11}$$

If we assume that the spontaneous supersymmetry breaking
is responsible for the observed mass splitting between
mesons and baryons, the order of magnitude of the constant a must be given by a hadronic mass, say the proton
mass,

$$\frac{1}{a} \sim m_p^2 \quad . \tag{12}$$

We find

$$\mu \sim (\kappa m_p) m_p , \tag{13}$$

where $\kappa m_p \sim 10^{-19}$. The mass of the spin 3/2 field is
very small, but we have hadronic mass splittings of
reasonable magnitude and zero cosmological constant[*].

GEOMETRY OF SUPERSPACE

We shall now describe how supergravity can be obtained from the geometry of superspace. If one takes
the differential geometry of superspace to be (super)

---

[*]The compensation between cosmological terms of opposite
sign was considered by Freedman and Das[7,9] in a specific
model. In that model difficulties seem to arise when
one attempts to complete the combined Lagrangian to a
locally supersymmetric invariant (private communication
from D. Freedman).

Reimannian[10,11], the connection with the space-time
formulation of supergravity[6] is not very direct and requires a limiting process in superspace[12]. This is due
to the fact that the field equations in Riemannian superspace do not admit as solution the flat superspace[4,13]
of ordinary global supersymmetry. It was for this very
reason that a different differential geometry in superspace was introduced by the Wess and the author[14] and,
independently, by Akulov, Volkov and Soroka[15]. The
superspace of global supersymmetry is a special case of
this kind of superspace. Furthermore, the equations of
supergravity take a very simple form. We describe this
below, following a recent paper by Wess and the author[16].

We begin by considering a general affine superspace.
Its points are parametrized by coordinates $z^M \equiv (x^m, \theta^\mu)$
where the x^m are the commuting space-time coordinates
while θ^μ are anticommuting variables. More precisely,
x^m are even and θ^μ odd elements of a Grassmann algebra.
Latin letters will denote vectorial (bosonic), Greek
letters spinorial (fermionic) indices. The supervierbien
matrix $E_M{}^A(z)$, where $A = (a, \alpha)$ and its inverse $E_A{}^M$, can
be used to transform world tensors into tangent space
tensors and vice versa. The submatrices $E_m{}^a$ and $E_\mu{}^\alpha$
consist of bosonic, $E_m{}^\alpha$ and $E_\mu{}^a$ of fermionic elements.
The superconnection $\Phi_{M,A}{}^B$ and the supervierbien can be
viewed as the coefficients of two one-forms

$$E^A = dz^M E_M{}^A, \qquad \Phi_A{}^B = dz^M \Phi_M{}_{,A}{}^B. \qquad (14)$$

Here the differentials dz^m are taken to anticommute with
each other and with the $d\theta^\mu$, while the $d\theta^\mu$ commute with
each other. Similarly, E^a anticommute with each other
and with E^α, while E^α commute with each other. Under a

linear transformation in the tangent space

$$\delta v^A = v^B X_B{}^A, \quad \delta u_A = - X_A{}^B u_B, \qquad (15)$$

the supervierbien $E_M{}^A$ transforms like the vector v^A, while the connection transforms as

$$\delta \Phi_A{}^B = \Phi_A{}^C - X_A{}^C \Phi_C{}^B - d X_A{}^B, \qquad (16)$$

and one can define the covariant differentials

$$\mathcal{D}v^A = dv^A + v^B \Phi_B{}^A, \quad \mathcal{D}u_A = du_A - \Phi_A{}^B u_B. \quad (17)$$

The notation of differential forms is compact and convenient and takes automatically into account all the sign changes due to the Grassmann nature of our variables. Many properties of Cartan forms generalize in a simple and natural way to our differential forms with Grassmann variables[14]. Because we write the differentials on the left, the differentiation operator $d = dz^M \partial/\partial z^M$ operates on a product starting from the right. For instance, if Ω_2 is a p-form, $d(\Omega_1\Omega_2) = \Omega_1 d\Omega_2 - (-1)^P d\Omega_1\Omega_2$. The theorem $d(d\Omega) = 0$ is valid and its inverse also applies with obvious restrictions on the topology of the domain of variation of the bosonic variables. In terms of co-variant derivatives, defined by $\mathcal{D} = dz^M \mathcal{D}_M$, the formulae (17) become

$$\mathcal{D}_M v^A = \frac{\partial}{\partial z^M} v^A + (-)^{bm} v^B \Phi_{M,B}{}^A,$$

$$\mathcal{D}_M U_A = \frac{\partial}{\partial z^M} U_A - \Phi_{M,A}{}^B U_B, \qquad (18)$$

where the sign factor $(-)^{bm}$ is defined by the convention that m = 0 if M is vectorial and m = 1 if M is spinorial, and similarly for B. The torsion and the curvature are defined by

$$T^A = dE^A + E^B \phi_B{}^A = \frac{1}{2} dz^N dz^M T_{MN}{}^A = \frac{1}{2} E^C E^B T_{BC}{}^A,$$

$$(19)$$

$$R_A{}^B = d\phi_A{}^B + \phi_A{}^C \phi_C{}^B = \frac{1}{2} dz^N dz^M R_{MN,A}{}^B = \frac{1}{2} E^C E^D R_{DC,A}{}^B.$$

As a consequence of their definition, they satisfy the Bianchi identities

$$dT^A + T^B \phi_B{}^A - E^B R_B{}^A = 0,$$

$$(20)$$

$$dR_A{}^B + R_A{}^C \phi_C{}^B - \phi_A{}^C R_C{}^B = 0.$$

Written out in terms of coefficients tensors the Bianchi identities take the form

$$E^C E^B E^A (\mathcal{D}_A T_{BC}{}^D + T_{AB}{}^{C'} T_{C'C}{}^D - R_{AB,C}{}^D) = 0,$$

$$E^C E^B E^A (\mathcal{D}_A R_{BC,D}{}^F + T_{AB}{}^{C'} R_{C'C,D}{}^F) = 0,$$

$$(21)$$

where

$$\mathcal{D}_A = E_A{}^M \mathcal{D}_M.$$

In order to give more structure to our superspace we must specialize the group in the tangent space. The simplest would be to require the tangent space to be a numerical tensor, $\eta_{AB} = (-)^{ab} \eta_{BA}$ and, with the further restriction of vanishing torsion, would lead to a

Riemannian superspace. Except for the use of vierbein and connection, the geometry would be equivalent to that described in[10,11] directly in terms of the metric tensor. For the reasons explained above, we choose a more restricted group in the tangent space and require the existence of a basis in which the matrices $X_A{}^B$ satisfy the relations

$$X_a{}^\beta = X_\alpha{}^b = 0, \quad X_a{}^b = L_a{}^b(z),$$

$$X_\alpha{}^\beta = \frac{1}{2} L_a{}^b (\Sigma_b{}^a)_\alpha{}^\beta, \tag{22}$$

where $L_a{}^b$ is an infinitesimal Lorentz matrix, $L_{ab} = -L_{ba}$ and $\Sigma_b{}^a = 1/4[\gamma_b, \gamma^a]$. In words, $L_\alpha{}^\beta$ describes the same Lorentz transformation as $L_a{}^b$ when applied to spinors. Our group consists therefore of ordinary Lorentz transformations, but dependent on both x and θ. Since the connection and the curvature $R_A{}^B$ are matrices belonging to the algebra of the tangent space group they both satisfy the same restrictions as $X_A{}^B$. In particular $R_{CD,\alpha\beta} = R_{CD,\alpha b} = 0$,

$$R_{CD,ab} = -R_{CD,ba}, \quad R_{CD,\alpha\beta} = \frac{1}{2} R_{CD,ab} (\Sigma^{ab})_{\alpha\beta}.$$

The equations of supergravity can be stated as simple restrictions on the torsion tensor in superspace. We take

$$T_{\alpha\beta}{}^c = 2i(\gamma^c)_{\alpha\beta}, \quad T_{\alpha\beta}{}^\gamma = 0, \tag{23}$$

$$T_{a\beta}{}^c = T_{a\beta}{}^\gamma = 0, \quad T_{ab}{}^c = 0, \tag{24}$$

while we leave the components $T_{ab}{}^{\gamma}$ which correspond to
the gauge invariant Rarita-Schwinger field, undetermined.
A number of relations can be immediately obtained by
combining (23) and (24) with the Bianchi identities (21).
Among them are

$$R_{\alpha\beta,c}{}^{d} = 0, \qquad (25)$$

$$R_{ab,\beta}{}^{\delta} + R_{\beta b,\alpha}{}^{\delta} - 2i(\gamma^{c})_{\alpha\beta} T_{cb,}{}^{\delta} = 0, \qquad (26)$$

$$R_{ab,c}{}^{d} - R_{ac,b}{}^{d} - 2i(\gamma^{d})_{\alpha\beta} T_{bc,}{}^{\beta} = 0, \qquad (27)$$

$$\mathcal{D}_{\alpha}R_{\beta b,c}{}^{d} + \mathcal{D}_{\beta}R_{ab,c}{}^{d} + 2i\ (\gamma^{c'})_{\alpha\beta} R_{c'b,c}{}^{d} = 0. \qquad (28)$$

Furthermore, with a little algebra, one can see that (26)
actually implies

$$(\gamma^{c})_{\alpha\beta} T_{cb,}{}^{\beta} = 0, \qquad (29)$$

and therefore also, from (27),

$$R_{ab,c}{}^{b} = 0. \qquad (30)$$

This in turn, through (28) tells us that

$$R_{ab,c}{}^{b} = 0. \qquad (31)$$

Finally, combining (26) and (27) and remembering the
relation

$$R_{ab,\beta}{}^{\delta} = \frac{1}{2} R_{ab,c}{}^{d}(\Sigma_{d}{}^{c})_{\beta}{}^{\delta},$$

one finds that

$$R_{ab,cd} = 2i(\gamma_b)_{\alpha\beta} T_{cd,}{}^{\beta} . \tag{32}$$

The simplest way to extract field equations from our superfield equations is to observe that it is possible to choose a gauge such that, as $\theta \to 0$, the superconnection $\Phi_{m,ab}$ becomes the usual connection in four-space, the supervierbein becomes*

$$E_m{}^a = e_m{}^a(x) , \quad E_m{}^{\alpha} = \tfrac{1}{2} \psi_m{}^{\alpha}(x),$$

$$E_{\mu}{}^a = 0 , \quad E_{\mu}{}^{\alpha} = \delta_{\mu}{}^{\alpha} , \tag{33}$$

where $e_m{}^a$ is the usual vierbein in four-space and $\psi_m{}^{\alpha}$ the Rarita-Schwinger field. It follows that, in the same limit,

$$T_{\mu\nu}{}^{\alpha} = T_{\mu n}{}^{\alpha} = R_{\mu\nu,a}{}^b = 0,$$

$$R_{m\nu,a}{}^b = e_m{}^c \delta_{\nu}{}^{\beta} R_{c\beta,a}{}^b , \tag{34}$$

while $T_{mn,}{}^a$ and $R_{mn,a}{}^b$ become the four-space torsion and curvature tensor, which we shall denote by $C_{mn,}{}^a$ and $R_{mn,a}{}^b$. Using (33) and the relation $T_{MN,}{}^C = E_M{}^A E_N{}^B T_{BA}{}^C$ we see that (23) and (24) imply the connection between the torsion and the spin density

---

*The higher powers in θ are also expressible in terms of the physical fields and their derivatives, provided one fixes the gauge appropriately. The coordinate transformations in superspace which preserve the choice of gauge can be expressed in terms of a coordinate transformation in ordinary space time and an x-dependent supersymmetry transformation (local supersymmetry transformation).

$$T_{mn,}{}^{c} - \frac{i}{2} \psi_{m}{}^{\alpha} (\gamma^{c})_{\alpha\beta} \psi_{n}{}^{\beta} = 0. \qquad (35)$$

On the other hand (29) gives the Rarita-Schwinger equation in the form

$$(\gamma^{c})_{\alpha\beta} T_{cd,}{}^{\beta} = 0, \quad T_{cd,}{}^{\beta} = e_{c}{}^{m} e_{d}{}^{n} (\mathcal{D}_{m} \psi_{n}{}^{\beta} - \mathcal{D}_{n} \psi_{m}{}^{\beta}), \quad (36)$$

where D_{m} is the covariant derivative used in the second of Ref. 5, and (31) gives the Einstein equation in the form

$$R_{ab,c}{}^{b} + \frac{i}{2} e^{bm} \psi_{m}{}^{\alpha} (\gamma a)_{\alpha\beta} T_{bc,}{}^{\beta} = 0. \qquad (37)$$

The equations (36), and (37) are equivalent, but not identical, to those given in Ref. 5. To establish the equivalence of the two forms of the Einstein equation observe that the Rarita-Schwinger equation implies that

$$(\gamma_{5})_{\alpha\beta} T_{ab,}{}^{\beta} = \frac{1}{2} \epsilon_{ab}{}^{cd} T_{cd,\alpha} . \qquad (38)$$

Instead of taking all of (23) and (24) as basic equations, one can take some of them and some of the equations we have derived from them through the Bianchi identities. A particularly interesting choice includes (23) and (25) among the basic equations.

Finally, let us observe that the above formalism permits the construction of superspace scalars. On the other hand, one knows how to transform a scalar into a density. One multiplies it by the det $E_{M}{}^{A}$, where the determinant of a matrix with commuting and anticommuting elements is defined in Ref. 11. By this procedure one can construct invariant actions in superspace.

REFERENCES

1. S. Deser and B. Zumino, in preparation. B. Zumino, Lectures at the 1976 Scottish Universities Summer School, August 1976.

2. D. V. Volkov and V. A. Soroka, JETP Letters $\underline{18}$ (1973) 312.

3. W. Bardeen, unpublished; B. de Wit and D. Z. Freedman, Phys. Rev. Letters $\underline{35}$ (1975) 827.

4. D. V. Volkov and V. P. Akulov, Phys. Letters $\underline{46B}$ (1973) 109.

5. J. Iliopoulos and B. Zumino, Nuclear Phys. $\underline{B76}$ (1974) 310. A. Salam and J. Strathdee, Phys. Letters $\underline{49B}$ (1974) 465. P. Fayet and J. Iliopoulos, Phys. Letters $\underline{51B}$ (1974) 461. L. O'Raifertaigh, Nuclear Phys. $\underline{B96}$ (1975) 331.

6. D. Z. Freedman, P. van Nieuwenhiuzen and S. Ferrara, Phys. Rev. D $\underline{13}$ (1976) 3214. S. Deser and B. Zumino, Phys. Letters $\underline{62B}$ (1976) 335.

7. D. Z. Freedman and A. Das, Stony Brook preprint ITP-SB-76-64.

8. S. MacDowell and F. Mansouri, private communication.

9. D. Z. Freedman, Stony Brook preprint ITP-SP-76-50.

10. R. Arnowitt and P. Nath, Phys. Letters $\underline{56B}$ (1975) 117.

11. R. Arnowitt, P. Nath and B. Zumino, Phys. Letters $\underline{56B}$ (1975) 81.

12. P. Nath and R. Arnowitt, Phys. Letters $\underline{65B}$ (1976) 73.

13. A. Salam and J. Strathdee, Nuclear Phys. $\underline{76B}$ (1974) 477. S. Ferrara, J. Wess and B. Zumino, Phys. Letters $\underline{51B}$ (1974) 239.

14. See, B. Zumino in Proc. of the Conf. on Gauge Theories and Modern Field Theory, Northeastern University, September 1975, eds. R. Arnowitt and P. Nath (MIT Press).

15. V. P. Akulov, D.V. Volkov and V. A. Soroka, JETP
    Letters 22 (1975) 396.

16. J. Wess and B. Zumino, University of Karlsruhe pre-
    print, to be published in Phys. Letters.

QUANTUM FLAVORDYNAMICS*

H. Fritzsch

CERN - Geneva, Switzerland

California Institute of Technology, Pasadena, CA

Charm is more than beauty.

Yiddish Proverb

I. INTRODUCTION

At particle physics conferences in the past the
organizers took usually great care in dividing the
various phenomena into "strong interaction physics,"
"electromagnetic interactions," "weak interactions," and
"gravitational interaction." This meeting is the first
one which breaks with this tradition and in which
electromagnetism and the weak interactions are treated
together - in a session on flavor interactions (quantum
flavordynamics, QFD). Although the final scheme of QFD
is still unknown, it is meanwhile clear, due to the
combined efforts of experimentalists and theorists during
the last years, that electromagnetism and weak interactions
have to be treated together in one single framework. This

---
*Work supported in part by the U.S. Energy Research and
Development Administration under Contract E-(11-1)-68

is, no doubt, a great advance towards a full understanding of all particle interactions, which, perhaps,
will be achieved some day. For the time being, let
us be content with the existence of three yet disconnected
fields in particle physics:

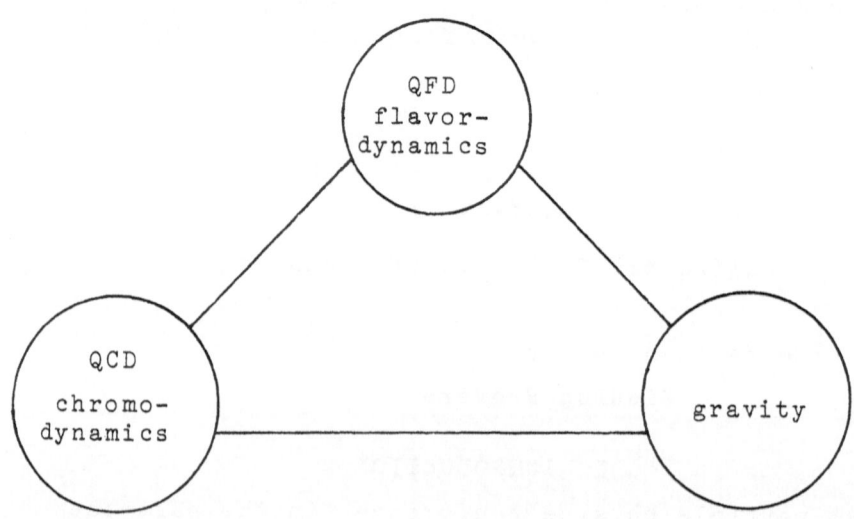

     Before we hear more details about the various
phenomenological aspects of flavor interactions in the
subsequent talks, I shall describe the theoretical landscape of QFD, which has opened up in front of the eyes
of theorists during the last year. Unfortunately, or
rather fortunately for theorists who feel themselves
young enough, this landscape contains, at the moment,
not only well-defined structures and contours, but also
a lot of rather foggy subjects, surely even some
delusions. To decide what is real and what is not, and
to investigate the still unexplored depths and abysses
of this landscape, remains one of the most challenging
tasks of experimentalists and theorists alike in the

years to come.

II. MYTH

The recent discovery of new particles which are stable with respect to the strong interactions[1], and the properties of their weak decay pattern suggest strongly the existence of charm which is needed for constructing $SU_2 \times U_1$ type gauge theories[2] of QFD. Although it is by no means sure that the charm changing weak current is the current[3] $[\bar{c}(s \cos\theta_c - d \sin\theta_c)]_L$ (L: lefthanded, θ_c: Cabibbo angle), it is quite possible that this is the case. Thus it seems useful to start the discussion with what I would like to call MYTH ("mynimal theory"): minimal in the sense that it incorporates all currents and degrees of freedom absolutely needed from the present phenomenological point of view, and nothing more.

MYTH in the strict sense I shall use it, is based on the four quark flavors u,d,s, and c, and on the four observed leptons. The gauge group is a direct product of the QCD and QFD gauge groups: $SU_3^c \times SU_2 \times U_1$, the fermion representations are given by

$$\begin{pmatrix} u & c \\ d' & s' \end{pmatrix}_L (u_R, d_R, s_R, c_R) \begin{pmatrix} \nu_e & \nu_\mu \\ e^- & \mu^- \end{pmatrix}_L (e^-_R, \mu^-_R) \, ,$$

$d' = d \cos\theta_c + s \sin\theta_c$, $(\theta_c$: Cabibbo angle) ,

$s' = -d \sin\theta_c + s \cos\theta_c$.

The neutrinos are supposed to be lefthanded Weyl particles. The minimal model contains four intermediate vector bosons (W^+, W^-, Z, γ). The symmetry breaking is

generated by one doublet of scalar bosons[4] the neutral
component of which generates a nonvanishing vacuum
expectation value. Below I list some of the important
features of MYTH:

a) The phenomenological weak Hamiltonian is

$$H^{weak} = \frac{4G_F}{\sqrt{2}} \left(j_\mu^{~+} j_\mu^{~-} + j_\mu^n j_\mu^n \right) , \tag{2.1}$$

$$j_\mu^{~+} = \bar{\nu}_e \gamma_{\mu L} e^- + (e \to \mu) + \bar{u} \gamma_{\mu L} d' + \bar{c} \gamma_{\mu L} s' ,$$

$$\left(\gamma_{\mu L} = \gamma_\mu \frac{1+\gamma_5}{2} \right) ,$$

$$j_\mu^n = \frac{1}{2} \left(\bar{\nu}_e \gamma_{\mu L} \nu_e - \bar{e}^- \gamma_{\mu L} e^- \right) + \frac{1}{2} \left(\bar{\nu}_\mu \gamma_{\mu L} \gamma_\mu - \bar{\mu}^- \gamma_{\mu L} \mu^- \right)$$

$$+ \frac{1}{2} \left(\bar{u}\gamma_{\mu L}u - \bar{d}\gamma_{\mu L}d \right) + \frac{1}{2} \left(\bar{c}\gamma_{\mu L}c - \bar{s}\gamma_{\mu L}s \right) - \sin^2\theta_w j_\mu^e .$$

(j_μ^e: electromagnetic current, θ_w: $SU_2 \times U_1$ mixing angle)

$$\frac{4G_F}{\sqrt{2}} = \frac{g^2}{2M_w^2} , \quad M_w = \frac{37.3 \text{ GeV}}{\sin\theta_w} , \quad e/g(SU_2)=\sin\theta_w, M_z=M_w/\cos\theta_w .$$

b) The theory is free of anomalies; all fermion charges
add to zero.

c) The photon mass is zero due to the requirement that
only the neutral component of the scalar doublet is allowed
to have a nonvanishing vacuum expectation value (charge
conservation). The fact that the scalar fields transform
as a doublet leads to an "isospin" relation among the
massive boson masses: $M_w = M_z \cos\theta_w$. Thus the charged
current and neutral current Fermi constants (which, in
general, are unrelated) are equal if we choose the

normalizations of the currents used above. Not only
the ratio of neutral current cross sections, e.g., the
total cross section ratio $\sigma^{\bar{\nu}\bar{\nu}}/\sigma^{\nu\nu}$ ($\sigma^{\nu\nu}$: neutrino-hadron
neutral current cross sections), which depends only on
the algebraic form of the neutral current, can be
predicted as a function of θ_w, but also the ratios
$\sigma^{\nu\nu}/\sigma^{\nu\mu-}$, $\sigma^{\bar{\nu}\bar{\nu}}/\sigma^{\bar{\nu}\mu+}$. The neutrino production data are
in good agreement with the prediction for $\sin^2\theta_w \approx 0.3...0.4$.

d) Parity violation in atomic physics.

In MYTH the neutral current is a superposition of
a vector and an axial vector current. Consequently Z
boson exchange leads to parity violation, in particular
in atomic physics[5]. The present status of the
experimental situation in this field will be discussed
later by Bouchiat. Let me only mention a few theoretical
points. The leading parity violation effect in case of
atoms with large A will arise due to the following term
in the n.c. effective Hamiltonian:

$$H^{n.c.} = \frac{4G_F}{\sqrt{2}} \cdot j_\mu^{\ n}(\text{quarks},\text{vector}) \cdot j_n^{\ \mu}(\text{electron},\text{axial vector}).$$

$$(2.2)$$

This term is the dominating one since $j_\mu^{\ n}$(quarks,vector)
acts coherently on nucleons. In the limit of nonrelativistic nuclear motion the parity violating atomic
physics Hamiltonian reduces to

$$H_{p.v.} = \frac{G}{\sqrt{2}} \cdot \frac{Q_w}{2} \int \rho(\vec{x})\ \psi_e^+ \gamma^5 \psi_e d^3x ,$$

$$(2.3)$$

($\rho(\vec{x})$: nuclear charge density, ψ: electron wave function)
where Q_w is the nucleus expectation value of the neutral
vector current density:

$$Q_w = 4 <\text{nucleus}| \int j_0^{\ n}(\text{quarks})\ d^3x |\text{nucleus}> ,$$

which can easily be expressed in terms of the number
of u and d quarks in the nucleus, or alternatively in
terms of the nuclear charge Z and atomic mass number
A (or neutron number N):

$$Q_w = (1 - 8/3 \sin^2\theta_w)(\text{number of u-quarks}) + (-1 + 4/3 \sin^2\theta_w)$$

$$\cdot \text{ number of d-quarks}$$

$$= Z (1 - 4 \sin^2\theta_w) - N$$

$$= Z (2 - 4 \sin^2\theta_w) - A \quad . \tag{2.4}$$

Some preliminary experimental results have been
publicized already[6]: Those experiments were done on
$^{209}_{83}$ Bi. In MYTH the effective charge is always negative,
and for values of $\sin^2\theta_w$ between 0.25 and 0.4 (this is
the relevant range indicated by the neutrino results if
MYTH is correct) Q_w varies between -126 and -176. It is
important to check both sign and magnitude of Q_w by
experiment.

e)

The quark and lepton masses as well as the Cabibbo
angle are parameters of the theory, put in from outside
in the form of the suitably chosen Yukawa interaction
coupling constants. The neutral current is parity
violating and flavor diagonal, independent of the values
for the various parameters of the theory: The absence
of $|\Delta S| = 2$ neutral currents is "natural" (see Ref. 7).

f) Quark masses

It is easy to see that in the minimal theory the
effective quark mass matrix depends only on five parameters,
which we can choose to be the four quark masses
m_u, m_d, m_s, m_c, and the Cabibbo angle. No mass relations

exist, nor does there exist any theoretical understanding
of the mass pattern. The phenomenology of the observed
hadron spectrum gives $m_u \lesssim m_d$, $m_s \gg (m_u + m_d)/2$,
$m_c \gg m_s$. Applying PCAC for the pseudoscalar mesons
π and K one finds specifically[8]

$$(m_d - m_u)/(m_d + m_u)/2 \simeq 0.6, \quad (m_d + m_u)/2m_s \simeq m_\pi^2/2m_K^2 \simeq 1/26.$$

The effective quark masses are not well-defined
numbers; in QCD they are renormalization group dependent
parameters. One may argue that the PCAC relations above
are valid, if one chooses a renormalization point μ such
that the lowest loop approximation to the renormalization
group equation is applicable, in which case one can
assign actual numbers (in MeV) to the quark masses. For
$\mu \simeq$ a few GeV the typical values would be[9] $m_u \simeq 4$ MeV,
$m_d \simeq 7.5$ MeV, $m_s \simeq 150$ MeV. The mass of the charmed quark
is ~ 1600 MeV.

We emphasize the very big "isospin violation"
exhibited by these quark masses: $(m_d - m_u)/(m_d + m_u)/2$, but
by the ratio $(m_u - m_d)/M_0$, where M_0 is the typical strong
interaction mass scale supposedly generated by the
introduction of a mass scale in QCD via the renormalization
point. Thus isospin symmetry is <u>not</u> a consequence of
nearly exact mass degeneracy between m_u and m_d, as often
stated, but a consequence of the <u>smallness</u> of the u-d
quark masses compared to M_0. <u>The strong interactions
generate isospin symmetry</u>.

In the kaon system the u-d quark mass difference
dominates the "electromagnetic" mass difference $M_{K^o} - M_{K^+}$.
Here the quark mass contribution is positive, but the
electromagnetic contribution is negative. The situation
is different for the mass difference $M_{D^+} - M_{D^o}$; here both
the quark mass contribution and the electromagnetic one
(the latter probably dominated by the Coulomb contribution)

are positive. Hence one expects $M_{D^+} - M_{D^o} > M_{K^o} - M_{K^+}$.
More detailed calculations lead to the result
$M_{D^+} - M_{D^o} \simeq 6.5 \ldots 8$ Mev[10], a value which is consistent
with the present experimental data. It would be quite
useful to determine ΔM_D very accurately. Here is a
place in charm spectroscopy where one can learn something about the ordinary quarks.

g) Baryon and lepton number are conserved in MYTH.
However in addition electron number and muon number are
also conserved. The latter follows simply because the
MYTH Lagrangian is invariant under U_1 gauge transformations acting separately on the $(\nu_e \ e^-)$ and on the
$(\nu_\mu \ \mu^-)$ system. The only way to break this symmetry
is to add a term which mixes the muonic and electronic
system, e.g., by introducing a leptonic version of the
Cabibbo angle: $\nu_e \rightarrow \cos\theta \ \nu_e + \sin\theta\nu_\mu$, $\nu_\mu \rightarrow -$. However,
the neutrinos are massless-such a rotation has no physical
meaning: the rotated like the unrotated states are
eigenvalues of the mass matrix. The masslessness of the
neutrinos insures the separate conservation of electron
and muon number. In particular the decay $\mu \rightarrow e\gamma$ is
absolutely forbidden.

The presence of the Cabibbo angle in the quark
system insures that there exist no two separately conserved
baryon numbers (u-d number and c-s number). (This would
be the case if either $m_u = m_c$ or $m_d = m_s$). Thus in
particular the transitions $s \rightarrow d + \gamma$ (the quark analog
of $\mu \rightarrow e\gamma$) occurs. For free quarks the amplitude
describing this process is[11] approximately
$e \ G_F/\sqrt{2} \cdot 1/8\pi^2 \ m_s \cdot (m_c^2 - m_u^2)/M^2 \cdot \log (M_w^2)/m_c^2 \ \sin\theta_c \cos\theta_c$.
The branching ratio for the radiative decay is

$$\frac{\Gamma(s \to d + \gamma)}{\Gamma(s \to d + e^- + \bar{\nu}_e)} \simeq 2/3 \cos^2\theta_c \ \alpha/\pi \left[\frac{(m_c^2 - m_u^2)}{M_W^2} \log\frac{M_W^2}{m_c^2} \right]^2$$

$$\sim 10^{-8} . \tag{2.5}$$

A slight modification of MYTH would be to give up the interpretation of the neutrinos as Weyl fields and to introduce righthanded neutrino field components as well. In this case there is no reason to have no mass terms for the neutrinos, and it would be most natural to have (small) neutrino masses as well as a leptonic version of the Cabibbo angle. (The present experimental limits on the neutrino masses are $m_{\nu_e} \lesssim 30$ eV and $m_{\nu_\mu} \lesssim 0.5$ MeV). In addition there exists an astrophysical bound[12] which implies here: $m_{\nu_e} + m_{\nu_\mu} < 16$ eV. Let us denote the eigenvalues of the neutrino mass matrix by ν_1, ν_2. The lepton system can be written as

$$\begin{pmatrix} \cos\theta\nu_1 + \sin\theta\nu_2 & , & -\sin\theta\nu_1 + \cos\theta\nu_2 \\ & & \\ e^- & , & \mu^- \end{pmatrix}_L \tag{2.6}$$

(θ: leptonic analog of Cabibbo angle).

There are two interesting consequences of such a scheme.

$\underline{\mu \to e\gamma:}$

The decay $\mu \to e\gamma$ occurs; muon and electron number are not separately conserved. The branching ratio is

$$B = \frac{\Gamma(\mu \to e\gamma)}{\Gamma_\mu^{tot}} = \frac{3\alpha}{32\pi} \left(\sin\theta \ \cos\theta \ \frac{m_{\nu_1}^2 - m_{\nu_2}^2}{M_W^2} \right) . \tag{2.7}$$

This branching ratio is, in any case, extremely small.
Even if we take m_1 = 0.5 MeV, $m_2 \simeq 0$, and $\sin\theta = 1/\sqrt{2}$,
which would give an upper limit, one finds $B \simeq 10^{-26}$.
For $m_1 \sim m_2 \sim$ a few eV (see astrophysical bound), one
has $B \sim 10^{-43}$. In any case: Even if a large violation
of muon number conservation is introduced by a large
leptonic mixing angle θ, the induced decay $\mu \to e\gamma$ is
by far too small to be measurable.

Neutrino oscillations

The existence of the leptonic mixing angle θ leads
to neutrino oscillations[13] with an oscillation length

$$L = \frac{p \, [\text{MeV}]}{4 \, |m_1^2 - m_2^2| \, [\text{eV}^2]} \cdot 10\text{m} \quad . \tag{2.8}$$

(p: neutrino momentum).

This formula is made suitable for application by
use of the relevant units. The best limit on $\nu_e - \nu_\mu$
oscillations at present comes from the Gargamelle
neutrino experiments using the ν_μ neutrino beam of the
CERN-PS. For maximal interference (θ = 45°) one has
$|m_1^2 - m_2^1| < 1$ eV^2.

New experiments, either with ν_μ beams in accelerator
laboratories or with $\bar{\nu}_e$'s from reactors, now being
prepared by several groups, will be able to improve this
limit considerably or, perhaps, will establish the
existence of neutrino oscillations.

h) Weak decays

One analysis about the relevance of MYTH for the
weak interactions, as observed in nature, brings up once
again all the odd things about the nonleptonic weak
interaction. There are two separate types of problems
related to the nonleptonic weak decays of strange particles.

1) In MYTH as in the conventional current-current
Hamiltonian, the nonleptonic weak interaction Hamiltonian
for strange particle decay consists of a $|\Delta I| = 1/2$ and
$|\Delta I| = 3/2$ part; the strong $|\Delta I| = 1/2$ enhancement as
observed in nature is not evident. The decay $K_s \to \pi\pi$
is forbidden in the SU_3 limit; nevertheless it is one
of the fastest weak decays observed. 2) The strength of
the nonleptonic $|\Delta I| = 1/2$ dominated $|\Delta S| = 1$ weak
interaction is much larger (by about a factor 15...20
in amplitude) than the strength of the $|\Delta S| = 1$ semi-
leptonic interaction. Strange particles almost never
decay semileptonically, (branching ratio $\sim 10^{-3}$ in case
of hyperons) unless they are inhibited to decay via the
$|\Delta I| = 1/2$ interaction like the K^+.

It has been shown that higher order QCD corrections
enhance the $|\Delta I| = 1/2$ term relatively to the $|\Delta I| = 3/2$
term. The effect can be calculated in the regime where
the QCD coupling constant is small such that the lowest
loop approximation to the renormalization group equation
can be applied. If one chooses this region to start at
1 GeV, (which may be too optimistic), the enhancement is[14]
approximately 3 for $g^2/4\pi$ (1 GeV) $\simeq 0.4$. This is much
too small to account for the observed $|\Delta I| = 1/2$
enhancement (~ 20 in amplitude). However, it must be
emphasized that this calculation accounts only for gluon
exchanges with relatively large momenta involved
($q^2 > M^2 \simeq 1$ GeV^2). Although gluon exchanges with
$q^2 < M^2$ cannot be treated quantitatively by perturbation
theory (this constitutes the long range part of the
problem) it may be that this long range contribution is
the relevant contribution enhancing the $|\Delta I| = 1/2$ term
by a factor approximately 20 compared to the $|\Delta I| = 3/2$
term.

Perhaps one way to understand the long range contribution of the $|\Delta I| = 1/2$ enhancement in case of hyperons is the argument of Ref. 15 where it was shown that the baryon matrix elements of the $|\Delta I| = 3/2$ term vanish in case of simple nonrelativistic SU_6 wave functions. This argument amounts to the assumption that the dominating reaction in a hyperon decay is the annihilation of a diquark involving s and the subsequent recreation of a diquark containing no s quark (see figure below). Thus $q\bar{q}$ creation diagrams are forbidden.

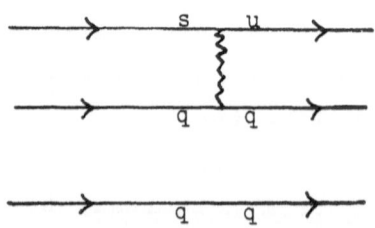

The diquark annihilation argument would have a good chance for being a correct explanation of the $|\Delta I| = 1/2$ rule for hyperon decays if one could show in addition that the weak decay described in the figure above indeed dominates over the ones where $q\bar{q}$ pairs are created. Since the annihilation of the (sq) diquark occurs at very short distances ($\sim 1/M_W$), one has to know the baryon wave function at the origin in order to make the argument more quantitative. Using PCAC the s-wave hyperon decay amplitudes can be expressed in terms of the baryon wave function at the origin:
$A \sim 1/F_\pi \ G_F \ \sin\theta_c \cdot |\psi_{sq}(r_{sq} = 0)|^2.$

Numerical estimates have been made recently by Donoghue, Holstein and Golowich[16] and C. Schmid[17]. In Ref. 16 $\psi(0)$ is calculated on the basis of the MIT bag approach. One finds only a relatively small enhancement

of the nonleptonic decay. However, the correct hyperon
decay amplitude is reproduced if radiative gluon
corrections are taken into account. Schmid uses the
observed charge radius of the nucleon in order to
estimate $\psi(0)$. He finds very good agreement with the
observed magnitude for Coulomb-type wave functions[17].

Summarizing the present situation for nonleptonic
strange particle decay one can say that the $|\Delta I| = 1/2$
rule, the $K_s \rightarrow \pi\pi$ problem, and the problems of nonleptonic weak decays are still problems which are not
fully understood. However the situation does not look
hopeless: an understanding of the mysteries of $|\Delta S| = 1$
nonleptonic decays may well be possible within MYTH.

The problem about the $|\Delta I| = 1/2$ enhancement of
the nonleptonic $|\Delta S| = 1$ Hamiltonian leads us to the
question about a possible enhancement of nonleptonic
decays of charmed particles. It seems now that charmed
particles have a fairly large semileptonic branching ratio
(of the order of 30% [see the following talk by F. Gilman]).
In fact, the semileptonic branching ratio is about the one
predicted by naive counting of degrees of freedom in free
quark theory (two lepton doublets, three colored quark
doublets), namely 40%. Thus essentially no enhancement
of nonleptonic decay amplitudes seems to be present; and
one is invited to speculate that the analog of octet
($|\Delta I| = 1/2$) enhancement for SU_4 (20 dominance over 84)
does not take place for the charm sector. Thus far no
real satisfactory solution for this paradox has been
offered[18]. However, it is quite possible that the long
range $|\Delta I| = 1/2$ enhancement discussed above is absent in
the charm case, and only the rather minor short range
QCD enhancement is left over.

i) Dimuons, trimuons...

If MYTH is correct, charm is the only new degree
of freedom which can be produced in high energy
neutrino-hadron scattering. No new leptons, no yet
further types of quarks are supposed to exist. It is
well-known that the dimuon events can be explained in
terms of charm excitation, assuming a semileptonic
branching ratio for charm decay of the order of 30-40%.
In MYTH there is only one source for the "wrong sign"
dimuons, namely charm-anticharm associated production
(D_0-$\bar{D}_0$ mixing is suppressed), which necessarily leads
also to trimuon events which have to be found. Taking
the data of Ref. 19, the "wrong" sign dimuon rate is of
the order of 0.1%, implying a cross section for $c\bar{c}$
associated production of the order of 0.5% for B = 0.2,
and a trimuon rate[20] of $2 \cdot 10^{-4}$.

Unfortunately there are other sources of trimuon
events, e.g., electromagnetic production of muon pairs,
which give rise to a trimuon rate in neutrino scattering
of the same order of magnitude. (In hadronic collisions
the prompt μ/π ratio is approximately 10^{-4}. It may be
that the prompt muon production is due to electromagnetic
dimuon production, in which case one is led to a trimuon
rate of order 10^{-4} in ν_μ and $\bar{\nu}_\mu$ scattering).

The new neutrino experiments now under way at CERN
and FNAL will, hopefully, clarify which picture of the
multilepton events in (anti)neutrino scattering is correct.

III. PROBLEMS FOR MYTH

MYTH has been and is an excellent model to confront
the real world with. But is it really true, or at least
approximately true? Let me list all those problems, or
possible problems, experimentalists and theorists have

collected as possible ammunition to execute MYTH
eventually.

a) Who understands CP-violation?

CP violation is mysterious for two reasons: 1) Why
is it there at all? and 2) Why is it so small? As
emphasized above, MYTH in its ground state discussed
thus far cannot accommodate CP violations. One possible
way out of this dilemma was suggested recently, namely
to expand MYTH to (MYTH)' by adding more Higgs scalars[21].
That way more space is generated to accommodate phase
parameters.

Such an approach can give, as especially emphasized
by Weinberg, some reason why the observed CP violation
is so small. The point is the Higgs boson exchange leads
to an effective Fermi interaction among the quarks and
leptons with an effective Fermi constant of order
$G_F \, m^2_{fermion}/m^2_{Higgs}$, which for reasonable values of the
mass parameters is typical of milliweak strength. Thus
even if the Higgs scalar interaction violates CP in a
maximal way, we would not be able to see this at low
energies, apart from milliweak effects. In MYTH the
term of the milliweak interaction introduced by Higgs
boson exchange which is relevant for $|\Delta S| = 1$ transitions
is

$$H^{|\Delta S|=1} \simeq G_F/m^2_H \, e^{i\delta} \, m_s \sin\theta_c \cos\theta_c$$

$$(m_c \bar{s}_R c_L \bar{c}_R d_L + m_u \cdot \bar{s}_R u_L \bar{u}_R d_L) + h.c. \, , \quad (3.1)$$

where δ is an unknown but generally big phase parameter
(for maximal CP violation $\delta = \pi/2$).

This Hamiltonian above consists of a $|\Delta I| = 1/2$ and
$|\Delta I| = 3/2$ term. In order to be consistent with experiment,

or in order to obtain the same type of predictions as
the superweak theory, one has to assume that the
$|\Delta I| = 1/2$ term dominates. <u>A new $|\Delta I| = 1/2$ rule is
required</u>. A possible solution to this problem has
recently been discussed by Minkowski[22], who argued that
the second term in the Hamiltonian (3.1) which is pure
$|\Delta I| = 1/2$ might actually dominate. (Matrix elements
of the second term taken between ordinary hadrons are
proportional to the number of $\bar{c}c$ pairs in their wave
functions and hence suppressed). He uses a similar method
as described in Ref. 23 where it was shown that in QCD
new terms involving gluon fields are generated. Specifically here the term

$$H^{induced} \sim \left(G_F / 4\pi^2 m_H^2 \right) e^{i\delta} \sin\theta_c \cos\theta_c m_c^2 \cdot$$

$$\cdot g_{st} \bar{s}_R \sigma_{\mu\nu} G^{\mu\nu} d_L + h.c.$$

($G_{\mu\nu}$: color octet gluon field strength) is generated whose
matrix elements are clearly enhanced compared to the
m.e. of the first term in eq. (3.1) (enhancement factor
$\sim m_c/m_u$).

As emphasized by Weinberg, at least three Higgs
doublets are needed to generate spontaneously a CP
violating phase by the scalar self interaction. <u>Thus a
modified MYTH accommodating several Higgs doublets is the
minimal theory to describe all known facts about particle
physics (except gravity)</u>.

Finally, we should like to add that even the $|\Delta I| = 1/2$
rule (and the dominant nonleptonic weak interaction as a
whole) might be due to Higgs boson exchange. (This point
was studied recently by C. Hill[24]).

b)   Trouble with antineutrinos

The recent experimental results in antineutrino-
hadron scattering at high energies indicate strong
violations of scaling[25], in particular a substantial
rise of the ratio $\sigma^{\bar{\nu}\mu^+}/\sigma^{\nu\mu^-}$ and the average value of
the variable y. Qualitatively such an effect is
expected in QCD, where the $\bar{q}q$ contribution to the
nucleon wave function increases with increasing energy[27].
However, it has been argued on the basis of detailed
calculations that the effect is too big to be
accommodated solely by QCD, and a new righthanded
coupling of a new heavy quark of charge -1/3 b to the
u-quark $(\bar{u}b)_R$, or a new righthanded coupling of a
new heavy quark N of charge -4/3 $(\bar{a}d)_R$, is needed in
order to explain the observed effect by the production
of a new quark flavor. Which of these two explanations
(scaling violation - new quark), if any, is correct
cannot be decided on the basis of present data. Of
course, if it turns out that a new flavor is produced,
MYTH would be wrong; in a year from now we shall know
the answer.

c)   Slow birth of a heavy lepton

In March 1977 it will be two years, after the
first rumors about possible heavy lepton physics at
SPEAR reached the physics community. Meanwhile the
μ-e events at SPEAR, which are supposed to originate
from the disintegration of a pair of new charged leptons
U_1 (mass ∿ 1.7...2 GeV) have become real physics,
studied both at SPEAR[28] and DORIS[29]. Thus the U particle
seems to exist. Is it indeed a heavy lepton? If yes,
an extension of MYTH to include new leptons and possibly
new quarks would be necessary.

d)   Is the muon partly an electron?

In the fifties, the decays $\mu \to e\gamma$ were looked for, and not found on the level predicted by theorists[30] (branching ratio $\alpha/\pi \sim 10^{-3}$). Subsequently a new particle (ν_μ) was invented to suppress this decay. Meanwhile the limit on the branching ratio for the $\mu \to e\gamma$ decay has become approximately $2 \cdot 10^{-8}$, and there are new experiments going on at the various pion factories to improve this limit.[*] Suppose this decay is found at the level of 10^{-9} or 10^{-10}. Is MYTH in trouble? It seems that this is the case. Remember: even a large violation of muon number conservation by the neutrino mass matrix as discussed above, does not lead to an appreciable $\mu \to e\gamma$ rate. Above we mentioned that there is only one possible source of CP violation in MYTH - the Higgs sector. We also learned that there have to exist several Higgs doublets for producing CP violation. Can we generate an appreciable $\mu \to e\gamma$ decay rate the same why? The answer is yes, as given recently by Bjorken and Weinberg[31]. If there is more than one Higgs doublet, not only CP violation, but also muon number nonconservation can occur. The largest contribution to the decay $\mu \to e\gamma$, in particular, is of order $(\alpha/\pi)^3 \sim 10^{-8}$ in branching ratio, which is close to the observed limit.

_____

[*] Our interest in the $\mu \to e\gamma$ decay was stimulated by the experiment currently going on at SIN where a signal has been observed which can be interpreted as a signal for the decay $\mu \to e\gamma$ (estimated branching ratio $\sim 10^{-9}$). (I am indebted to Drs. Eichenberger and Povel for this private communication).

IV. BEYOND MYTH

a) How to live with the U lepton?

What kind of modification of MYTH would be necessary if the U particle is a new lepton? This question cannot be answered without having detailed additional information about this new lepton. Some we already have:

a) Mass $\sim$ 1.7...2 GeV.

b) U^- decays part of the time via a three-body decay into e^- and μ^-: $U^- \rightarrow e^-(\mu^-) + N_1 + N_2$, where N_1, N_2 are basically noninteracting light neutral objects, (observed branching ratio $\sim$ 15...20%)[28,29].

The U decay is likely to be a weak decay. Thus one of the neutral objects is likely to be $\bar{\nu}_e$ ($\bar{\nu}_\mu$), the other we may identify with the light neutral lepton associated with U^- (ν_μ).

It has often been incorrectly stated, that the addition of a new charged lepton to the set of observed leptons required the existence of other quarks, since otherwise in a gauge theory approach the existence of anomalies would render the theory nonrenormalizable. A few counter examples:

1. We add the doublets $\begin{pmatrix} \nu_U \\ U^- \end{pmatrix}_L \begin{pmatrix} \nu_U \\ U^- \end{pmatrix}_R$; ν_U is a Fermi-Dirac neutrino. The weak current involving U^- is a vector current.

2. We add a triplet: $\begin{pmatrix} U^+ \\ \nu_U \\ U^- \end{pmatrix}_L$.

In both cases no anomalies are introduced.

b) Sequential QFD

Centainly the most naive extension of MYTH would be to add more lefthanded lepton and quark doublets (altogether n lepton and quark doublets)[32]:

$$
\begin{pmatrix} u & c & t & \dots \\ d' & s' & b' & \dots \end{pmatrix}
\qquad
\begin{pmatrix} \nu_e & \nu_\mu & \nu_U & \dots \\ e^- & \mu^- & U^- & \dots \end{pmatrix}
$$

$$
\begin{pmatrix} d' \\ s' \\ b' \\ \cdot \\ \cdot \\ \cdot \end{pmatrix}
= U
\begin{pmatrix} d \\ s \\ b \\ \cdot \\ \cdot \\ \cdot \end{pmatrix}
$$

U: unitary matrix.

There are n^2 real parameters describing the unitary matrix U of which $2n-1$ can be absorbed by fixing the relative phases of the 2n quark fields. Thus there are $(n-1)^2$ real parameters. We can rewrite the matrix U as a modified orthogonal matrix where the elements are multiplied by phases $e^{i\delta}$. One has $1/2\, n(n-1)$ generalized Euler angles and $1/2\,(n-1)(n-2)$ phases. In MYTH there is, of course, only one angle (Cabibbo angle), and no phase parameter. No CP violation via weak mixing parameters is possible in MYTH.

In the three doublet case

$$
\begin{pmatrix} u & c & t \\ d' & s' & b' \end{pmatrix}_L
\qquad
\begin{pmatrix} \nu_e & \nu_\mu & \nu_U \\ e^- & \mu^- & U^- \end{pmatrix}_L
$$

one is dealing with three "Euler-Cabibbo angles" and one phase[33].

If we generalize MYTH to n =3, one faces one important problem. In general there are three rotation angles, and not only one. From experiment we know $d' \simeq d \cos\theta_c + s \sin\theta_c$. This Cabibbo-type universality starts to become an accident; it becomes "unnatural." Though nature is special, and it may well be that it has chosen to mix d only with s in the weak current, and only very little with b. <u>If there are more than four quarks, the Cabibbo universality of the weak currents is expected not to be exact</u>. To search for a violation of universality would be a useful task for experimentalists. <u>CP violation</u>[33]: The six-quark six-lepton model of the type discussed above allows CP violation: one phase is available. The resulting picture of CP violation is very similar to the super-weak theory. Since in the six-quark model it can be arranged that the phase parameter causing CP violation appears only in weak currents connecting the light with the heavy quarks, the CP violation dominates in the $|\Delta I| = 1/2$ channel. This leads in particular to $|\epsilon'/\epsilon| \ll 1$. The ϵ-parameter for the K_0-$\bar{K}_0$ system is of the order $\sin\delta \cdot \sin\theta_3$ where θ_3 is the (s-b) mixing angle (which has to be relatively small, if we want to preserve the Cabibbo universality ($\sin\theta_3 \lesssim 0.25$)). Thus the smallness of CP violation is not really understood (as in most other models of CP violation). However, it is related to the s-b mixing angle. If one could find a theoretical reason why this angle must be small ($\sin\theta_3 \lesssim 10^{-2}$), the smallness of CP violation would become plausible.

The dipole moment of the neutron is very small. For $m_t \sim m_b \sim 5 - 10$ GeV it is of order 10^{-30} cm.

c) Vectorlike weak interactions[34]

A vectorlike theory is a theory in which the parity violation is not intrinsic, but due to the presence of parity violating terms in the fermion mass matrix. In the limit where the fermion masses are set to zero the theory is a pure vector theory, and the gauge currents are vectorial. If the fermion mass matrix is turned on, it acquires both scalar and pseudoscalar terms, and parity is violated. The unitary transformation in the fermion space which diagonalizes the fermion mass matrix and eliminates its pseudoscalar part (in the absence of CP violation) introduces apparent axial vector currents.

In the conventional gauge theory framework the fermion masses are generated by the coupling of the fermion fields to scalar fields which develop non-zero vacuum expectation values ("spontaneous symmetry breaking").

In a vectorlike theory the violation of parity is part of the spontaneous symmetry breaking, in contrast to MYTH where the parity violation is an intrinsic property of the field equations.

d) Vectorlike Models based on SU_2^W doublets only

A radical possibility to construct vectorlike schemes is to place all leptons and quarks in SU_2^{weak} doublets. The smallest vectorlike scheme of this sort is one based on six lepton flavors and six quark flavors[34]:

$$\begin{pmatrix} u & t & c \\ d' & b & s' \end{pmatrix}_L \qquad \begin{pmatrix} u & t & c \\ b & d'' & s'' \end{pmatrix}_R$$

$$\begin{pmatrix} \nu_e & N_E & \nu_\mu \\ e^- & E^- & \mu^- \end{pmatrix}_L \qquad \begin{pmatrix} \nu_e & N_E & N_\mu \\ E^- & e^- & \mu^- \end{pmatrix}_R \quad . \qquad (4.1)$$

Here t and b are new flavors of quarks ("top" and "bottom") with charges 2/3 and -1/3, respectively. The lefthanded quarks are rotated by the Cabibbo angle θ_c into d' = d $\cos\theta_c$ + s $\sin\theta_c$, and s' = d $\sin\theta_c$ + s $\cos\theta_c$. The observed universality of the weak interactions requires that there is no (or only little) mixing of d' with b. There may, however, be substantial mixing between b and s', i.e., the fields b and s' in the scheme above may be replaced by the rotated fields $\tilde{b}$ = b $\cos\tilde{\theta}$ + s' $\sin\tilde{\theta}$ and $\tilde{s}'$ = -b $\sin\tilde{\theta}$ + s' $\cos\tilde{\theta}$. For simplicity we have not displayed this rotation in the scheme above.

Among the righthanded quarks, the observed V-A structure of the weak currents connecting the u quark with the d' combination forbids any appreciable coupling of u_R to either d_R or s_R. Consequently its partner must be pure or nearly pure b. The linear combinations d" = d $\cos\theta"$ + s $\sin\theta"$ and s" = -d $\sin\theta"$ + s $\cos\theta"$ are, for the moment, left arbitrary, (one can argue in favor of $\theta" \approx 0$)[34].

The scheme contains, in addition to the conventional leptons, one new charged lepton M^-, one new massive neutral lepton N_M and the righthanded partner $(\nu_e)_R$ of the electron neutrino as well as the massive Majorana partner of the muon neutrino ν_μ. We assume that the neutral lepton N_M is a Fermi-Dirac particle (its mass term conserves lepton number). The fields $(\nu_e)_L$, $(\nu_e)_R$ and $(\nu_\mu)_L$ are assumed to be massless in the absence of the weak interactions. The violation of lepton number conservation occurs only via the N_μ mass term. Thus neutrinoless double β decay does not occur in lowest order of the weak interaction. However, there exists the decay $K^- \rightarrow \pi^+ + \mu^- + \mu^-$, which occurs in second

order of the weak interactions. (This decay was
estimated in Ref. 35). The expected branching ratio
$K^- \to \pi^+ + \mu^- + \mu^- / K^- \to$ all is of the order of 10^{-14} for
$M_{N_\mu} \sim$ several GeV , i.e., more than ten orders of
magnitude smaller than the experimental limit. It
remains to be seen, if the accuracy of the experiments
can be improved such that an experimental test of the
lepton number violation of the kind we are discussing
becomes feasible.

In the scheme (4.1) all lepton and quark fields
take part in the weak interaction. Consequently the
neutral current is a vector current:

$$J_\mu^{neutral} = (-\tfrac{1}{2}+z)\bar{e} + (\tfrac{1}{2} - \tfrac{2}{3}z)\bar{u}\gamma_\mu u +$$

$$(-\tfrac{1}{2} + \tfrac{1}{3}z)\bar{d}\gamma_\mu d (z = \sin^2\theta) \qquad (4.2)$$

+ terms, involving the other fermions (θ: $SU_2 \times U_1$
mixing angle).

A direct generalization of the six-flavor scheme
(4.1) is the following eight-flavor scheme:

$$\begin{pmatrix} u & t & c & \nu \\ d' & b & s' & h \end{pmatrix}_L \quad \begin{pmatrix} u & t & c & \nu \\ b & d & h & s \end{pmatrix}_R$$

$$\begin{pmatrix} \nu_e & N_E & \nu_\mu & N_M \\ e^- & E^- & \mu^- & M^- \end{pmatrix}_L \quad \begin{pmatrix} N_E & \nu_e & N_\mu & \nu_\mu \\ e^- & E^- & \mu^- & M^- \end{pmatrix}_R , \qquad (4.3)$$

where we have not indicated the possibility to have weak
angles between the various leptons and quarks. The

eight-flavor scheme contains two new charged leptons E^-, M^-, and two new massive leptons N_E, N_M (Fermi-Dirac spinors), and five new quark flavors. The essential difference between the six-flavor and the eight-flavor scheme is that there is no need to introduce massive Majorana neutrals. The lepton mass matrix need not be lepton number violating.

e) SU_2^W doublets and singlets

Apart from the pure vector model discussed above, many other vectorlike theories can be constructed, in particular models with contain both SU_2 doublets and singlets. Examples[34,36]:

Quarks

	Doublets		New heavy quarks of charge		
			2/3	-1/3	-4/3
A	$\begin{pmatrix} u & c \\ d' & s' \end{pmatrix}_L$	$\begin{pmatrix} u & t \\ b & d \end{pmatrix}_R$	c,t	b	-
B	$\begin{pmatrix} u & c \\ d' & s' \end{pmatrix}_L$	$\begin{pmatrix} u & c \\ b & h \end{pmatrix}_R$	c	b,h	-
C	$\begin{pmatrix} u & c \\ d' & s' \end{pmatrix}_L$	$\begin{pmatrix} u & c \\ b & s \end{pmatrix}_R$	c	b	-
D	$\begin{pmatrix} u & c \\ d' & s' \end{pmatrix}_L$	$\begin{pmatrix} t & c \\ d & s \end{pmatrix}_R$	c,t	-	-
E	$\begin{pmatrix} u & c \\ d' & s' \end{pmatrix}_L$	$\begin{pmatrix} t & v \\ b & h \end{pmatrix}_R$	c,t,v	b,h	-
F	$\begin{pmatrix} u & c \\ d' & s' \end{pmatrix}_L$	$\begin{pmatrix} t & c \\ b & s \end{pmatrix}_R$			

$$G \begin{pmatrix} u & c \\ d' & s' \end{pmatrix}_L \begin{pmatrix} d & s \\ a & b \end{pmatrix}_R \qquad c \qquad - \qquad a,b$$

$$H \begin{pmatrix} u & c \\ d' & s' \end{pmatrix}_L \begin{pmatrix} t'' & c \\ d & s \end{pmatrix}_R \qquad c,t \qquad -$$

$$I^{7,37)} \begin{pmatrix} u & c & t \\ d' & s' & b' \end{pmatrix}_L \begin{pmatrix} u'' & c'' & t'' \\ b & s & d \end{pmatrix}_R (v'') \qquad c,t,v \quad b \qquad -$$

Leptons:

	Doublets	New leptons of charge	
		0	−1

$$J \begin{pmatrix} \nu_e & \nu_\mu \\ e^- & \mu^- \end{pmatrix}_L \begin{pmatrix} \nu_e & \nu_\mu \\ E^- & M^- \end{pmatrix}_R \qquad\qquad\qquad E^-, M^-$$

$$K \begin{pmatrix} \nu_e & \nu_\mu \\ e^- & \mu^- \end{pmatrix}_L \begin{pmatrix} E^o & M^o \\ e^- & \mu^- \end{pmatrix}_R \qquad E^o, M^o \qquad\qquad -$$

$$L \begin{pmatrix} \nu_e & \nu_\mu \\ e^- & \mu^- \end{pmatrix}_L \begin{pmatrix} E^o & M^o \\ E^- & M^- \end{pmatrix}_R \qquad E^o, M^o \qquad\qquad E^-, M^-$$

$$M \begin{pmatrix} \nu_e & \nu_\mu \\ e^- & \mu^- \end{pmatrix}_L \begin{pmatrix} E^+ \\ E^o \\ e^- \end{pmatrix}_R \begin{pmatrix} M^+ \\ M^o \\ \mu^- \end{pmatrix}_R$$

f)   Comparison with experiment

If we compare any of the models above with experiment, we have to take into account expecially the following facts:

1)   Apart from the c-quark there is no evidence at SPEAR for the existence of further quarks, which would manifest themselves by new narrow $\bar{q}q$ states in the e^+e^- channel.   These new types of heavy quarks must have an effective mass > 4 GeV.   (This constraint is somewhat marginal in case of charge -1/3 quarks).

2)   One, perhaps two, new charged leptons may exist with masses $\sim$ 2 GeV.

3)   The results of the inclusive neutral current experiments on isoscalar targets are

	Gargamelle[38]	HPWF[39]	CITF[40]
$R^{\nu N}$	0.25 ± 0.04	0.31 ± 0.06	0.24 ± 0.02
$R^{\bar{\nu} N}$	0.39 ± 0.06	0.39 ± 0.10	0.39 ± 0.06

In order to obtain information about the structure of the neutral current it is important to know the ratio $\sigma^{\bar{\nu}\nu}/\sigma^{\nu\nu} = (R^{\bar{\nu}}/R^{\nu})\sigma^{\bar{\nu}\mu^+}/\sigma^{\nu\mu^-}$.   In order to avoid the problems associated with the observed energy dependence of $\sigma^{\bar{\nu}\mu^+}/\sigma^{\nu\mu^-}$ at very high energies we base ourselves on the Gargamelle data which give $\sigma^{\bar{\nu}\nu}/\sigma^{\nu\nu} = 0.59 \pm 0.14$.

4)   The neutral current cross section ratio $R_e = \sigma^{\nu\mu e^-}/\sigma^{\bar{\nu}\mu e^-}$ is measured to be $\sim$ 1/2 (with large uncertainty, the case $R_e = 1$ is still not excluded[41]).

The Gargamelle group reports a cross section[42] $\sigma^{\bar{\nu}\mu e^-} \simeq (1.3 \pm 0.8) \, 10^{-42} \, E_\nu [\text{GeV}] \text{cm}^2$ (this result is corrected for loss of efficiency on the basis of an essentially constant y-distribution (within the minimal

$SU_2 \times U_1$ theory and $\sin^2\theta_w = 0.4$). An admixture of
$(1-y)^2$ in the y distribution will increase the cross
section slightly, but not significantly). The correct
cross section reported by the AAchen-Padova group is
in agreement with the Gargamelle value[43].

The results of the experiment of Reines et al.[44],
for the process $\bar{\nu}_e e^- \to \bar{\nu}_e e^-$ are also consistent with the
Gargamelle result, if one applies μ-e universality to
the neutral current interaction.

I. The Neutral Current

We study first the hadronic cross sections under
the simplifying assumption that the $\bar{q}q$ pairs in the
nucleon wavefunction can be neglected (valence quark
approximation), and discuss the following four distant
cases neglecting all weak doublets which do not contain
u or d quarks and setting $\theta_c = 0$.

		α	β
I	$\begin{pmatrix} u \\ d \end{pmatrix}_L$	0	0
II	$\begin{pmatrix} u \\ d \end{pmatrix}_L \begin{pmatrix} u \\ b \end{pmatrix}_R$	1	0
III	$\begin{pmatrix} u \\ d \end{pmatrix}_L \begin{pmatrix} t \\ d \end{pmatrix}_R$	0	1
IV	$\begin{pmatrix} u \\ d \end{pmatrix}_L \begin{pmatrix} u \\ b \end{pmatrix}_R \begin{pmatrix} t \\ d \end{pmatrix}_R$	1	1 .	(4.4)

The effective Hamiltonian is given by

$$H^{weak} = \frac{4G}{\sqrt{2}}(j_\mu^+ j_\mu^- + \rho \cdot j_\mu^n j_n^\mu) \quad \rho = \frac{M_w^2}{M_Z^2 \cos^2\theta}$$

$$j_\mu^n = j_\mu^3 - \sin^2\theta_w j_\mu^e \tag{4.5}$$

(Note: Here we allow the Z mass to be unconstrained
and introduce another free parameter ρ; in MYTH one
has $\rho = 1$ due to the Higgs doublet structure).

The various possibilities can be rewritten in
compact notation by the following doublet:

$$\binom{u}{d}_L \cdot 1 + \binom{u}{b}_R \alpha + \binom{t}{d}_R \beta \quad , \tag{4.6}$$

where the parameters α, β take the values $(0,1)$ and are
given for the various cases in the above table.   In the
valence quark approximation $(\bar{u}=\bar{d}=s=\bar{s}=0)$ one finds

$$R^\nu = \rho^2 \left[\frac{1}{2} + \frac{1}{12}\alpha^4 + \frac{1}{12}\beta^4 - z(1 + \frac{2}{9}\alpha^2 + \frac{1}{9}\beta^2) + \frac{20}{27}z^2\right],$$

$$R^{\bar{\nu}} = \rho^2 \left[\frac{1}{2} + \frac{3}{4}\alpha^4 + \frac{3}{4}\beta^4 - z(1 + 2\alpha^2 + \beta^2) + \frac{20}{9}z^2\right] ,$$

$$\frac{\sigma^{\bar{\nu}\bar{\nu}}}{\sigma^{\nu\nu}} = \frac{\frac{1}{6} + \frac{1}{4}(\alpha^4+\beta^4) - z(\frac{1}{3} + \frac{2}{3}\alpha^2 + \frac{1}{3}\beta^2) + \frac{20}{27}z^2}{\frac{1}{2} + \frac{1}{12}(\alpha^4+\beta^4) - z(1 + \frac{2}{9}\alpha^2 + \frac{1}{9}\beta^2) + \frac{20}{27}z^2} \quad . \tag{4.7}$$

The ratios $\sigma^{\bar{\nu}\mu\bar{\nu}\mu}/\sigma^{\nu\mu\nu\mu}$ for the various possibilities
I through IV are displayed in Figure 1.
We conclude: Case I (MYTH and models E,F) is in good
agreement with the data for $\sin^2\theta_w \simeq 0.3\ldots0.4$.   Case II
(models B,C) is in good agreement with the data for
$\sin^2\theta_w \simeq 0.0\ldots0.6$; note this wide range.   Case III (model
D) is in agreement with the data for $\sin^2\theta_w < 0.05$.   If
the $\bar{q}q$ sea is added, the curve of case III moves slightly
above the dashed area, thus for $\sin^2\theta_w \lesssim 0.05$ the deviation
of $\sigma^{\nu\mu\bar{\nu}}/\sigma^{\nu\mu\nu\mu}$ from the measured value is slightly more
than one standard deviation.   Case IV (model A and pure
vector model) are excluded by three standard deviations.

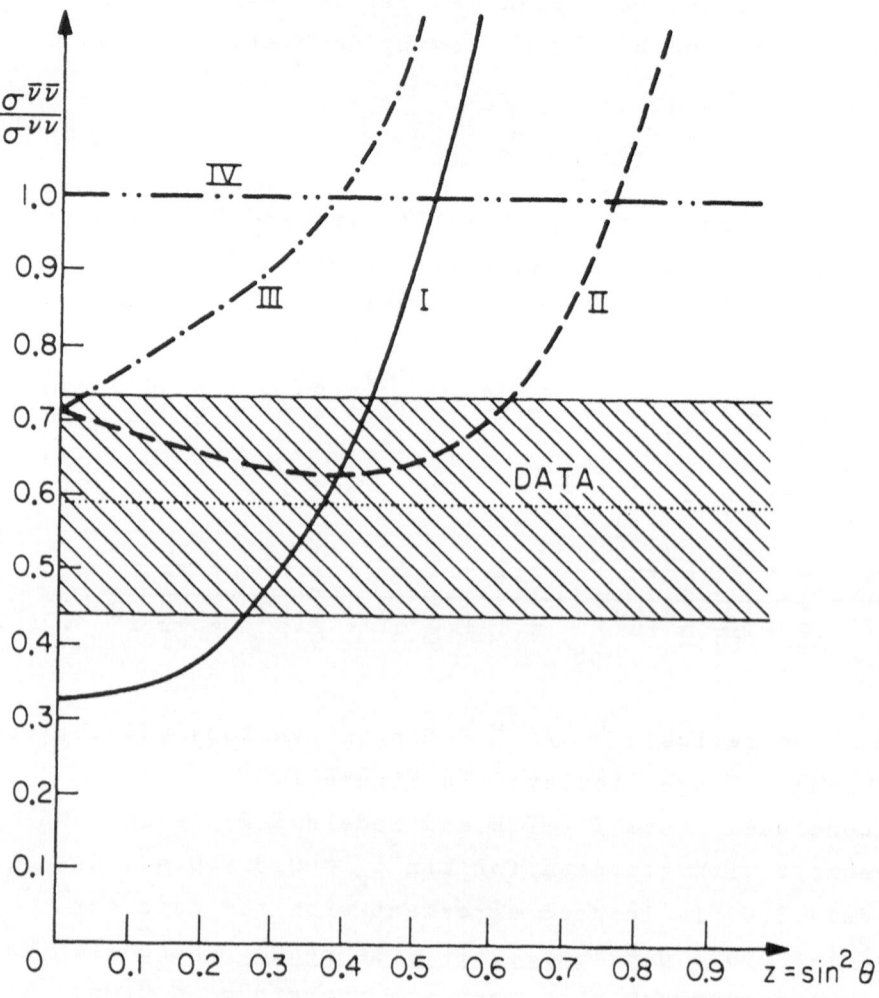

Figure 1. $\dfrac{\sigma^{\bar{\nu}\bar{\nu}}}{\sigma^{\nu\nu}}$ for possibilities I - IV in text.

II. Leptonic Neutral Current Cross Sections

Studying the implications of the various $SU_2 \times U_1$ models for leptons, we have to keep in mind that the leptons cannot be treated independently of the hadrons, at least not at the present stage where the data very scarce. The cross section $\sigma^{\bar{\nu}_\mu e^-}$ depends on ρ and $\sin^2\theta_w$. In order to get rid of the strength parameter ρ one has to use the hadronic data in addition. I have carried out such an analysis for the case where the lefthanded electron acts as the $I_z^{weak} = -1/2$ member of a SU_2^{weak} doublet, and the righthanded electron has an arbitrary I_z^{weak}. The result is as follows: All values of I_z^{weak} (e^-_R) are excluded except I_z^{weak} (e^-_R) = 0 (the minimal case) or I_z^{weak} (e^-_R) = -1/2 (the vector case).

Thus, in particular SU_2^{weak} triplet assignments for e^-_R are excluded; such possibilities were discussed within unified theories based on SU_3^{weak}[34,45) or the exceptional groups[46). The lepton assignment as given in Ref. 45 gives $I_z^{weak}(e^-_R) = -1/2$; this is still consistent with experiment, however, it is threatened by the result of the Aachen-Padova group[41).

III. New Charged Currents

Many of the $SU_2 \times U_1$ theories involving new leptons and quarks, especially the vectorlike theories, introduce new righthanded currents carrying a valence quark u or d into a new heavy quark $((\bar{t}d)_R, (\bar{u}b)_R)$. At the moment there is no evidence of a $(\bar{t}d)_R$ current. The so-called $\bar{\nu}$-anomaly can be understood if the current $(\bar{u}b)_R$ is present and if $m_b \simeq 5$ GeV[25). However, I think at the moment one has to reserve judgement with respect to this question - the quoted experimental errors in the $\bar{\nu}_\mu$ experiments are substantial, and fits including QCD corrections to scaling can describe the experimental results within the quoted errors[47]. Only new experiments with much better statistics can tell if b quark excitation takes place in $\bar{\nu}_\mu$ scattering or not.

New currents add new terms to the nonleptonic weak Hamiltonian which may be relevant for the nonleptonic weak decays of strange (and charmed) particles. However, I expect a four-quark operator like $\bar{s}h\bar{h}d$ (h: heavy quark) to contribute very little to the nonleptonic strange particle decay, due to the absence of heavy quark flavors in normal hadrons. Recently it has been emphasized that new right-handed currents involving heavy quarks can lead in QCD to the appearance of a new term in the nonleptonic Hamiltonian which may be important and even dominating, namely the quark bilinear $\bar{s}\sigma_{\mu\nu}G^{\mu\nu}d$ ($G^{\mu\nu}$: gluon field strength, summation over color understood). For example, in scheme C the new right-handed current $(\bar{c}s)_R$ is present, which generates the new term[48]

$$H^{\text{nonlept.}} = - G/\sqrt{2} \sin\theta \cdot m_c \frac{g}{4\pi^2} \bar{s}\sigma_{\mu\nu}G^{\mu\nu}\frac{1+\gamma_5}{2}d + \text{h.c.}$$

(it corresponds to the decay s → d + gluon, g: gluon-quark coupling constant). The important features of this new term are the following:

i)   It contains no heavy quark fields. Thus its matrix elements between normal hadrons can be big.

ii) It is enhanced by m_c.

iii) It is pure $|\Delta I| = 1/2$ and has all the desired properties for a nonleptonic weak Hamiltonian.

A consistent picture of all nonleptonic decays of
strange particles emerges if one assumes that the term
in the equation above dominates the decays. Specific
wave function calculations performed recently within
the MIT bag model indicate that this assumption may be
consistent[49].

We emphasize that the new term described above is
enhanced by the large value of m_c. A similar enhancement
would not occur in the case of charmed particle decay.
Thus charmed particles are supposed to have a large
semileptonic decay mode, in agreement with recent
estimates on the basis of the observed dimuon events
in ν_μ hadron scattering.

Independent of the $|\Delta I| = 1/2$ rule argument, it is
important to check if the current $(\bar{c}s)_R$ is present in
the decays of charmed particles. It would, for example,
influence the nonleptonic decays of charmed particles
(especially the F meson decay) as well as the semileptonic decays of charmed baryons.

IV. The New Lepton Physics

1. A righthanded current in U lepton decay?

One particular question which arises with respect
to the new charged U lepton is the one about the
structure of its decay amplitude. If U is a sequential
heavy lepton (i.e., essentially a heavy muon), it
would decay via a lefthanded current. However, in many
vectorlike theories the U lepton decays via a right-
handed current. A careful study of the e^- - μ events
and of the cross section for inclusive μ production in
e^+e^- annihilation will allow us to answer that question
in the near future.

2. Heavy neutral leptons

Heavy neutral leptons are incorporated in many

vectorlike theories. Certainly they are the most
peculiar states in the zoo of new elementary fermions
theorists speculate about, and the ones most difficult
to see. The present experimental limits on the
existence of heavy neutral leptons are very poor. If
such leptons are coupled by a charged weak current to
the electron or muon (e.g. by $\begin{pmatrix} N \\ e^- \end{pmatrix}_R$, $\begin{pmatrix} N \\ \mu^- \end{pmatrix}_R$), the
limits are $m_N > 500$ MeV or $m_N > 400$ MeV respectively
(from kaon decay). Such neutral leptons could be
produced in the weak decays of charmed particles or
in the U particle decay, and would manifest themselves by multilepton events; experimentalists should
be aware of such events.

    If massive neutral leptons exist, but are not
(or only very weakly) coupled by a charged current to
e^- and μ^-, there are at present essentially no
experimental limits. There could exist a massive neutral
lepton of mass 50 MeV, which might either be stable or
decay slowly, and we would not have seen it. There is
only one possibility to discover such leptons, namely
in e^+e^- annihilation via the weak neutral current. If
a neutral lepton N belongs, e.g., to an SU_2^{weak} doublet
representation, the $\bar{N}N$ pair production at the highest
PETRA or PEP energies will be approximately 20...30% of
one unit in R. If N is not too massive and is coupled
to μ^- (e.g., $\begin{pmatrix} N \\ \mu^- \end{pmatrix}_R$), events of the type $e^+e^- \rightarrow (\mu^- + \pi^+)_{jet}$
$+ (\mu^+ \pi^-)_{opposite\ jet}$ $+$ no missing energy would occur.
Most interesting would be the production of Majorana
neutrals. Since a Majorana mass term violates lepton
number conservation maximally, one is led to rather
spectacular events, e.g., $e^+e^- \rightarrow (\mu^- \pi^+)_{jet} + (\mu^- \pi^+)_{opposite\ jet}$. [34)]
    A neutral lepton which is entirely uncoupled from
e^- and μ^-, is either stable, or decays into a lower-lying

member of its SU_2^W representation, which then might find its way to disintegrate into hadrons or light leptons.

3. Radiative decay of the U lepton

It has been emphasized by various authors[50], that the U lepton might have a large radiative decay mode, e.g., $(U^- \to e^-\gamma, U^- \to \mu^-\gamma)$. These decays occur typically in vectorlike theories, involving heavy neutral leptons. For example the scheme $\begin{pmatrix} \nu_\mu & N \\ \mu^- & U^- \end{pmatrix}_L$ $\begin{pmatrix} \nu_\mu & N \\ U^- & \mu^- \end{pmatrix}_R$ would lead to the decay $U^- \to \mu^-+\gamma$ (via the process described in the figure below.

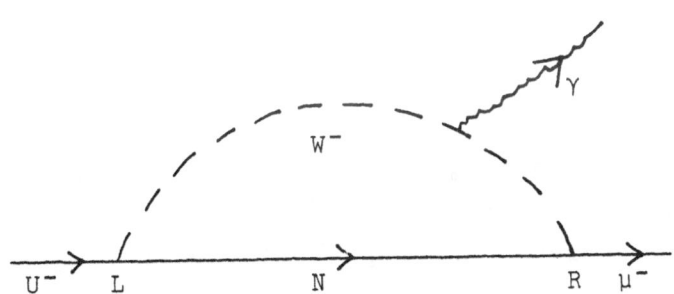

The ratio $\rho_\mu = \Gamma(U^- \mu^-\gamma)/\Gamma(U^- \to \nu_{\mu R} + \mu^-+\nu_{\mu L})$ is $6\,\alpha/\pi \cdot (m_N/m_{U^-})^2$, i.e., very sensitive to the mass of the neutral lepton N (see table below):

m_{U^-} = 1.9 GeV

m_N [GeV]	1	2	5	10	15
ρ_μ	0.004	0.01	0.09	0.37	0.82

If m_N is indeed several GeV, the radiative U decay constitutes a sizable fraction of the U^- leptonic decay. It is important to look for radiative decay modes in U decay.

V.   Muon Number Nonconservation

In the presence of new heavy leptons the possible decay $\mu \to e\gamma$ and muon number nonconservation in general have to be studied again. If U^- is a sequential heavy lepton, and ν_{U^-} massless (or very light), the situation with respect to muon number nonconservation is basically the same as in the minimal four-lepton scheme. However the situation changes considerably if U^- comes together with a new massive neutral lepton:

$$\begin{pmatrix} \nu_e{}' & \nu_\mu{}' & N' \\ e^- & \mu^- & U^- \end{pmatrix}_L .$$

This case was recently studied by the author[51]. If $m_N > m_{U^-}$, the U^- decay can only proceed via (small) weak mixing angles, implying a relatively large lifetime for U^- ($\tau > 10^{-11}$ sec). By the same mixing the decay $\mu \to e\gamma$ will occur, with a branching ratio of order $10^{-7} \cdot (m_N/M_w)^4$, which is e.g., approximately 10^{-9} for $m_N \simeq 15$ GeV.

In many vectorlike schemes[34] the righthanded electron and muon are coupled to heavy neutral leptons, which in principle can mix with each other (scheme C above):

$$\begin{pmatrix} \nu_e & \nu_\mu \\ e^- & \mu^- \end{pmatrix}_L \begin{pmatrix} N_e{}' & N_\mu{}' \\ e^- & \mu^- \end{pmatrix}_R \quad \begin{aligned} N_e{}' &= \cos\theta N_e + \sin\theta N_\mu \\ N_\mu{}' &= \perp \end{aligned} .$$

Such a scheme will lead to muon number non-conservation, as emphasized recently by Cheng and Li[52], and the author[53]. The branching ratio for the $\mu \to e\gamma$ decay is of the order of $(\alpha/\pi)(\cos\theta \, \sin\theta \cdot \dfrac{m_{N_e}^2 - m_{N_\mu}^2}{M_W^2})^2$, which is in the region accessible by experiment ($\gtrsim 10^{-11}$) for m_N^2 approximately a few GeV^2.

We should like to add one important remark about the scheme discussed above. The massive righthanded neutral leptons N_e, N_μ are coupled to the righthanded charged fermions. In principle it is possible that there exists also lefthanded couplings. If we want to preserve the univerality of the weak interactions, such couplings must be small. Suppose we consider an admixture of $N_{e(\mu)L}$ to the neutral weak partner of e_L or μ_L, say the scheme

$$\begin{pmatrix} \nu_e + \varepsilon & N_e & \nu_\mu \\ & & \\ e^- & & ,\mu^- \end{pmatrix}_L \begin{pmatrix} N_e' & N_\mu' \\ & \\ e^- & \mu^- \end{pmatrix}_R. \qquad (\varepsilon \lesssim 0.1)$$

Such a scheme would lead immediately to an unacceptably large $\mu \to e\gamma$ rule. The branching ratio for the $\mu \to e\gamma$ decay can easily be calculated (see Ref. 50). One finds

$$B = 6\left(\frac{\alpha}{\pi}\right) \varepsilon^2 \cdot \left(\frac{m_N}{m_\mu}\right)^2 \cdot \sin^2\theta \quad,$$

which would imply $\varepsilon \lesssim 10^{-4}$ for $m_N \sim 2$ GeV and $\sin^2\theta \sim O(1)$. This constraint on the smallness of ε is so powerful that it can hardly be so small by accident. We have to require that all LR contributions cancel automatically. As noted recently [54], this is the case in one interesting situation, namely if the e- and μ-mass terms arise by the vacuum expectation value of an SU_2 singlet Higgs scalar. In this case the fermion mass term can be written as

$$m_\mu \, (\mu^-_R{}^*\mu_L + N^*_\mu{}'_R \nu_{\mu L}) + h.c. + e \rightarrow \mu + \bar{N}N\text{-terms}.$$

The off-diagonal terms $\bar{N}\nu$ lead to N-admixtures in the neutrino states. The diagonalization of the neutral lepton mass matrix yields:

$$\begin{pmatrix} \nu_e + \cos\theta \cdot \dfrac{m_e}{m_{N_e}} \cdot N_e + \sin\theta \cdot \dfrac{m_e}{m_{N_\mu}} \cdot N_\mu & \nu_\mu - \sin\theta \cdot \dfrac{m_\mu}{m_{N_e}} N_e + \cos\theta \dfrac{m_\mu}{m_{N\mu}} N_\mu \\[2ex] e^- & \mu^- \end{pmatrix}_L .$$

It can be seen easily that the scheme above leads to an exact cancellation of the L-R terms up to order G^2. (Note: The L-R amplitudes are proportional to m_N, the N admixtures to the neutrinos are proportional to $1/m_N$. Thus the final result is independent of m_N. However, in the limit $m_{N_e} = m_{N_\mu}$ the angle θ becomes undefined and can be set to zero, thus in this case the $\mu \rightarrow e\gamma$ amplitude vanishes. Since, on the other hand, the result is mass independent, the cancellation remains intact even for $m_{N_e} \neq m_{N_\mu}$).

The scheme discussed above leads within an $SU_2 \times U_1$ theory to a vectorial leptonic neutral current, a possibility which is at the edge of being excluded by experiment [41].

5. Is Nature "Natural?"

Glashow and Weinberg, and in some modified form, Paschos, have recently proposed a new principle of QFD:

the absence of $|\Delta S| = 1$ neutral currents of order G and
$G\alpha$ is required to be "natural," i.e., independent of
the special choice of parameters in the theory[7]. This
principle requires in particular, that all quarks of
charge $-1/3$ must have the same values of T_3^W and $\vec{T}^2$. A
similar requirement may hold for the quarks of charge
$2/3$, although here the phenomenological situation is
less clear.

If we apply "naturalness" for quarks of charge
$-1/3$, very few models besides the minimal one are left
for serious consideration, namely scheme (D,H) and
extensions of this scheme, e.g., the seven-quark
extension of the six-quark vector theory (scheme I)

$$\begin{pmatrix} u' & c' & t' \\ d & s & b \end{pmatrix}_L \begin{pmatrix} u'' & c'' & t'' \\ d & s & b \end{pmatrix}_R (v_L', v_R'')$$

(v: heavy quark of charge $2/3$, u'...denote linear
combinations of the various quark mass eigenstates).

The apparent advantage of the seven-quark theory
compared to the pure vector theory is that it may offer
an explanation of the $\bar{\nu}$-anomaly (b-quark excitation)
without giving $\sigma^{\bar{\nu}\bar{\nu}} = \sigma^{\nu\nu}$. However, a more quantitative
analysis shows that this is not the case, in particular
the b-quark mass comes out unacceptably low[37,55].

Should one really impose "naturalness" as a
constraint? The following argument shows that this is
rather doubtful. It is well-known from quantum mechanics
that states which are close to each other in energy mix
stronger as states which are far apart. Thus it might
be that very heavy quarks mix very little with light
quarks such that no danger for $|\Delta S| = 1$ neutral current

transitions results. The following example shows that
such a situation is not unrealistic. Let me consider
one new quark of charge -1/3 (b) beside d,s, and let me
assume that the mixing of the various quarks of charge
-1/3 with respect to the weak interaction is described
by some universal parameter λ; as shown in the following
matrix:

$$\begin{pmatrix} m_d & \lambda & \lambda \\ \lambda & m_s & \lambda \\ \lambda & \lambda & m_b \end{pmatrix} .$$

The Cabibbo angle is (in lowest order of λ):

$$\theta_c = \frac{\lambda}{m_s - m_d} .$$

Similarly the (db) and (sb) mixing angles are

$$\theta^{db} \approx \theta^{sb} \approx \lambda/m_b - m_{d(s)} \approx \lambda/m_b ,$$

which gives $\theta_c/\theta_{db} \approx m_b/m_s$. It is easy to see that in
particular the $K_L \to \mu^+\mu^-$ decay is of fourth order in θ_{db}:

$$\frac{\Gamma(K_L \to \mu^+\mu^-)}{\Gamma(K_+ \to \mu\nu)} \approx 10^{-8} \approx \theta_{db}^4 ,$$

thus it follows $\theta_{db} < 10^{-2}$ and $m_b > 3$ GeV. This constraint
is not very strong; note that $m_b > 5$ GeV, according to
the ν_μ scattering and the SPEAR results.

We conclude "Naturalness" in the scheme of Ref. 7
should not be used as a constraint in QFD, if very heavy

quarks (m > 3 GeV) are involved. Only a final theory
of QFD, including a theory of the weak mixing angles,
can give a definite answer to the question what kind of
scheme is allowed and what is not.

6.   A Look Beyond $SU_2 \times U_1$

The recent experimental results in neutrino scattering
and the results of the experiments designed to look for
parity violation in atoms, if taken seriously, have
created a contradiction within $SU_2 \times U_1$ theories. On the
one hand it seems that parity is violated in the leptonic
neutral current coupling[41], on the other hand no trace
of a parity violation in atoms is seen[6]. At the
moment it is still possible that the contradiction is not
real; the experiments are extremely difficult, and in
particular the interpretation of the atomic physics
experiments with respect to weak interaction theory is
not yet completely understood. Nevertheless it might
be wise to think for a moment about the consequences, if
the experimental results both in neutrino-electron
scattering and in atomic physics are confirmed.

We should like to emphasize (see also Bouchiat's
talk) that the neutrino scattering experiments do not
really test parity violation in the neutral current
Hamiltonian; they do not measure a pseudoscalar quantity.
The only firm conclusion one can draw from the neutrino
cross section data is that the neutral current Hamiltonian
consists of at least two terms:

1. V·V (V: vector current)

2. A·A (A: axial vector current).

The presence of a V·A interference term is not
required. Of course, with an $SU_2 \times U_1$ theory the only
way to obtain a V·V and an A·A term at the same time is
to have a neutral current of the form $\alpha \cdot V + B \cdot A(\alpha, B$

real parameters ≠ 0) in which case a V·A interference
term results automatically.

However, if there are several Z bosons, say two,
this is not true. Theories of this type were considered
in Refs. 56,57. For example, it might be that there is
a boson coupled to a vector current (Z_V), and another
one coupled to an axial vector current (Z_A). Two years
ago such a situation has been discussed within the
framework of a larger unified theory[56]. The remarkable
feature of the scheme discussed in Ref. 56 was that the
neutral current interaction is not only of the form
V·V + A·A (i.e., parity conserving), but in addition
the effective neutral current interaction relevant for
neutrino scattering is identical to the minimal theory.
A similar situation was discussed recently by Mohapatra
and Sidhu[58].

7. Outlook

There are many indications today, both in experiment
and theory, that QCD in the form of the quark-gluon
Yang-Mills theory is the correct theory of the strong
interactions. A similar statement cannot be made for
QFD; although a more or less definite pattern of QFD has
emerged during the last years, it is still unclear what
the final theory of flavor dynamics is going to be. Too
many questions are yet unanswered, in particular the
questions of charge quantization and of the details of the
symmetry breaking (mass generation etc). Are we really
close to having discovered (either directly or indirectly)
all basic degrees of freedom (leptons, quarks, bosons),
or is it just a small corner of QFD, we have seen thus
far? Theorists who are looking for a complete understanding of nature in the form of one grand unified theory
tend to believe that the latter is the case, and many yet

undiscovered new "elementary" particles and interactions
(leptoquarks, diotons, new unconventoonal types of leptons
and quarks) exist[56]. If this is indeed the case, a
final theory of QFD will not be at hand very soon - there
are stilly many things to do.

Acknowledgement

This talk was written during my stay at Caltech
in January 1977. I am indebted to the members of the
High Energy Physics group at Caltech for many discussions.
I would especially like to think Professor M. Gell-Mann
for stimulating conversations and the kind and generous
hospitality extended to me.

REFERENCES

1. For a thorough discussion of new particle production
    and references, see F. Gilman's talk (these Proceedings).

2. S. L. Glashow, Nucl. Physics $\underline{22}$, 579 (1961);
    A. Salam and J. C. Ward, Phys. Lett. $\underline{13}$, 168 (1964);
    S. Weinberg, Phys. Rev. Lett. $\underline{19}$, 1264 (1967).

3. Y. Hara, Phys. Rev. $\underline{134}$, B701 (1964);
    S. L. Glashow, J. Iliopoulos, and L. Maiani, Phys.
    Rev. $\underline{D2}$, 1285 (1970).

4. S. Weinberg, See Ref. 2.

5. For a detailed discussion and references see Bouchiat's
    talk (these Proceedings).

6. P. Soreide et a., "The search for parity nonconserving
    optical rotation in atomic bismuth," (Seattle-Oxford
    preprint (1976).

7. S. L. Glashow and S. Weinberg, Harvard preprint 1976,
    to be published in Phys. Rev. See also: E. Paschos,
    BNL preprint (1976).

8. M. Gell-Mann, R. L. Oakes, and B. Renner, Phys. Rev.
    $\underline{175}$, 2195 (1968); S. L. Glashow and S. Weinberg,
    Phys. Rev. Lett. $\underline{20}$, 224 (1968).

9. See e.g., H. Leutwyler, Physics Letters $\underline{48B}$, 45 (1974).

10. H. Fritzsch, Phys. Lett. $\underline{63B}$, 419 (1976); K. Lane and
    S. Weinberg, Phys. Rev. Lett. $\underline{37}$, 717 (1976); See
    also: A. DeRujulá, H. Georgi and S. L. Glashow, Phys.
    Rev. Lett. $\underline{37}$, 398 (1976).
    S. Ono, Phys. Rev. Lett. $\underline{37}$, 655 (1976).

11. See e.g., M. K. Gaillard and B. W. Lee, Phys. Rev.
    $\underline{D10}$, 897 (1974).

12. R. Cowsik and J. M. McClelland, Phys. Rev. Lett. $\underline{29}$,
    669 (1972).

13. For recent discussions see:
    S. M. Bilenky and B. Pontecorvo, Phys. Lett. $\underline{61B}$, 248

(1976); H. Fritzsch and P. Minkowski, Phys. Lett. 62B, 72 (1976); S. Eliezer and D. Ross, Phys. Rev. D10, 3088 (1974).

14. G. Altarelli and L. Maiani, Phys. Lett. 52B, 351 (1974); M. K. Gaillard and B. W. Lee, Phys. Rev. Lett. 33, 108 (1974); A. I. Vainshtein, V. I. Zakharov, M. A. Shifman, ITEP preprint 64 (Moscow 1976).

15. R. P. Feynman, M. Kislinger, and F. Ravndal, Phys. Rev. D3, 2706 (1971); J. C. Pati and C. H. Woo, Phys. Rev. D3, 2920 (1973).

16. J. F. Donoghue, B. Holstein, and E. Golowich, to be published in Phys. Rev.

17. C. Schmid, ETH preprint (December 1976).

18. See also; J. Ellis, M. K. Gaillard, and D. V. Nanopoulos, Nucl. Phys. B100, 313 (1975).

19. A. Benvenuti, Phys. Rev. Lett. 35, 1199, 1203, 1249 (1975).

20. The CFR group has found recently several trimuon events; the estimated trimuon production rate is $\sim 3 \cdot 10^{-4}$ (B.C. Barish et a., preprint CALT-68-567).

21. S. Weinberg, Phys. Rev. Lett. 37, 657 (1976); P. Sikivie, Maryland preprint 1976.

22. P. Minkowski, University of Bern preprint (November 1976).

23. H. Fritzsch and P. Minkowski, Phys. Lett. 61B, 275 (1975).

24. C. Hill, to be published.

25. See the subsequent talk by Barnett, and references therein.

26. See e.g., G. Altarelli, G. Parisi, and R. Petronzio, Phys. Lett. 63B, 183 (1976).

27.   M. Barnett, H. Georgi, and H. D. Politzer, Phys.
      Rev. Lett. $\underline{37}$, 1313 (1976); J. Kaplan and
      F. Martin, to be published;
      B. W. Lee and R. E. Shrock, FNAL preprint 76/61-THY.

28.   M. L. Perl et a., Phys. Rev. Lett. $\underline{35}$, 1489 (1975);
      M. L. Perl, et a., Phys. Lett. $\underline{63B}$, 466 (1976);
      G. J. Feldman et al., Phys. Rev. Lett. $\underline{38}$, 117 (1977).

29.   H. Meyer, these Proceedings.

30.   R. P. Feynman and M. Gell-Mann, Bull. Am. Phys.
      Soc. $\underline{2}$, 391 (1957); G. Feinberg, Phys. Rev. $\underline{110}$,
      1482 (1458); N. Cabibbo and R. Gatto, Phys. Rev.
      $\underline{116}$, 1334 (1959).

31.   J. D. Bjorken and S. Weinberg, to be published in
      Phys. Rev. Letters.

32.   See e.g., H. Harari, Les Houches lecture notes
      (1976), Weizmann Institute preprint WIS-76/54 PH.

33.   M. Kobayashi and K. Maskawa, Prog. of Theoret.
      Phys. $\underline{49}$, 652 (1973); L. Maiani, Phys. Lett. $\underline{62B}$,
      183 (1976); S. Pakvasa and H. Sugawara, to be
      published; J. Ellis, M. K. Gaillard, and D. V.
      Nanopoulos, Nucl. Phys. $\underline{B109}$, 213 (1976).

34.   For a review and references see:
      H. Fritzsch, Proceedings, of the Int. Neutrino
      Conf., Aachen 1976 (CERN preprint TH-2198, July 1976).

35.   T. P. Cheng, Phys. Rev. $\underline{D14}$, 1367 (1976).

36.   I am discussing here the various models from a
      special point of view, which is not always shared
      by other theoreticians. Therefore I shall not give
      single references to each model. Most of the models
      discussed below can be found in one of the following
      references: P. Fayet, Nucl. Phys. $\underline{B78}$, 14 (1974);
      H. Fritzsch, Copenhagen lectures (Caltech preprint
      CALT-68-524 (1975); Y. Achiman, K. Koller, and

T. F. Walsh, Phys. Lett. 59B, 261 (1975);

R. M. Barnett, Phys. Rev. D13, 671 (1976),

SLAC-PUB-1821 and 1850 (1976); P. Ramond, Nucl.

Phys. B110, 214 (1976); F. Gürsey and P. Sikivie,

Phys. Rev. Lett. 36, 775 (1976); H. Fritzsch and

P. Minkowski, Phys. Lett. 63B, 99 (1976);

J. Kandaswamy and J. Schechter, preprint COO-3533-77

(July 1976); G. Segré and J. Weyers, Phys. Lett.

65B, 243 (1976); R. L. Kingsley, Rutherford

preprint RL-76-113 (1976); Y. Achiman and T. F.

Walsh, DESY preprint 76146 (1976); V. Barger and

D. V. Nanopoulos, preprint COO-562 (1976);

K. Kang and J. Kim, Phys. Lett. 64B, 93 (1976);

T. C. Yang, DESY preprint 1976.

37. H. Fritzsch, to be published in Phys. Lett. (CERN
preprint TH2249).

38. P. Musset, report given at the Tbilisi Conference
(1976).

39. A. Benvenuti et al., Phys. Rev. Lett. 37, 1039 (1976).

40. B. C. Barish et al., Caltech preprint CALT-68-544
(1976).

41. H. Faissner, report given at the Tbilisi Conference
(1976).

42. F. J. Hasert et al., Phys. Lett. 46B, 121 (1973).
J. Blietschau et a., CERN preprint EP/PHYS/76-42
(June 1976), to be published in Nucl. Phys.

43. H. Faissner, private communication.

44. See e.g., F. Reines, Proceedings of the Int. Neutrino
Conf. (Aachen 1976).

45. H. Fritzsch and P. Minkowski, Phys. Lett. 63B, 99
(1976).

46. F. Gürsey and P. Sikivie, Phys. Rev. Lett. 36,
775 (1976); P. Ramond, Nucl. Phys. B110, 214 (1976).

47. A detailed analysis has recently been done by
     G. Fox (unpublished).

48. H. Fritzsch and P. Minkowski, Phys. Lett. 61B,
     275 (1976); R. K. Ellis, to be published in Nucl.Phys.

49. J. F. Donoghue, E. Golowich, and B. Holstein,
     Amherst preprint (1976), to be published in Phys. Rev.

50. H. Fritzsch and P. Minkowski, Caltech preprint
     CALT-68-538 (1976) unpublished; K. Fujikawa, to
     be published in Phys. Lett; F. Wilczek and A. Zee,
     Nucl. Phys. B106, 461 (1976).

51. H. Fritzsch, Caltech preprint CALT-68-583
     (January 1977).

52. T. P. Cheng and L. F. Li, preprint (December 1976).

53. H. Fritzsch, unpublished.

54. J. D. Bjorken, K. Lane, and S. Weinberg, unpublished.

55. R. M. Barnett, to be published in Phys. Rev.

56. H. Fritzsch and P. Minkowski, Annals of Physics 93,
     193 (1974) and: Nucl. Phys. B103, 61 (1976).

57. R. N. Mohapatra and J. C. Pati, Phys. Rev. D11, 566,
     2558 (1975); G. Senjanovic and R. N. Mohapatra,
     Phys. Rev. D12, 1502 (1975).

58. R. N. Mohapatra and D. P. Sidhu, preprint
     (December 1976).

PRODUCTION OF NEW PARTICLES IN ELECTRON-POSITRON ANNIHILATION*

Frederick J. Gilman

Stanford Linear Accelerator Center

Stanford University, Stanford, California 94305

I. INTRODUCTION

At last year's conference in this series, I recall speaking[1] in a session on "Psychotherapy and Neuroses." But at that time, with charmed particles still to be directly observed and with a number of important puzzles to be solved, in many ways I felt like someone presenting problems rather than cures. With the experimental progress in the year since then, the present discussion should be much more reassuring.

In fact, most of the major components of the "new physics" in electron-positron annihilation appear to have fallen into place. In particular, we have a whole spectroscopy of "bound charm" states followed at higher energies by the production of pairs of particles manifesting "naked charm." So far there is no indication that the properties of the new, weakly decaying mesons deviate from those expected for charmed mesons, properties

---

* Work supported by the Energy Research and Development Administration.

which were deduced, or deducible, from the classic
Glashow, Iliopoulos, and Maiani paper[2] of 1970.[3]

So, with no great discovery in electron-positron
annihilation for six months, I will review a number
of areas where there is important progress, but of a
more detailed quantitative nature. In doing so, I will
touch on the charmonium states, on charmed mesons, and
on the evidence for a charged heavy lepton. A number
of topics discussed at some length in my talk[4] at the
Brookhaven APS Meeting will not be covered here.

Therefore, this talk may be viewed in some regards
as trying to tie up some of the loose ends that were
left over from a year ago. As such, it is also
symptomatic of an era I think has already begun of
filling in details within the broad outline of the physics
of electron-positron annihilation which is now established.

II. CHARMONIUM

The available experimental evidence points strongly
to the new narrow states below ~ 3.7 GeV being the
ground state plus orbital and radial excitations of a
fermion-antifermion system. The general situation with
respect to this "charmonium" system has been recently
reviewed elsewhere[4] and we concentrate here on more
specialized subjects where there have been very recent
developments.

There is increasing evidence that these states are
singlets not only with respect to SU(2), i.e., have
zero isospin, but also with respect to SU(3). For
some time we have known[5] that only upper limits were
set for ψ decay into $K_S K_L$, $K^* \bar{K}^*$, $K^{**} \bar{K}^{**}$, and $K \bar{K}^{**}$, but
that decays into KK* and K*K** are clearly seen.[6] Such a
pattern of unseen and observed decay modes corresponds

exactly to that of the forbidden and allowed decays of
an SU(3) singlet with odd charge conjugation.[7]

The recent paper of Vannucci et al.[8] on mesonic
decays of the ψ contains branching ratios (or upper
limits) for these and many other modes. This permits
further tests of the SU(3) singlet assignment of the ψ
such as the relative branching fractions into $\pi\rho$, $K\bar{K}*$,
and $\eta\phi$, or into ρA_2, $K*\bar{K}**$, ωf, and $\phi f'$.

Tests for the SU(3) singlet character of each of
the three states between ψ and ψ' identified with the
^3P charmonium levels come from their relative decays[9]
to $\rho^0\pi^+\pi^-$ vs $K*^0K^-\pi^+$. For $\chi(3414)$ there is also the
ratio of the $\pi^+\pi^-$ to K^+K^- branching ratios.[9]

All these measurements are consistent with SU(3)
singlet assignments for the corresponding states. In
particular, the results reported by Vannucci et al.[8]
and corresponding measurements at DESY[10] are consistent,
and point toward such an assignment for the $\psi(3095)$.
There seems to be some symmetry breaking present in the
decay mechanism for $\psi \rightarrow$ pseudoscalar meson plus vector meson where Vannucci et al. report relative branching
ratios for $\pi^+\rho^-$: $K^+\bar{K}*^-$: $\eta\phi = 2.1 \pm 0.5 : 0.8 \pm 0.15 :$
0.5 ± 0.3, whereas theory (for an SU(3) singlet ψ with
p-wave phase space) would give 1 : 0.85 : 0.5. However,
if expressed in terms of octet and singlet amplitudes
for the final state, this corresponds to a 10 to 20%
ratio of octet to singlet amplitude. Such a level of
symmetry breaking is roughly the same as found in some
"ordinary" hadron decays,[11] and should therefore cause
no alarm as to the basic conclusion that the ψ and other
charmonium states are consistent with being SU(3) singlets.

The small widths of all the charmonium states below
~ 3.7 GeV are conventionally explained by the Okubo-Zweig-

Iizuka rule,[12] for the charmed quark and antiquark
composing the charmonium do not appear in the hadrons
making up the decay products and the corresponding
quark diagram is disconnected. A striking result taken
from the list of branching ratios[8] for ψ decays is
that

$$\Gamma(\psi \to \omega f)/\Gamma(\psi \to \omega f') \gtrsim 10$$

(only an upper limit exists[8] for $\psi \to \omega f'$). Note that
in terms of quark diagrams $\psi \to \omega f'$ (with ω-ϕ and f-f'
ideally mixed) corresponds to a doubly disconnected one.
Thus we see a substantial extra suppression in rate in
the case of a decay corresponding to a doubly rather
than singly disconnected quark diagram. In the same
vein, $\psi \to \phi f'$ is seen, but $\psi \to \phi f$ is not. An extra
disconnection gives extra suppression!

This statement may not appear to jibe with the
milder suppression (of $\sim 1/5$) of $\phi\pi^+\pi^-$ (which
corresponds to a doubly disconnected quark diagram)
compared to $\omega\pi^+\pi^-$ (which can correspond to a singly
disconnected one) reported earlier.[5] However, I would
argue this ratio of overall rates gives a misleading
impression of what is happening. A look (Fig. 1) at
the $\pi^+\pi^-$ mass spectrum produced in association with
an ω or ϕ shows them to be entirely different. There are
almost no $\pi^+\pi^-$ events above 1 GeV produced along with
a ϕ, whereas a majority of dipions associated with an
ω lie there. Just below 1 GeV, where the $\pi^+\pi^-$ mass
distribution associated with a ϕ is concentrated, that
associated with an ω has a valley.

A possible mechanism to understand this, which is
consistent with the experimental observations, is as
follows: The final state ϕ is made together with a

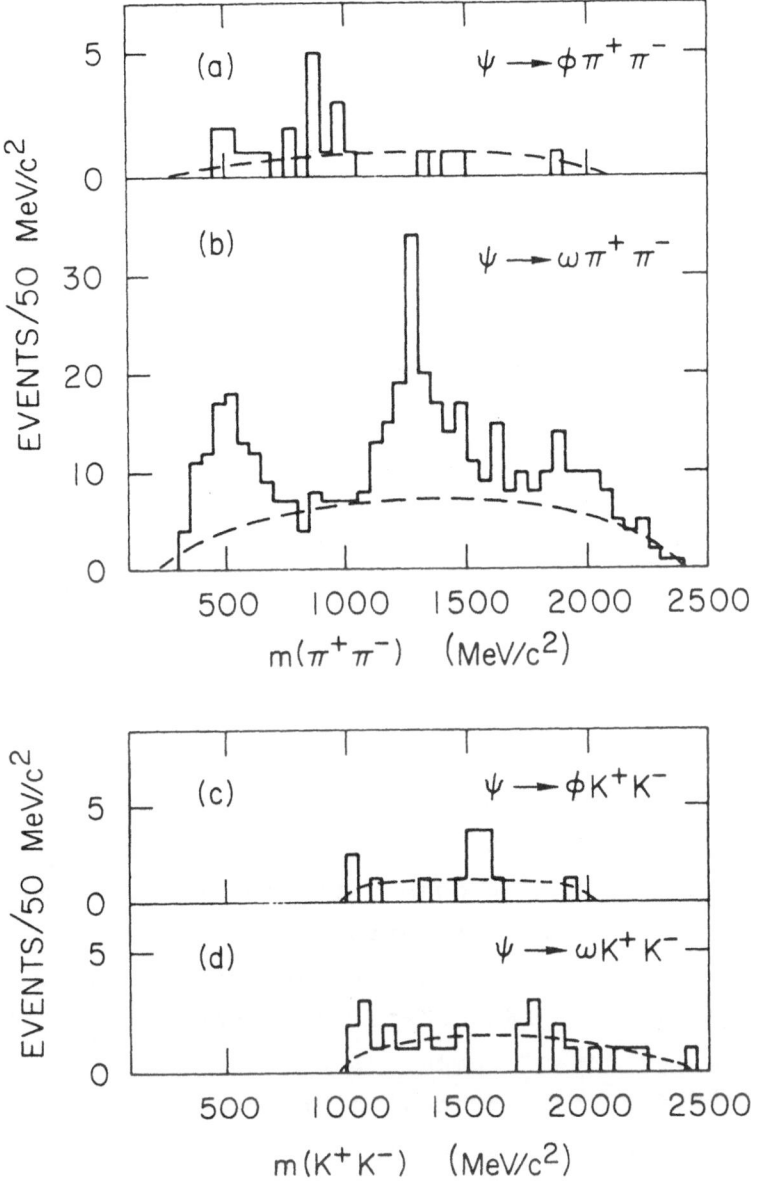

Fig. 1 Invariant mass of the $\pi^+\pi^-$ system in the decays
(a) $\psi \to \phi\pi^+\pi^-$, (b) $\psi \to \omega\pi^+\pi^-$. Invariant mass of the
K^+K^- system in the decays (c) $\psi \to \phi K^+K^-$, (d) $\psi \to \omega K^+K^-$.
The dashed lines indicate the shape predicted by phase
space corrected for detection biases (from Ref. 8).

resonance containing a strange quark-strange antiquark
($s\bar{s}$) pair. Such a process corresponds to a singly
disconnected diagram, since the ϕ is almost entirely
$s\bar{s}$. If the mass of the $s\bar{s}$ system is less than about
1 GeV (e.g., for the ϵ or S*) it decays into $\pi\pi$, since
$K\bar{K}$ is kinematically forbidden. On the other hand, if
the mass is greater than 1 GeV, and particularly if the
state is dominantly composed of $s\bar{s}$ (e.g., the f'), it
decays to $K\bar{K}$. In other words, the second disconnection
associated with the $\phi\pi\pi$ final state takes place at low
mass, "inside" a resonance, and is at least partly due
to phase space inhibiting the $K\bar{K}$ mode. At high mass,
the ratio of $\phi\pi\pi$ to $\omega\pi\pi$ events and the absence of
$\psi \rightarrow \omega f'$ and $\psi \rightarrow \phi f$ indicates a considerably larger
suppression of decays corresponding to doubly disconnected
diagrams.

The list of observed mesonic decays of the ψ now
available, plus the constraints of zero isospin and
isospin conservation in the decay process, permit
putting bounds on the branching ratio for decays into
final channels with the same particle types and multiplicty
but different charge combinations. Thus, the observations
of ψ decay modes with particular charge assignments in
the 3π, 5π, 7π, 9π, $K\bar{K}\pi$, $K\bar{K}2\pi$, $K\bar{K}3\pi$, $K\bar{K}4\pi$, and $2K2\bar{K}$
channels allow establishment of a lower bound[8] of
21.6 ± 2.4% for the ψ branching fraction into all charge
assignments for these channels. A model in which all
isospin states are populated statistically gives
30.2 ± 3.3% for the sum of these channels on the basis
of the same observations[8] of modes with particular
charge assignments.

These newer, more detailed, numbers for
individual modes, and the corresponding bounds or

estimates for the sum of both observed and unobserved
decays, agree quite well with the overall picture of
ψ decays discussed previously.[1] When added to the
known branching ratios for e^+e^-, $\mu^+\mu^-$, decay into
hadrons through a virtual photon, and estimates of
modes involving higher K and π multiplicity of baryons,
a large majority of all ψ decays are accounted for.

An important change in accounting for the ψ'
decays has occurred, however. The present status[13]
of this is as follows:

TABLE I

ψ' Branching Ratios

Mode	Branching Ratio (%)
$\psi' \rightarrow e^+e^-$, $\mu^+\mu^-$	2
$\rightarrow \gamma_v \rightarrow$ hadrons	3
$\rightarrow$ hadrons(direct decays)	~ 10
$\rightarrow \pi\pi\psi$	~ 50
$\rightarrow \eta\psi$	4
$\rightarrow \gamma + {}^3P_J(J=0,1,2)$	18-28
$\rightarrow \gamma + \chi(3455)$	$\gtrsim 1$
Total	88-98

The estimated width for ψ' direct decays to hadrons is
simply scaled[1] from the corresponding width for the
ψ by the ratio of $\Gamma(\psi' \rightarrow e^+e^-)$ to $\Gamma(\psi \rightarrow e^+e^-)$. What
is mostly new here are the gamma decay rates into the
${}^3P_J$ states between the ψ and ψ' taken from the results
of the MP^2 S^3D collaboration[14] at SPEAR as well as the
SLAC-LBL magnetic detector collaboration.[15] The individual
branching ratios for these decays have probable values

of 7 to 8%, and are at or even above the previous
upper limits.[16] As a result the "missing" decays of
the ψ' may now be in place. There is no longer any
need to find other major ψ' modes, although presently
undetectable ones at the 10% level cannot be ruled out.

    With the new knowledge of, for example,
BR($\psi' \rightarrow \gamma\chi(3414)$)), together with older measurements
of $\psi' \rightarrow \gamma\chi \rightarrow \gamma$ + hadrons we can extract branching ratios
for various χ decays. Using isospin invariance we then
may convert branching ratios for particular modes into
lower bounds on general modes of the same multiplicity
or, using a statistical model, into estimates of the
rate for the general mode. For the $\chi(3414)$ this
exercise yields the following:[17]

TABLE II

$\chi(3414)$ Branching Ratios

Mode	Branching Ratio (%)	
	Lower Bound	Statistical Model
$\pi\pi$	1.4 ± 0.5	1.4 ± 0.5
4π	8.0 ± 1.5	10.7 ± 2.0
6π	4.1 ± 1.5	9.8 ± 3.7
$K\bar{K}$	1.9 ± 0.6	1.9 ± 0.6
$K\bar{K}\pi\pi$	10.8 ± 2.7	14.4 ± 3.6
Total	26.2 ± 3.5	38.2 ± 5.6

Thus at least ~ 25% of all $\chi(3414)$ decays are accounted
for, and probably almost 40%. We have come a long way
since the first information on the hadronic decays of
the ψ!

III. CHARMED MESONS

The neutral charmed meson D^0 with mass $\sim$ 1865 MeV is observed[18] to decay into $K^-\pi^+$, $K_s^0\pi^+\pi^-$, and $K^-\pi^+\pi^-\pi^+$ from e^+e^- annihilation experiments,[18,19] and possibly $K_s^0\pi^+\pi^-\pi^+\pi^-$ in photoproduction.[20] Its charged partner D^+ has been observed[21] in the nonleptonic mode $K^-\pi^+\pi^+$. Work on establishing cross section times branching ratio values for these and other modes, or providing upper bounds, is in progress.

It is of considerable interest to know the actual branching ratio of the D^0 into, say, $K^-\pi^+$. Since all that is measured at the moment is the product of production cross section and branching ratio at some particular e^+e^- center-of-mass energy, this now requires a guesstimate of the total production cross section of the D^0. Taking charm production at $\sqrt{Q^2}$ = 4.028 GeV as being 10 to 15 nb, between 0.5 and 1.0 D^0 per event containing charm (there is also D^+ production in these events), and a cross section times branching ratio[22] for $e^+e^- \rightarrow (D^0 \rightarrow K^-\pi^+)+\ldots$ of $\sim$ 0.25 nb, one finds that $BR(D^0 \rightarrow K^-\pi^+)$ lies between approximately 1.5 and 5%. The branching ratios for other nonleptonic modes may all be estimated by comparison to $K^-\pi^+$.

An estimate of the semileptonic decay rates of charmed particles from e^+e^- annihilation data also can be obtained in much the same way as for the $K^-\pi^+$ nonleptonic decay of the D^0. Both the DASP[23] and PLUTO[24] groups working at DESY give peak cross sections[25,26] for inclusive anomalous $e^\pm$ production of $\sim$ 3 nb near 4.03 GeV e^+e^- center-of-mass energy. If, again, the inclusive charm production is 10 to 15 nb at this energy, then one deduces a branching ratio for semi-electronic decay of some weighted average of the D^0 and

D^+ of 10 to 15%.

An analogous estimate can be made from the recent data[27] obtained using the muon tower and magnetic detector at SPEAR on anomalous muon production. For momenta above 910 MeV and an average e^+e^- center-of-mass energy of 6.9 GeV, it is found that ~ 2.2% of charged tracks are muons in events with $\geq$ 3 observed charged prongs. Taking a charged multiplicity of 5, assuming that the ratio of 2.2% for anomalous muons to charged tracks is true for momenta below 910 MeV also, and the guesstimate that 1/3 of the events with four or more charged particles at such center-of-mass energies contain charmed particles, one deduces[28] a semimuonic branching ratio of ~ 17%. Note that at these energies one is likely producing a mix of not only D^0 and D^+, but also the F^+ and charmed baryons, all of which may have different branching ratios.

From these estimates, as well as the rate for dilepton production in neutrino-induced events,[29] it now seems very likely that the semielectronic and semimuonic branching ratios for some charmed particles are each in the 10 to 20% range, for a total semileptonic branching ratio of 20 to 40%. This implies that there is little or no enhancement of the rate for nonleptonic as compared to semileptonic decays above the naive level obtained by assuming the charmed quark decays as if it were free into $s + e^+\nu$, $s + \mu^+\nu$, $s + u\bar{d}$ in the ratios 1:1:3.

From study of D production near 4 GeV we know that it occurs principally via two-body or quasi-two-body modes, e.g., $e^+e^- \to D^0\bar{D}^{*0}$. If one then observes the decay of $D^0 \to K^-\pi^+$, the angular distribution of the K (or π) relative to the beam and D line-of-flight directions may be studied. Calling ψ the angle between the D line of flight and the incident e^+ beam direction, and θ,ϕ the polar and azimuthal angles of the K relative to the $e^+e^- \to D\bar{D}^*$ production plane with z axis along the D line of flight, one can show that the angular distribution of the K has the form[30,31]

$$W(\psi,\theta,\phi) \propto \frac{1+\cos^2\psi}{2}W_T(\theta) + \frac{1-\cos^2\psi}{2}\cos 2\phi W_P(\theta) + \frac{1-\cos^2\psi}{2}W_L(\theta)$$

$$+ \frac{\cos\psi\sin\psi}{\sqrt{2}}\cos\phi W_I(\theta) . \qquad (1)$$

The structure functions $W_T(\theta)$, $W_P(\theta)$, $W_L(\theta)$, and $W_I(\theta)$ have forms which are characteristic of the D and D* spins and are generally interdependent in a nontrivial way for any particular set of spin assignments. This should make it possible, at least for low values of the spins of D and D*,[30,31] to use the data to rule out all but a single assignment for the D and D*.

There appears to be substantial D production in e^+e^- annihilation in the 4 GeV region, and in particular at the bumps in the cross section at 4.028, ~ 4.11, and 4.414 GeV. Also, from SPEAR data there has been no evidence of $e^+e^- \to \psi + \dots$, with upper bounds on the cross section for this process previously reported[32] as being in the range of 1% of $\sigma(e^+e^- \to \text{hadrons})$. Note that, from our experience with $\psi' \to \pi\pi\psi$, if the vector resonances in the 4 GeV region are higher mass relatives of the ψ and ψ', then we might well expect widths for these states to decay to $\pi\pi\psi$, $4\pi\psi$, $\eta\psi$, etc., to be tens to hundreds of keV. Since the total width of these vector states lies in the tens of MeV range, branching ratios into $\psi + \dots$ might be expected to be 10^{-3} to 10^{-2}. Anything much larger might well be an indication that some

of these states are not simply further radial excitations
of the $c\bar{c}$ system, decaying primarily in a Zweig rule
allowed manner into pairs of charmed mesons, but
possibly $c\bar{q}\bar{c}q$ states, e.g., "molecular charmonium."
Such an assignment of some of the states in the 4 GeV
region has been considered by a number of authors,[33,34,35]
including the possible decay chain $c\bar{q}\bar{c}q \rightarrow c\bar{c} + q\bar{q}$. The
smallness of any signal[32] for inclusive ψ production
in the 4 GeV region would seem to be evidence against
such an assignment of states, together with a decay
chain leading to charmonium levels among the final
particles.

A good ideal of the information we presently have
on D masses and D spectroscopy comes from a careful
study of the recoil mass spectra recoiling against a
detected D. Equivalently, at a fixed e^+e^- center-of-mass
energy, one studies the D momentum or kinetic energy
spectrum.

The recoil mass spectrum against a detected D^0 for
all data from 3.9 to 4.6 GeV is shown in Fig. 2. The
peaks near 1.87 GeV and 2.01 GeV are indicative of $D^0\bar{D}^0$
and $D^0\bar{D}^{*0}$ production (as well as kinematic reflections
from $D^*\bar{D}$ with D^* decay into a detected D with emission
of a pion or photon). The peak near 2.15 GeV is likely
due to $D^*\bar{D}^*$ production with detection of a D^0 from
D^* decay.[36]   The collection of events above $\sim$ 2.4 GeV
indicates yet other D^0 production mechanisms, possibly
including higher D resonances in this mass region.

The D^0 momentum or kinetic energy spectrum at
$\sqrt{Q^2}$ = 4.028 GeV has been particularly useful in extracting
properties of the D^0 and D^{*0} from the peaks due to the
channels $D\bar{D}$, $D\bar{D}^* + \bar{D}D^*$, and $\bar{D}^*D^*$. Preliminary results
from this effort are as follows:[22]

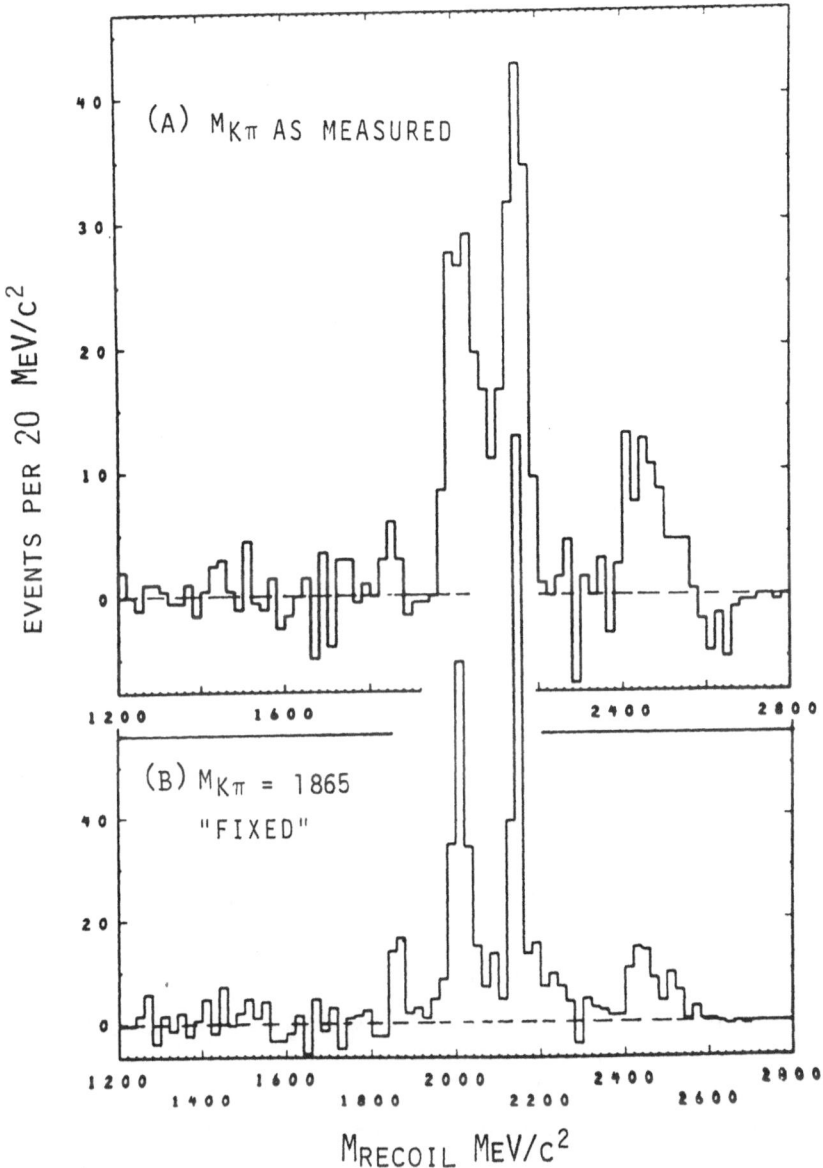

Fig. 2 (a) M_{recoil} distribution for the $K^{\pm}\pi^{\mp}$ signal as measured. (b) M_{recoil} distribution for the $K^{\pm}\pi^{\mp}$ signal for fixed $M_{K\pi}$ = 1865 MeV. Each distribution is background subtracted. Data are from $e^{+}e^{-}$ annihilation from 3.9 to 4.6 GeV (from Ref. 46).

$$M_{D*^O} = 2005.5 \pm 1.5 \text{ MeV}, \tag{2}$$

which is measured relative to $E_{beam} = 2014$ MeV. The
Q-value for $D*^O \to \pi^O D^O$ is 3 ± 2 MeV, which gives

$$M_{D^O} = 1867 \pm 3 \text{ MeV}. \tag{3}$$

Furthermore, with such a small Q-value, $D*^O \to \gamma D^O$ is
expected to be competitive with $D*^O \to \pi^O D^O$; and it is:[22]

$$\frac{BR\ (D*^O \to \gamma D^O)}{BR\ (D*^O \to \pi^O D^O)} = 1.0 \pm 0.3. \tag{4}$$

The D^+ spectrum has more background and it is harder to
get unique fits. However, the peak from $D^+ \bar{D}*^-$ is
rather clear, and it gives[22]

$$M_{D^+} + M_{D*^+} = 3880 \pm 5 \text{ MeV}. \tag{5}$$

Subtracting the previously listed D^O and $D*^O$ masses,
one finds

$$(M_{D^+} - M_{D^O}) + (M_{D*^+} - M_{D*^O}) = 7 \pm 6 \text{ MeV}. \tag{6}$$

Since both mass differences are expected to be positive
and of the same order, each is probably 2 to 8 MeV.[4]
Such values are at the lower end of the range of
theoretically predicted ones.

IV. HEAVY LEPTONS

We start the discussion of heavy leptons in
antihistorical order with the work on anomalous muon

production in e^+e^- annihilation using the muon tower
and magnetic detector at SPEAR.[27] In particular we focus
on events with only two charged particles detected,
with a cut on the angle between the planes containing
the charged tracks and the beam to be greater than $20°$
and with a cut on the missing mass squared (against
the two charged prongs) to be greater than 1.5 GeV^2.
Both cuts serve to limit QED backgrounds. Again, the
momentum of the potential muon must be greater than
910 MeV, to distinguish it (statistically) from hadrons
on the basis of its penetration of the matter in the
muon tower.

After subtraction of remaining QED background,
muons from pion and kaon decay, and hadron "punch through"
the muon tower, a significant anomalous muon signal
remains. In particular, for e^+e^- center-of-mass energies
of 5.8 to 7.8 GeV (average 6.9 GeV), the cross section
for anomalous muon production in two prong events with
the above cuts[27] is (Fig. 3):

$$\sigma(e^+e^- \to \mu^\pm + \text{one charged track} + ...) = 212 \pm 49 \text{ pb}.$$

Although a cross section of a few hundred picobarns
sounds negligibly small, in fact it is _big_ on the scale
of electron-positron interactions at these energies. To
see this we do the following _rough_ calculation. First
we make a correction for the cuts, particularly those
for the muon momentum and the coplanarity angle. A
rough correction for both of these multiplies the cross
section by about a factor of three to obtain an estimated
cross section without cuts for $\mu^\pm$ + one charged track+ ...
of $\sim$ 640 pb.

Now, such anomalous muons must arise from either
production of a new hadron, carrying a quantum number

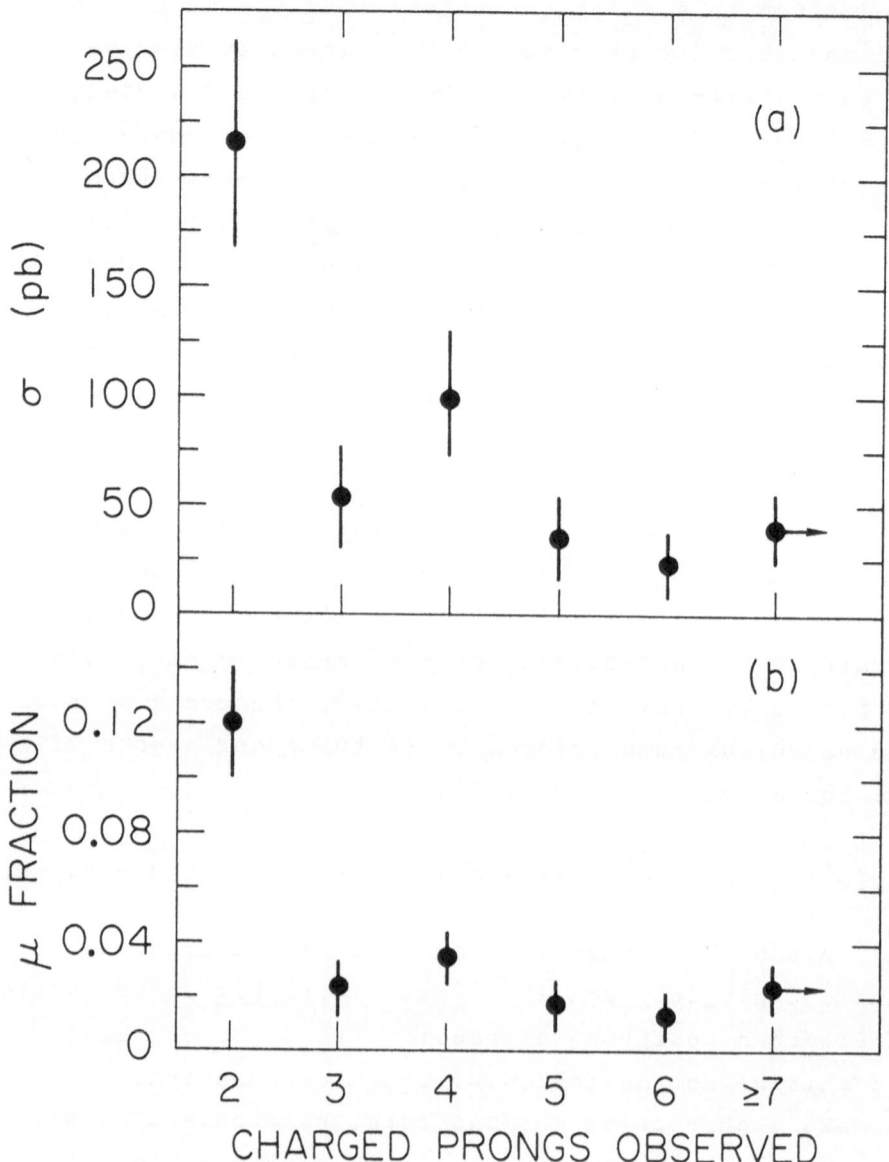

Fig. 3 (a) Anomalous muon production cross section vs the number of charged prongs observed. (b) Ratio of anomalous muons to candidates vs the number of charged prongs observed in the $E_{c.m.}$ range 5.8 to 7.8 GeV (from Ref. 27).

conserved in strong and electromagnetic interactions so that it decays semileptonically, or the production of a heavy lepton. In either case we have pair production of these new particles along with their antiparticle. In the case of a new hadron it is likely that additional hadrons are produced:

$$e^+e^- \rightarrow U + \bar{U} + \ldots ,$$

whereas a heavy lepton would be pair produced with no other particles. We take the production cross section for this process to be that for muon pairs, i.e., the point fermion cross section, which is ~ 1.6 nb at these energies. Recall that the inclusive production of hadrons containing a new charge 2/3 (-1/3) quark is expected to be $\frac{4}{3}\sigma_{point}$ ($\frac{1}{3}\sigma_{point}$), while pair production of charged heavy leptons (well above threshold) is given by σ_{point}.

Putting the factors together, we have[37]

$$2 \times \sigma(e^+e^- \rightarrow U+\bar{U}+\ldots) \times BR(U \rightarrow \mu+\ldots) \times BR(\bar{U}+\ldots \rightarrow \text{one charged prong}$$
$$+\ldots) \simeq 640 \text{ nb}.$$

With $\sigma(e^+e^- \rightarrow U+\bar{U}+\ldots) = 1.6$ nb, we deduce that

$$BR(U \rightarrow \mu+\ldots) \geq 0.2$$

by using $BR(\bar{U}+\ldots \rightarrow \text{one charged prong}+\ldots) \leq 1$. By μ-e universality we then must have a new particle[38] with a branching ratio into an electron or muon plus neutrals ~ 40%!

Furthermore, the momentum spectrum of the detected muons is hard (Fig. 4), and different from the (soft) spectrum seen in events with ≥ 3 charged particles detected.

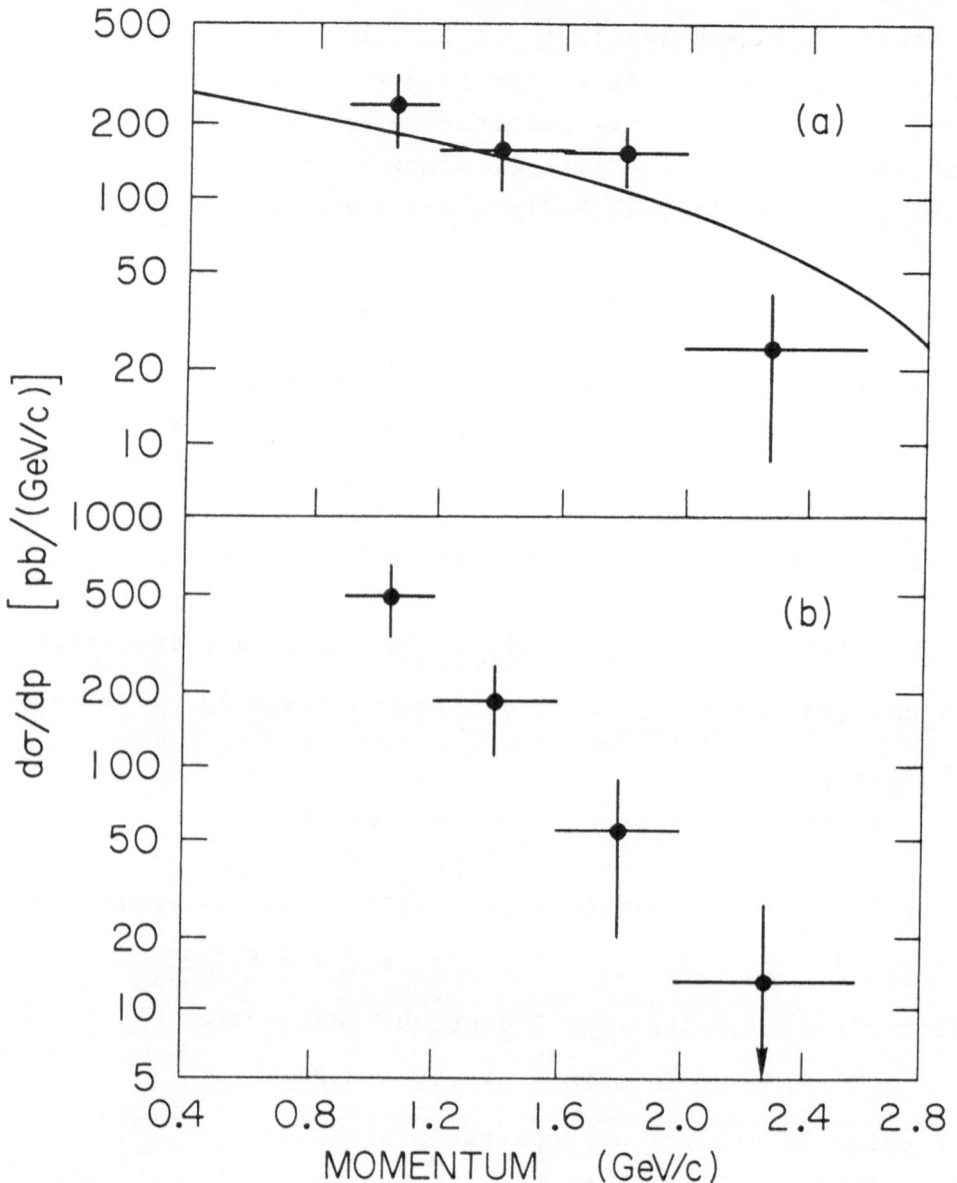

Fig. 4   Differential cross section for anomalous muon
production versus momentum for (a) two prong events and
(b) multiprong events in the $E_{c.m.}$ range 5.8 to 7.8 GeV
(from Ref. 27).

This, together with the estimated branching ratio given
above, makes it very difficult to associate the U with
a hadron, and in particular with charmed particles. For
our experience with D's teaches us that their momentum
spectrum in e^+e^- annihilation is not hard and that the
momentum spectrum of the leptons in their semileptonic
decay is quite soft. Folding together these two spectra
then gives an inclusive anomalous lepton spectrum in
e^+e^- annihilation from D production which is soft.[39]
We expect this to be true generally for leptons originating
from decays of new hadrons. Furthermore, all known D
decays involve K mesons, as expected theoretically.
These, together with the pions expected to be produced
in association with <u>any hadron</u> well above threshold,
would typically yield several additional charged prongs
in each event containing new hadrons, thus throwing it
out of the μ + one charged particle topology. Finally,
for the case of charmed particles we already have
accounted for a 20 to 40% semileptonic branching ratio
from $\geq$ 3 prong events. We would have difficulty
accommodating another ~ 40% from 2 prong events!

   Thus, of the explanations offered, the least
strained and outlandish is that we have the production
of a charged heavy lepton plus antilepton. This single
explanation is consistent with the anomalous muon
production data[40] and all other data up to this time.
These include:

1. Anomalous eμ Events. Historically events of the
    form $e^+e^- \rightarrow e^\pm \mu^\mp$ + undetected neutrals were the
    first indication[41] of anomalous events which might
    indicate the existence of a charged heavy lepton.
    The energy dependence of the observed cross section
    is consistent with that expected for pair production

of a point fermion-antifermion. Analysis[42] of the
distribution of the angle between the electron and
muon and its energy dependence gives evidence for
the source of the events being production of a
pair of fixed mass particles, with the Lorentz
transformation from the heavy lepton rest frame
to the lab resulting in the e and μ being thrown
increasingly back-to-back at higher e^+e^- energies.
Further, the momentum spectrum of the leptons
indicates[42] a three body rather than two body decay
of the U. From these distributions one also deduces
the mass to lie between 1.6 and 2.0 GeV, with values
in the upper half of this range more likely.[43]
Assuming that the decays involving a muon or electron
are $\nu\mu\bar{\nu}_\mu$ and $\nu e\bar{\nu}_e$, respectively, and that these occur
with equal branching ratio, then the observed cross
sections give[42]

$$BR(U \rightarrow \mu\bar{\nu}\nu) = BR(U \rightarrow e\bar{\nu}\nu) = 17^{+6}_{-3}\% \ .$$

2. Anomalous ee and μμ Events. Although the QED
backgrounds are much higher, there is evidence[44]
for the analogs of the eμ events, but where the
leptons are e^+e^- or $\mu^+\mu^-$. They occur at a rate
consistent with the branching ratio given above
for $U \rightarrow \nu e\bar{\nu}_e$ or $U \rightarrow \nu\mu\bar{\nu}_\mu$.

3. $\mu^\pm$ hadron$^\pm$ + neutrals Events. A subset of the $\mu^\pm$ +
one charged prong +... events discussed at the outset
of this section contains sufficient identification
of the other charged particle to indicate that it
is not an electron or muon.[43,27] Such events
should occur when one heavy lepton decays to $\nu\mu\bar{\nu}_\mu$
and the other, for example, to $\nu\pi$ or $\nu\rho$. The rate

for these latter two decays relative to the $\nu\mu\bar{\nu}_\mu$
decay is calculable from otherwise known parameters
and the observed $\mu^{\mp}$ hadrons$^{\pm}$ + neutrals events
occur at a rate consistent with this.[43]

4. Two-Prong Topological Cross Section. Because the
great majority of decays of a ~ 2 GeV heavy lepton
involve a single charged prong ($\nu\mu\bar{\nu}_\mu$, $\nu e^-\bar{\nu}_e$, $\nu\pi^-$,
$\nu\rho^-$, $\nu\pi^-\pi^0\pi^0$, ...), pair production of heavy leptons
with a mass of ~ 2 GeV probably involves two
charged particles ~ 80% of the time. Crossing the
threshold for heavy lepton production shoud then
result in a jump in the two prong topological cross
section. As the e^+e^- center-of-mass energy
increases, the average multiplicity in hadronic
events rises and hadron production feeds the two
prong topology less and less; but the heavy lepton
decays are fixed in character and so its contribution
will increasingly dominate the two prongs, which
in turn will increasingly stick out in the distribution
of the cross section into topologies. This provides
an amusing analog to the situation in high energy
hadron collisions where diffraction dissociation
feeds (relatively fixed) low multiplicities while the
multiperipheral mechanism results in increasing
average multiplicity as the energy rises.

Assuming the existence of a heavy lepton, not only
are the exact values of the parameters associated with
it of great interest, but it may be used to gain
information on other physical parameters or particle
properties. For example, assuming the heavy lepton has
its own neutrino, ν_U, so that its decay involving a
muon is $U^- \rightarrow \mu^-\bar{\nu}_\mu\nu_U$, the muon momentum spectrum already
restricts[42] the mass of ν_U to be less than 700 MeV.

Limits (or observation!) on modes like $U \rightarrow e\gamma$ or $\mu\gamma$, and precise branching ratios for observed modes will provide important restrictions on the existence and properties of the spectroscopy and couplings of leptons.

In several ways the existence of a charged heavy lepton is the most exciting of all the spectacular discoveries made in electron-positron annihilation during the past few years. Unlike the charmed quark and its associated hadronic spectroscopy, nobody had a good reason for proposing such a particle beforehand. And particularly with the discovery of charm, neatly closing the books at four quarks and four leptons, there seemed every reason not to want such a particle. Now, more than ever, the pattern of quark and lepton masses is a mystery, and there is a high likelihood that still more fundamental fermions remain to be found.

REFERENCES

1. F. J. Gilman, in New Pathways in High Energy Physics II, ed. Arnold Perlmutter (Plenum Press, New York, 1976), p. 1.

2. S. L. Glashow, J. Iliopoulos, and L. Maiani, Phys. Rev. D 2, 1285 (1970).

3. See, for example, M. K. Gaillard, B. W. Lee, and J. L. Rosner, Rev. Mod. Phys. 47, 277 (1975) and J. Rosner, invited talk at this conference, and references to other work therein.

4. F. J. Gilman, invited talk at the 1976 Meeting of the Division of Particles and Fields of the APS, Brookhaven National Laboratory, 6-8 Oct 1976, and SLAC-PUB-1833 (1976), unpublished.

5. See, for example, G. Abrams, in Proc. 1975 Int. Symposium on Lepton and Photon Interactions at High Energies, Stanford University, 21-27 Aug 1975, ed. W. T. Kirk (SLAC, Stanford, California, 1975), p. 25.

6. We use K* to denote the vector resonance at ~ 890 MeV and K** to denote the tensor resonance at ~ 1420 MeV.

7. H. Harari, Stanford Linear Accelerator Report No. SLAC-PUB-1514 (1974). F. J. Gilman, in AIP Conference Proceedings, No. 26, High Energy Physics and Nuclear Structure - 1975, Los Alamos, 9-13 June 1975, ed. D.E. Nagle, R. L. Burman, B. G. Storms, A. S. Goldhaber, and C. K. Hargrave (American Institute of Physics, New York, 1975), p. 331. V. Gupta and R. Kogerler, Phys. Lett. 56B, 473 (1975).

8. F. Vannucci et a., SLAC-PUB-1862 (1976), unpublished.

9. G. Trilling, talk at the Topical Conference of the SLAC Summer Institute on Particle Physics, Stanford University, 2-13 Aug 1976 (unpublished).

10. W. Braunschweig et al., Phys. Lett. 63B, 487
    (1976); W. Bartel et al., Phys. Lett. 64B, 483 (1976).

11. N. P. Samios, M. Goldberg, and B. T. Meadows,
    Rev. Mod. Phys. 46, 49 (1974).

12. S. Okubo, Phys. Lett. 5, 165 (1963); G. Zweig,
    CERN preprints TH401 and TH412 (1964), unpublished;
    J. Iizuka, Suppl. Prog. Theor. Phys. 37-38, 21 (1966).

13. Branching ratios for the ψ, ψ', and other charmonium
    states are taken from the compilation of G. J.
    Feldman, lectures at the SLAC Summer Institute on
    Particle Physics, 2-13 Aug 1976, and SLAC-PUB-1851
    (1976), unpublished, unless otherwise indicated.

14. D. Coyne, talk at the American Physical Society
    Meeting at Stanford, California, 20-22 Dec 1976,
    on the results of the Maryland-Princeton-Padua-
    Stanford-SLAC-San Diego experiment at SPEAR, Bull.
    Am. Phys. Soc. 21, 1294 (1976).

15. J. S. Whitaker, W. Tanenbaum et al., Phys. Rev.
    Lett. 37, 1596 (1976).

16. J. W. Simpson et al., Phys. Rev. Lett. 35, 699 (1975).

17. We use $BR(\psi' \to \gamma\chi(3414)) = 7.5\%$ from Refs. 14 and
    15 and the branching ratios for $\psi' \to \gamma\chi \to \gamma +$
    hadrons reported in Ref. 9 and J. S. Whitaker,
    University of California Ph.D. Thesis and LBL
    preprint LBL-5518 (1976), unpublished.

18. G. Goldhaber, F. M. Pierre et al., Phys. Rev. Lett.
    37, 255 (1976).

19. R. F. Schwitters, invited talk at the 1976 Meeting
    of the Division of Particles and Fields of the APS,
    Brookhaven National Laboratory, 6-8 Oct 1976, and
    SLAC-PUB-1871 (1976), unpublished.

20. B. Knapp, invited talk at the 1976 Meeting of the
    Division of Particles and Fields of the APS,

Brookhaven National Laboratory, 6-8 Oct 1976
(unpublished).

21. I. Peruzzi, M. Piccolo, G. J. Feldman, H. K.
     Nguyen, J. E. Wiss et al., Phys. Rev. Lett. 37,
     569 (1976).

22. M. Breidenbach, talk at the American Physical
     Society Meeting, Stanford, California, 20-22 Dec
     1976, on results from the SLAC-LBL Magnetic
     Dectector Collaboration at SPEAR, Bull. Am. Phys.
     Soc. 21, 1283 (1976).

23. W. Braunschweig et al., Phys. Lett. 63B, 471 (1976).

24. J. Burmester et al., Phys. Lett. 64B, 369 (1976).

25. L. Criegee, talk at the American Physical Society
     Meeting, Stanford, California, 20-22 Dec 1976, on
     recent results from DORIS, Bull. Am. Phys. Soc.
     21, 1283 (1976).

26. H. Meyer, invited talk at this conference.

27. G. J. Feldman, F. Bulos, D. Lüke et al., Phys.
     Rev. Lett. 38, 117 (1977).

28. Similar estimates based on the muon tower data
     were first made by G. J. Feldman, whom I thank for
     discussions. We take only the data with $\geq$ 3
     charged prongs observed to make this estimate of
     semimuonic charm decays, since the corresponding
     two prong events likely arise dominantly from the
     decay of a charged heavy lepton. See Section IV
     of this talk.

29. See C. Baltay, H. Bingham, D. Cline, and O. Fackler,
     invited talks at the 1976 Meeting of the Division
     of Particles and Fields of the APS, Brookhaven
     National Laboratory, 6-8 Oct 1976 (unpublished).

30. F. J. Gilman, unpublished.

31. Some of the special low spin cases have also been

considered by J. D. Jackson (unpublished).

32. G. J. Feldman, invited talk at the Irvine Conference, December 1975 (unpublished). H. Meyer (Ref. 26) reports an inclusive ψ production cross section in the 4 GeV region which is a few tenths of a percent of the total cross section.

33. M. Bander et al., Phys. Rev. Lett. <u>36</u>, 695 (1976); C. Rosenzweig, Phys. Rev. Lett. <u>36</u>, 697 (1976); Y. Iwasaki, Prog. Theor. Phys. <u>54</u>, 492 (1975).

34. L. B. Okun and M. B. Voloshin, JETP Lett. <u>23</u>, 333 (1976).

35. A De Rújula, H. Georgi, and S. L. Glashow, Harvard preprint HUTP-76/A176 (1976), unpublished; S. Nussinov and D. P. Sidhu, Fermilab preprint FERMILAB-Pub-76/70-THY (1976), unpublished; P. H. Cox, S. Y. Park, and A. Yildiz, Harvard preprint HUTP-76/A182 (1976), unpublished.

36. A. De Rújula, H. Georgi, and S. L. Glashow, Phys. Rev. Lett. <u>37</u>, 398 (1976); K. Lane and E. Eichten, Phys. Rev. Lett. <u>37</u>, 477 (1976).

37. We assume that the muon and the other charged prong don't both come from the U(or $\overline{U}$).

38. If there are several new hadrons, then we are computing an average (weighted by relative production cross section) branching ratio, assuming the sum of their inclusive production cross sections is σ_{point}.

39. See, for example, A. Ali and T. C. Yang, DESY preprint DESY 76/39 (1976), unpublished; V. Barger, T. Gottshalk, and R. J. N. Phillips, University of Wisconsin preprint COO-881-569 (1976); unpublished; M. Gronau et al., DESY preprint DESY 76/62 (1976), unpublished.

40. Note that the previous <u>rough</u> calculation of BR $(U \rightarrow \mu + \ldots)$ is in fact an estimate of $BR(U^- \rightarrow \nu \mu^- \bar{\nu}_\mu)$ in the case U is a charged heavy lepton and it decays dominatly into one charged prong plus neutrals. The shape and magnitude of the inclusive muon spectrum of Ref. 27 agree with the heavy lepton hypothesis.

41. M. L. Perl et al., Phys. Rev. Lett. <u>35</u>, 1489 (1975).

42. M. L. Perl, G. J. Feldman et al., Phys. Lett. <u>63B</u>, 466 (1976).

43. See G. J. Feldman, lecture at the SLAC Summer Institute on Particle Physics, Stanford University, 2-13 Aug 1976, and SLAC-PUB-1852 (1976), unpublished.

44. See also H. Meyer, Ref. 26, for evidence from the PLUTO detector at DESY for μ + one charged prong and $\mu^\mp e^\pm$ events in the 4 GeV region at rates consistent with the cross sections measured at SPEAR (Refs. 41, 42, and 27).

45. M. L. Perl, invited talk at the International Neutrino Conference 1976, Aachen, Germany, 8-12 June 1976, and SLAC-PUB-1764 (1976), unpublished.

46. G. Goldhaber, talk at the Topical Conference of the SLAC Summer Institute on Particle Physics, Stanford University, 2-13 Aug 1976, and LBL preprint LBL-5534 (1976), unpublished.

RECENT RESULTS ON e^+e^- ANNIHILATION FROM PLUTO AT DORIS

Hinrich Meyer

University of Wuppertal

Gauss str. 20   D 5600 Wuppertal 1 West Germany

I. INTRODUCTION

A collaboration of DESY, the Universities of Hamburg, Siegen and Wuppertal* has used the magnetic detector PLUTO at the e^+e^- storage ring DORIS. Data have been taken at the two resonances J/ψ (3.1) and ψ' (3.7), just below the ψ' at $\sqrt{s}$ = 3.60 to 3.66 GeV and from $\sqrt{s}$ = 4.0 to 5.0 GeV.

A breakdown of the luminosities collected at the various energies is contained in table I. Data at the two narrow resonances have been used mostly for calibration purposes, since at the two resonances large

---

* Members of the PLUTO collaboration are: J. Burmester, L. Criegee, H.C. Dehne, K. Derikum, R. Devenish, J.D. Fox, G. Franke, G. Flügge, Ch. Gerke, G. Horlitz, Th. Kahl, G. Knies, M. Rössler, R. Schmitz, U. Timm, H. Wahl, P. Waloschek, G.G. Winter, S. Wolff, W. Zimmermann, V. Blobel, H. Jensing, B. Koppitz, E. Lohrmann, A. Bäcker, J. Bürger, C. Grupen, M. Rost, G. Zech, H. Meyer, K. Wacker

Table I

PLUTO 1976 runs
at the e^+e^- - storage ring DORIS

c.m. Energy	$\int Ldt$ nb^{-1}	month
3.1	50	Feb.
3.7	162	July, Sept.
3.6	604	July, Sept.
4.03	742	Sept.
4.1	716	May, June
4.2	247	Aug.
4.4	1169	March, Aug., Nov.
4.5	362	Nov.
4.6	434	Aug., Nov.
5.0	1375 +)	Nov., Dec.
Sum	5860 nb^{-1}	

+) not completely analyzed.

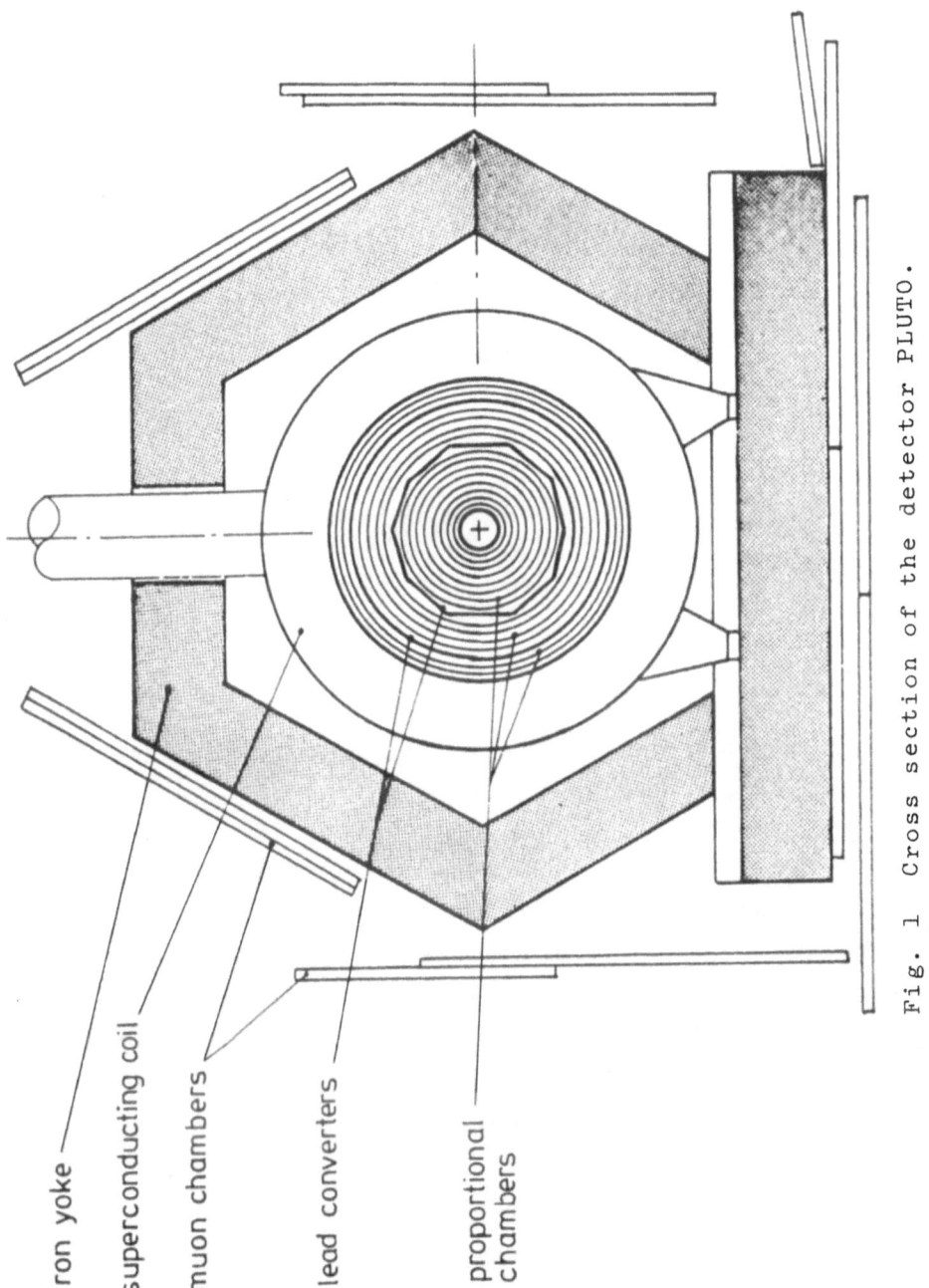

iron yoke

superconducting coil

muon chambers

lead converters

proportional chambers

Fig. 1 Cross section of the detector PLUTO.

data sets of hadronic events can be obtained with little
running time.

Our main emphasis was on data taking in the region
from $\sqrt{s}$ = 4.0 to 5.0 GeV, where the total cross section
has intriguing structures. Also in this energy region
the production of a pointlike new particle, a heavy
lepton with mass between 1.6 and 2.0 GeV has been
reported[1]. Whereas the heavy lepton would have very
smooth energy behaviour, one expects that the total
cross section structures are intimately related to
production of charmed particles.

The data taken at 3.6 GeV, below threshold for new
particle production, provide us with an extremely
useful reference point on "old physics".

Although preliminary in nature, we will have
important new information on the total cross section
for e^+e^- annihilation into hadrons, on inclusive
production of strange particles as well as the J/ψ (3.1)
resonance, on anomalous inclusive muon production, and
finally on μe - events.

II. THE DETECTOR

The magnetic detector PLUTO has been described
elsewhere[2]. Here we only present information on those
features which are most important for the physics topics
to be discussed. In Fig. 1 a cross section of the
dectector is shown looking down the positron beam. A
superconducting coil produces a uniform solenoidal field
of 20 KΓ in a volume of 140 cm diameter and 110 cm long.
Fourteen (14) proportional planes, with wires parallel
to the beam and cathode stripes oriented at 45° and 90°
with respect to the wires, provide charged particle
tracking. The detector is triggered by logical

combinations of proportional wire signals. Although
much more complicated in detail, it essentially
provides as the basic part a two particle trigger, both
particles have to meet momentum p > .240 GeV/c and
$|\cos\theta| \leq .87$. Since lepton (e,μ) and photon identification has been important we give some more experimental
details.

Two lead cylinders at radii of 37.5 cm (0.4 X_o)
and 59.4 cm (1.7 X_o) serve as photon and electron
detectors. Photons converted in the lead (or in the
material of the chambers and the beam pipe) can be
reconstructed at the few percent level from observed
e^+e^- pairs. At a higher level ($\sim$ 60%) only the
conversion point of photons is detected, but still
giving two constraints to reactions where the only
missing neutral particle is the photon. Electrons
develop showers in the lead cylinders, detected by
larger numbers of wires set in the chambers following
the lead. In what follows electrons are required to
produce > 10 hits after the second lead. Hadrons on
average produce 4.5 hits. The electron acceptance
given by the dimensions of the lead is - 0.55 $\leq \cos\theta$
$\leq$ 0.55 and all ϕ angles, at an average detection
efficiency of $\sim$ 55% for momenta larger than 300 MeV/c.
Hadrons sometimes produce more than 4-5 hits therefore
faking electrons. From hadronic events taken at 3.1
resonance the electron fake level of hadrons P(h $\rightarrow$ e)
has been determined to be 3.5 $\pm$ 0.5%.

The return flux iron yoke is used as a muon filter.
On average the particles have to traverse an equivalent
of $\sim$68 cm of iron (actually iron plus some material,
mostly copper and lead). Hadrons are filtered out by
interaction in the iron and muons of momentum greater

than 1 GeV/c penetrate. The muons are detected in
proportional tube planes with dimensions 108 × 180 cm^2.
A muon is required to hit the proper chamber within
108 × 25 cm^2 around the projected track point in the
chamber. Again - as in the electron case - data at
3.1 resonance provide the muon fake level of hadrons
P (h → μ) = 2.8 ± 0.7 % for momenta up to 1.5 GeV/c.

The geometrical acceptance is slightly momentum
dependent due to varying iron thickness, at high
momenta $\Delta\Omega$ = 0.43 × 4π.

From Monte-Carlo studies using the observed event
topologies as input and following an iterative procedure,
we find an event efficiency of ~80%, slightly depend on
$\sqrt{s}$ (see Fig. 2) including losses in the reconstruction
procedure. At 3.7 resonance, the efficiency is even
higher, certainly due to the cascade transition
ψ' → ππ J/ψ providing two additional pions for
triggering.

The luminosity is monitored by small angle Bhabha
scattering observed by a symmetrical arrangement of
four counter telescopes centered at a scattering angle
of ~ 130 mrad.

Fig. 3 shows a two prong event with two converted
photons. The material measured in units of radiation
length (X_o) that each particle from the interaction
region has to traverse is shown in the lower part of
the figure.

III. σ_{tot} $(e^+e^- \rightarrow$ HADRONS$)$

No attempt will be made to describe in detail the
analysis procedure which finally results in the total
cross section for hadron production. More details can
be found in a recent DESY-report[3]. Only a few

e⁺ e⁻ ANNIHILATION

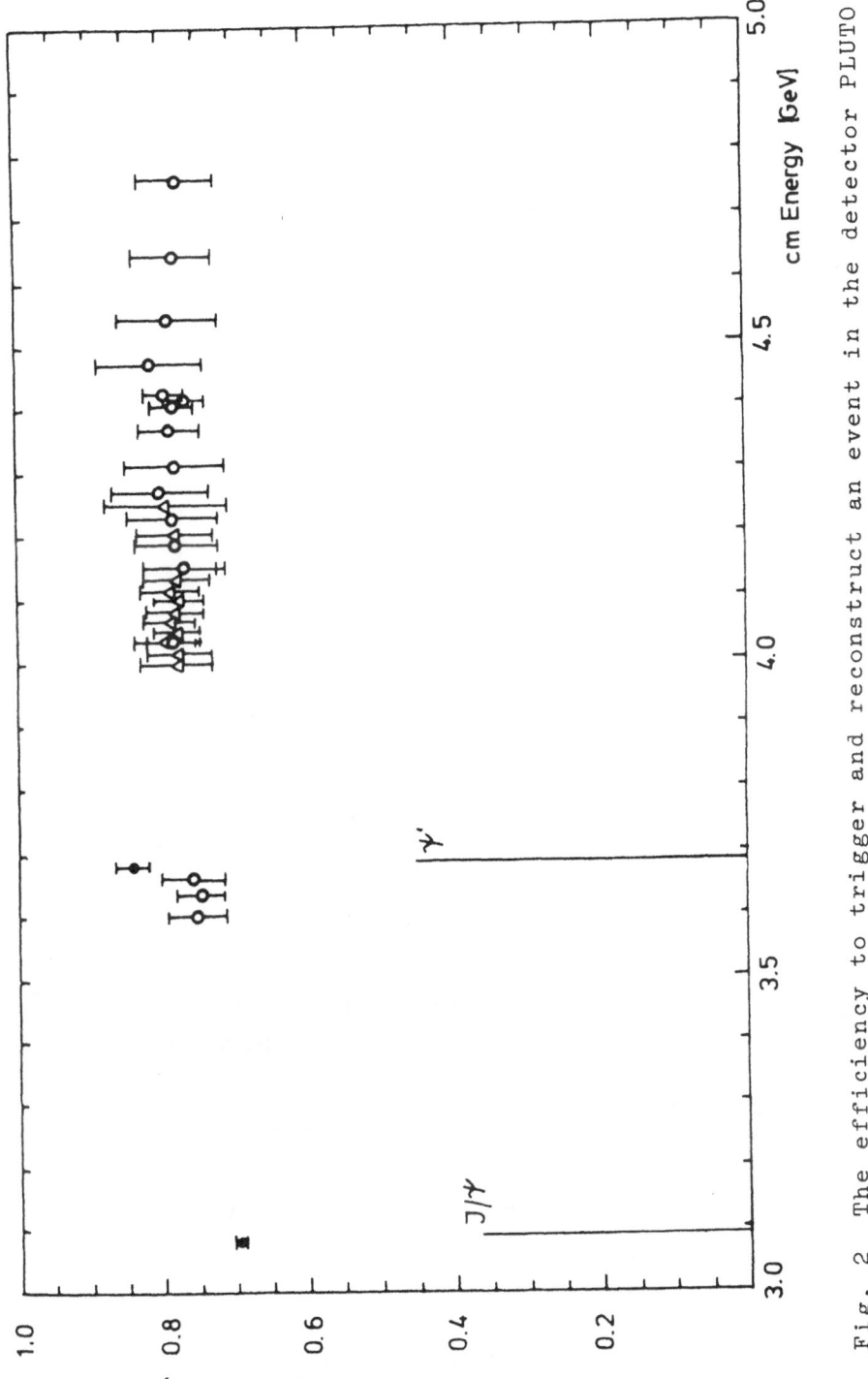

Fig. 2   The efficiency to trigger and reconstruct an event in the detector PLUTO.

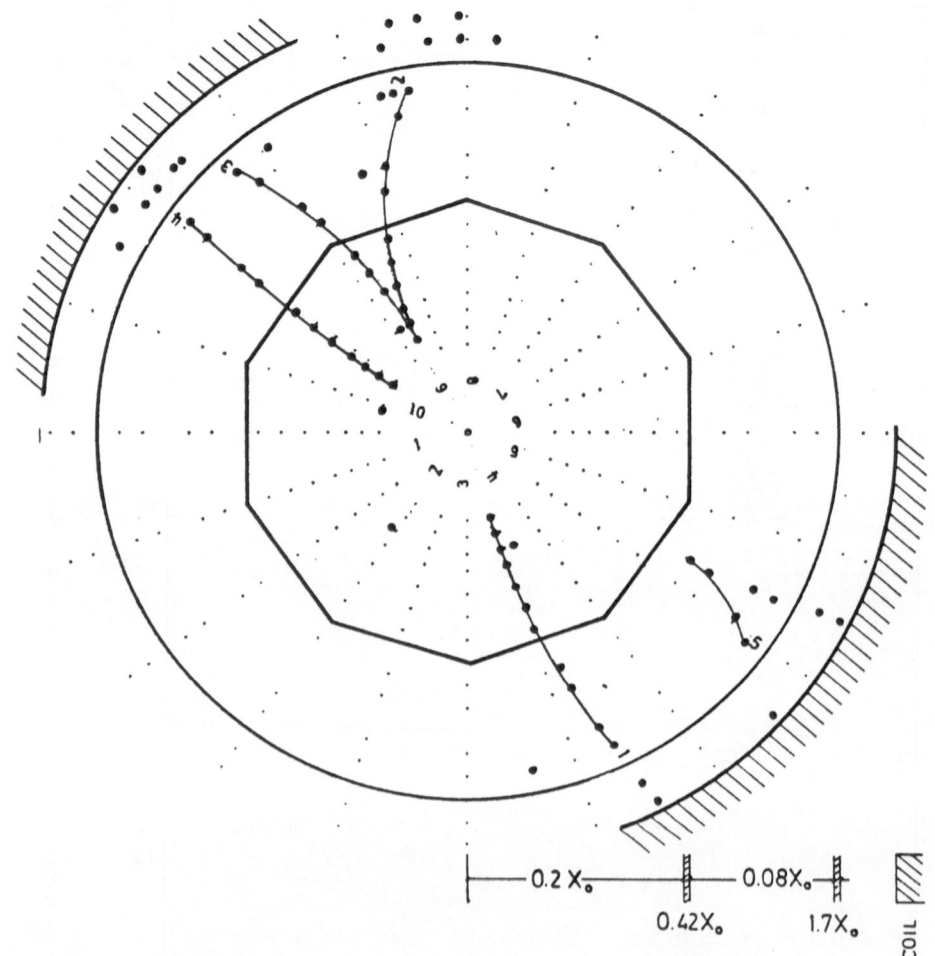

Fig. 3 A two prong event with two converted photons.
The material in units of radiation length (X_0) from the
center of the detector up to the coil is also shown.

important points will be mentioned.

a) Even with the large acceptance of the PLUTO
detector ($\Delta\Omega \simeq 0.90 \cdot 4\pi$) sizable corrections for
events lost in the trigger have to be made. So far
only an all pion phase space model has been used for
the extrapolation to full solid angle. The two charged
particle final state is most severely effected, it has
an efficiency of about 42% only, but it comprises a
significant fraction of the total cross section
($\sim$ 40% - see below).

b) The radiative corrections get fairly large
contributions from the tails of both narrow resonances,
even at $\sqrt{s}$ = 5.0 GeV, the total correction is of the
order of 15%, including the correction to the small
angle Bhabha monitor.

c) Purely leptonic final states from the well known
QED processes:

$$e^+ e^- \rightarrow e^+ e^- \tag{1}$$

$$\rightarrow \mu^+ \mu^- \tag{2}$$

$$\rightarrow \gamma\gamma \tag{3}$$

are excluded by various cuts from the hadronic event
sample. Since our detector is fairly thick (measured
in radiation length) process (1) presents some problems;
by radiation and conversion it even invades the higher
prong classes.

d) Decay products from pair produced heavy leptons have
not specifically been separated from the hadron sample.
As shown in the section on muon inclusive events, these
events so far can only partly be identified and, therefore, are counted as "hadronic" events, although the
production of point like sequential leptons (in Martin
Perls terminology) would be a QED process, followed by

weak decay of the heavy leptons and should be
subtracted from the hadronic events.

e) From higher order QED processes at $\sqrt{s}$ < 5 GeV one
expects large contribution only from

$$e^+e^- \rightarrow e^+e^- \, e^+e^- \tag{4}$$

and much less from

$$e^+e^- \rightarrow e^+e^- \, \mu^+\mu^- \tag{5}$$

The majority of events from (4) and (5) triggering our
detector produces a pair of low energy charged particles
that approximately balances transverse momentum. They
are largely removed by a cut in the relative azimuthal
angle $\Delta\phi$ < 150°.

Fig. 4 shows $R = \sigma_{tot}/\sigma_{\mu\mu}$ as a function of the
total center of mass energy $\sqrt{s}$. As compared to earlier
data from PLUTO[3] new data from the energy region
$\sqrt{s}$ > 4.3 have been added. The dotted curve just follows
the measured points. The main structures first observed
by the SLAC-LBL collaboration[4] are reproduced. At
4.0 GeV R steeply rises to a value of about 5 at 4.05 GeV.
A broader structure centered at 4.18 GeV follows. From
a valley at 4.3 GeV R again rises to ~ 5, at 4.415 GeV,
declining to a plateau of R ~ 4.6 up to 5 GeV, the
highest energy where we have taken data.

In detail our measured σ_{tot} differs from the
SLAC-LBL data[4]. At the resonances we are lower by
about one unit of R while otherwise there is good
agreement. This is not yet understood. Both experiments
quote possible systematic errors of ~ 12 %, the overall
comparison then certainly stays within this margin.

There is no complete separation into prong classes
available but a separation into 2-prong and multiprong
events was possible in this experiment. The percentage

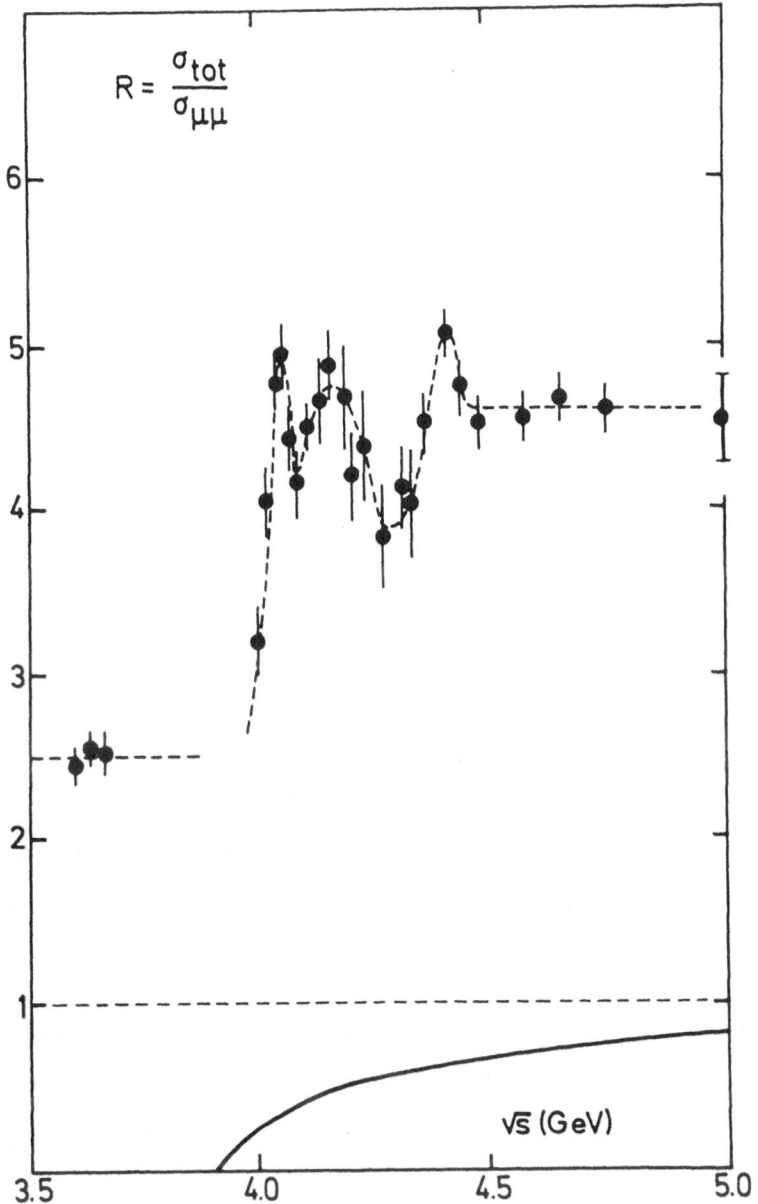

Fig. 4 The total cross section for e^+e^- annihilation into hadrons σ_{tot} divided by the muon pair cross section $\sigma_{\mu\mu}$, ●. The dotted line is used to guide the eye. The full line is the cross section for a pointlike heavy lepton of mass 1.95 GeV.

of 2-prong events is shown in Fig. 5. The surprising
result is the large value of about 35 % almost independent of energy. Since purely hadronic two-prong
events are likely to fall off fairly rapidly with
energy this result is very suggestive for the turn on
of a new process between 3.6 and 4.0 GeV, mostly
entering the two-prong channel. The heavy lepton in
fact provides a good candidate for this new process.
We come back to this question in the final part of
this talk.

The basic mechanism for hadron production is
assumed to be the production of a pair of pointlike
quarks, which subsequently fragment into hadrons. Each
quark type contributes a constant fraction to R and we
have

$$R \text{ (tot)} = R \text{ } (u,\bar{u}) + R \text{ } (d,\bar{d}) + R(s,\bar{s}) + R(c,\bar{c}) +$$

$$R \text{ (heavy lepton)} \tag{6}$$

Data at $\sqrt{s} = 3.6$ GeV measure the contribution from the
first three terms, while above $\sqrt{s} = 4.0$ GeV we have
additional contributions from the charmed quark and the
heavy lepton. The full curve in the lower part of
Fig. 4 indicates the heavy lepton contribution calculated
from

$$R(L) = \sigma(L^+L^-)/\sigma(\mu^+\mu^-) = \frac{3\beta - \beta^3}{2} \tag{7}$$

where β is the velocity of the heavy lepton.

Since charmed quarks (by invention) have to fragment
dominantly into strange particles, one should observe
above 4.0 GeV a new piece in R with two kaons per new
event. To find out what actually happens in the next
part data on inclusive production of K_S^o will be presented.

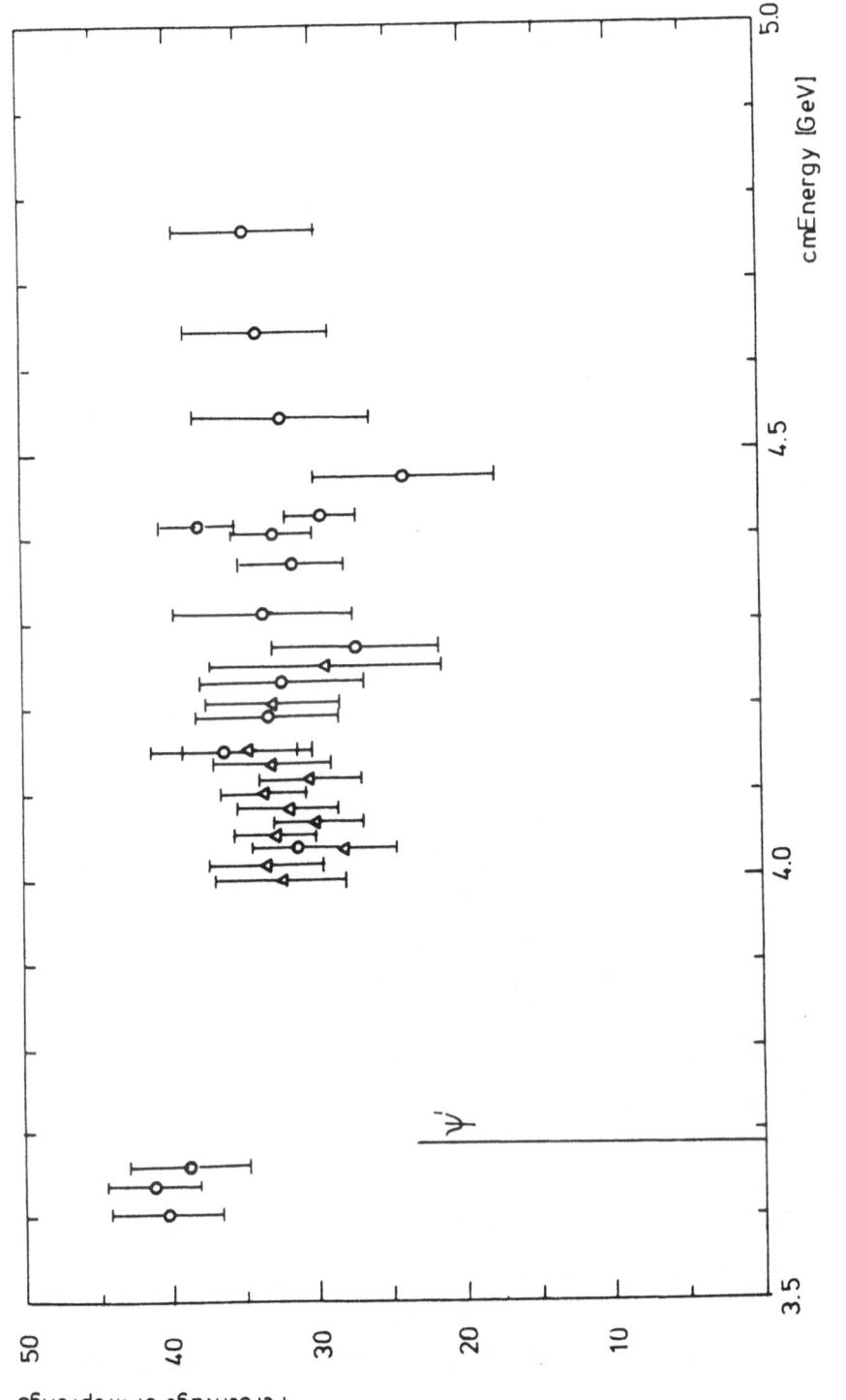

Fig. 5 The ratio of the two prong cross section to the total cross section as function of energy $\sqrt{s}$.

IV. INCLUSIVE PRODUCTION OF K_s^o

All events with more than 2 charged particles have been used for a search for K_s^o [14].

$$e^+e^- \rightarrow \pi^+\pi^- + \text{charged particles.} \qquad (8)$$

The effective mass of pairs of oppositely charged particles - assuming they are pions - has been calculated. In addition, a minimum decay length of 15 mm for the pair is required. Using all geometrical constraints and momentum conservation, we do find a strong K_s^o signal in the $\pi^+\pi^-$ mass distribution (see Fig. 6). Correcting for the decay mode $K_s^o \rightarrow \pi^o\pi^o$ and for the finite detection efficiency as function of momentum, we get the inclusive yield of K_s^o as function of the energy of the K_s^o normalized to the beam energy $X = E_K/E_{beam}$. There is a cut on the kaon momentum at 200 MeV/c, since below this momentum the detection efficiency becomes very small. The data are presented in Fig. 7, showing the inclusive cross section $S/\beta \cdot d\sigma/dx$, where β is the K_s^o velocity. Going from 3.6 GeV to 4.03 GeV, a dramatic rise in the inclusive K_s^o cross section is observed at $X < 0.5$ (Fig. 7a).

At higher energies (Fig. 7b) there is still a sizable excess at lower X, but at $X > 0.5$ the cross section approximately scales, as expected in quark fragmentation models.

The increase at $X < 0.5$ supports the hypothesis that the K_s^o come from the decay of a pair of charmed mesons. Not far above threshold both charmed mesons have small velocity but their total energy equals that of the beam and $E(k) < \frac{1}{2} E(beam)$ follows. Since the excess is largest at lowest X, we infer a dominance of

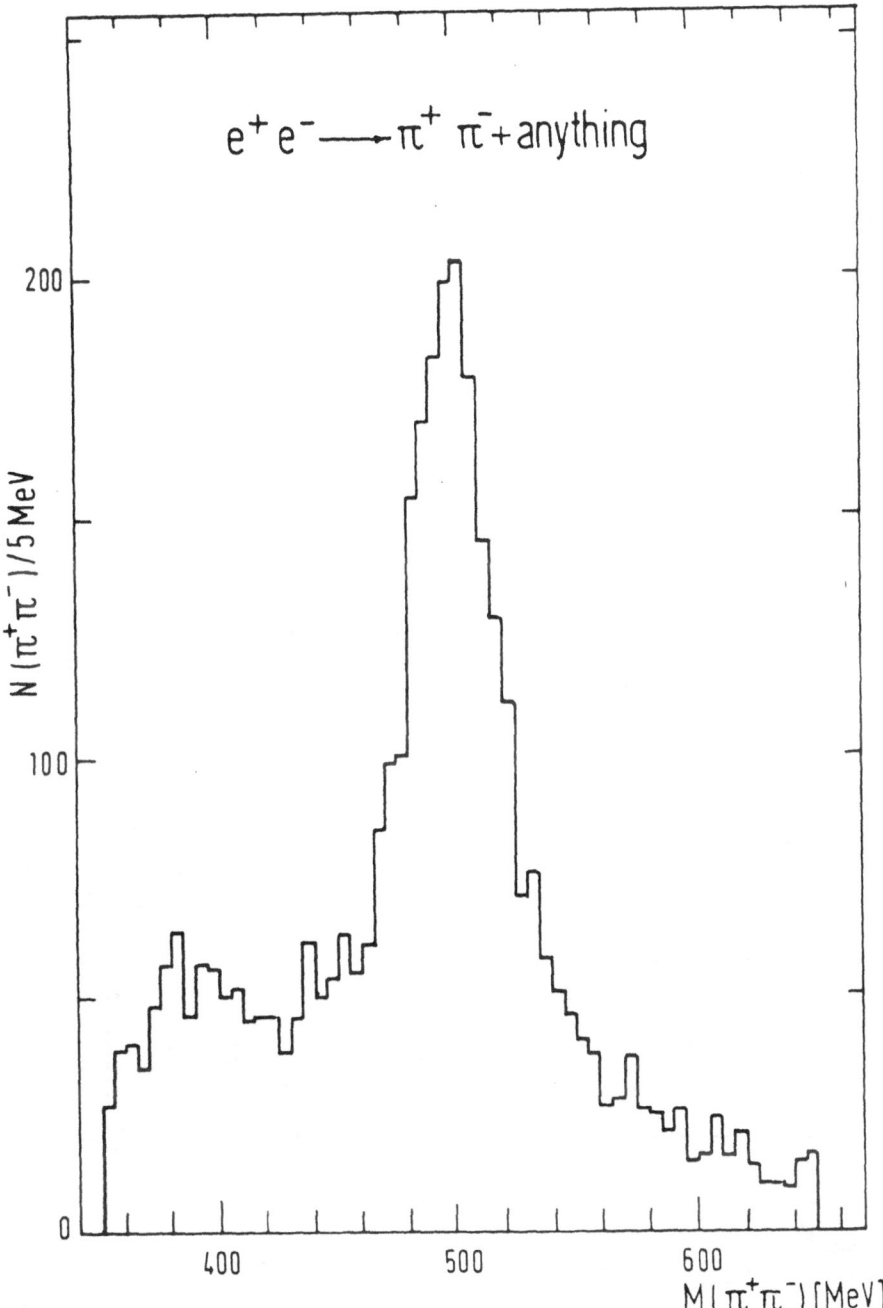

Fig. 6 The K_s^o signal in the effective mass of pion pairs.

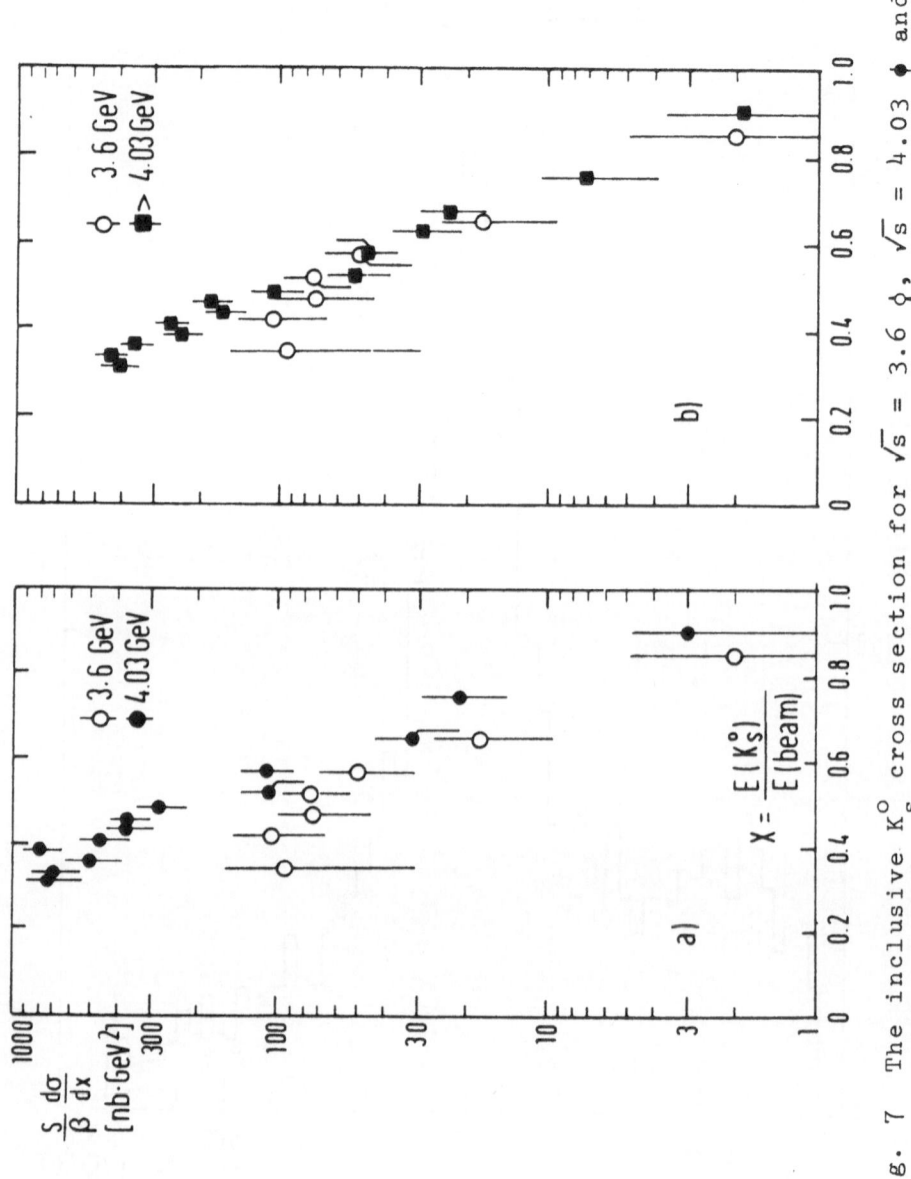

Fig. 7 The inclusive K_S^0 cross section for $\sqrt{s} = 3.6$ ◊, $\sqrt{s} = 4.03$ ♦ and $4.03 < \sqrt{s} < 4.80$ GeV ✚ .

multibody decays of the charmed particles.

The integrated cross section is shown in Fig. 8.
No correction for the K_s^o momentum cut $p(K_s^o) < 200$ MeV/c
has been applied. However, by extrapolating the
observed momentum spectrum to zero K_s^o momentum, we
estimate the loss to be of the order of 10 %. Also
radiative corrections have not yet been applied. To
determine the K_s^o yield per event, we therefore refer to
the total cross section n o t corrected for radiative
effects[3]. Below charm threshold we find ~ 0.08 K_s^o/event.
To find the extra contribution from new physics, we
subtract this value from the data above $\sqrt{s} = 4.0$ GeV.
The result is shown in Fig. 9 (open points). Since we
expect $N (K^+ + K^-) = N (K^o + \bar{K}^o) = 2N (K_s^o)$ supported by
recent results from the DASP group[5] we should find
0.5 K_s^o/new event while the open points give ~ 0.25 K_s^o/new
event.

This is where the heavy lepton comes in again. The
kaon yield from the heavy lepton is expected to be
suppressed due to the Cabibbo-angle and has been assumed
to be zero. Subtracting the heavy lepton contribution
from the new part of the total cross section one arrives
at the full point in Fig. 9 close to the value of 0.5.
[Also, the experimental points should move upwards once
the correction for the momentum cut has been applied.
This cannot be done in a model independent way.]

At $\sqrt{s} = 4.03$ and 4.415 we have taken fairly large
data sets, about 20.000 hadronic events each. When
comparing the K_s^o yield at the two energies, we observe
much less kaon per new event at 4.415 GeV, by about
a factor of three. At $\sqrt{s} = 4.03$ GeV we know about
very sizeable charm particle production[6]. Tentatively
one may conclude that the 4.415 structure is not very

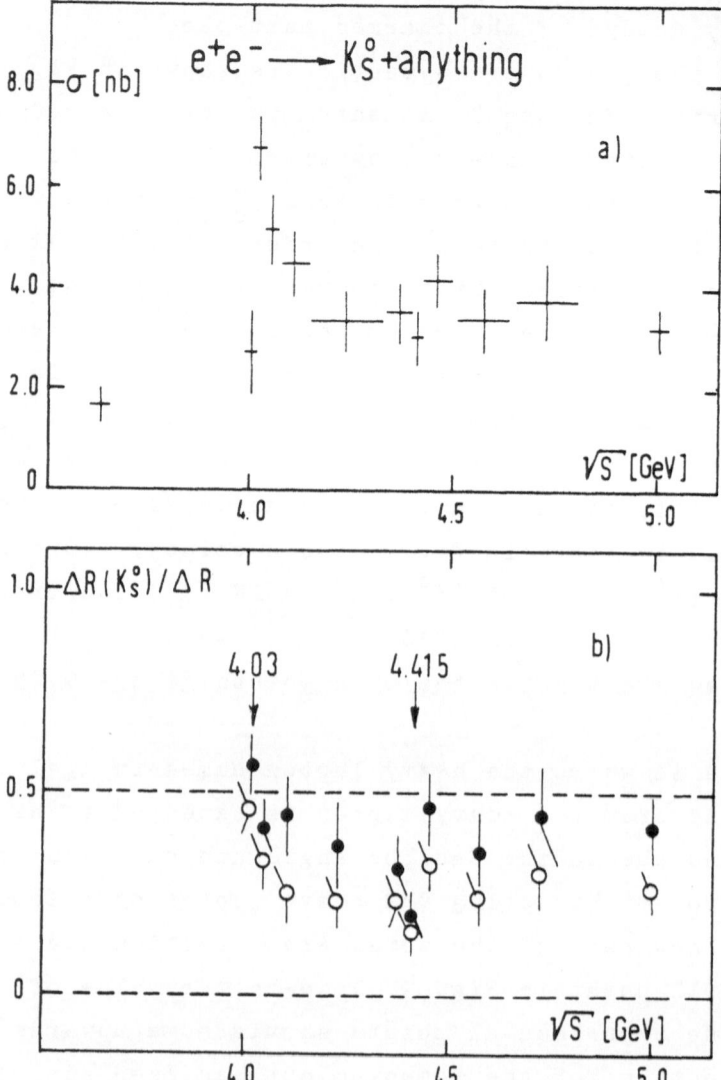

Fig. 8 The total inclusive K_s^o cross section for
P (K_s^o) > 0.2 GeV as function of $\sqrt{s}$ (GeV).

Fig. 9 The number of K_s^o (p > 0.2 GeV) per new event for
$\sqrt{s}$ > 4.0 GeV, ϕ. Accounting for the heavy lepton
contribution to new events above 4.0 GeV results in
the full points ϕ.

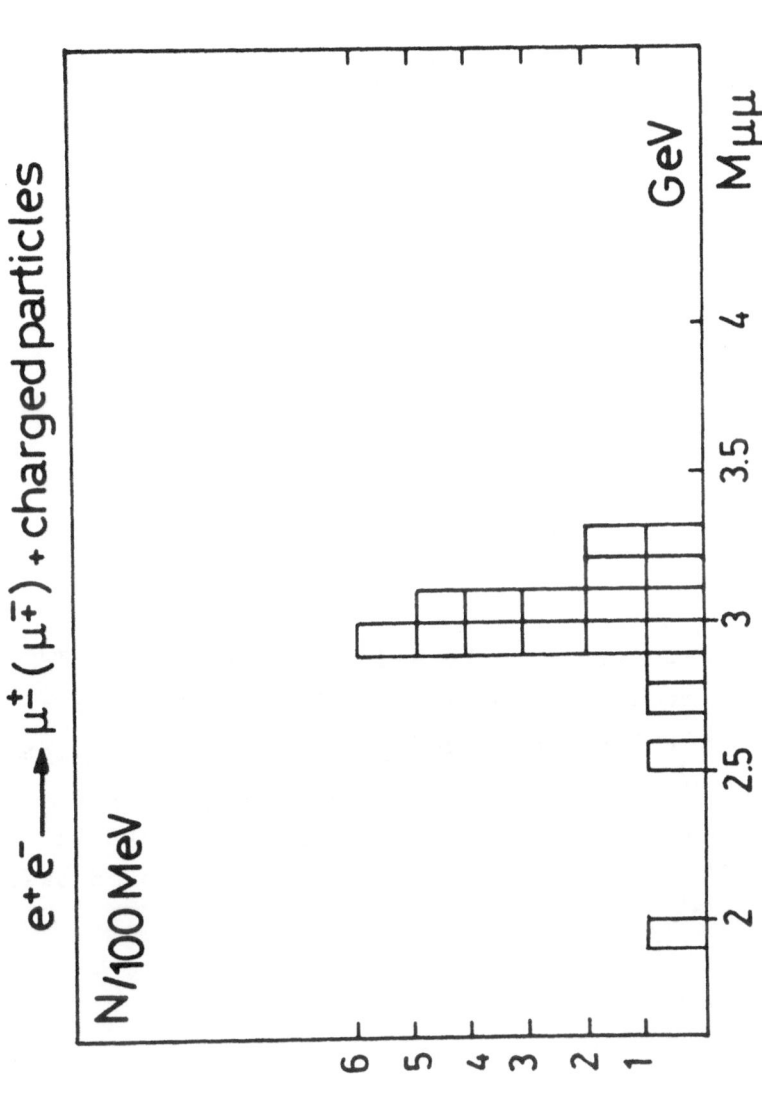

Fig. 10 The effective mass distribution for dimuon events in the multiprong events.

strongly connected with charm, at least not to D^+, D^o.
If the 4.415 were a $c\bar{c}$ enhancement the question arises
where is the charm content going to? It has been
suggested that more complicated quark structures should
exist in this energy region (named charm molecules) since
they supposedly consist of $c\bar{q}\bar{c}q$ systems[7]. It was
concluded that a fair fraction of the decay modes of
the "molecules" would involve a J/ψ in the final state
carrying the charm content of the molecule. Inclusive
J/ψ production can be studied in PLUTO and will be the
next topic to be discussed.

V.   INCLUSIVE J/ψ PRODUCTION

The $\mu^+\mu^-$ decay mode of the J/ψ provides the signal
for inclusive J/ψ production. Only one of both muons
is required to be identified. Events from

$$e^+e^- \rightarrow \mu^\pm + p^\mp + \text{charged particles} \qquad (9)$$

have been selected and the effective mass $M (\mu^\pm p^\mp)$ has
been calculated. The momentum of particle p has to be
larger than 1 GeV, and $|\cos\Theta| < 0.752$. All candidates
consistent with normal hadronic events have been rejected. The distribution for $M(\mu^\pm p^\mp)$ is shown in
Fig. 10. A clear clustering at the J/ψ mass is seen.
Events due to the radiative tail of the ψ'

$$e^+e^- \rightarrow \gamma\, e^+e^-$$
$$\psi' \rightarrow \pi^+\pi^- J/\psi \qquad (10)$$
$$\mu^-\mu^+$$

have been rejected using the kinematic constraints in
the reaction chain (10).

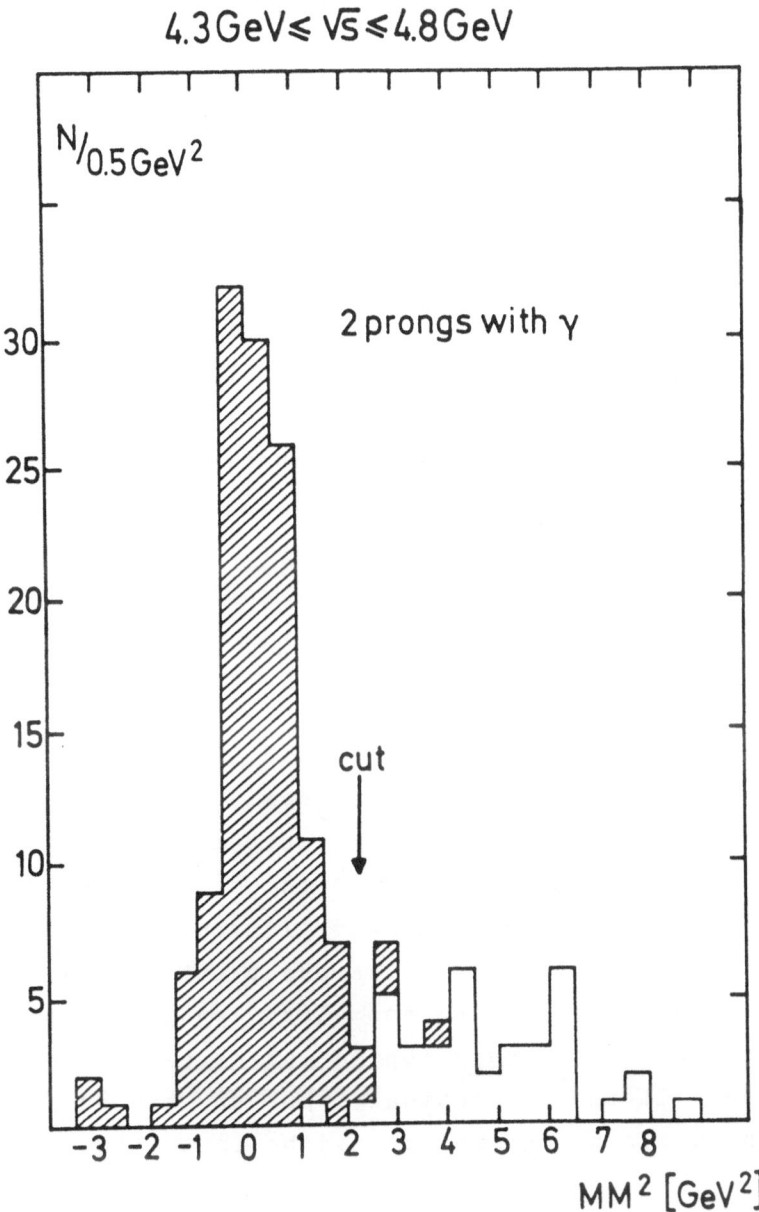

Fig. 11 Two prong events with at least one muon candidates and 1 or more photons. The hatched events have only one photon converted at a position compatible with the reaction $e^+e^- \to \mu^+\mu^-\gamma$.

Only 4 events in the J/ψ mass region remain
which may be due to the direct process

$$e^+e^- \to J/\psi + \text{hadrons}. \tag{11}$$

None of the four events is at 4.03 GeV or 4.415 GeV.
Correcting for acceptance and the $\mu^+\mu^-$ branching ratio
of the J/ψ we find a cross section for reaction (11)
of ~ 35 pb, averaged over $4.0 < \sqrt{s} < 5.0$ GeV.

The two s.d. upper limits at 4.03 GeV and 4.415 GeV
are ~ 100 pb, which is about 0.4 % of the total cross
section. One concludes that inclusive J/ψ production
is only a small fraction of the total cross section and
in fact much lower (by about a factor 20) than the
predictions from "charm molecule" models[7].

VI. ANOMALOUS INCLUSIVE MUON PRODUCTION

The new pieces in the total cross section, coming
from charm and heavy lepton, should be most easily
detected by observation of the (semi)-leptonic decay
modes of the new particles. One expects branching
ratios of the order of 15 % for these channels[8,9].

A distinction between heavy lepton and charm
signals is accomplished by the observation of additional
particles accompanying an anomalous lepton signal. The
decay modes of the heavy lepton can be estimated with
relatively little theoretical input and the result is
that the dominant channel should be in the two-prong
class[9]

$$e^+e^- \to L^+L^- \tag{12}$$
$$\llcorner\to \nu_L + \mu^- + \nu$$
$$\llcorner\to \nu_L + (1 \text{ charged particle})^+$$

with large missing energy and momentum since there
are at least three missing neutrinos. The charm
signal is expected to be dominantly in the multiprong
class

$$e^+e^- \rightarrow C \bar{C} \qquad\qquad\qquad (13)$$

Events with at least one track as a muon candidate
have been selected from our data. Tracks are accepted
as muon candidates if they fall into the muon-chamber
acceptance and have momentum larger than 1 GeV/c and
$|\cos\Theta| \le 0.752$.

The muon candidate sample contains

a) background from hadron punch through and decay;

b) Muons from known sources, mostly QED;

c) a possible signal from semi-leptonic decays of
new particles.

VIa) MULTIPRONG MUON EVENTS

We proceed in first discussing the multiprong
events,

$$e^+e^- \rightarrow \mu^\pm + \ge 2 \text{ charged hadrons.} \qquad (14)$$

The data sample is summarized in table II.

The background from source a) is estimated as
follows: all tracks that meet the requirements p > 1 GeV
and $|\cos\Theta| < 0.752$ and not reaching a muon chamber are
called hadrons. From the data at 3.1 resonance we
know that 2.8 ± 0.7 % of all hadrons appear as muons.
This fraction is being subtracted from the muon
candidates. Some QED events also enter the multiprong

Table II

Muliprong Events

E(GeV)	3.6	4.0 - 4.3	4.3 - 4.8
$L(nb^{-1})$	613	1660	2037
Hadrons	630	1684	2100
Muons	14	44	109
μ(corrected)	17.5	52.9	130.8
Hadron background	17.6	47.2	58.8
Events	0 ± 5	6 ± 12	72 ± 27
σ(pb)	< 36	< 40	79 ± 29

class due to radiative muon pair production with the
photon converted into an e^+e^- pair,

$$e^+e^- \rightarrow \mu^+\mu^-\gamma \qquad\qquad (15)$$
$$\rightarrow e^+e^-$$

and at least one of both muons identified. The events
are rejected due to a 1 constraint (if one electron
is not well measured) or 4 constraint fit. The only
other known source in the multiprongs is reaction
(11) [inclusive J/ψ production] with at least one
muon from J/ψ decay detected. This event class has
been discussed above, and has been subtracted on an
event by event basis.

Cross section are calculated assuming an isotropic
distribution of the muons and using the luminosities
and event numbers given in table II. No correction
for the momentum cut at 1 GeV has been applied.

From table II it is concluded that at energies
$\sqrt{s}$ = 3.6 GeV and $\sqrt{s}$ = 4.0 - 4.3 GeV no anomalous
signal remains. At the higher energy $\sqrt{s}$ = 4.3-4.8 GeV,
an effect is observed at the three standard deviation
level. The events may be due to charm pair production
(see (13)) with semi-leptonic decay of one of both
charm particles. There is no proof for this hypothesis
but the size of the signal is compatible with estimates
by Gronau et al.[10] which is based on the experimentally
observed inclusive electron spectrum by the DASP
group[11]. A small fraction of the multiprong events
is expected to come from heavy leptons[9].

VIb) TWO-PRONG MUON EVENTS

The largest fraction of two-prong muon events is
due to QED processes. In order to suppress events due

to elastic muon production

$$e^+e^- \rightarrow \mu^+\mu^-$$

a cut in the relative azimuthal angle $\Delta\Phi < 170^\circ$
has been applied. A large contribution is expected
from radiative muon pair production

$$e^+e^- \rightarrow \mu^+\mu^-\gamma \qquad\qquad (14)$$

A missing mass cut is applied. Since the resolution
in missing mass degrades with increasing $\sqrt{s}$, the cut
is energy dependent and events due to (14) are
rejected by

$$MM^2 > 1.4 \cdot (E_b/1.8)^2 \quad (GeV^2)$$

For about 60 % of the events (14) the photon converts
in the detector and the conversion point provides two
more constraints to the reaction (14).

In Fig. 11 the missing mass distribution is
shown for all events with one or more photons accompanying the two charged tracks. The hatched area
are events with only one photon observed and converted
at a position in the detector compatible with reaction
(14). This provides a check on the missing mass cut
applied on an event by event basis. Fig. 12 shows
the missing mass distribution for all two-prong events
with no other particles observed in the detector. Again,
the missing mass cut removes the contribution from
reaction (14). As a further check we examined the
direction of the missing momentum vector for the events
due to (14) which, in the majority of the cases points
to regions of the detector where little material for
conversion is available.

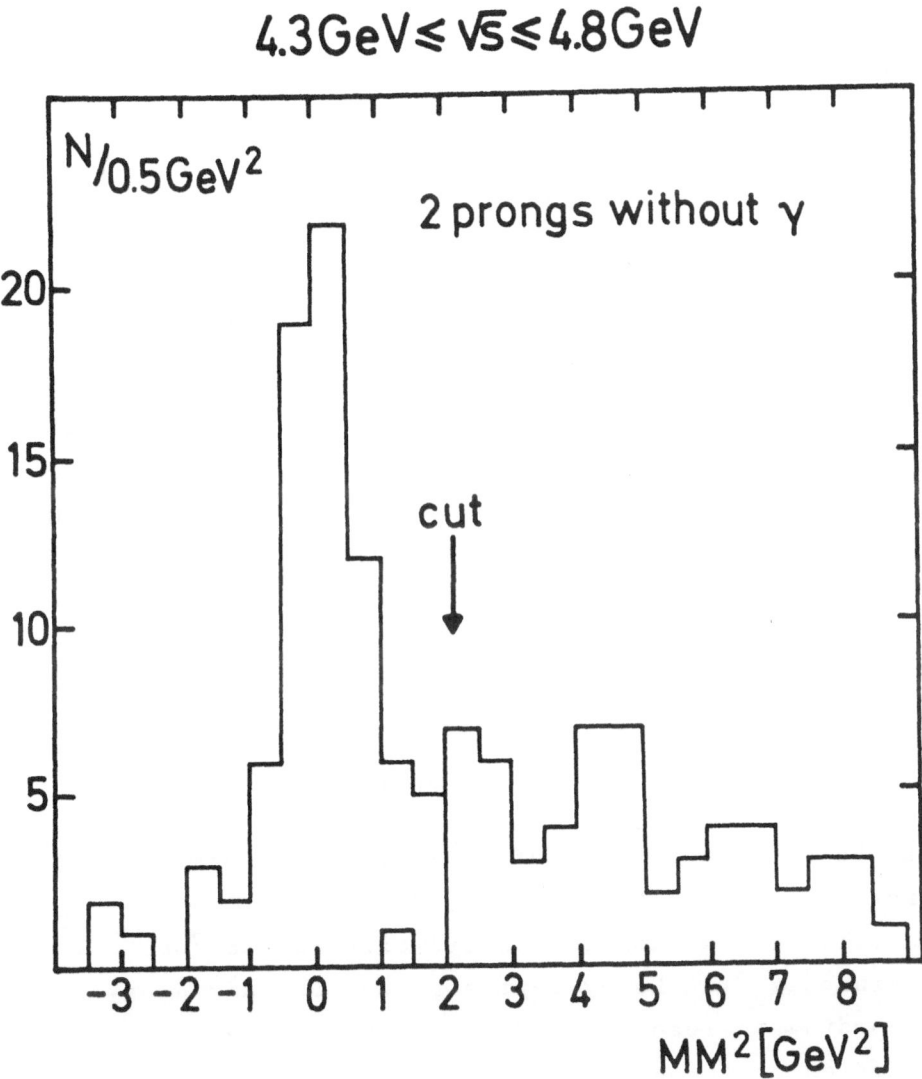

Fig. 12 Two prong muon events with no photon observed.

After subtraction of the QED events*, the background
from punch through and decay is estimated following
the procedure applied for the multiprong events. The
numbers are summarized in table III.

Cross section are calculated as for multiprongs,
again no correction for the momentum cut has been
applied. In addition no attempt has been made
correcting for the missing mass cut since it cannot
be done in a model independent way.

No anomalous signal is observed at $\sqrt{s}$ = 3.6 GeV
whereas sizeable signals are seen at $\sqrt{s}$ > 4.0 GeV.

A similar result has been found by the SLAC-LBL
group at SPEAR[12]. Fig. 13 shows a comparison where
I have scaled down the published cross sections from
Ref. 12 accounting for the different muon momentum
acceptances. Both experiments are in good agreement.
A threshold for the reaction

$$e^+e^- \to \mu^\pm + (1 \text{ charged particle})^\mp$$
$$+ (\text{missing energy}) \tag{15}$$

can be estimated at about $\sqrt{s}$ = 3.90 GeV from Fig. 13,
providing a mass M(L) of $\sim$ 1.95 GeV.

Two prong muon events are expected to come
dominantly from heavy lepton production and decay
according to reaction (12). If true, a certain
fraction of the events should be of the type,

$$e^+e^- \to L^\pm L^\mp \to \mu^\pm \nu\nu + e^\mp \nu\nu \tag{16}$$

_____

* Contributions from $\mu\mu\gamma\gamma$ final states have been
estimated by F. Gutbrod and T. Rek (private communication) and found to be small, less than 10 % of our
final results.

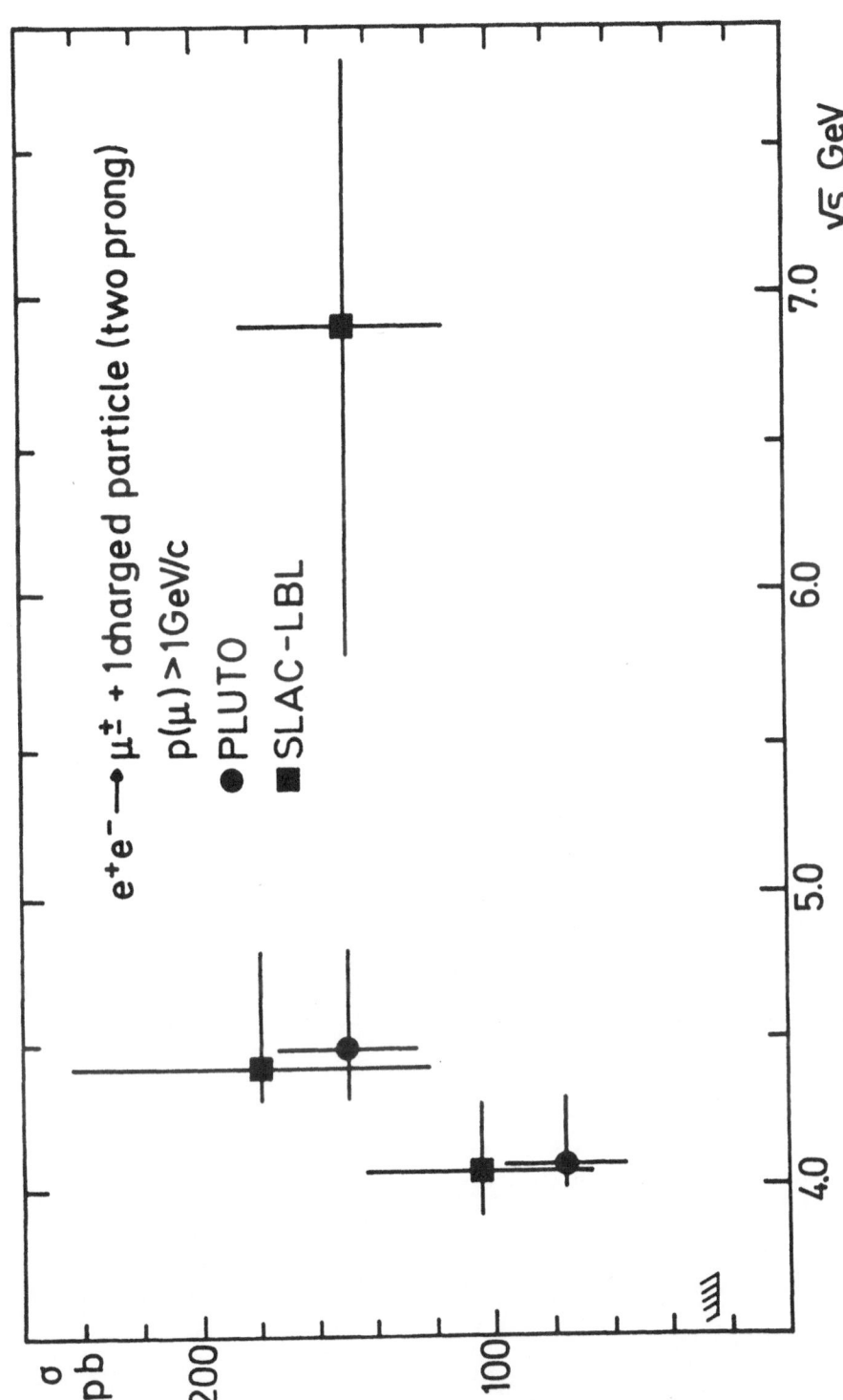

Fig. 13   The inclusive muon two prong cross section at function of √s.

Table III

Two Prong Events

E(GeV)	3.6	4.0 - 4.3	4.3 - 4.8
L(nb^{-1})	613	1660	2037
Hadrons	209	592	743
Muons	66	195	299
QED	60	153	206
μ(corrected)	7.3	53.1	108.8
Hadron background	5.9	16.6	20.8
Events	1.4 ± 2	36.5 ± 10	88 ± 15
σ(pb)	< 23	76 ± 21	149 ± 25

μe events with no other particles observed in the
detector.

The expected number of μe events can be estimated
using the following input:

a)   a heavy lepton mass of 1.95 GeV

b)   the production cross section is given by

$$\sigma(L^+ L^-) = \sigma_{\mu^+\mu^-} \cdot \frac{3\beta - \beta^3}{2}$$

c)   the decay mode $L \rightarrow \nu$ + hadrons gives 1 charged
     hadron in 87 % of all cases[9]. From Fig. 5 it
     can be seen that there is a large two prong
     cross section, in fact large enough to be
     consistent with the heavy lepton contribution
     being dominantly in the two prong channel

d)   taking account for the electron detection
     efficiency and geometrical acceptance.

One then expects 10 events of reaction (16).

VIc)   μe EVENTS

We do find 12 events of the type (16). Fig. 14
shows the missing mass squared vs. $\Delta\Phi$ for the events.
This is compatible with expectation.

An example of a μe event is shown in Fig. 15. In
this case the electron radiates a hard photon in the
first lead cylinder, the photon is converted in the
second lead cylinder, while the remaining low energy
electron spirals through the detector. This provides
a unique signal for the electron.

From the known electron and muon fake levels
(see part II of this talk), we estimate that less
than 1.5 events are background. This clearly

establishes a μe signal in our detector*. The electron
energy spectrum certainly is compatible with the
heavy lepton hypothesis as can be seen on Fig. 16.
The curve is a V-A leptonic decay spectrum folded with
the momentum dependent detection efficiency for
electrons. Since we know that charm particles are
produced in this energy region, and also semi-leptonic
decay modes have been observed[13], could the μe
signal be due to charm decay?

VId)   μe EVENTS IN THE MULTIPRONG CHANNEL

The μe events observed in the two prong class are
also compatible with the following charm decay mode

$$e^+ e^- \rightarrow \quad C^+ \quad + \quad C^- \tag{17}$$

$$K_L^0 \; \mu^+ \; \nu \qquad\qquad K_L^0 \; e^- \; \nu$$

Since the K_L^0 always excapes detection in PLUTO
this is the μe signature. Let us assume that all 12
μe events observed are due to this charm source. We
then expect 3 times more events where at least one K_L^0
is replaced by K_S^0 that decays to two pions almost
always close to the production vertex. These events
enter the multiprong class and we should expect about
30 events of this kind.

All multiprong events with a muon candidate have
been examined for electron candidates among the
additional tracks.

From the hadron electron fake level p (h → e) of
3.5 % we expect 7 (seven) events and find 6 (six). All

---

* Note added in proof: the number of μe events is doubled
now from analysis of the 5 GeV data.

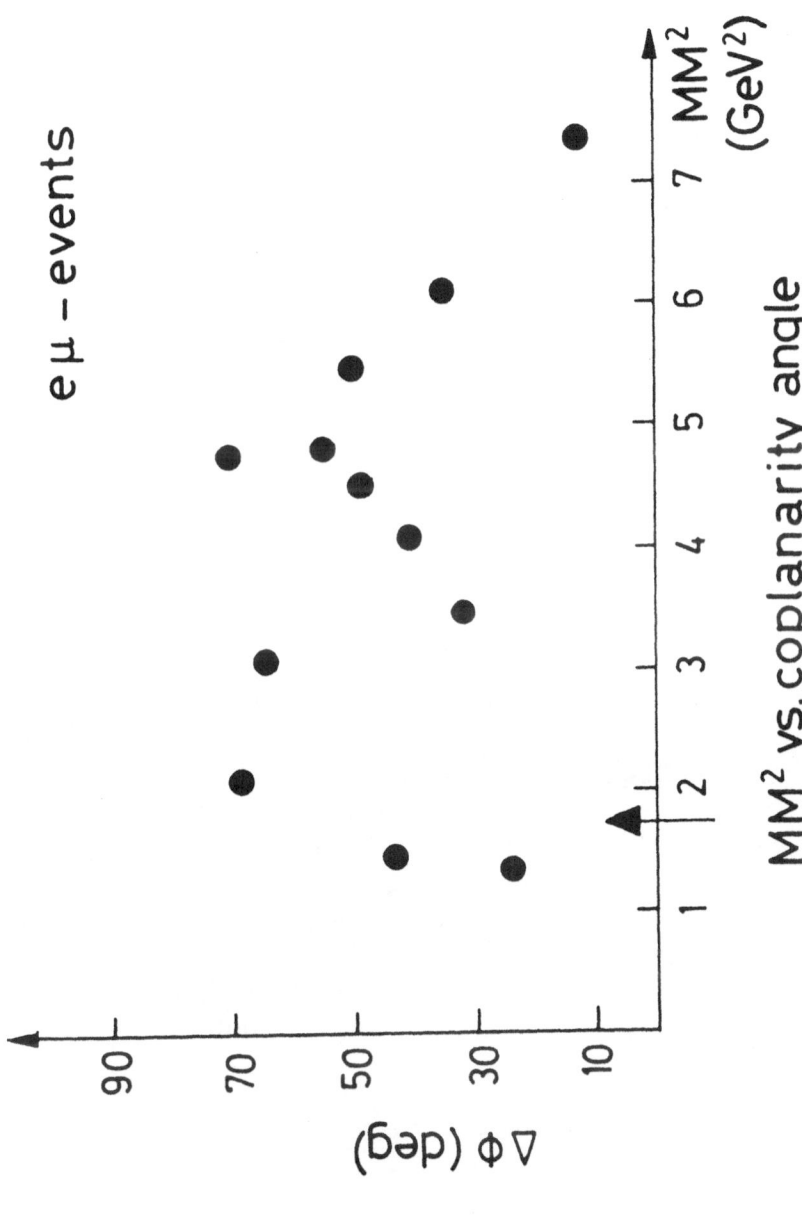

Fig. 14 The missing mass squared for eμ events vs. the difference in the azimuthal angle of the e$^-$ and μ$^-$tracks.

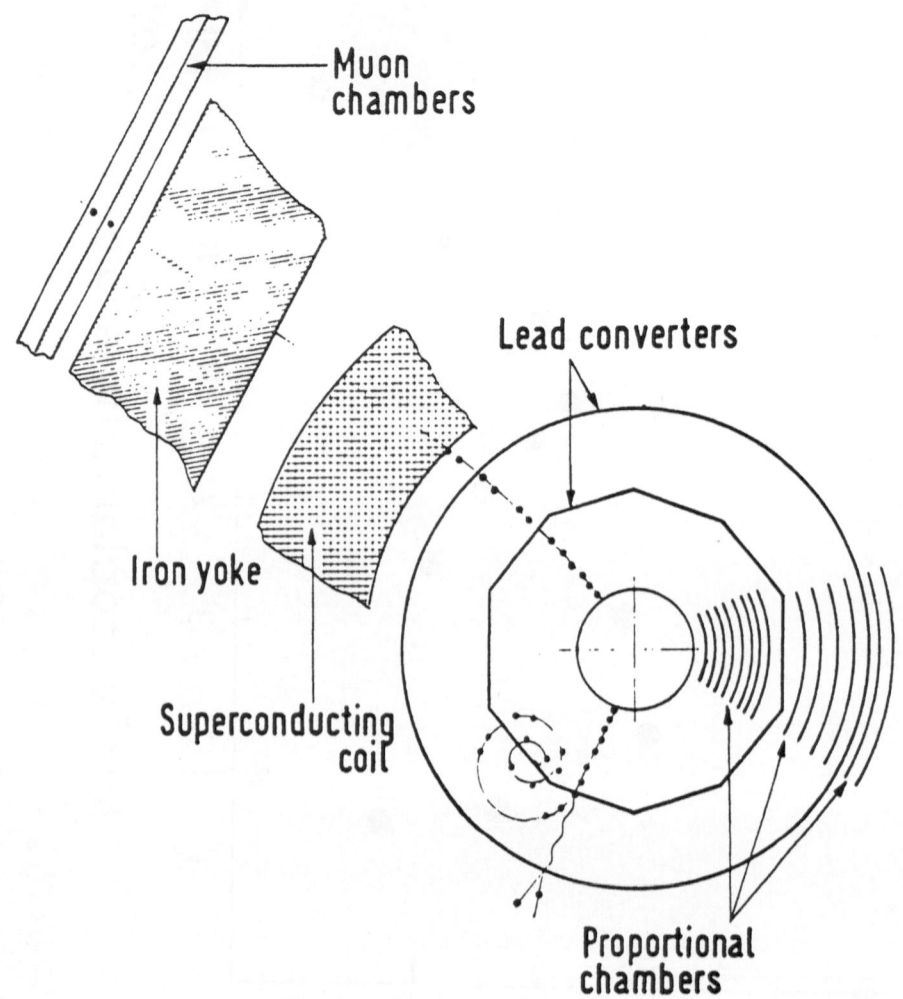

Fig. 15   An e μ event in PLUTO.

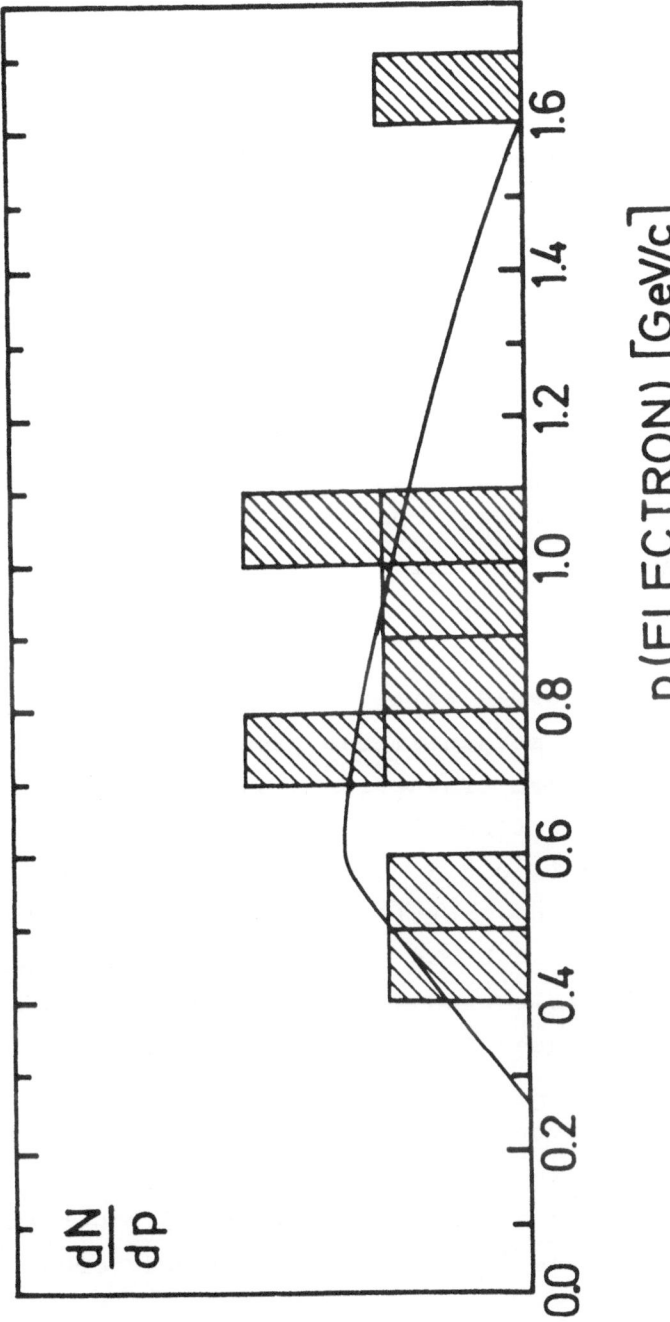

Fig. 16 The electron momentum spectrum for the 9 e μ events in the energy
region 4.3 < √s < 4.8 GeV. The curve is calculated for a V-A decay spectrum
of a 1.95 GeV heavy lepton including the electron detection efficiency as
function of momentum and normalized to the number of events observed.

6 events are therefore consistent with background. One
concludes that the μe events are very unlikely to come
from semileptonic charm decays*.

VII. EVIDENCE FOR A HEAVY LEPTON

This section summarizes all experimental facts,
that bear on the question of the existence of a heavy
lepton, and came out from the PLUTO data so far
analysed.

a) The contribution from the two prong channel

$$e^+e^- \rightarrow \text{two hadrons + missing energy}$$

to the total cross section does not fall off with
increasing energy (see Fig. 5). This exactly is
expected to happen if the heavy lepton has the
anticipated decay structure[9] leading dominantly to
the two prong final state.

b) We know, that charmed particles are produced above
4 GeV. From the GIM mechanism one expects about 2
kaons per event. Charm production should contribute
to the total cross section with ~ 1.5 R. The heavy
lepton contributing (~ 0.8 R) should not add to the
kaon yield since strange particle final states are
suppressed due to the Cabibbo angle. The K_s^o yield per
new event is compatible with the GIM mechanism only
if the heavy lepton does in fact not contribute to the
kaon yield (see Fig. 9).

c) Inclusive muon production is sizeable only in the
two prong channel. The inclusive μ two prong cross
section is large, has a threshold at about 3.9 GeV and
weak energy dependence (Fig. 13), as required for a

---

* This result is also confirmed from the 5 GeV data.

pointlike particle of mass ~ 1.95 GeV.

d) The proper number of μe events is observed with correct kinematic properties (Fig. 14).

e) No anomalous μe events together with charged hadrons are observed. This excludes charm as the source of the μe events.

Acknowledgements

I am grateful to F. Gutbrod, H. Fritzsch and T. Walsh for very useful discussions.

I also want to thank the DESY directorate for support.

REFERENCES

1. M. L. Perl et al., Phys. Rev. Lett. $\underline{35}$, 1489 (1975)
    M. L. Perl et al., Phys. Lett $\underline{63B}$, 466 (1976).

2. E. Lohrmann, Journal de Physique, Colloq. $\underline{C2}$, 37
    (1976).

3. J. Burmester et al., DESY 76/53, October 1976
    and Phys. Lett. $\underline{66B}$, 395 (1977).

4. J. Siegrist et al., Phys. Rev. Lett. $\underline{36}$, 700 (1976).

5. R. Brandelik et al., DESY 77/12, February 1977.

6. G. Goldhaber et al., Phys. Rev. Lett. $\underline{37}$, 255 (1976)
    I. Peruzzi et al., Phys. Rev. Lett. $\underline{37}$, 564 (1976).

7. C. Rosenzweig, Phys. Rev. Lett. $\underline{36}$, 697 (1976)
    M. Bander et al., Phys. Rev. Lett. $\underline{36}$, 695 (1976)
    A. De Rujula, H. Georgi, S. L. Glashow, Phys. Rev.
    Lett. $\underline{38}$, 317 (1977).

8. M. K. Gaillard, B. W. Lee, J. L. Rosner, Rev.
    Mod. Phys. $\underline{47}$, 277 (1975).

9. Y. S. Tsai, Phys. Rev. $\underline{D4}$, 2821 (1971)
    K. Fujikawa, N. Kawamoto, Phys. Rev. $\underline{D14}$, 59 (1976).

10. M. Gronau et al., DESY 76/62, November 1976.

11. W. Braunschweig et al., Phys. Lett. $\underline{63B}$, 471 (1976).

12. G. J. Feldman et al., Phys. Rev. Lett. $\underline{38}$, 117 (1977).

13. J. Burmester et al., Phys. Lett. $\underline{64B}$, 369 (1976).

14. J. Burmester et al., DESY 77/14, 1977 to be published
    in Phys. Lett.

THE SEARCH FOR HEAVY PARTICLES*

R. Michael Barnett

Stanford Linear Accelerator Center

Stanford, California 94305

ABSTRACT

Direct and indirect evidence for the existence of
a new heavy quark b and of new heavy neutral leptons N_e
and N_μ can be sought in neutrino and e^+e^- scattering.
These particles are expected to have right-handed currents. Discussion is given on the characteristics,
production and decay of hadrons such as $b\bar{b}$, $u\bar{b}$ and $d\bar{b}$,
and of the massive neutral leptons. Muon number viola-
tion with and without N_e and N_μ is considered.

I.   INTRODUCTION

Interest in gauge theories of the weak and electromagnetic interactions[1] has led many people in the last
few years to consider the possibility that new heavy
particles, both hadrons and leptons, may exist and might
have right-handed currents.[2] Two types of heavy particles

---

*Work supported by the Energy Research and Development
Administration.

will be considered here: (1) hadrons containing a quark, b,[3-7] of charge -1/3 with a right-handed coupling to u quarks, (2) massive neutral leptons[4-7], N_e and N_μ, with right-handed couplings to e and μ.

Several experimental results have motivated recent interest in new heavy particles. Originally the anomalous behavior of charged-current antineutrino scattering hinted at the possibility of a $(u,b)_R$ coupling. However, more substantial evidence concerning this coupling may be found in dilepton ($\mu^+\mu^-$ and μ^+e^-) antineutrino data at high energies, soon to be reported. Among the best evidence for the existence of b quarks would be the observation of $b\bar{b}$ states in e^+e^- scattering at CESR, PETRA and PEP colliding beams (located at Cornell, DESY and SLAC, respectively).

The first possible evidence for the couplings $(N_e,e)_R$ and $(N_\mu,\mu)_R$ arose from the atomic parity violation experiments. These experiments appear (given present experimental and theoretical uncertainities) to be inconsistent with the standard four-quark, four-lepton model with left-handed couplings; but they appear to be consistent with models where the electron has similar left- and right-handed couplings. More attention for these couplings occured when reports (discussed at this conference) were heard indicating that the decay μ→eγ _may_ have been observed. This decay would be expected in models with the above couplings. Also intriguing is the possibility that the reported $\mu^-\mu^-$ and $\mu^-\mu^+\mu^-$ events in neutrino scattering data (of which much more will soon be available) can be understood in terms of decays to N_e and N_μ. At PETRA and PEP there will be clear means of observing N_e and N_μ and determining their mass and mixing.

Many factors enter into the analysis of the relevant data. Among these are the scaling violations expected in asymptotically free gauge theories, and the branching ratios to muons of hadrons with b quarks. Also relevant to the possibility of b quarks and N_e and N_μ leptons with right-handed couplings are the various neutral-current neutrino interactions (νN deep inelastic, $\nu_\mu e$, $\bar{\nu}_\mu e$, $\bar{\nu}_e e$, νp, $\bar{\nu} p$, etc.). The presence of such couplings in gauge theories alters the expected behavior of neutral-current reactions (even at low energies).

The above subjects, which have been discussed by many different authors, will be summarized in this report. For simplicity and clarity, these topics will be discussed in the context of (only) three SU(2) × U(1) gauge models of the weak and electromagnetic interactions, shown in Table I. The Weinberg-Salam-Glashow-Iliopoulos-Maiani (WS-GIM) model[1,8] has no right-handed couplings. The CHYM model[3-6] has the $(u,b)_R$, $(N_e,e)_R$ and $(N_\mu,\mu)_R$ couplings, but no $(t,d)_R$ coupling. Other variations of the leptonic couplings are possible, and in particular, the versions[6] of this model found from the exceptional group E_7 have slightly different couplings. The vector model[7] has, in addition to the above couplings, the $(t,d)_R$ coupling. While the vector model is in some disrepute at the present time, it is useful for comparison purposes.

One is by no means limited to considerations of SU(2) × U(1) models. However, the larger groups can give more parameters and make it easier to fit data. Among alternative models are the $SU(2)_L \times SU(2)_R \times U(1)$ models of De Rujula, Georgi, Glashow[9] and Mohapatra,

TABLE I

WS-GIM Model

$$\begin{pmatrix} u \\ d \end{pmatrix}_L \quad \begin{pmatrix} c \\ s \end{pmatrix}_L \qquad\qquad u_R \quad d_R \quad c_R \quad s_R$$

$$\begin{pmatrix} \nu_e \\ e \end{pmatrix}_L \quad \begin{pmatrix} \nu_\mu \\ \mu \end{pmatrix}_L \qquad\qquad e_R \quad \mu_R$$

CHYM Model

$$\begin{pmatrix} u \\ d \end{pmatrix}_L \quad \begin{pmatrix} c \\ s \end{pmatrix}_L \quad b_L \ g_L \qquad \begin{pmatrix} u \\ b \end{pmatrix}_R \quad \begin{pmatrix} c \\ g \end{pmatrix}_R \quad d_R \ s_R$$

$$\begin{pmatrix} \nu_e \\ e \end{pmatrix}_L \quad \begin{pmatrix} \nu_\mu \\ \mu \end{pmatrix}_L \quad \begin{pmatrix} \nu_U \\ U \end{pmatrix}_L \qquad \begin{pmatrix} N_e \\ e \end{pmatrix}_R \quad \begin{pmatrix} N_\mu \\ \mu \end{pmatrix}_R \quad \begin{pmatrix} N_U \\ U \end{pmatrix}_R$$

Vector Model

$$\begin{pmatrix} u \\ d \end{pmatrix}_L \quad \begin{pmatrix} c \\ s \end{pmatrix}_L \quad \begin{pmatrix} t \\ b \end{pmatrix}_L \qquad \begin{pmatrix} u \\ b \end{pmatrix}_R \quad \begin{pmatrix} c \\ s \end{pmatrix}_R \quad \begin{pmatrix} t \\ d \end{pmatrix}_R$$

$$\begin{pmatrix} \nu_e \\ e \end{pmatrix}_L \quad \begin{pmatrix} \nu_\mu \\ \mu \end{pmatrix}_L \quad \begin{pmatrix} \nu_U \\ U \end{pmatrix}_L \qquad \begin{pmatrix} N_e \\ e \end{pmatrix}_R \quad \begin{pmatrix} N_\mu \\ \mu \end{pmatrix}_R \quad \begin{pmatrix} N_U \\ U \end{pmatrix}_R$$

Sidhu.[10] But no discussion of these models is given here.[11]

II.   EVIDENCE FOR b QUARK IN $\nu N \to \mu X$

A.   b Quarks vs. Scaling Violation

The anomalous behaviors with energy of the ratio, R_c, of $\bar{\nu}$ to ν total cross sections and of the anti-neutrino average y (where $y \equiv (E_\nu - E_\mu)/E_\nu$) have been interpreted[3-5,12-14] as evidence for the right-handed coupling of u quarks to a quark, b (of charge -1/3 and of mass 5-6 GeV). This argument has been clouded by the alternative possibility that the anomalies could be explained as scaling violations expected in asymptotically free gauge theories.[12-16] These anomalies have been well discussed in the literature and the old arguments will not be repeated here.

It would be useful to find tests of scaling violation which are independent of the $(u,b)_R$ coupling, and tests of the $(u,b)_R$ coupling which are independent of asymptotic freedom corrections. F. Martin and I[17] have sought such tests (as described below).

There are two types of scaling violation expected in asymptotically free theories. The dominant effect is the logarithmic change with Q^2 (and therefore with E) of the relative amounts of each type of quark, antiquark and gluons. The valence quarks (u and d) decrease and sea quarks ($\bar{u}$, $\bar{d}$, s, $\bar{s}$, c, $\bar{c}$, etc.) increase with Q^2. Since sea quarks are concentrated at small x, this causes the <x> to decrease with Q^2. A second effect (called shrinkage) is that the <x> of valence quarks decreases with Q^2; however, this effect (unlike the first) has little effect on R_c and <y>.

B.   Scaling Violation Independent of Model

These scaling violations can be measured experimentally by the quantity

$$R_x \equiv \frac{\sigma(x<0.15)}{\sigma(x>0.15)} , \qquad (2.1)$$

shown[17] in Fig. 1. Clearly for E<80 GeV this quantity is independent of the $(u,b)_R$ coupling. If the secondary effect of decreasing <x> for valence quarks were included, the rise of R_x from 5 to 80 GeV would be about 30 percent greater than that shown on the curve labelled AF. In order for scaling violation to be a viable explanation of the antineutrino anomalies, R_x would have to rise even more quickly than as shown in Fig. 1 (with label "AF").

C. $(u,b)_R$ Independent of Scaling Violation

Are there tests which are independent of the asymptotic freedom corrections (or any similar scaling violations)? Since the sea contributions are concentrated at small x, any effects due to increasing sea can be eliminated by considering only events at large x (defined here as x>0.15). In Fig. 2 the <y> for x>0.15 is shown. As anticipated, the WS-GIM model gives little rise with E in contrast to the CHYM model (with a $(u,b)_R$ coupling) which shows a significant rise. Similarly, one could examine R_c at x>0.15.

D.   Tests with Dileptons

Significant tests for a quark, b, with coupling $(u,b)_R$ might be found in data for

$$\nu N \rightarrow \mu^- \mu^+ X$$

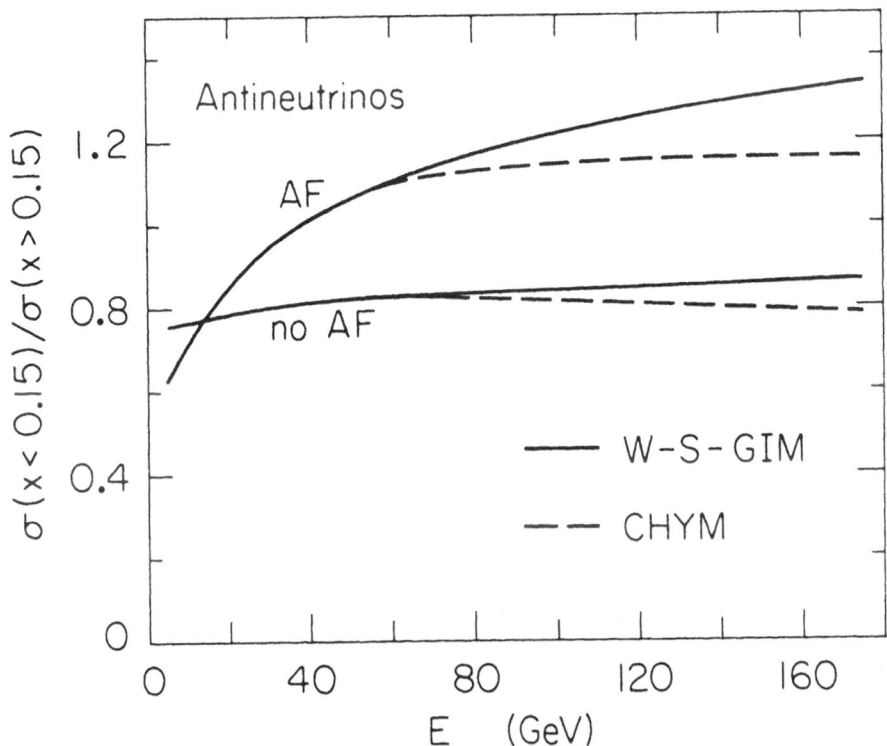

Figure 1. The energy dependence of R_x, the ratio of small x to large x cross sections, for antineutrino scattering. The calculated R_x are shown for the cases with and without asymptotic freedom (AF) corrections. Effects of shrinkage (discussed in text) are not included. For the CHYM model, m_b = 5 GeV.

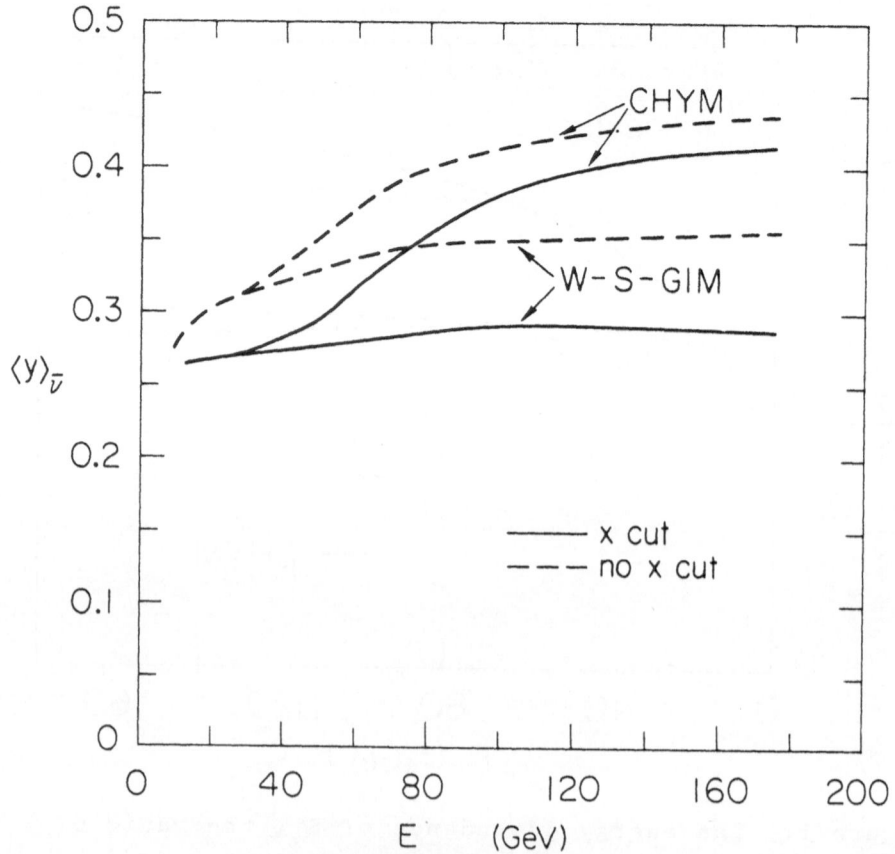

Figure 2. The average value of y for x>0.15 and for all x in antineutrino scattering. The efficiencies and cuts of the HPWF experiment (1976) are included. For the CHYM model, m_b=5GeV. In all cases asymptotic freedom cor-rections are incorporated.

and $\hspace{6cm}$ (2.2)

$$\bar{\nu}N \rightarrow \mu^+\mu^-X$$

(or with μ^-e^+ and μ^+e^-). An especially useful quantity[17,18] should be the following "ratio of ratios":

$$R_r \equiv \frac{\sigma(\bar{\nu}\rightarrow\mu\mu)}{\sigma(\bar{\nu}\rightarrow\mu)} \Big/ \frac{\sigma(\nu\rightarrow\mu\mu)}{\sigma(\nu\rightarrow\mu)} \ . \hspace{2cm} (2.3)$$

This ratio allows one freedom from knowing the branching ratio of charm to muons (assuming mesons and baryons are similar). However, some input is required on the relative $b\rightarrow\mu$ and $c\rightarrow\mu$ branching ratios, and this will be discussed shortly. Clearly the relative ν and $\bar{\nu}$ cross sections, which are difficult to determine, cancel out of R_r. While the separate ν and $\bar{\nu}$ ratios are very sensitive to the detection efficiency for the secondary (slow) muon, the ratio of ratios is found to be rather insensitive to this problem.[19,20] However, R_r should clearly not be calculated with cross sections from different experiments with different cuts. Somewhat more sensitivity (but still small) is found for detection efficiency for initial (fast)muons; however, this can be accounted for theoretically by model-independent means. There is some sensitivity to the amount of strange sea quarks relative to $\bar{u}$ and $\bar{d}$ sea quarks, but at higher Q^2 (and E), it is reasonable to assume SU(3) symmetry.

This ratio, R_r, is however very dependent on asymptotic freedom corrections as is shown[17] in Fig. 3 (where $s=\bar{s}=\bar{u}=\bar{d}$, $m_b=5$ GeV, and $\Gamma(b\rightarrow\mu)=\Gamma(c\rightarrow\mu)$ are assumed). Since b production is a valence process, an increasing sea to valence ratio goes against dimuon production through b quarks. It should be emphasized that the $(u,b)_R$ prediction (labelled CHYM here) would decrease if $m_b>5$ GeV

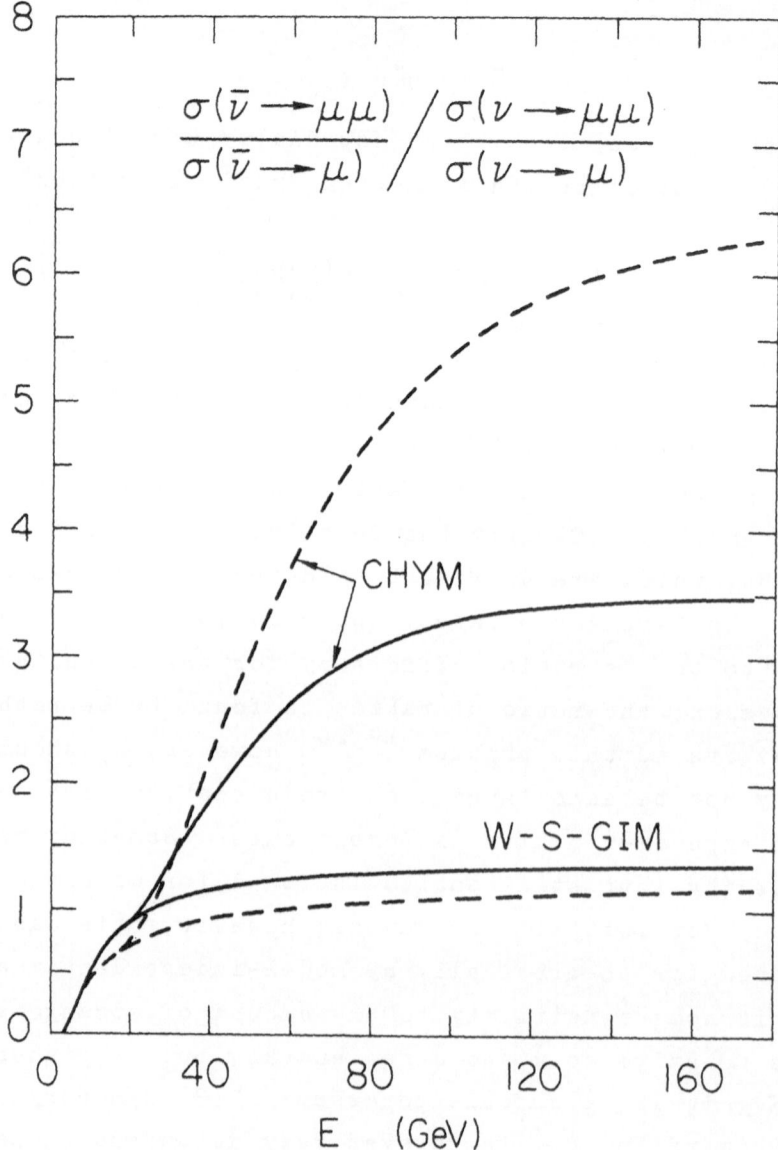

Figure 3. The antineutrino dimuon to single muon ratio divided by that ratio for neutrinos. The solid (dashed) curves include (exclude) asymptotic freedom corrections. For the CHYM model, it is assumed m_b = 5 GeV, B(b→µ) = B(c→µ), and s quarks = $\bar{u}$ quarks.

and if $\Gamma(b\rightarrow\mu)<\Gamma(c\rightarrow\mu)$ whereas the WS-GIM prediction cannot easily be increased.

      E.   <u>Branching Ratios of Y Mesons to Muons</u>

     In considering dilepton events (as discussed above), it is necessary to know the relative branching ratios of $b\rightarrow\mu$ and $c\rightarrow\mu$. This has been discussed by R. Cahn and S. Ellis,[18] and I will summarize their results. In this discussion I define the mesons containing b quarks as

$$Y^+ \equiv u\bar{b}, \quad Y^o \equiv d\bar{b} . \qquad (2.4)$$

There are two decay modes possible for Y^+ and two for Y^o as shown in Fig. 4:

    Modes (4a) and (4b): The width is

$$\Gamma(b\rightarrow u\mu\nu) = \frac{G^2 m_b^5}{192\pi^3} . \qquad (2.5)$$

With phase space (assuming simple three-body decays) included, the naive ratios for X in $b\rightarrow u+X$ are

$e\nu$	$\mu\nu$	$U\nu$	ud	cs	
1	1	0.5	3	1.5	(2.6)

(quarks have a factor 3 for three colors). However, one must remember that both c and U, formed in b decay, also decay into muons. Then given the branching ratio $B(U\rightarrow\mu)\approx20\%$, one finds the branching ratio

$$B_{a,b}(b\rightarrow\mu)\approx B(c\rightarrow\mu). \qquad (2.7)$$

This result can be obtained even if some of the above assumptions are relaxed.

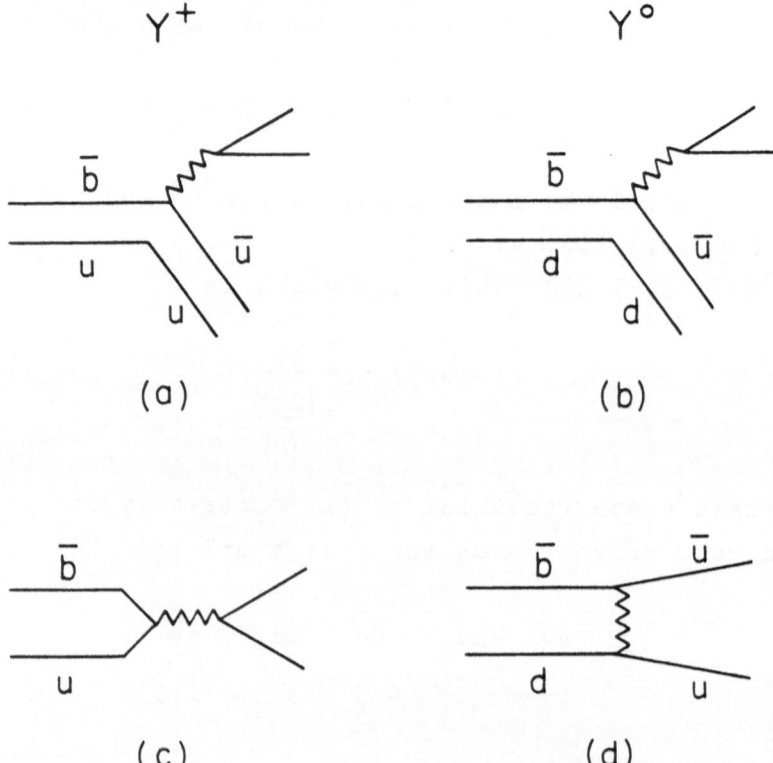

Figure 4.   The decay modes of $Y^+ \equiv u\bar{b}$ and $Y^0 \equiv d\bar{b}$.

Mode (4c): The widths for these annihilation diagrams are

$$\Gamma(Y^+ \to c\bar{s}) = 9 \frac{G^2 m_c^2}{2\pi} |\psi(0)|^2,$$

$$\Gamma(Y^+ \to U\nu) = 3 \frac{G^2 m_U^2}{2\pi} |\psi(0)|^2.$$

$$(2.8)$$

Since the width of the pseudoscalar Y^+, in this mode, is proportional to produced mass, the other possible products are negligible. Clearly we again find

$$B_c(b \to \mu) \approx B(c \to \mu). \qquad (2.9)$$

Note that this mode is Cabibbo-suppressed for D^+ mesons.

Mode (4d): The width for this mode which has left-right couplings (unlike D_0 which has left-left couplings) is

$$\Gamma(Y^0 \to u\bar{u}) = 2 \frac{G^2 m_b^2}{\pi} |\psi(0)|^2. \qquad (2.10)$$

Obviously

$$B_d(b \to \mu) = 0. \qquad (2.11)$$

In order to compare the relative rates, Cahn and Ellis needed to estimate the wave function at the origin $|\psi(0)|^2$. Since its dependence is on the reduced mass, one can compare with estimates of $|\psi(0)|^2$ for charmed mesons. Lane and Weinberg[21] use $|\psi(0)|^2 = 0.02$ and De Rujula, Georgi, Glashow[22] use $|\psi(0)|^2 = 0.55$. From these numbers one obtains (with $m_b = 5$ GeV)

$$\frac{B[(Y^{+}+Y^{0})\to\mu]}{B[(D^{+}+D^{0})\to\mu]} \approx \begin{cases} 0.97 \text{ (LW)} \\ 0.67 \text{ (DGG)} \end{cases} \qquad (2.12)$$

In conclusion, one finds that the b quark may have a slightly smaller branching ratio to muons than does the c quark.

III.   NEUTRAL CURRENT NEUTRINO SCATTERING

Charged-current scattering provides direct tests for heavy particles. The neutral-current neutrino interactions, while presumably lacking such direct production, do allow one to examine the structure of the theory even at low energies. The neutral-current, deep-inelastic scattering gave the earliest evidence of problems for the vector model of weak interactions.

Deep-inelastic scattering through neutral currents has been widely discussed,[23] so only brief remarks will be added here. F. Martin and I have investigated[17] the effects of corrections expected in asymptotically free theories. To the extent that one considers only the ratios of neutral to charged currents

$$R_{\nu} \equiv \frac{\sigma(\nu N \to \nu X)}{\sigma(\nu N \to \mu X)} , \qquad (3.1)$$

the effects tend in general to cancel. By including all energy dependent effects (asymptotic freedom corrections, experimental cuts, new currents, etc.) in theoretical calculations of the numerator and denominator of R_{ν}, we could determine the best $\sin^2 \theta_W$ for each model (considering the three experiments which occur at different energies); this is shown later in this section.

With this determination of $\sin^2 \theta_W$, we[17] could address the "problem" that rising $\sigma(\bar{\nu}N \to \mu X)/E$ and "constant"
$R_{\bar{\nu}}$ (comparing three experiments) implies $\sigma(\bar{\nu}N \to \bar{\nu}X)/E$ must
be rising (suggesting, perhaps, charm-changing neutral-
currents). In fact there is no problem for the WS-GIM
or CHYM models (see Fig. 5): (a) Any rise in $\sigma(\bar{\nu}N \to \mu X)$
due to asymptotic freedom corrections is approximately
matched in $\sigma(\bar{\nu}N \to \bar{\nu}X)$; (b) Accounting for experimental
cuts[24-26] would lower both high energy points by 20-30%
(from values shown) so $R_{\bar{\nu}}$ is not really constant; (c)
The error bars are large.

There are three other types of neutral-current experiments against which models can be tested. The νp
elastic scattering experiments[27,28] appear reasonably
consistent[23,29] with the WS-GIM and CHYM models and inconsistent with the vector model. In the next few months,
greatly increased statistics and new calculations of
background should make this experiment[30] a crucial test.

In the elastic scattering[31-33] of ν and e and in the
atomic parity violation experiments,[34] the neutral currents also probe the weak interactions of the electron.
ν_e elastic scattering, as discussed elsewhere,[23,25] is
consistent with the electron having (in addition to (ν_e,
e)$_L$) the couplings e_R or (N_e,e)$_R$ but not (E^+,E^o,e^-)$_R$ or
(e^-,E^{--})$_R$, etc. So the WS-GIM and CHYM models (as described in Section I) are consistent with present data.

The search for parity violation in atomic physics
via weak neutral-currents is potentially a critical test
of the electron coupling. Present experiments on bismuth
still have large systematic uncertainties and some people[36]
argue that the atomic-nuclear theory is also uncertain.
If the above complaints are ignored, the apparent lack
of parity violation in the Seattle and Oxford experiments[34]

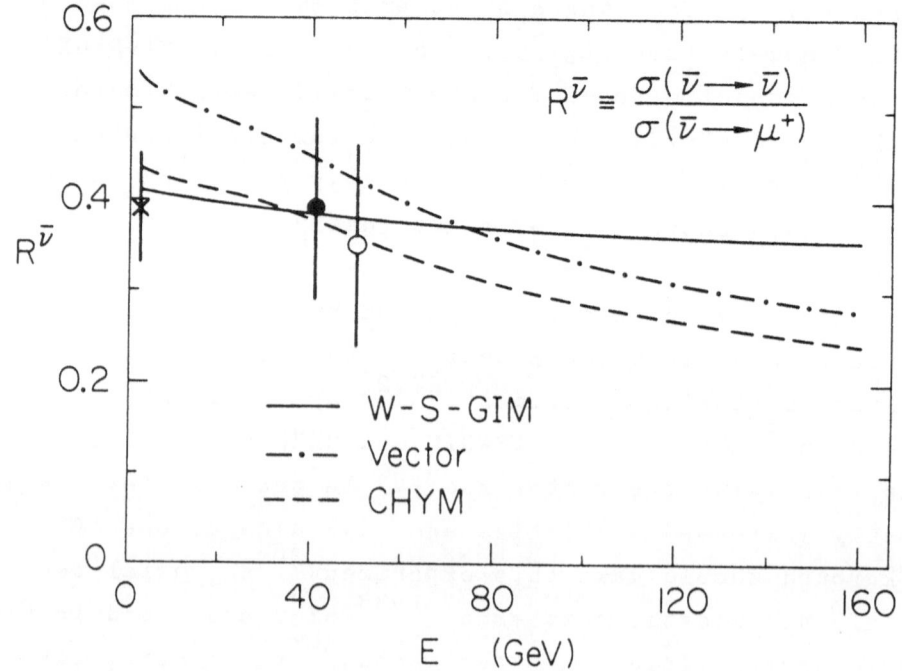

$$R^{\bar{\nu}} \equiv \frac{\sigma(\bar{\nu} \longrightarrow \bar{\nu})}{\sigma(\bar{\nu} \longrightarrow \mu^+)}$$

—— W-S-GIM
—·— Vector
--- CHYM

$R^{\bar{\nu}}$

E   (GeV)

Figure 5.   The ratio $R_{\bar{\nu}}$ of neutral to charged current antineutrino scattering.   The curves include asymptotic freedom corrections.   The point at E = 2 GeV from Gargamelle[26] has been corrected for experimental cuts, since at E = 2 GeV it is model-independent (assuming scaling). The points at E≈40 GeV (HPWF[24]) and at E≈50 GeV (CF[25]) have not been corrected.   In these models, these points would be lowered by 20-30% by correcting for cuts.

casts doubt upon the WS-GIM model and is suggestive of models (such as CHYM) where the electron's neutral-currents are pure vector, assuring no parity violation irrespective of $\sin^2 \theta_W$. Perhaps, more decisive will be the proposed parity-violation searches on hydrogen.[37]

The N_e and N_μ suggested by νe elastic scattering and by the atomic parity-violation experiments are discussed in later sections.

All of these neutral-current results are summarized in Figs. 6-8.

IV. THE b$\bar{\text{b}}$ MESONS

A. Spectroscopy of b$\bar{\text{b}}$

An important test of the existence of a b quark will be the search for b$\bar{\text{b}}$ states in e^+e^- scattering. Let us define β to be the lowest-lying vector b$\bar{\text{b}}$ state (the equivalent of ψ). Eichten and Gottfried (EG)[38] have estimated the β, β',... and $Y(\equiv u\bar{\text{b}})$ masses in a linear-Coulombic potential of the form

$$V(r) = -\frac{4}{3}\alpha_s \left(\frac{1}{r}\right) + \frac{r}{a^2} \qquad (4.1)$$

(while I will consider only the case of b quarks and masses of 5-6 GeV, the work of EG is more general). From charm, EG found a = 2.2 GeV^{-1}, $\alpha_s(m^2=10)=0.2$ and $m_c = 1.35$ GeV. The β mass is calculated from

$$M_\beta(m_b) = 2m_b + E_0(m_b) + \Delta(m_b), \qquad (4.2)$$

where E_0 is the ground state eigenvalue of the Schrödinger equation and Δ is the "zero of energy" term (approximated by EG as $\Delta(m_b)=\Delta(m_c)(m_c/m_b)$, $\Delta(m_c)=-225$ MeV). The result

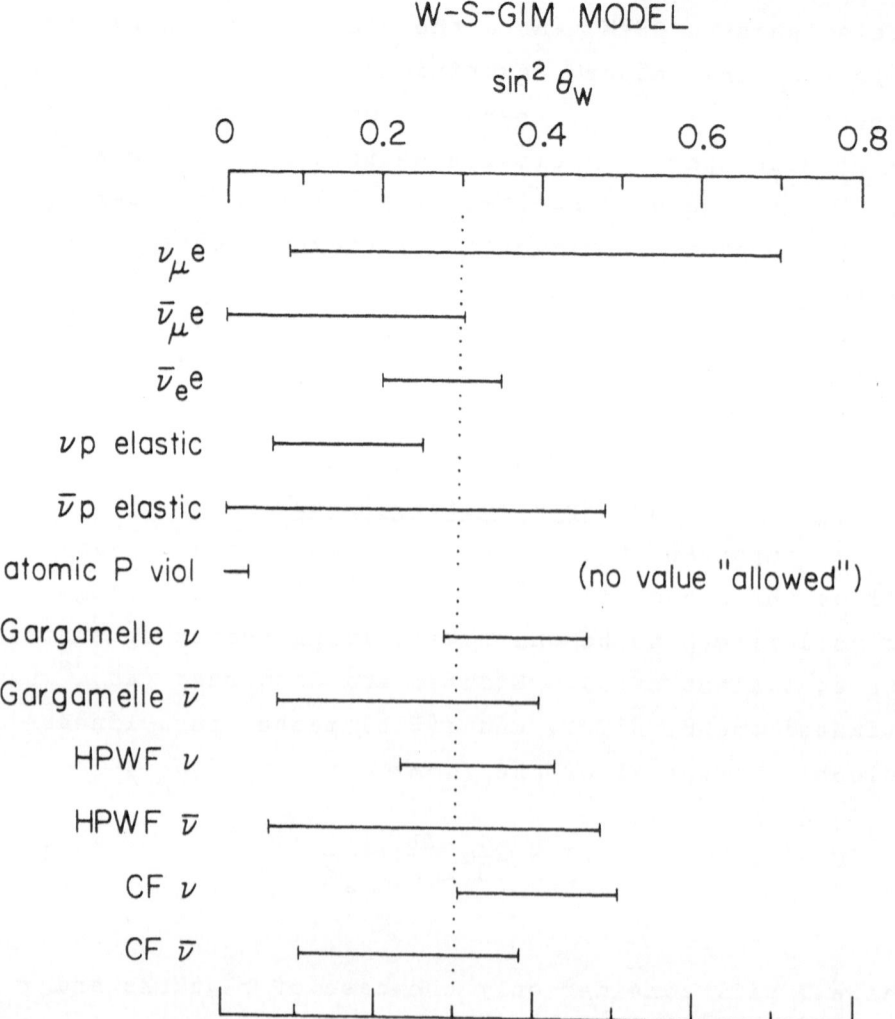

Figure 6. The allowed values of $\sin^2 \theta_W$ for various
neutral-current experiments. The lines show the regions,
for the WS-GIM model, within one standard deviation (two
for νe experiments) of experiment. The bottom six experiments are the deep-inelastic neutrino experiments
where theory includes asymptotic freedom corrections and
experimental cuts.

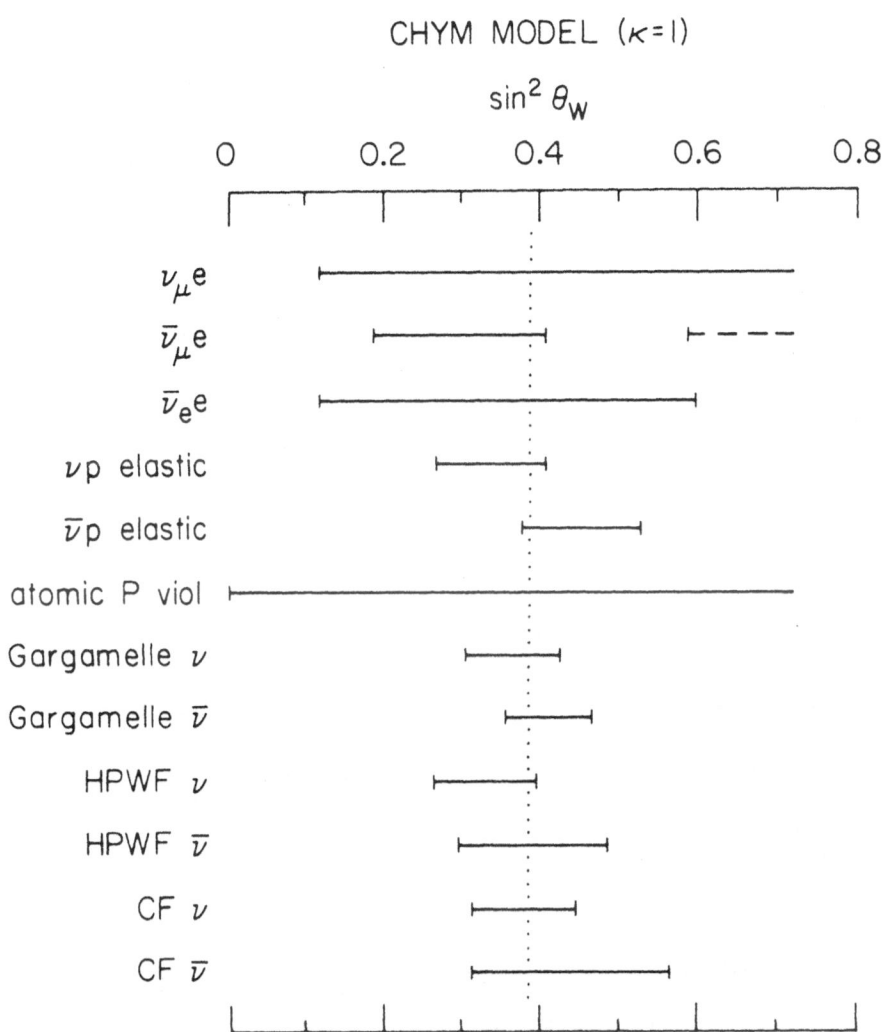

Figure 7. The same as Fig. 6, but for the CHYM model.

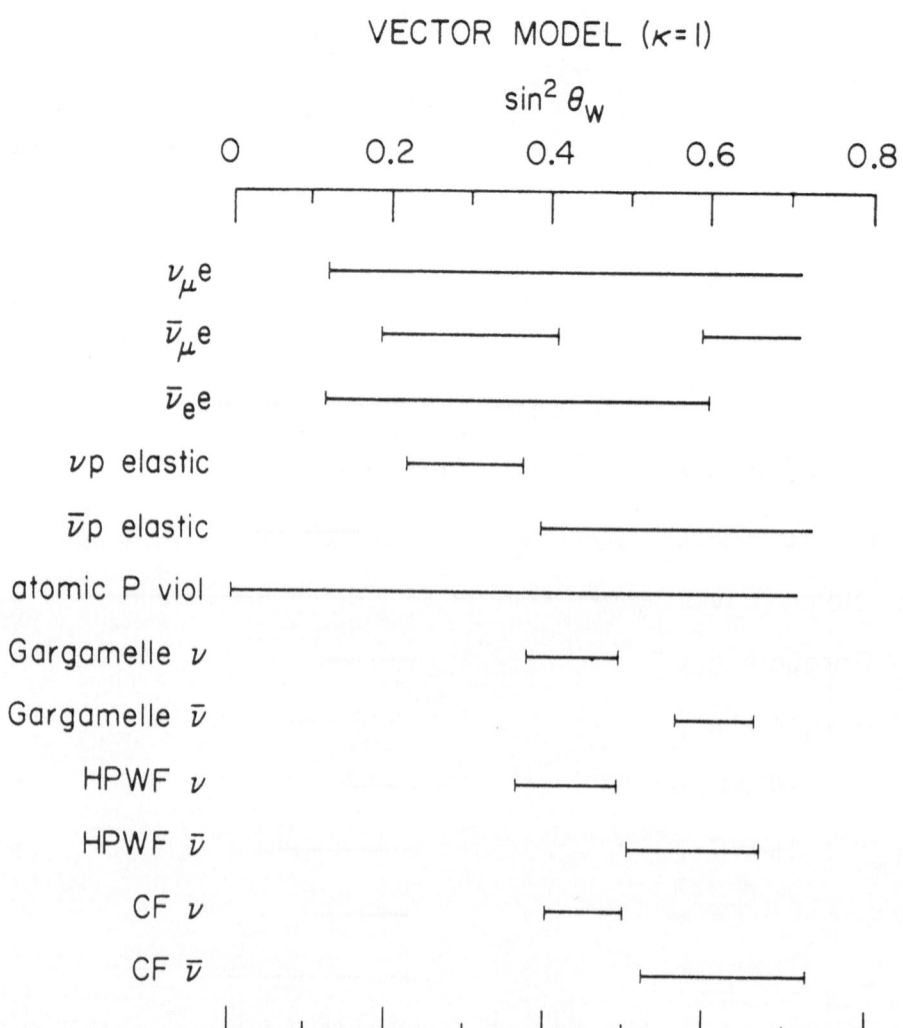

Figure 8.   The same as Fig. 6, but for the vector model.

of this calculation is (for $m_b = 5$ GeV)

$$m_\beta = 10.3 \text{ GeV}, \qquad m_{\beta'} = 10.7 \text{ GeV}, \qquad m_{\beta''} = 11.05 \text{ GeV}.$$
$$(4.3)$$

Similarly the mass of $Y \equiv u\bar{b}$ and $d\bar{b}$ can be calculated from

$$m_Y = m_D + (m_b - m_c) + \frac{3}{4} \left(1 - \frac{m_c}{m_b}\right) (m_{D^*} - m_D). \qquad (4.4)$$

The last term is, of course, a simple means of estimating the hyperfine contribution. EG find for $m_b = 5$ GeV

$$m_Y = 5.6 \text{ GeV} . \qquad (4.5)$$

It is now clear, since $2m_Y = 11.2$ GeV, that not only are β and β' below threshold for $Y\bar{Y}$ production, but also β'' (and β''' if $m_b \gtrsim 6$ GeV which is quite possible). In adtion there are two sets of p-wave states below threshold. This is shown in Fig. 9 (from EG[38]). The enormous number of gamma and hadron decays which result are shown in Figs. 10 and 11 (also Ref. 38).

B. Can β Be Seen at the New Accelerators?

It will be quite difficult to see the β states (assuming they exist) in e^+e^- scattering[39] at the new accelerators PETRA (at DESY), CESR (at Cornell) and PEP (at SLAC). The reasons for this are: (a) the b charge is half the c charge; (b) Γ_{ee} is proportional to m_b^{-2} and therefore the integrated cross section is also; (c) the expected experimental resolution will be much worse than at SPEAR; (d) the production is proportional to m_β^{-2}.

The integrated area under a resonance in e^+e^- scattering is given by

$$\Sigma_\beta = \frac{6\pi^2}{m_\beta^2} B_{had} \Gamma_{ee} \ . \tag{4.6}$$

B_{had} is the branching ratio to hadrons and should be close to 1.0. Γ_{ee} is proportional to

$$\Gamma_{ee} \propto \frac{|\psi(0)|^2}{m_b^2} Q^2 \ . \tag{4.7}$$

Eichten and Gottfried[38] found $\Gamma_{ee}(\beta)$ by comparison with $\Gamma_{ee}(\psi)$. Using potential (4.1) (which gives $|\psi(0)|^2$ in-creasing by a factor of ten from c to b), they found

$$\Gamma_{ee}(\beta) = 0.7 \text{ keV} \tag{4.8}$$

which gives for $m_b = 5$ GeV

$$\Sigma_\beta = 130 \text{ nb-MeV} \tag{4.9}$$

(and $\Sigma_{\beta'} = 90$, $\Sigma_{\beta''} = 70$ nb-MeV). Compare this with $\Sigma_\psi = 10,000$ nb-MeV.

Background can be estimated as follows. If $R(e^+e^-) \approx 5.3$ at $\sqrt{s} = 10.3$ GeV, then the usual cross section is $\sigma_{had} = 4.3$nb. The experimental resolution at PEP is

$$\Delta m = 16 \text{ MeV at } \sqrt{s} = 10.3 \text{ GeV}, \tag{4.10}$$

giving a background area of

$$\sigma_{had}\Delta m = 70 \text{ nb-MeV} \ . \tag{4.11}$$

From Eqs. (4.9) and (4.11) we see that signal to background is only 2 for β of mass 10.3 GeV (1.3 for β' and
1.0 for β'') in comparison with 250 for ψ above background.

Given that the expected luminosity for PEP and PETRA
at $\sqrt{s}$ = 10 GeV is the same as at SPEAR (10^{31} cm^{-2} sec^{-1}),
one can compare with the present upper limits on Σ for
narrow resonances.[40] For most of the lower energy region
the limits are not good, but at higher energies much more
stringent limits are set:

$\sqrt{s}$(GeV)	Σ(nb-MeV)
3.2-5.4	1000
5.4-7.0	100
7.0-7.4	30

The conclusion (comparing with Eq. (4.9)) is then that if
experiments at PEP (etc.) scan at the same level as the
early SPEAR scans (the lower energies), they will definitely miss all β states (and even states of 2/3 charge
quarks). However, if enough time for a careful and fine
scan is alloted, then the β states can be found.

One can also estimate the maximum value (above
background) which $R(e^+e^-)$ will reach, given the resolution (4.10). In the Breit-Wigner approximation

$$R_{max} \approx 169 \, \frac{\Gamma_{ee}(keV)}{\Delta(MeV)} \, B_{had} \approx 7, \qquad (4.12)$$

compared to 300 for ψ.

In conclusion, experimentalists should be aware that
the search for $\beta \equiv b\bar{b}$ states will be difficult.

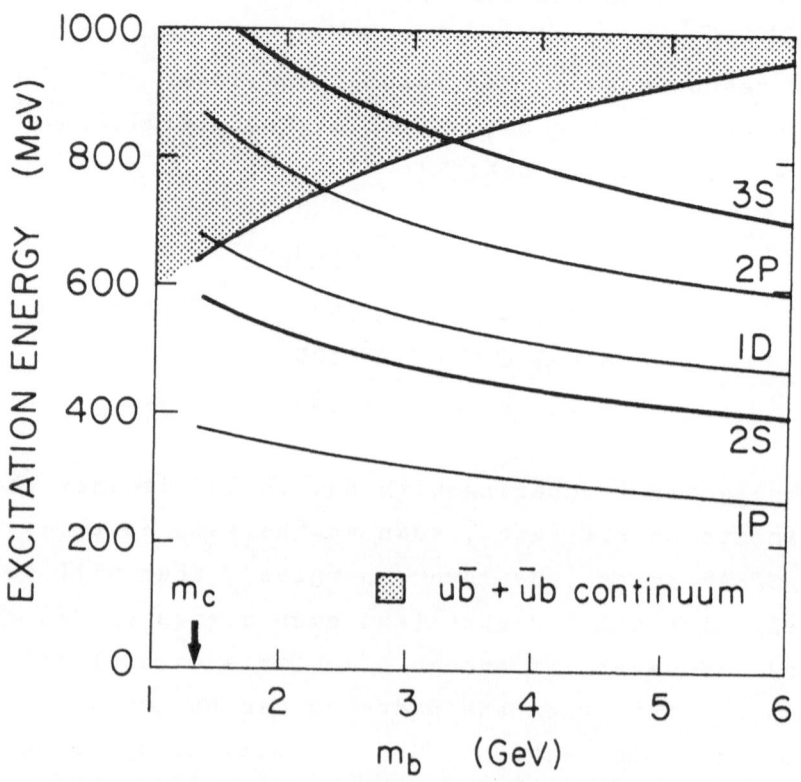

Figure 9. The splitting between the lowest vector $b\bar{b}$ state and the radial excitations and p-wave states, as a function of m_b. The region where $Y\bar{Y}$ production can occur is shaded ($Y\equiv u\bar{b}$ and $d\bar{b}$). This figure was taken from Ref. 38.

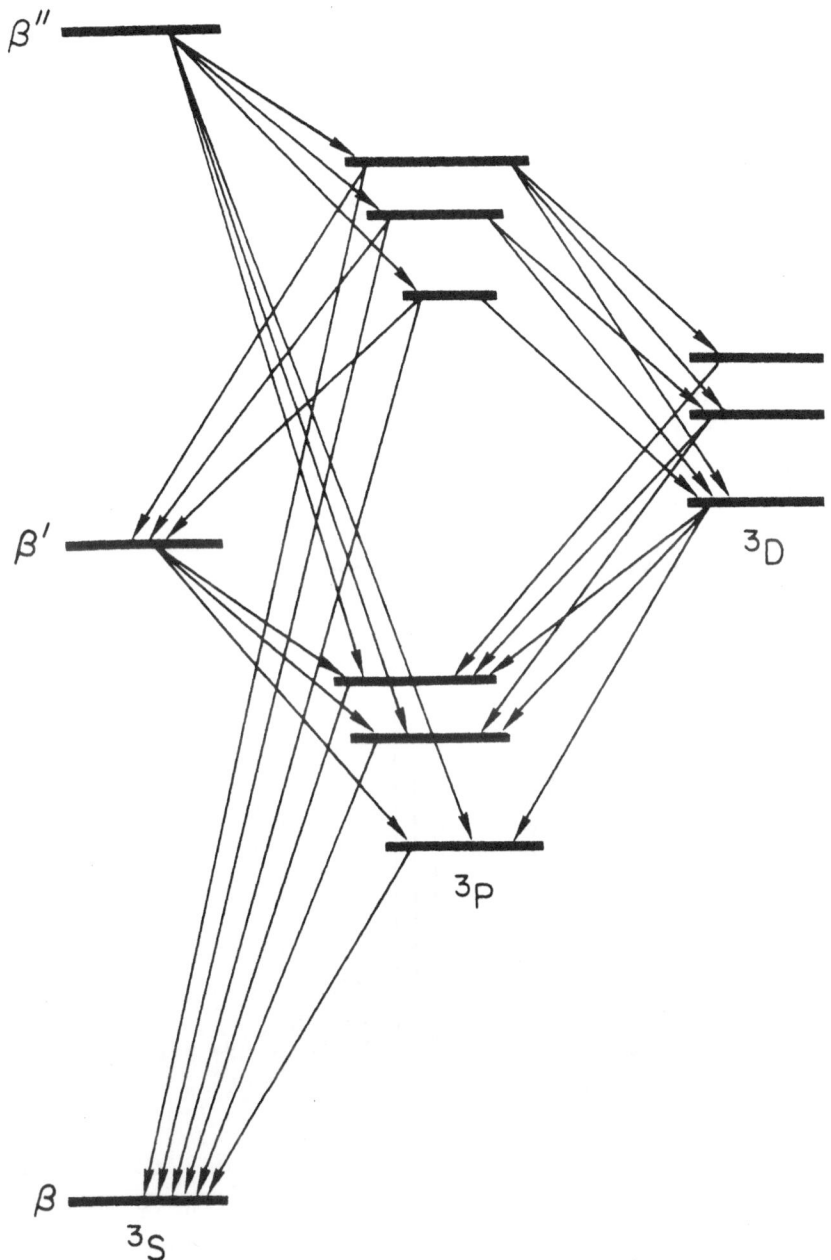

Figure 10. The allowed gamma transitions for $\beta (\equiv b\bar{b})$ states if m_b = 5 GeV. This figure was taken from Ref. 38.

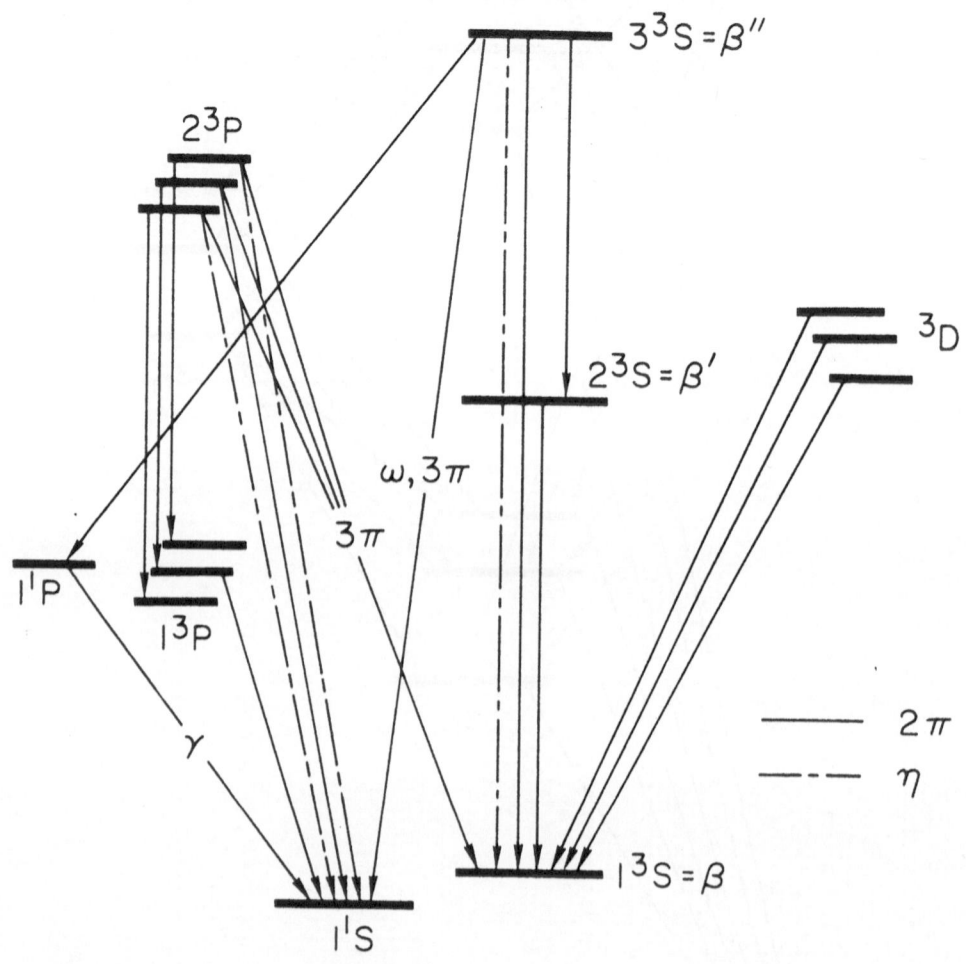

Figure 11. The allowed hadronic transitions for $\beta(\equiv b\bar{b})$ states if $m_b = 5$ GeV. This figure was taken from Ref. 38.

C.   Branching Ratios for β

Before considering β, let us consider the branching ratios[41] of ψ:

$$\Gamma(\psi \rightarrow \text{hadrons}) = 59 \text{ keV,} \qquad (4.13)$$

$$\Gamma(\psi \rightarrow e^+ e^-) = \Gamma(\psi \rightarrow \mu^+ \mu^-) = 4.8 \text{ keV,} \qquad (4.14)$$

so the $e^+ e^-$ branching ratio is

$$B(\psi \rightarrow e^+ e^-) = 7\% \text{ .} \qquad (4.15)$$

The decay of ψ to hadrons can occur two ways: through gluons and through a photon. Given that $\Gamma_{ee} = 4.8$ keV and $R(e^+ e^-) \approx 2.3$ near ψ, one can estimate

$$\Gamma(\psi \rightarrow \gamma \rightarrow \text{hadrons}) \approx 11 \text{ keV,} \qquad (4.16)$$

so that

$$\Gamma(\psi \rightarrow \text{gluons} \rightarrow \text{hadrons}) \approx 48 \text{ keV.} \qquad (4.17)$$

For β we have from Eq. (4.8) $\Gamma_{ee}(\beta) = 0.7$ keV. Assuming $R(e^+ e^-)$ is still approximately 5.3 near β we find[39]

$$\Gamma(\beta \rightarrow \gamma \rightarrow \text{hadrons}) \approx 3.6 \text{ keV.} \qquad (4.18)$$

Following the arguments of Appelquist and Politzer,[42] the assumptions that gluon decay occurs through three gluons and that $\alpha_s \approx 0.2$ at ψ (both of which can be disputed) would lead to

$$\Gamma(\beta \to \text{gluons} \to \text{hadrons}) \approx 16 \text{ keV} \qquad (4.19)$$

and

$$B(\beta \to e^+ e^-) \approx 4\%. \qquad (4.20)$$

V.   MUON NUMBER VIOLATION
A. General Remarks

A brief discussion will be given here of three models (all in SU(2) × U(1) which lead to violations of muon number in processes such as $\mu \to e\gamma$, $K_L \to \mu e$, $\mu \to 3e$ and $\bar{\mu}$ + nucleus $\to$ e + nucleus. Needless to say, recent interest in this subject has been sparked by the unofficial reports from SIN in Zurich that several events have been observed which look like $\mu \to e\gamma$ and are above the expected background. But it need not concern theorists (yet) whether this particular experiment has or has not observed $\mu \to e\gamma$. The question is whether we expect muon number violation. These reports have reminded us that such experiments provide us with an interesting tool for understanding the structure of the weak and electromagnetic interactions.[43]

B. Higgs Exchange

In context of the WS-GIM model (although it is applicable elsewhere) Bjorken and Weinberg[44] consider the interactions of leptons with Higgs scalars:

$$H = -g_1 \overline{\begin{pmatrix} \nu_\mu \\ \mu^- \end{pmatrix}_L} \begin{pmatrix} \phi_1^+ \\ \phi_1^o \end{pmatrix} \mu_R^- - g_2 \overline{\begin{pmatrix} \nu_e \\ e^- \end{pmatrix}_L} \begin{pmatrix} \phi_2^+ \\ \phi_2^o \end{pmatrix} \mu_R^-$$

$$-g_3 \overline{\begin{pmatrix} \nu_\mu \\ \mu^- \end{pmatrix}_L} \begin{pmatrix} \phi_3^+ \\ \phi_3^o \end{pmatrix} e_R^- - g_4 \overline{\begin{pmatrix} \nu_e \\ e^- \end{pmatrix}_L} \begin{pmatrix} \phi_4^+ \\ \phi_4^o \end{pmatrix} e_R^- + \text{H.C.} , \qquad (5.1)$$

where the ϕ_i are linear combinations (not necessarily
independent) of several scalar fields of definite mass.
Since the μ and e are defined as the physical states
found in the diagonalization of the mass matrix, if
there is only one Higgs doublet (as is sometimes assumed),
then g_2 and g_3 must be zero. However, if there are more
than one Higgs doublet, then in general it is possible
that g_2 and/or g_3 are nonzero and virtual Higgs scalars
will give physical transitions between μ and e such as
shown in Fig. 12. Because the Higgs coupling to the
light leptons are so weak, the two loop diagrams (Fig.
12b), in general, dominate one loop diagrams (Fig. 12a):

$$\frac{1 \text{ loop}}{2 \text{ loops}} \sim \frac{2\pi}{\alpha} \left(\frac{m_\mu}{m_H}\right)^2 . \qquad (5.2)$$

Bjorken and Weinberg roughly estimate[44]

$$\frac{\mu \to e\gamma}{\mu \to e\nu\bar\nu} \lesssim 10^{-8} , \qquad (5.3)$$

where the present experimental limit is 2.2×10^{-8}. The
decay $\mu \to 3e$ occurs by a very small tree graph. $K_L \to \mu e$ is
forbidden in lowest order (or one would get strangeness-
changing neutral-currents). They also predict

$$\frac{\mu^- N \to e^- N}{\mu^- N \to \nu N'} \sim 4\times10^{-9} , \qquad (5.4)$$

where N is a nucleus and the experimental limit is 1.6
$\times10^{-8}$.

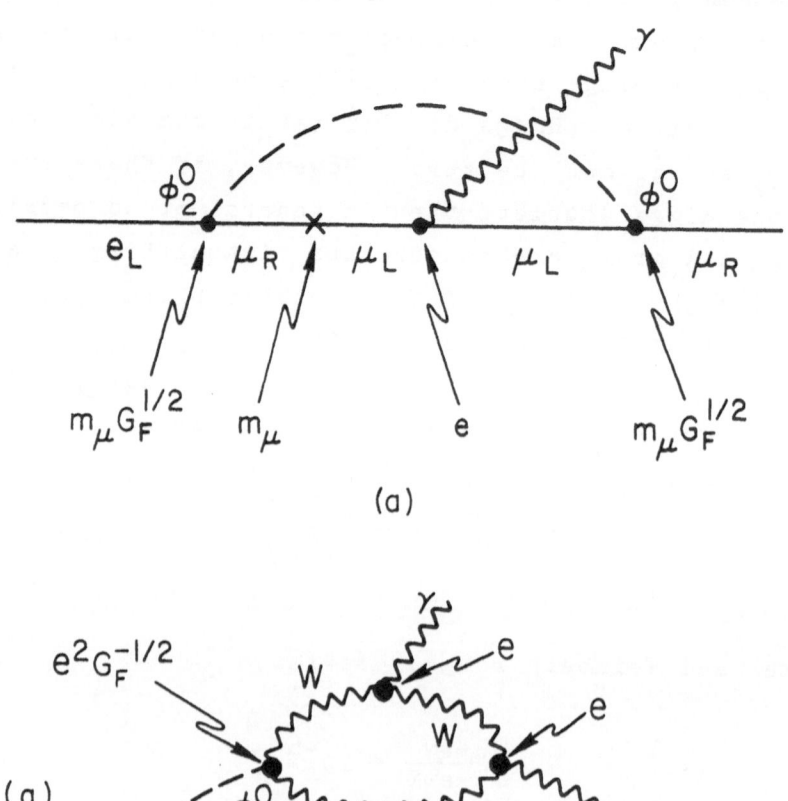

(a)

(b)

Figure 12.   One (a) and two (b) loop diagrams in which
virtual Higgs exchange leads to the decay μ→eγ. This
figure was taken from Ref. 44.

C. Mixing Between N_e and N_μ

In the context of models such as the CHYM and vector models (see Section I), Cheng and Li[45] considered the mixing of massive neutral leptons which have right-handed couplings to the electron and muon. Greater detail of their work is given in T. P. Cheng's talk at this conference, but a brief summary follows. In analogy with the Cabibbo mixing of the d and s quarks, they suggest

$$N_e' = N_e \cos \phi + N_\mu \sin \phi ,$$
$$N_\mu' = -N_e \sin \phi + N_\mu \cos \phi .$$
(5.5)

Then clearly if one considers the simple one-loop diagram of Fig. 13, there will be a GIM-like cancellation. The cancellation is not complete to the extent that N_e and N_μ have unequal masses; the amplitude for this $\mu \rightarrow e\gamma$ process is proportional to

$$\cos \phi \sin \phi \left(m_{N_\mu}^2 - m_{N_e}^2 \right) .$$
(5.6)

Bjorken, Lane and Weinberg[46] argue that the Higgs couplings which give masses and lead to the above mixing also cause small but finite mixing of the left-handed singlet parts of N_e and N_μ with ν_e and ν_μ. This mixing is order m_μ/m_N. There are, as a result, left-right diagrams in addition to the right-right diagram, Fig. 13. In amplitude they find if right-right terms give +1, left-right terms give -6; but essential features of the Cheng-Li work remain unchanged. If the term Eq. (5.6) is 1 GeV^2, then

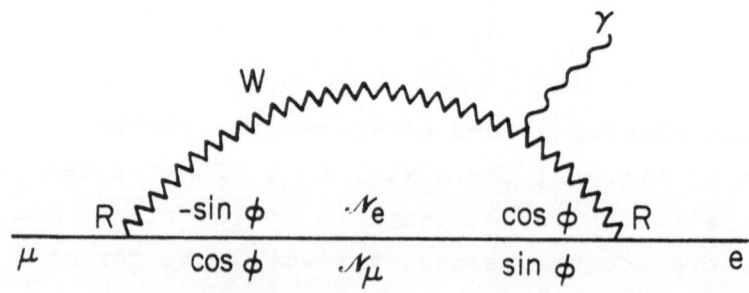

Figure 13.   One of the diagrams in which N_e and N_μ exchange leads to the decay $\mu \to e\gamma$.   This approach was suggested by Cheng and Li.[45]

$$\frac{\mu \to e \gamma}{\mu \to e \nu \bar{\nu}} \sim 10^{-9}. \qquad (5.7)$$

The processes $K_L \to \mu e$, $\mu \to 3e$, $\mu^- N \to e^- N$ are allowed, but below present experimental limits.

D. Mixing Between ν_e, ν_μ and N_U

Glashow[47] and Fritzsch[48] have shown that muon number can be violated without right-handed currents and with only one Higgs doublet. They propose that the charged heavy lepton has a left-handed coupling to a massive neutral lepton which can mix with ν_e and ν_μ

$$\begin{pmatrix} \nu'_e \\ e \end{pmatrix}_L \qquad \begin{pmatrix} \nu' \\ \mu \end{pmatrix}_L \qquad \begin{pmatrix} N'_U \\ U \end{pmatrix}_L \qquad (5.8)$$

Decays such as $\mu \to e \gamma$ occur then in the same fashion as proposed by Cheng and Li[45] (see above) where Δm^2 is replaced with $m^2_{N_U}$.

The mixed states can be written as:

$$\nu'_e = \nu_e \cos \theta + N_U \sin \theta ,$$

$$\nu'_\mu = \nu_\mu \cos \phi + (-\nu_e \sin \theta + N_U \cos \theta) \sin \phi, \quad (5.9)$$

$$N'_U = (N_U \cos \theta - \nu_e \sin \theta) \cos \phi - \nu_\mu \sin \phi.$$

Both angles can be shown to be small by the need to maintain universality (seen through μ and β decay) and by the lack of ν_e in ν'_μ (muon neutrinos do not produce electrons in scattering). If $\mu \to e \gamma$ is observed at the 10^{-10} - 10^{-9} level and is to be explained in this manner, then the smallness of the angles θ and ϕ requires that N_U have a much larger mass than U. As a consequence, the charged heavy lepton U can decay only through the mixing of N_U with ν_e and ν_μ.

It would be possible to rule out this method of muon number violation by measuring τ_U carefully, but the present limits on τ_U are not restrictive enough. Given the mixing of neutral leptons, the charged lepton U could be produced in ν_μ scattering although it would be difficult to observe at that level.

VI. OBSERVATION OF N_e AND N_μ

J. D. Bjorken and I have considered the production of massive neutral leptons, N_e and N_μ, which have right-handed couplings to e and μ (see the CHYM and vector models, Section I). Some of the following discussion originally appeared in Ref. 49 (Bjorken and Llewellyn Smith).[50]

A. Direct Production in Deep Inelastic Scattering

In the process μp or $ep \rightarrow (N_\mu$ or $N_e) +$ hadrons, as shown in Fig. 14, one could look for decays such as $N_\mu \rightarrow \mu\mu\nu$ which would probably have a branching ratio of about 10%. Since this is a weak process, the highest energies are desirable. At E=100 GeV, $\sigma \approx 10^{-36}$ cm^2 so such dimuons would be hard to detect unless strong cuts are made such as: (a) require missing energy (the neu-trino)·; (b) require that the muon pair have large $p_\perp$ (since this is unlikely in other sources of muons); (c) observe m ($\mu\mu$); (d) use the Pais-Treiman relations[51] for $p_{\mu+}$ vs. $p_{\mu-}$.

The process $\nu N \rightarrow N+$ hadrons shown in Fig. 14 occurs because of the mixing described by Bjorken, Lane and

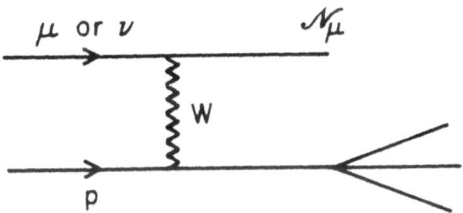

Figure 14.   Direct production of N_μ in either muon or
neutrino deep-inelastic scattering off a nucleon.

Weinberg.[46] Since the mixing is proportional to m_μ/m_N,
the production of N (with subsequent decay to $\mu\mu\nu$) relative to the usual μ production is (if $m_N \approx 2$ GeV):

$$\left(\frac{\mu}{N_\mu}\right)^2 \, 0.1 \approx 5\times10^{-4} \, , \qquad (6.1)$$

compared to charm decay which gives dimuons at the 10^{-2}
level. Perhaps with cuts as discussed above, this signal
might be visible.

    B.   Indirect Production in Neutrino Scattering

    The following process (also discussed by J. Rosner
at this conference) is possible if $N(\equiv N_e$ or $N_\mu)$ is light
enough (see Fig. 15)

$$\nu N \to \mu^- + D + \text{hadrons}$$
$$\hookrightarrow K\mu^+ N$$
$$\hookrightarrow \mu^-(\mu^+\nu) \, . \qquad (6.2)$$

For F production (which may be 10% of D production) the
same process occurs, but without the K. According to
the reported estimates of Rosner, an N mass of 1.15 GeV
maximizes the branching ratio for $F\to\mu N$ (at about 20%).

    Since N_e and N_μ mix, one doesn't know whether the
production will be in association with a μ (as shown)
or with an e. Similarly, the decay can be to μ or to
e. And, of course, the virtual W (at the final decay)
gives $\mu\nu$ only 20% of the time. Since counter experiments
do not distinguish electrons from hadrons, if the D decay
is to KeN(with $N\to\mu X$), a "same sign dimuon" could be observed. Otherwise a trimuon event can result, and at
20% of that level quadramuons (although muon detection

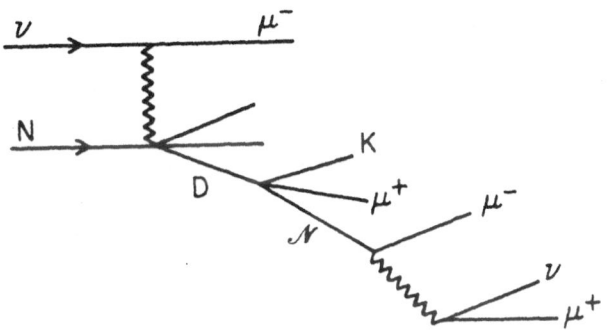

Figure 15.   Multiple muon production in neutrino-nucleon
scattering through the production of a charmed meson D
which decays into a neutral heavy lepton N.   For F meson
production and decay, simply remove the K meson.

efficiency hurts the chances of seeing such events).
If N is light enough, it is possible that $B(D \to K\mu N) \approx 5\%$
and that (e.g.) $B(N \to \mu X) \approx 30\%$ so that

$$\frac{\mu^- \mu^-}{\mu^+ \mu^-} \approx \frac{1}{10} , \qquad\qquad (6.3)$$

which is the rate observed experimentally.[52] The trimuon
rate (corrected for efficiency) could be as large.

In trimuon events the two secondary muons could
have a relatively low invariant mass (less than 2 GeV).
The secondary muon in same-sign dimuons should be relatively slow.

If the N_e and N_μ are too heavy for D or F decay,
then perhaps they occur as products of $u\bar{b}$ and $d\bar{b}$ decay.
In any case the subject of $\mu^- \mu^-$ and trimuon events is
an interesting one to pursue, since there is no obvouis
explanation of them at this time.

C. Direct Production in $e^+ e^-$ Scattering

As energy increases, the weak interactions begin
to become competitive with the electromagnetic interactions. Then the process (see Fig. 16)

$$e^+ e^- \to \nu_e \; \substack{N_e \\ \big\downarrow \\ e^+\pi^- \text{ or } \mu^+\pi^-} \qquad (6.4)$$

should be considered (it is allowed in models such as
CHYM and vector). The cross section to a good approximation is

$$\sigma(e^+ e^- \to \bar{N}_e \nu_e \text{ or } N_e \bar{\nu}_e) = \frac{G^2 s}{\pi} . \qquad (6.5)$$

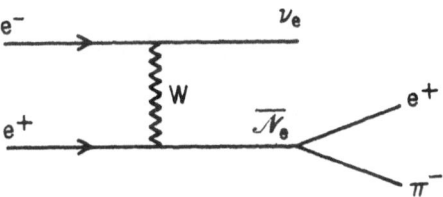

Figure 16. The direct production in e^+e^- scattering of ν_e and $\bar{N}_e$ by W exchange. The $\bar{N}_e$ then decays into $e^+\pi^-$.

At PEP or PETRA

$$\sigma(s=1000 \text{ GeV}^2) = 3 \times 10^{-35} \text{ cm}^2, \qquad (6.6)$$

and since luminosity will be 10^{32} cm^{-2} sec^{-1}, they should produce ten $N_e \nu_e$ events per hour. The branching ratio to the eπ mode can be taken from charged lepton estimates in Thacker and Sakurai[53] and Tsai.[54] This ratio varies with mass giving about 70% for 0.6 GeV, 30% for 1 GeV, 8% for 2 GeV, and 4% for 3 GeV.

Even at SPEAR it is conceivable that such events could have been produced. If $m_N \leq 1$ GeV, then at SPEAR

$$B\sigma \approx 0.25 \times 10^{-36} \text{ cm}^2 \quad . \qquad (6.7)$$

Since the total running at high SPEAR energies is 20×10^{36} cm^{-2}, there would be five such events of $e^+e^- \rightarrow \nu e \pi$ or $\nu \mu \pi$.

Ordinarily one would think there is no possibility to separate five events from background; 40,000 charged leptons have probably been produced but only 100 seen (and indirectly); and many charmed mesons were produced before their discovery. However, these events are quite unique. One could first examine only two charged prong events in which both prongs go in the same general direction. Next require that the pair's momentum equal that of the beam. Then since this is a three particle event, one can reconstruct and find that the missing particle is massless. At this point little background should remain, and one can eliminate most by separating $\pi^+\pi^-$, $\mu^+\mu^-$ and e^+e^- from the sample. If any events remain, the eπ or $\mu\pi$ pairs should have the same invariant mass (that of N_e).

Similarly the process $e^+e^- \rightarrow Z^0_{virtual} \rightarrow N_e \bar{N}_e$ is allowed
but the cross section is smaller, and the signatures of
the decay modes are not quite so distinct. However, an
important point is that this process is allowed in a
greater range of theories than is $\nu_e \bar{N}_e$ production. At
PEP and PETRA the resulting decay product $e\pi$, $\mu\pi$, $\mu\mu\nu$,
$\mu e\nu$ (etc.), will be produced in two distinct jets (one
for each N); within each jet one could make cuts on invariant mass.

D. Indirect Production in e^+e^- Scattering

As discussed in Section VI.B, N_e and N_μ can be the
decay products of D, F, Y ($\equiv u\bar{b}$) and U (heavy charged
leptons) _if_ the N masses are light enough. For the heavy
leptons U produced in e^+e^- scattering at high energies,
the products of each decay will be in jets and therefore
easier to find. The resulting events could contain 4μ,
$4e$, 3μ $1e$, etc. Or one could again look for $e\pi$ and $\mu\pi$
pairs. While production of N_e and N_μ in this fashion
could be greater than through Z (as above), the signal
is confused by the added presence of accompanying hadrons
or neutrinos.

Acknowledgements

The preparation of this report was greatly aided
by discussions with V. Barger, B. Barish, J. Bjorken,
R. Cahn, T. P. Cheng. D. Cline, S. Ellis, G. Feldman,
H. Fritzsch, H. Georgi, F. Gilman, S. Glashow, K. Lane,
T. Y. Ling, A. Mann, F. Martin, D. Nanopoulos, D.
Politzer, J. Rosner, and S. Weinberg. I would like to
thank the organizers of Orbis Scientiae-1977 for in-
viting me to the conference.

REFERENCES

1. S. Weinberg, Phys. Rev. Lett. 19, 1264 (1967); A.
    Salam, in Elementary Particle Physics: Relativistic
    Groups and Analyticity (Nobel Symposium No. 8),
    edited by N. Svarthholm (Almquist and Wiksell,
    Stockholm, 1968).

2. Earlier papers considering right-handed currents
    include: R. Mohapatra, Phys. Rev. D 6, 2023 (1972);
    M. A. B. Beg and A. Zee, Phys. Rev. D 8,1460 (1973);
    A. Zee, Phys. Rev. D 9, 1772 (1974); A. De Rújula et
    al., Rev. Mod. Phys. 46, 391 (1974).

3. M. Barnett, Phys. Rev. Lett. 34, 41 (1975) and Phys.
    Rev. D 11, 3246 (1975).

4. M. Barnett, Phys. Rev. D 13, 671 (1976).

5. P. Fayet, Nucl. Phys. B 78, 14 (1974); Y. Achiman
    et al., Phys. Lett. 59B, 261 (1975).

6. F. Gursey and P. Sikivie, Phys. Rev. Lett. 36, 775
    (1976); and P. Ramond, Nucl. Phys. B 110, 214 (1976).

7. A. De Rújula et al., Phys. Rev. D 12, 3589 (1975).
    H. Fritzsch et al., Phys. Lett. 59B, 256 (1975).
    R. L. Kingsley et al., Phys. Rev. D 12, 2768 (1975).
    S. Pakvasa et al., Phys. Rev. Lett. 35, 703 (1975).

8. S. L. Glashow, J. Iliopoulos and L. Maiani, Phys.
    Rev. D 2, 1285 (1970).

9. A. De Rújula et al., Harvard preprint HUTP-77/A002
    (1977).

10. R. N. Mohapatra and D. P. Sidhu, preprint CCNY-HEP-
    76/14 (1976).

11. H. Fritzsch and P. Minkowski, Nucl. Phys. B 103, 61
    (1976).

12. M. Barnett et al., Phys. Rev. Lett. 37, 1313 (1976);
    Phys. Rev. Lett. 36,1163 (1976); and Phys. Rev. D 14,
    70 (1976).

13. J. Kaplan and F. Martin, Nucl. Phys. B <u>115</u>, 333 (1976).

14. B. W. Lee and R. E. Shrock, preprint Fermilab-Conf-76/61-THY (1976).

15. G. Altarelli, G. Parisi and R. Petronzio, Phys. Lett. <u>63B</u>, 183 (1976). A good review of this topic: V. I. Zakharov, preprint ITEP-127, Rapporteur's talk at the XVIII Int. Conf. on High Energy Physics, Tbilisi, July 1976.

16. Scaling violations are discussed in various contexts in: G. Shaw <u>et al</u>., University of California, Irvine preprint 77-7; P. H. Frampton and J. J. Sakurai, preprint UCLA/77/TEP/3; R. Budny, Rockefeller preprint COO-2232B-113; R. C. Casella, NBS preprint (1976); R. Kögerler and D. Schildknecht, preprint TH. 2247-CERN.

17. M. Barnett and F. Martin, Stanford Linear Accelerator Center preprint "Distinguishing Scaling Violations from New Currents" (1977).

18. R. N. Cahn and S. D. Ellis, University of Michigan preprint UM HE 76-45.

19. B. C. Barish <u>et al</u>., Caltech preprint CALT 68-567, comparision of data with different E_μ cuts indicates ν and $\bar{\nu}$ are affected similarly.

20. V. Barger, T. Gottschalk, and R. J. N. Phillips, University of Wisconsin preprint COO-564.

21. K. Lane and S. Weinberg, Phys. Rev. Lett. <u>37</u>, 717 (1976).

22. A. De Rújula <u>et al</u>., Phys. Rev. Lett. <u>37</u>, 398 (1976).

23. M. Barnett, Phys. Rev. D <u>14</u>, 2990 (1976); V. Barger and D. V. Nanopoulos, Phys. Lett. <u>63B</u>, 168 (1976); C. H. Albright <u>et al</u>., Phys. Rev. D <u>14</u>, 1780 (1976).

24. A. Benvenuti et al., Phys. Rev. Lett. 37, 1039 (1976).

25. B. C. Barish, Caltech preprint CALT-68-544; and L. Stutte, in Proceedings of the International Conference on Production of Particles with New Quantum Numbers, edited by D. B. Cline and J. J. Kolonko (University of Wisconsin, Madison, 1976), p. 388.

26. J. Blietschau et al., preprint CERN/EP/PHYS 76-55 (1976).

27. D. Cline et al., Phys. Rev. Lett. 37, 252 and 648 (1976); and L. Sulak et al., Harvard preprint, presented at 1976 Neutrino Conference, Aachen.

28. W. Lee et al., Phys. Rev. Lett. 37, 186 (1976).

29. See Refs. 23 and D. P. Sidhu, Phys. Rev. D 14, 2235 (1976); E. Fischbach et al., Phys. Rev. Lett. 37, 582 (1976).

30. L. Sulak, private communication.

31. F. J. Hasert et al., Phys. Lett. 46B, 121 (1973) and J. Blietschau et al., preprint CERN/EP/PHYS 76-42 (1976).

32. F. Reines et al., Phys. Rev. Lett. 37, 315 (1976).

33. Aachen-Padua Collaboration (unpublished).

34. P.E.G. Baird et al. (Oxford and Washington collaborations), Nature 264, 528 (1976).

35. M. Barnett, Stanford Linear Accelerator Center preprints SLAC-PUB-1821 (to be published in Phys. Rev. D, February 1, 1977) and SLAC-PUB-1850 (Proceedings of 1976 Meeting of the Division of Particles and Fields of the APS, Brookhaven National Laboratory, October 1976).

36. S. Meshkov, at this conference.

37. G. P. Lepage, Stanford Linear Accelerator Center preprint SLAC-PUB-1859 (1976); R. N. Cahn and G. L. Kane, Michigan preprint UM He 77-1 (1977).

38. E. Eichten and K. Gottfried, Cornell preprint CLNS-350 (1976).

39. A discussion similar to that in Section IV. B and C can be found in Ref. 18 where $pp \to \beta \to \mu^+ \mu^-$ is also considered.

40. G. J. Feldman, Stanford Linear Accelerator Center preprint SLAC-PUB-1851 (1976) (SLAC Summer Institute on Particle Physics, August 1976).

41. G. S. Abrams, in <u>Proc. 1975 Int. Symposium on Lepton and Photon Interactions at High Energies</u>, edited by W. T. Kirk (Stanford Linear Accelerator Center, Stanford University, Stanford, California, 1975), p. 25.

42. T. Appelquist and H. D. Politzer, Phys. Rev. Lett. <u>34</u>, 43 (1975).

43. See for example: A. Pais, Rockefeller preprint COO-2232B-118.

44. J. D. Bjorken and S. Weinberg, Stanford Linear Accelerator Center preprint SLAC-PUB-1868 (1977).

45. T. P. Cheng and L. -F. Li, Carnegie-Mellon preprint (1977); see also their report to this conference, and S. M. Bilenkii <u>et al</u>., Dubna preprint JINR-E2-10374 (1977).

46. J. D. Bjorken, K. Lane and S. Weinberg, Stanford Linear Accelerator Center preprint in process.

47. S. L. Glashow, Harvard preprint HUTP-77/A008.

48. H. Fritzsch, Caltech preprint CALT-68-583.

49. J. D. Bjorken and C. H. Llewellyn Smith, Phys. Rev. D <u>7</u>, 887 (1973).

50. V. Barger and D. V. Nanopoulos have also discussed these topics in Wisconsin preprint COO-583.

51. A. Pais and S. B. Treiman, Phys. Rev. Lett. <u>35</u>, 1206 (1975).

52. A. Benvenuti et al., Phys. Rev. Lett. 35, 1199, 1203 and 1249 (1975).

53. H. B. Thacker and J. J. Sakurai, Phys. Lett. 36B, 103 (1971).

54. Y. -S. Tsai, Phys. Rev. D 4, 2821 (1971).

PARITY VIOLATION IN ATOMIC PHYSICS AND

NEUTRAL CURRENTS

C. Bouchiat

Laboratoire de Physique Théorique de l' Ecole

Normale Supérieure

Paris, France

I. INTRODUCTION

Using atomic physics experiments as a way to get information about electron-hadron neutral currents seems, at first sight, a hopeless enterprise. It is only for momentum transfer q of the order of a few hundred GeV that the neutral current electron-hadron amplitude A_w can complete with the electromagnetic one A_{EM}. The ratio A_w/A_{EM} for momentum transfer typical of atomic processes, i.e. for $q \sim m_e \alpha$, turns out to be exceedingly small:

$$A_w/A_{EM} \sim G_F / \frac{\alpha}{(m_e \alpha)^2} = G_F m_e^2 \alpha \sim 10^{-14}. \tag{1}$$

Indeed, as a first estimate of the rotation of the plane of polarization of visible light propagating in an optically inactive medium, in 1958 Zeldovich gave a value of 10^{-13} radian/m. At that time, his conclusion was that such an effect, obviously could not be observed[1].

In 1973 we decided to have a new look at the
problem and in particular at what we shall call,
following de Rujula[2], the "Dull Atomic Physics" or more
briefly the D.A.P.. At the risk of boring the audience,
I shall spend some time on the D.A.P. factor. If one
wishes to increase the ratio A_w/A_{EM}, one obvious sug-
gestion is to work with heavy atoms (Z>50) since we
expect a coherent effect of the nucleons similar to
the one operating in neutrino astrophysics[3]. The previous estimate Zeldovich is multiplied by the "weak
charge" Q_w of the nucleus, which in the Weinberg-Salam[4]
model is given by $Z(1-4\sin^2\theta_w)-N$. Although an appreciable
enhancement of about two orders of magnitude is obtained
in this way, it is certainly not enough to bridge the
feasibility gap. A little to our surprise, we found
that the D.A.P. factor which we assumed to be of the
order of unity in the previous consideration could do
the job. The D.A.P. factor combined with the nuclear
coherence effect gives rise to a Z^3 law[5] for the parity
mixing amplitudes in heavy atoms. Enhancements of the
order of 10^6 to 10^7 are readily obtained.

Furthermore, the dye laser technology opens a new
field of experiments in atomic spectroscopy and, in
particular, allows the study of highly forbidden radiative transitions where a weak electron-nucleus interaction has the best chance to show up giving rise to
asymmetries of the order of 10^{-4}-10^{-3}. In the last
three years, various tests of parity violation induced
by neutral currents in electronic atoms and muonic atoms
have been proposed, and some of them have led to experiments now under completion[5-10]. The level of accuracy
at which the gauge models of weak interactions can be
tested will certainly be reached in the near future.

II. THE PARITY VIOLATING ELECTRON-NUCLEUS
INTERACTION INDUCED BY NEUTRAL CURRENTS

The parity violating electron-nucleon interaction will be described by a Lagrangian density of the current-current type. We shall restrict ourselves to vector and axial-vector neutral currents of the "first class". The parity violating Hamiltonian $H_{p.\nu.}$ can then be written as:

$$H_{p.\nu} = H_{VA} + H_{AV} ,$$

$$H_{VA} = G_F/\sqrt{2} \int \bar{e}(x)\gamma_\mu e(x)A_h^\mu(x)d^3x,$$ (2)

$$H_{AV} = G_F/\sqrt{2} \int \bar{e}(x)\gamma_\mu \gamma_5 e(x)V_h^\mu(x)d^3x.$$

In the above expression, $e(x)$ is the Dirac field of the electron; $A_h^\mu(x)$, $V_h^\mu(x)$ are axial vector and vector neutral hadronic currents respectively. If we assume that $A_h^\mu(x)$ and $V_h^\mu(x)$ conserve strangeness and eventually charm and colour, we are led to currents "diagonal" in the quark fields

$$V_h^\mu(x) = \sum_i a_{vi} \bar{q}_i(x)\gamma^\mu q_i(x),$$

$$A_h^\mu(x) = \sum_i a_{Ai} \bar{q}_i(x)\gamma^\mu \gamma_5 q_i(x).$$ (3)

The hermiticity of $H_{p.\nu}$ implies the reality of the dimensionless constants a_{vi} and a_{Ai}. From this, it follows immediately that H_{AV} and H_{VA} are automatically even under time reflection. Within the same theoretical framework a Hamiltonian $H_{p.\nu.}$ constructed with "scalar" and "pseudoscalar" currents would violate T automatically;

although the constants a_{Vi}, a_{Ai} are the relevant ones for a phenomenological description of high energy scattering, they are not well suited to the study of very low energy phenomena. We shall replace $H_{p.\nu.}$ by the effective Hamiltonian $H_{p.\nu.}^{eff.}$ in which the nucleons are treated as point-like particles:

$$H_{p.\nu.}^{eff.} = G_F/\sqrt{2} \int \bar{e}(x)\gamma^\mu e(x)\Big(C_{Ap}\bar{p}(x)\gamma_\mu\gamma_5 p(x)$$

$$+ C_{An} \bar{n}(x)\gamma_\mu\gamma_5 n(x)\Big)d^3x$$

$$+ G_F/\sqrt{2}\int\bar{e}(x)\gamma^\mu\gamma_5 e(x)\Big(C_{Vp}\bar{p}(x)\gamma_\mu p(x) \qquad (4)$$

$$+ C_{Vn}\bar{n}(x)\gamma_\mu n(x)\Big)d^3x.$$

Now $p(x)$ and $n(x)$ are the Dirac fields associated with the proton and neutron. The real constants C_{Ap}, C_{An}, C_{Vp}, C_{Vn} are the weak vector and axial vector charges of the proton and neutron.

The Weinberg-Salam model makes definite predictions about the constants C_A and C_V:

$$C_{Vp} = -\frac{1}{2} + 2\sin^2\theta_w, \quad C_{Vn} = \frac{1}{2},$$

$$C_{Ap} = -C_{An} = \lambda(2\sin^2\theta_w - \frac{1}{2}). \qquad (5)$$

θ_w is the now familiar weak interaction angle, which, from the analysis of neutrino experiments, is restricted to the following values:

$$0.3 \lesssim \sin^2\theta_w \lesssim 0.4 . \qquad (6)$$

$\lambda = 1.25$ is the ratio between the vector and axial coupling constant in neutron β decay.

Although the latest neutrino data on neutral currents seems to favour the Weinberg-Salam model and to
exclude a pure V (or pure A hadronic currents), the
knowledge of the coupling constants C_{Vp}, C_{Vn} etc.....
is crucial to discriminate between various gauge models
of weak interactions.

The experiments of the first generation described
below will yield a kind of weak vector charge of the
nucleus conventionally defined as[5]

$$Q_w = -2[C_{Vp}Z + (A-Z)C_{Vn}]. \tag{7}$$

It is of interest to discuss the values taken by
Q_w in the various gauge models which have been discussed
recently. We shall first restrict ourselves to the
models associated with an $SU_2 \otimes U_1$ gauge group. The left-
handed doublets of the standard model are usually kept
unchanged. One adds[11] a certain number of new singlets,
doublets or even triplets involving the right-handed
quarks U_R, d_R, C_R and right-handed leptons e_R... together
with new quarks t, b, d' and new leptons E^-, N^o,...
Let us give some examples of right-handed doublets:

$$\begin{pmatrix} U_R\cos\alpha + U_R'\sin\alpha \\ \\ b_R \end{pmatrix}, \begin{pmatrix} t_R \\ \\ d_R\cos\beta + d_R'\sin\beta \end{pmatrix}, \begin{pmatrix} N_R^o \\ \\ \bar{e}_R\cos\gamma + E_R^-\sin\gamma \end{pmatrix}.$$

It is a straightforward matter to compute the new
values of C_{Vp} and C_{Vn} in terms of the mixing angles

α, β, γ, and the values of the weak isospin 3-component T_{3R}^u, T_{3R}^d, T_{3R}^e attributed to the right-handed u_R, d_R, e_R fields:

$$C_{Vp} = \frac{1}{2}(c-1)(1+2a-b-4\sin^2\theta_w)k^{-2},$$

$$(8)$$

$$C_{Vn} = \frac{1}{2}(c-1)(-1+a-2b)k^{-2},$$

with a, b, c, given by:

$$a = 2\ T_{3R}^u\ \cos^2\alpha,$$

$$b = -2\ T_{3R}^d\ \cos^2\beta,$$

$$(8bis)$$

$$c = -2\ T_{3R}^e\ \cos^2\gamma.$$

The factor $k = M_{z_0}\cos\theta_w/M_w$ allows Higgs mechanisms different from the one originally proposed by Weinberg and Salam.

Experiments involving elastic scattering of V_μ on nucleons and electrons can be used to restrict the values of the constants a,b,c[12]. In view of large experimental uncertainties a wide range of values of Q_w is allowed. For the particular case of B_i (Z=83, A=221), values of Q_w smaller than the Weinberg-Salam value by an order of magnitude are not excluded within the $SU_2 \otimes U_1$ scheme. But if the ratios

$$p_N = \frac{\sigma(V_\mu+N\to V_\mu+N)}{\sigma(\bar{V}_\mu+N\to \bar{V}_\mu+N)} \quad \text{and} \quad p_e = \frac{\sigma(V_\mu+\bar{e}\to V_\mu+\bar{e})}{\sigma(\bar{V}_\mu+\bar{e}\to \bar{V}_\mu+\bar{e})}$$

remain different from 1, it is clear that, with only

one neutral vector boson, the effective charge Q_w cannot be zero for arbitrary values of Z and A.

There is, of course, also the possibility that the $SU_2 \otimes U_1$ weak gauge group is only part of the truth and that more complicated schemes involving several neutral heavy vector bosons associated, for instance, with a $SU_2 \otimes SU_2 \otimes U_1$ gauge group are relevant[13]. Within such a framework the results $p_N \neq 1$ and $p_e \neq 1$ do not imply that parity is violated in neutral current physics. The V_μ used in neutral current experiments-assumed to be massless-are of the left-handed type. The parity violating Hamiltonian

$$H^{p.v} = G_F/\sqrt{2}[\bar{V}_\mu \gamma_p (1+\gamma_5) V_\mu \bar{e} (g_v \gamma^e + g_A \gamma^e \gamma_5) e] \qquad (9)$$

gives rise to the same predictions as the parity conserving Hamiltonian

$$H^{p.c} = G_F (g_v \bar{V}_\mu \gamma_e V_\mu \bar{e} \gamma^p e + g_A \bar{V}_\mu \gamma_p \gamma_5 V_\mu \bar{e} \gamma p \gamma_5 e) \qquad (10)$$

if the incident massless neutrino beam is in a pure helicity (-1) state. In other words, if the one Z_0 hypothesis is not made, with neutrino physics experiments only, it is not possible to tell whether or not parity is conserved in neutral currents interactions[14].

In order to discuss the physics of neutral currents in atoms, it is convenient to derive from the second quantized effective Hamiltonian $H_{p.v}^{eff.}$ a nonrelativistic electron-nucleus potential $V^{p.v}(\vec{r})$:

$$V^{p.\nu}(\vec{r}) = -G_F/\sqrt{2}\left\{\frac{\vec{s}\cdot\vec{p}}{m_e}\ \delta^3(\vec{r})[ZC_{Vp}+(A-Z)C_{Vn}]\right.$$

$$+\left(\frac{\vec{p}}{m_e} - 2i\ \frac{\vec{s}_\wedge\vec{p}}{m_e}\right)\delta^3(\vec{r})[\vec{S}_pC_{Ap}+\vec{S}_nC_{An}]$$

$$\left. + \text{hermitian conjugate}\right\}.$$

$\vec{r}$, m_e, $\vec{s}$, $\vec{p}$ are respectively the position, mass, spin
$\vec{s} = \frac{1}{2}\vec{\sigma}$ and momentum operator of the electron. $\vec{S}_p$ and
$\vec{S}_n$ are the <u>total</u> spin operators of the protons and
neutrons in the nucleus. If one wishes to include scalar
and pseudoscalar neutral currents (or tensor pseudotensor
currents), one has simple to add an imaginary part to
the constants C_{Vp} and C_{Vn} (or C_{Ap} and C_{An}).

III. THE MATRIX ELEMENT OF $V^{p.\nu.}$ BETWEEN VALENCE STATES
OF ATOMS. VARIATION WITH ATOMIC NUMBER: THE Z^3 LAW

Under the influence of the parity violating potential, parity mixing occurs between atomic wave functions.
For high Z atoms, the term independent of the nuclear
spin associated with the hadronic vector current will
give the dominant contribution. In the dependent particle
model of the atom, because of its zero-range nature, will
mix only s and p states.

The matrix element of $V^{p.\nu.}$ is given in terms of
the radial functions $R_{n\ell}(r)$ by:

$$\langle n \ s_{1/2} | V^{p.v.} | n' \ p_{1/2} \rangle =$$

$$\frac{-3i}{32\pi m_e} G_F/\sqrt{2} [C_{Vp} Z + C_{Vn}(A-Z)] \qquad (12)$$

$$x \frac{d}{dr} R_{n1}(0) x R_{no}(0).$$

This matrix element presents some similarity with the expression giving the hyperfine splitting of atomic states, which is proportional to $|R_n(0)|^2$. There exists in the literature a semiempirical formula due to Fermi and Segré giving the wave function at the origin $R_{no}(0)$, which only requires the knowledge of the energy spectrum of the s valence electron.

In Ref. (5), the derivation of the Fermi-Segré formula has been improved and extended in order to give the starting coefficient of the radial wave function at the origin, $\lim_{r \to 0} r^{-\ell} R_{n\ell}(r)$, for any value of the orbital angular momentum ℓ.

Let us quote the result

$$\lim_{r \to 0} r^{-\ell} R_{n\ell}(r) = \frac{2^{\ell+1} Z^{\ell+\frac{1}{2}}}{V_n^{3/2} a_0^{\ell+3/2} (2\ell+1)!} [1 + \delta_\ell(\varepsilon_n, Z)],$$

$$(13)$$

with

$$\delta_\ell(\varepsilon_n, Z) = -\varepsilon_n^{3/2} \frac{d\mu_{n\ell}}{d\varepsilon} - \zeta \frac{\ell(\ell+1)(2\ell+1)}{6}$$

$$+ 0(\varepsilon_n^2) + 0(\zeta^2). \qquad (14)$$

a_0 is the Bohr radius, $\varepsilon_n = -V_n^{-2}$ is the binding energy in atomic units. $\mu_{n\ell}(\varepsilon)$ is the interpolated quantum defect $(\mu_{n\ell}(\varepsilon_n) = n-V_n)$; the constant ζ is given in terms of the average potential acting on the valence electron by

$$\zeta = \frac{a_0}{Z^2 e^2} \frac{d}{dr} [rV(r)] \Big|_{r=0}.$$

For $\varepsilon_n \ll 1$ (small binding energy) and high Z, $\delta_{0,1}$ is at most of the order of 10%.

The remarkable feature of this formula, besides its simplicity, is the fact that, practically, it only requires a good knowledge of the energy spectrum and it is independent of the detailed shape of the electronic potential.

The mathematical accuracy of the above formulae for $\ell = 0,1$ has been checked with numerical computations and found to give the correct results with errors smaller than ten per cent.

When used to compute the hyperfine separation of atoms and ions with one s valence electron, the Fermi-Segré formula appears to be remarkably accurate provided relativistic corrections are included properly. As an illustration, we give a table obtained from results quoted in the book of H.G. Kuhn[15]:

	Li	Na	K	Ca+	Rb	Cs	Au
$\dfrac{\text{Fermi-Segré}}{\text{experiment}}$:	0.97	0.88	0.92	0.96	0.94	1.00	0.95

Since the mathematical correctness of formula (13) has been verified independently the above table gives

in fact a test of the single particle approximation.
Using the results given in formula (14) to evaluate the
matrix element of $V^{p.\nu}$ one arrives at the final expression:

$$\langle n\ s_{1/2}|V^{p.\nu}|n'p_{1/2}\rangle =$$

$$-i\ \hbar\ G_F\ Z^3\ \frac{[1+\delta_0(\epsilon_n)+\delta_1(\epsilon_n')](C_{vp}+(\frac{A}{Z}-1)C_{vn})}{2\sqrt{2}\pi m_e c\ a_0^4\ v_n^{3/2}\ v_{n'}^{3/2}}\ .$$

$$(15)$$

From the above expression, it is apparent that the
matrix element follows a Z^3 law. It is possible to give
simple physical interpretations of the three powers of
Z; one power of Z expresses <u>the coherent effect of the</u>
<u>nucleons</u> (the weak vector charge of the nucleus is the
sum of the vector charges of the constitutents). The
second power of Z reflects the fact that the <u>density of</u>
<u>the valence electron at the nucleus increases</u> like Z,
because of the Coulomb attraction of the nucleus; the
third power follows from the proportionality of $V^{p.\nu}$
to the <u>velocity of the electron near the nucleus which</u>
<u>grows like Z</u>.

The matrix element of $V^{p.\nu}$ expressed in Eq. (15)
has still to be corrected for relativistic effects.
The relativistic correction factor K_r is quite substantial
due to the fact that the radial Dirac wave functions of
an $s_{1/2}$ and $p_{1/2}$ electron moving in the Coulomb field of
a point charge are infinite at the origin. K_r has been
obtained in Ref. (5) in terms of the nuclear radius R
and gives rise to a <u>further increase with Z</u>:

$$K_r = \left[\frac{\Gamma(3)}{\Gamma(2\sqrt{1-Z^2\alpha^2}+1)} \left(\frac{2ZR}{a_0}\right)^{\sqrt{1-Z^2\alpha^2}-1} \right]^2 .$$

(16)

For example, the value of K_r is 2.8 for Cs (Z = 55) and 9.0 for Pb(Z=82).

Up to now, our discussion has been limited to valence electron states which give rise to radiative transitions involving visible light. One may think of transitions in the X ray range involving shells K and L. For heavy atoms the matrix element of $V^{p.\nu}$ between K and L states obeys a Z^5 law. However, negative powers of Z associated with the energy denominators and the ratio E_1/M_1 between electric and magnetic dipole amplitudes appear in the computation of the observable quantities and they almost completely cancel the factor Z^5 of the matrix element of $V^{p.\nu}$. The same phenomena do not occur with valence states and the observable parity violation effects will reflect the variation with Z of the matrix element of $V^{p.\nu}$.

In the hydrogen atom (or hydrogen-like ions), there is a possibility of getting an enhancement of the parity violating effect using the fact that n $s_{\frac{1}{2}}$ and n $p_{\frac{1}{2}}$ are almost degenerate (by applying a magnetic field, the real part of the energy difference can even by canceled).

However, the observation of the strongly suppressed n $s_{\frac{1}{2}} \rightarrow n's_{\frac{1}{2}}$ transition in the presence of the nearby allowed n $s_{\frac{1}{2}} \rightarrow n'p_{\frac{1}{2}}$ transitions may give rise to serious background problems. From a theoretical point of view, the hydrogen-deuterium experiments are evidently of very

great interest first, because no uncertainty will affect
the atomic physics computations, and secondly the experiments could, in principle, lead to a complete determination of the four coupling constants C_{Vp}, C_{Vn}, C_{Ap},
C_{An} [16].

IV. PHYSICAL IMPLICATIONS OF PARITY MIXING IN ATOMIC WAVE FUNCTIONS

Γ -invariance forbids static electric dipole moment,
but permits electric dipole transition moment between
states of same parity. The basic principle of most experiments proposed so far is the observation of the
interference between the abnormal E_1 amplitude and the
normal magnetic dipole amplitude M_1 or the electric dipole amplitude $E_{ind.}$ induced by an external static electric field.

Before going to a description of the possible experimental methods let us give order of magnitude estimates of E_1 and M_1. We define E_1 and M_1 as the matrix
elements of the z -component of the electric dipole d_z
and of the magnetic dipole operator μ_z.

The $E_1^{p \cdot \nu}$ amplitude between two s-states (or two
p-states of same parity) is given to first order by
an expression of the form:

$$E_1^{p \cdot \nu \cdot} = \sum_{n''} \frac{\langle ns | V^{p \cdot \nu} | n''p \rangle \langle n''p | dz | n's \rangle}{E_n - E_{n''}}$$

$$\tag{17}$$

$$+ \frac{\langle ns | dz | n''p \rangle \langle n''p | V^{p \cdot \nu} | n's \rangle}{E_{n'} - E_{n''}},$$

where the allowed electric dipole amplitudes $\langle ns | dz | n's \rangle$
are typically of the order of $|e| \, a_0$. Up to factors of

the order of unity, the modulus of the matrix element of $V^{p.\nu}$ can be expressed in unit of $\frac{1}{2} m_e c^2 \alpha^2$ (Ryberg) as:

$$\frac{|<V^{p.\nu}>|}{\frac{1}{2} m_e c^2 \alpha^2} \sim \frac{\alpha^2 G_F m_e^2}{\pi \sqrt{2}} \; Z^2 A K_r \bar{C}_V \; ,$$

where we have defined $\bar{C}_V$ as:

$$\bar{C}_V = \frac{Z}{A} C_{Vp} + (1 - \frac{Z}{A}) C_{Vn} \; .$$

Assuming that there is no large cancellation in the sum over the intermediate states, the following expression can be used to get an order of magnitude of $|E_1^{p.\nu}|$:

$$|E_1^{p.\nu}| \sim \frac{\alpha^2 G_F m_e^2}{\pi \sqrt{2}} \; Z^2 A K_r \bar{C}_V |e| a_0 \; . \qquad (18)$$

Using the numerical value $G_F m_e^2 \simeq 3 \times 10^{-12}$, one finally obtains

$$|E_1^{p.\nu}| \sim 3.6 \times 10^{-17} \; Z^2 A K_r \bar{C}_V |e| a_0 \; . \qquad (19)$$

Taking for instance the case of Bi(Z=83, A=209), $K_r \simeq 9$ $|E_1^{p.\nu}| \approx 4.7 \times 10^{-10} |e| a_0 C_V$.

By going to heavy atoms Z>50, a factor of the order of 10^6-10^7 is gained. For a normal magnetic dipole transition, the magnetic dipole amplitude is of the order of one Bohr magneton: $|M_1| \sim |\mu_B| = \frac{\alpha}{2} C |e| a_0$. One introduces the dimensionless ratio:

$$\frac{|E_1^{p \cdot \nu}|}{|M_1|/c} \sim \frac{\sqrt{2}}{\pi} \alpha \, G_F \, m_e^2 \, Z^2 A K_r \bar{C}_V \sim 10^{-14} Z^2 A K_r \bar{C}_V.$$

(20)

Taking the case of Bi again, one finds:

$$\frac{|E_1^{p \cdot \nu}|}{|M_1|/c} \sim 1.3 \times 10^{-7} \bar{C}_V.$$

(21)

These order of magnitude estimates have been con-firmed by explicit computations which have been performed by using essentially two methods. The first one involves an explicit summation over the intermediate states[5]. The matrix elements of $V^{p \cdot \nu}$ are those given by the semi-empirical formula. The electric dipole amplitudes $<n's|d_z|n"p>$ are taken either directly from experiments or from phenomenological calculations. The summation is usually dominated by a finite number of states correspond-ing to low excitation energies. The big advantage of this method is that the important ingredients are all accessible to experimental measurements. The second method, perhaps mathematically more elegant, uses a Green function technique[5,17,18] to perform the summation over the intermediate states, but it requires a detailed knowledge of the potential. One introduces the wave function $\Phi(E, E_n)$ corresponding to an angular momentum $\ell = 1$, defined as:

$$\Phi(E, E_n) = (E - H_1)^{-1} d_z |ns> .$$

(22)

H_1 is the one-particle Hamiltonian for p states. The radial part of $\Phi(E, E_n)$ is the solution of an inhomogeneous second order differential equation which can be solved by numerical methods.

Knowing $\Phi(E, E_n)$, one obtains the electric dipole amplitude $E_1^{p \cdot \nu}$ by the following expression:

$$E_1^{p \cdot \nu} = <\Phi(E_n, E_n') \mid V^{p \cdot \nu} \mid ns> + <n's \mid V^{p \cdot \nu} \mid \Phi(E_{n'}, E_n)>. \tag{23}$$

The two methods have been used to compute the 6s-7s transition of cesium, and they give results which are in very good agreement with each other[5,19].

For atoms with several valence electrons, the Green's function method which makes the assumption that the energy denominators in formula (17) can be approximated by single particle values could lead to appreciable erros if $E_1^{p \cdot \nu}$ is denominated by low excitation intermediate states.

V.   EXPERIMENTS

Many experiments have been suggested which could lead to an experimental demonstration of the existence of weak neutral currents in atoms. It is not possible to review here all the proposals. I will concentrate my discussion on those which, to my knowledge, have materialized as experimental projects now in a well-advanced stage or near completion. The two types of experiments I shall discuss proceed from two different philosophies and both have their advantages and their drawbacks.

The first type of experiment involves the excitation of twice forbidden magnetic dipole transitions in heavy atoms. One looks for an asymmetry of the excitation cross sections with respect to the circular polarization of the incident light (circular dichroïsm). Because of the hindrance factor affecting the M_1 amplitude, asymmetries

of the order of 10^{-4} are expected. In these experiments
the elimination of the systematic errors is not a priori
too difficult, but because of the very low cross sections
involved, the difficulties will lie in the signal to
noise ratio.

The second type of experiment is a search for an
optical rotatory power of an atomic Bismuth vapor for
wavelengths in the vicinity of a normal magnetic transition[9,20,21]. The choice of normal transitions is dictated by practical considerations. In such an experiment
there is no intensity problem, but the rotation angle
to be observed is of the order of $|cE^{p \cdot \nu}/M_1| \sim 10^{-7}$
radians, and consequently, the problem of systematic
errors is certainly the most serious one.

We shall first discuss the optical rotation of
Bismuth experiments since the preliminary results published by the two groups involved have received much attention.

A. Optical rotation of Bismuth vapor

The physics of optically active molecules tells us
that a very sensitive way to detect an interference between magnetic dipole and electric dipole transition
moments is to measure the optical rotation of light near
the mixed transition. Circular dichroism and optical
power are closely related phenomena since they are respectively associated with a difference between the
imaginary and real parts of the indices for right and
left circularly polarized light.

The rotation of the plane of polarization of light
of wave length λ transmitted through a unit length of
matter having different indices $n_R (n_L)$ for right (left)
circularly polarized photon is given by:

$$\phi = - \frac{\pi}{\lambda} R_e(n_R - n_L). \tag{24}$$

Assuming a simple Breit-Wigner shape, the optical rotation ψ can be written in terms of κ_{max}, absorption coefficient at the line peak as:

$$\phi(\omega) = 2 \kappa_{max}. \frac{\dfrac{\omega - \omega_0}{\Gamma/2}}{1 + \left(\dfrac{\omega - \omega_0}{\Gamma/2}\right)^2} \; Im\left\{\frac{cE_1^{p\cdot\nu}}{M_1}\right\}, \tag{25}$$

$\omega = 2\pi c/\lambda$ is the frequency of light; ω_0 and Γ are respectively the frequency and the width of the line. It would seem, looking at this formula, that one should work with the twice forbidden M_1 transition in order to have a ratio ϕ/κ_{max}. as large as possible. However, in twice forbidden transitions like the $6s_{\frac{1}{2}} \rightarrow 7s_{\frac{1}{2}}$ of Cesium the angle $\phi \, \ell$ being proportional to $E_1^{p\cdot\nu} \, M_1$ is far too small for any reasonable length ℓ of traversed matter, and one has to restrict to normal magnetic transition with $|M_1| \sim |\mu_B|$. A good choice seems to be the transitions between the ground state of Bi and the first excited states (all these states belong to the configuration $6p^3$) which fall in the visible region where high power dye lasers are now available. The ratio $R = Im\left\{E_1^{p\cdot\nu}/M_1\right\}$ has been computed by several authors using different technqiues and approximations. The results obtained are given in the following table together with the preliminary experimental results of the Washington[24] and Oxford[25] groups:

OPTICAL ROTATION IN ATOMIC BISMUTH

$^{4}S_{3/2} \rightarrow {}^{2}D_{3/2}(\lambda=878$ nm)		$^{4}S_{3/2} \rightarrow {}^{2}D_{5/2}(\lambda=648$nm)
	$R/10^{-8}$	$R/10^{-8}$
(17)	-32	-43
(17)	-24	-32
Theory(18)	-35	
(W.S)(22)	-24	-32
(23)	-16	-21
Experiments:	-8±3	10±8
Washington[24,26]		Oxford[25,26]

In the two experiments the systematic errors ΔR affecting R are estimated by the authors to be:

$$\Delta R = \pm~10 \times 10^{-8}$$

and, according to them, not yet fully understood. One may also raise questions about the calibration procedure which is not fully explained in the existing publications.

It is difficult to draw any definite conclusions from the data given above. The only thing one can say is that there is still no evidence for a violation of parity. In view of the rather large systematic errors and the theoretical uncertainty illustrated by the dispersion of the results of various computations, it seems to me somewhat risky to conclude that there is a definite contradiction with the Weinberg-Salam predictions.

B. Circular Dichroïsm in twice forbidden magnetic transitions

The experiments on forbidden magnetic transitions when data is available (hopefully in the next few months) will certainly help to clarify the situation. As we have already said they involve a much larger value of the ratio

$$R = Im\{E_1^{p.\nu}/M_1\}.$$

Furthermore, the elements involved in experiments, Cesium and Thallium, have a much simpler atomic structure and the theoretical predictions are more reliable especially for the Cesium where the theoretical uncertainty is believed not to exceed 15%.

In the Cesium experiment under completion at E.N.S. at Paris, the forbidden magnetic transition: photon + $6s_{\frac{1}{2}} \rightarrow 7s_{\frac{1}{2}}$ is excited by a dye laser. The transition is detected through the fluorescence radiation associated with the transition

$$7s_{\frac{1}{2}} \rightarrow 6p_{\frac{1}{2}} \ .$$

Instead of directly measuring the interference $E_1^{p.\nu} M_1$ it appears more convenient to observe a polarization effect which is characteristic of an interference between $E_1^{p.\nu}$ and the electric dipole transition moment $E_1^{ind.}$ induced by a static electric field $\vec{E}_0$. With this technique, it is possible to suppress completely the rather <u>important background</u> associated with the $6s_{\frac{1}{2}} \rightarrow 7s_{\frac{1}{2}}$ radiative transitions induced by collisions:

$$photon + Cs + Cs \rightarrow Cs + Cs^*.$$

In Ref. (5), it was shown that the interference between the mixed $M_1 + E_1^{p.\nu}$ produces an electronic

polarization in the final state $\vec{P}_e^{(1)}$ given by the
following expression:

$$\vec{P}_e^{(1)} = \frac{8}{3} \frac{F(F+1)}{(2I+1)^2} [R_e\{\frac{M_1}{E_1^{ind.}}\} + \xi_i \; Im\{\frac{E_1^{p \cdot \nu}}{E_1^{ind.}}\}]\hat{k}_i \wedge \hat{E}_0. \tag{27}$$

F is the hyperfine quantum number of the lower as
well as the upper state ($\Delta F = 0$ transition). I denotes
the nuclear spin. $\hat{k}_i$ and $\hat{E}_0$ are unit vectors along the
incident photon momentum and the static electric field.
ξ_i is the circular polarization of the incident photon.
The vector $\hat{k}_i \wedge \hat{E}_0$ has the transformation properties of
an angular momentum under space and time reflection. The
presence in $\vec{P}_e^{(1)}$ of a term of the form $\xi_i \; \hat{k}_i \wedge \vec{E}_0$ is a
clear indication that the atomic Hamiltonian contains a
P - odd T - even piece.

The amplitude $E_1^{p \cdot \nu \cdot}$ being imaginary and the amplitude
$E_1^{ind \cdot}$ real, the ratio $R = Im\{E_1^{p \cdot \nu}/M_1\}$ can be obtained
from the observation of ξ_i dependence of the polarization
$\vec{P}_e^{(1)}$.

Simple considerations of invariance predict the
existence of a component of $\vec{P}_e$ along $\hat{k}_i$ when the incident
beam is circularly polarized:

$$\vec{P}_e^{(2)} = p(F) \; \xi_i \; \hat{k}_i , \tag{28}$$

where $p(F)$ is a number of the order of 10^{-1} which can be
obtained from independent spectroscopic measurements.
It is important to note that $\vec{P}_e^{(2)}$ and $\vec{P}_e^{(1)}$ can be unambiguously distinguished, first because of their dif-
ferent directions but also owing to the fact that $\vec{P}_e^{(1)}$
reverses with the electric field while $\vec{P}_e^{(2)}$ is independent

of $\vec{E}_0$. In fact the detection of the known polarization $\vec{P}_e^{(2)^0}$ offers a very convenient way of calibrating the electronic polarization in terms of optical signals.

In this kind of experiment the first step is the measurement of the M_1 amplitude which arises from rather complex relativistic effects hard to compute directly in a reliable way. From the observation of ξ_i independent part of $\vec{P}_e^{(1)}$, the E.N.S. group has obtained the following value of M_1[27]:

$$M_1 = <6sm = \tfrac{1}{2}|\mu_z|7sm = \tfrac{1}{2}>$$

$$= -4.24 \times 10^{-5}|\mu_B| . \tag{29}$$

Using the value of electric dipole amplitude $E_1^{p \cdot \nu}$ induced by the parity violating potential of the Weinberg-Salam model with $\sin^2\theta_w$:

$$E_1^{p \cdot \nu} = -|e|<6sm = \tfrac{1}{2}|z|7sm = \tfrac{1}{2}>$$

$$= -i1.7 \times 10^{-11} |e|a_0 , \tag{30}$$

the following ratio R is predicted:

$$R = Im\{E_1^{p \cdot \nu}/M_1\} = 1.1 \times 10^{-4}.$$

Up to now only an upper limit on R has been published[28]:

$$|R| \leq 1 \times 10^{-2}.$$

This very preliminary result will hopefully be improved by more than two orders of magnitude in the next few months.

A similar experiment suggested in Ref. (5) is under way at the University of California, Berkeley. It involves the excitation of the $6p_{\frac{1}{2}} \rightarrow 7p_{\frac{1}{2}}$ (292nm) transition in Thallium. The M_1 amplitude has been measured by using the same technique as in the Cesium case and found to be[29]:

$$M_1 = -(2.11\pm0.30)10^{-5}|\mu_B|.$$

The electric dipole amplitude $E_1^{p.\nu} = <7p_{\frac{1}{2}}|d_z|6p_{\frac{1}{2}}>$ has been computed first in Ref. (5) and subsequently by others[30,31] and contrary to what is stated in Ref. (31) the different results agree within 20% leading to the following value of R:

$$R = \mathrm{Im}\{\frac{E_1^{p.\nu}}{M_1}\} \simeq 10^{-3}.$$

REFERENCES

1. Ya.B. Zel'dovich, Zh Eksperim, i. Theor. Fiz $\underline{36}$,
    964 (1959) (transl. Soviet. Phys. JETP $\underline{9}$, 682 (1959).

2. De Rujula, Report at the Tiflis Conference.

3. D. Freedman, Phys. Rev. $\underline{D9}$, 1389 (1974).

4. S. Weinberg, Phys. Rev. Lett. 19, 1264 (1967)
    A. Salam, "Elementary Particle Physics" ed. by N.
    Svartholm p. 367 (1968). The first proof of the
    renormalizability of the Weinberg-Salam model was
    given by: G.'t Hooft, Nucl. Phys. B $\underline{33}$, 173 (1971);
    ibid. B $\underline{35}$, 167 (1971).

5. M.A. Bouchiat and C.C. Bouchiat, Phys. Letters $\underline{48B}$,
    111 (1974) ibid. J. Physique $\underline{35}$, 899 (1974); ibid.
    J. Physique $\underline{36}$,493 (1975).

6. G. Feinberg and M.Y. Chen, Phys. Rev. D $\underline{10}$, 190
    (1974).

7. J. Bernabeu et al. Phys. Letters $\underline{50B}$, 467 (1974)

8. A.N. Moskalev, Zh E.T.F. Pis. Red. $\underline{19}$, 229, 394 (1974)
    (JETP Lett. $\underline{19}$, 141, 216 (1974).

9. I.B. Khriplovich, Zh Eksperim, I. Theor. Fiz. Pis.
    Red. $\underline{20}$ 689 (1974). (Transl. Sov. Phys. JETP Lett.
    $\underline{20}$, 315 (1974).

10. R.R. Lewis and W.L. Williams, Phys. Lett. B $\underline{59}$, 70
    (1974).

11. P. Fayet, Nuclear Physics B $\underline{78}$, 14 (1974). A.
    De Rujula, H. Georgi, S.L. Glashow, Phys. Rev. Lett.
    $\underline{35}$, 69 (1975). F. Wilczek, A. Zee, R. Kingsley and
    S.B. Treiman, Phys. Rev. D $\underline{12}$, 2768 (1975). H.
    Fritzch, M. Gell-Mann and P. Minkowski, Phys. Lett.
    $\underline{53B}$, 256 (1975). S. Pakvasa, W.A. Simmons and S.F.
    Tuan, Phys. Rev. Lett. $\underline{35}$, 702 (1975). R.M. Barnett,
    Phys. Rev. Lett. $\underline{34}$, 41 (1975). Y. Achiman, K. Koller
    and T.F. Wash, Phys. Lett. $\underline{59}$B 261, (1975). F. Gursey

and P. Sikivie, Phys. Rev. Lett. 36, 775 (1976).
V. Barger and D.V. Nanopoulos, Phys. Lett. B63,
168 (1976).

12. V. Barger and D.V. Nanopoulos, Preprint University
of Wisconsin COO-562 (1976). K.P. Das and N.G.
Deshpande, Preprint University of Oregon OITS
1962 (1976). R.M. Barnett, SLAC Preprint 1821 (1976).
B. Kayser, Preprint.

13. R. Incoul and J. Nuyts, Nuclear Physics B98, 429
(1975). H. Fritzch and P. Minkowski, Nuclear Phys.
B103, 61 (1976). R.N. Mohapatra and D.P. Sidhu,
BNL Preprint CCNY 76/14 E. Ma, Phys. Lett. B65,
468 (1976).

14. L. Wolfenstein, Proceedings of the 1975 International
Symposium on Lepton and Photon Interactions at High
Energy Stanford University August 1975.

15. H.G. Kuhn, "Atomic Spectra", p. 342-346, Longman 1969.

16. R.R. Lewis and W.L. Williams, Phys. Lett. 59B, 70
(1975). R.N. Cahn and G.L. Kane, Preprint University
of Michigan UMHE 77/1.

17. M. Brinicombe, C.E. Loving and P.G.H. Sandars, J.
Phys. B 9, L237 (1976).

18. E.M. Henley and L. Wilets, Phys. Rev. A14, 1411
(1976).

19. C.E. Loving and P.G.H. Sandars, J. Phys. B, Atom.
Mol. Phys. 8L, 336 (1975).

20. D.C. Soreide and E.N. Fortson, Bull. Am. Phys. Soc.
20,491 (1975).

21. P.G.H. Sandars in "Atomic Physics" IV, ed. G. Zu
Putlitz, E.M. Weber, A. Winnacker, Plenum Press
New York (1975).

22. I.P. Grant, N.C. Pyper and P.G.H. Sandars, to be
published.

23. I.B. Kriplovich, to be published.

24. D.C. Soreide et al. Phys. Rev. Letters $\underline{36}$, 352
    (1976).

25. P.E.G. Baird et al. Invited talk to the Fifth
    International Conference on Atomic Physics,
    Berkeley (1976).

26. E.N. Forston and P. Baird et al. Nature 264, 528
    (1976).

27. M.A. Bouchiat and L. Pottier, J. Physique Lettres
    $\underline{36L}$, 189 (1975).

28. M.A. Bouchiat and L. Pottier, Phys. Letters $\underline{62B}$,
    327 (1976).

29. S. Chu, E.D. Commins and R. Conti, Berkeley Preprint
    and Communication to the Fifth International Con-
    ference on Atomic Physics, Berkeley 1976.

30. D. Neuffer, to be published.

31. O.P. Sushkov, V.V. Flambaum, and I.B. Kriplovich,
    to be published.

CORRELATIONS IN HADRON COLLISIONS AT HIGH TRANSVERSE MOMENTUM

Richard P. Feynman

California Institute of Technology

Pasadena, California 91125

Unfortunately it is not my intention to give anything like a general summary report on the situation in high transverse momentum collisions. All I want to do is give a report on work on some particular model described in a paper by Field and Feynman[1]. The ideas have now been applied to correlation experiments. The results have been described by Fox[2] at the APS meeting in Brookhaven. What I want to do here is just summarize the ideas of these papers and to tell you what the situation now appears to be.

We use a specific model of how the high transverse momentum particles arise. We suppose that they result from direct collisions between quarks. The quarks find themselves as partons in the original colliding hadrons. After a pair of quarks collide they come out at large $p_\perp$. As they leave the collision they generate, by a kind of cascading process, a jet of hadrons of various momenta in the general direction of the original quark. This is supposed to happen in exactly the same way as it presumably happens in deep inelastic scattering of electrons

on protons; or again as the quark pairs generated e^+e^-
collision separate and produce hadron jets. This quark-
quark collision idea was one of the first proposed to explain the large $p_\perp$ particles but it was early pointed out
that on any simple model of this collision, cross sections
should scale as $p_\perp^{-4}$. However, experimentally, the behavior is much more like $p_\perp^{-8}$. This has suggested that some
mechanism other than direct quark-quark collision may be
the origin of the phenomenon. The view we wish to test
is the possibility that the collisions are indeed quark-
quark collisions, but for reasons unknown the energy dependence of the collision varies inversely as the eighth
power of the momentum.

In order to analyze this we need three kinds of
information. a) The quark distribution in the original
hadron. For protons this is directly determinable by
electron and neutrino scattering. For large x Drell-
Yan have given arguments (from the proton form factor)
that the behavior should be as $(1-x)^3$. For the pion we
have no information. The Drell-Yan type of analysis
suggests that at large x the distribution should be flat,
$(as(1-x)^0)$. We have assumed this. This means that pions
should be more effective than protons for experiments
requiring large momentum quarks. The idea is confirmed
by the experimental[3] results shown in Fig. 1. We have
used these experiments, in fact, to give us a still more
detailed idea of the quark distribution in a pion. The
second thing we shall have to know for our model is b) The
way that quarks make the hadron jet. This should
come directly from experiments with leptons and Fig. 2
shows the distribution of hadrons produced by a quark
in different lepton experiments[4]. It is seen that the
various experiments substantially agree and we have used
this curve to make predictions as to what to expect when

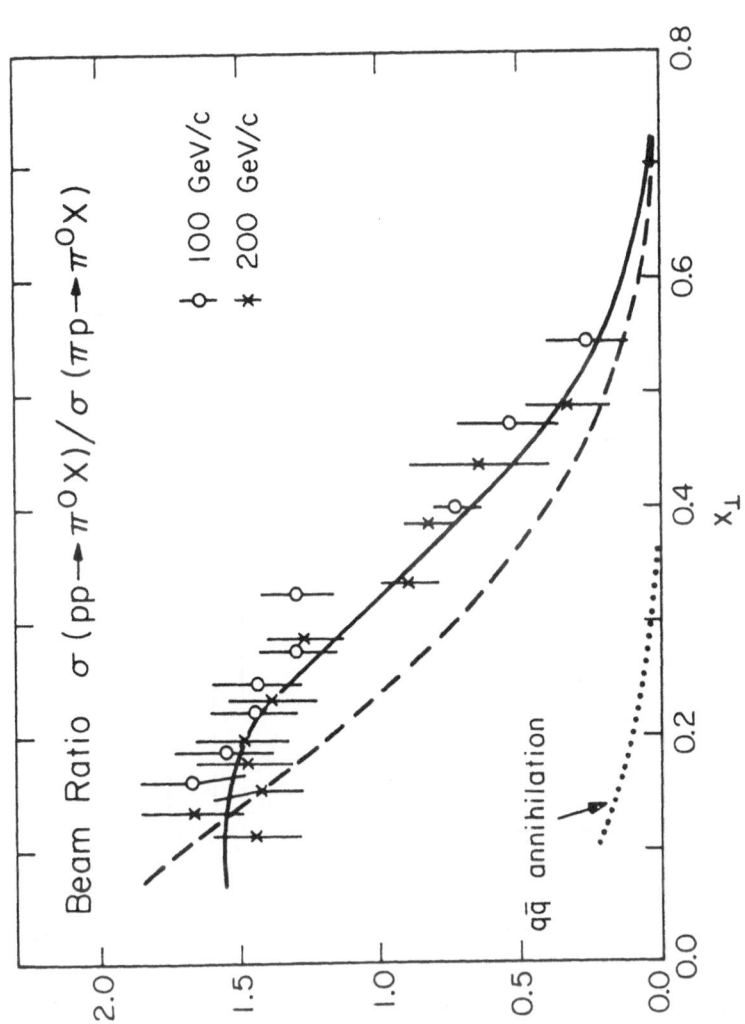

Figure 1: (From Ref. 1) The ratio of the invariant cross section $\sigma(pp \to \pi^0 + X)$ /$\sigma(\pi^+ p \to \pi^0 + X)$ at $\theta_{cm} = 90°$ and $P_{lab} = 100$ and 200 GeV (Donaldson et al., Phys. Rev. Letters 36, 1110 (1976)). The fit with the solid line was used to determine the distribution of quarks in the pion, relative to the distribution in the proton.

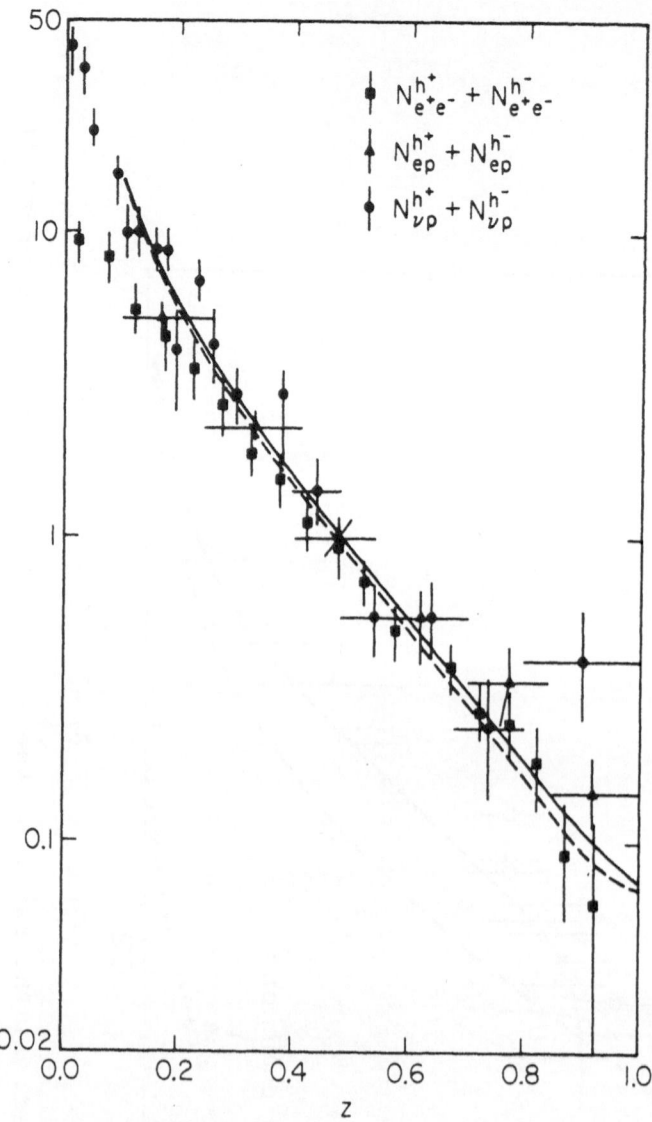

Figure 2: Distribution of hadrons seen in various lepton
experiments (Ref. 4) in the direction of the presumably
recoiling quark. They should be the same for each experiment, except that the flavor makeup of the average
quark differs. The solid line gives the fit used in Ref.
1 for νp; the dotted line gives the fit for $e^- e^+$. The
difference is the effect of differences in the number
of strange quarks.

the quarks come out of a hadron collision. For our present purposes please note that as z approaches 1 we have
supposed that the distribution approaches a constant
(rather than 0, which might also be consistent with experiment). We expect this because the probability that
a quark might be nearly a pure pion plus a little low
momentum debris might be nearly the same as the probability that a pion is nearly a pure quark. We have assumed this is a constant (times dx) as we discussed
under a) and have taken, therefore, a constant for the
large z behavior of b). Finally, we shall need c) The
cross section that two quarks will scatter one another.
The scaling experiments indicate that this varies inversely as the eighth power of momentum, we have to
guess the angular dependence is the same as the determination of how the cross section varies with the ratio
t/s (t is the momentum transfer, s the energy squared
of the quark collision). We have chosen empirically

$$\frac{d\sigma}{dt} = \frac{2300 \text{ mb}}{s(-t)^3}$$

by trying to get agreement with hadron data. The agreement is shown in Fig. 3 where the solid line gives the
predictions for the above formula. Not all the points[5]
on Fig. 3 were available when the choice was made. The
points with crosses were the result of new data[6] from
the CCHK collaboration at the ISR. We were gratified to
see that our model has some predictive power.

     We do not believe this cross-section need represent
any fundamental law. It is only determined empirically
to fit experiment over the range of observations ($p_\perp$ from
2 to 7 GeV). Other models unlike ours are capable of

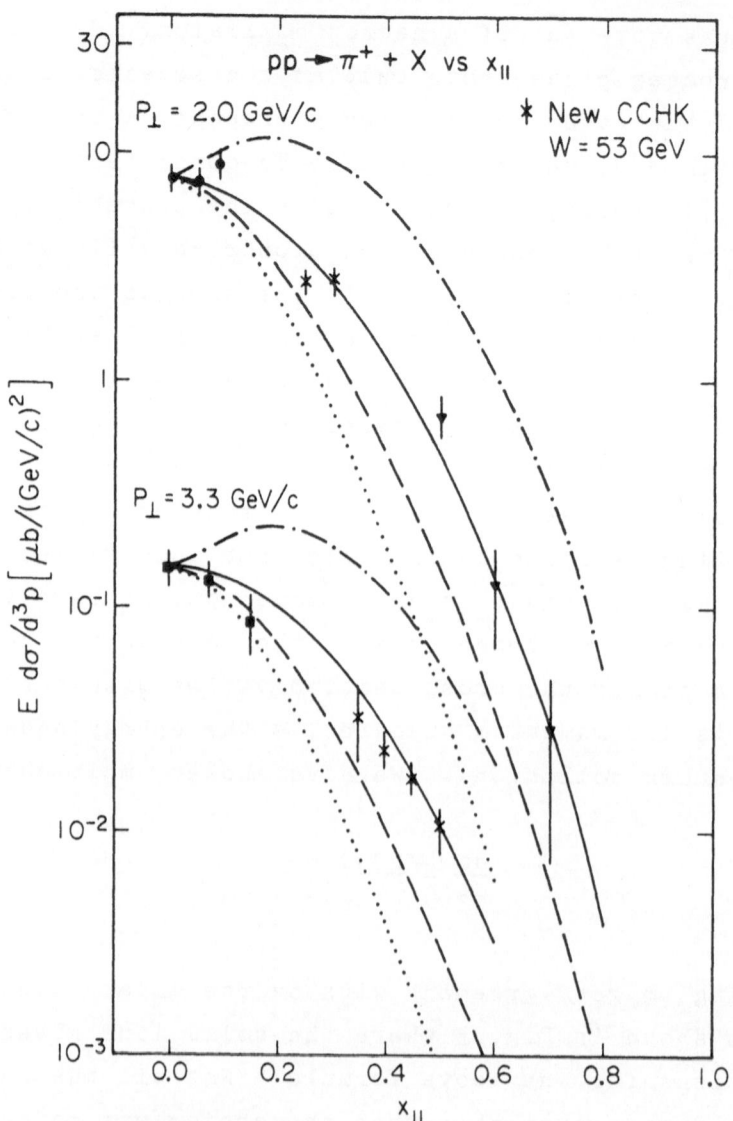

Figure 3: (From Ref. 1). The angular dependence of the
invariant cross sections at W = 53 GeV and $p_\perp$ = 2.0 and
3.3 GeV/c, resulting from various choices for the quark-
quark scattering differential cross-section $d\hat{\sigma}/d\hat{t}$:
$(\hat{s}^2+\hat{u}^2)/(\hat{s}^2\hat{t}^4)$ (dash-dot), $1/(-\hat{s}\hat{t}^3)$ (solid), $1/(\hat{s}^2\hat{t}^2)$
(dashed), $1/\hat{s}^4$ (dotted). New data[6] have been added as
crosses to the data[5] originally included in Ref. 1.
$1/(-\hat{s}\hat{t}^3)$ is preferred.

predicting the cross section; the CIM model, in fact, obtaining just this mathematical form. If fundamental quark coupling occurs as asymptotic freedom expects, the cross section should fall as $p_\perp^{-4}$ (divided by some logarithms) but is smaller than the above for $p_\perp < 7$ GeV if Politzer's coupling constants are used. Thus, in their view, at higher momentum the fall-off might become less rapid than $p_\perp^{-8}$.

The cross-section need not be fundamental in another sense, it need not be pure quark-quark scattering. Along with the quarks there may be some kind of gluon bremsstrahlung - as long as this bremsstrahlung for quark-quark collisions is nearly the same as any gluon bremsstrahlung that may exist in a quark-lepton scattering. That is because we take lepton data to determine the distribution of hadrons generated from what we call a quark.

I would like to take this opportunity to show some further progress made since the publication of our paper[1]. A sensitive test of our model is the ratio of positive to negative pions predicted under various circumstances, for this depends on the number of u and d quarks in the proton and verifies that the outgoing particles reflect the charge properties of the presumed colliding quarks. We were asked by the Chicago-Princeton group to predict results for their experiment of protons on protons or on neutrons at 400 GeV laboratory momentum. We sent them the bands indicated by the grey areas of Fig. 4 to represent our predictions with their uncertainties. They obliged us, as you see, by the experimental points indicated by the heavy dots[7]. I believe these are still preliminary data, but it does look like we are on the right track.

With all these functions determined we set out to

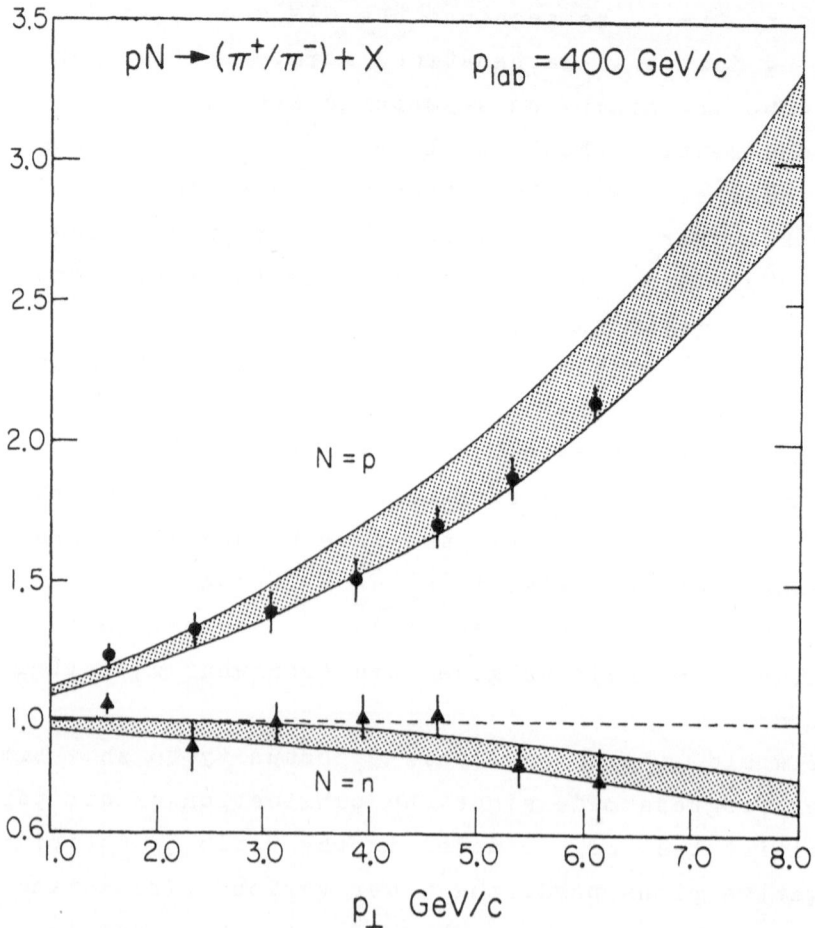

Figure 4: The ratio of numbers of π^+ to π^- expected in inclusive proton collisions are plotted vs. $p_\perp$ for proton-proton and proton-neutron collisions for $\theta_{cm} = 97°$. The theory of Ref. 1 including uncertainties predicted that the ratio should lie within the dotted bands. The results of subsequent experiments[7] are given by the heavy dots. The large ratios observed in the proton-proton case for large $p_\perp$ are interpreted as due to the large ratios of u and d quarks in the proton at large x, and confirm that it is not primarily gluons which generate the observed pions.

try to fit all the correlation experiments with no new parameters. But, as you will see, we had a certain difficulty.

To understand how our model produces its effects in correlation experiments, I would like to emphasize a salient feature. It is most important for the comparison of experiments done with a jet trigger and with a single particle trigger. (By a jet trigger I mean an experiment which is triggered when several particles together have a large total $p_\perp$.) Theoretically the total momentum of the transverse jet is of course the momentum of the quark which made it $P_q = P_{jet}$. But when a single particle is measured it has only a fraction z_c of the total momentum of the quark $z_c \cdot P_q$. The point is however, that z_c is large ranging from .8 to .9. Its variation in different circumstances is shown by the triangles in Fig. 5 (from Ref. 1), the lines (like error bars) indicate the range of z_c values which give important contributions.

If a particle of a given momentum is found, in principal it will come in one of two ways: It may have come from a jet of very much higher momentum which disintegrated in a normal way to give relatively low momentum hadrons. Or, on the other hand, it may have been from a jet which disintegrated in a very rare mode in which 85% of the quark momentum appears in just one particle. In fact, it is the latter which occurs most often. The reason is that the very rapid $p_\perp^{-8}$ fall-off of the quark collision cross section makes it so very difficult to get high momentum quarks that events are biased to come from the lowest quark momentum possible. This effect has been noticed by almost everyone who has worked on these problems. This z_c is numerically as

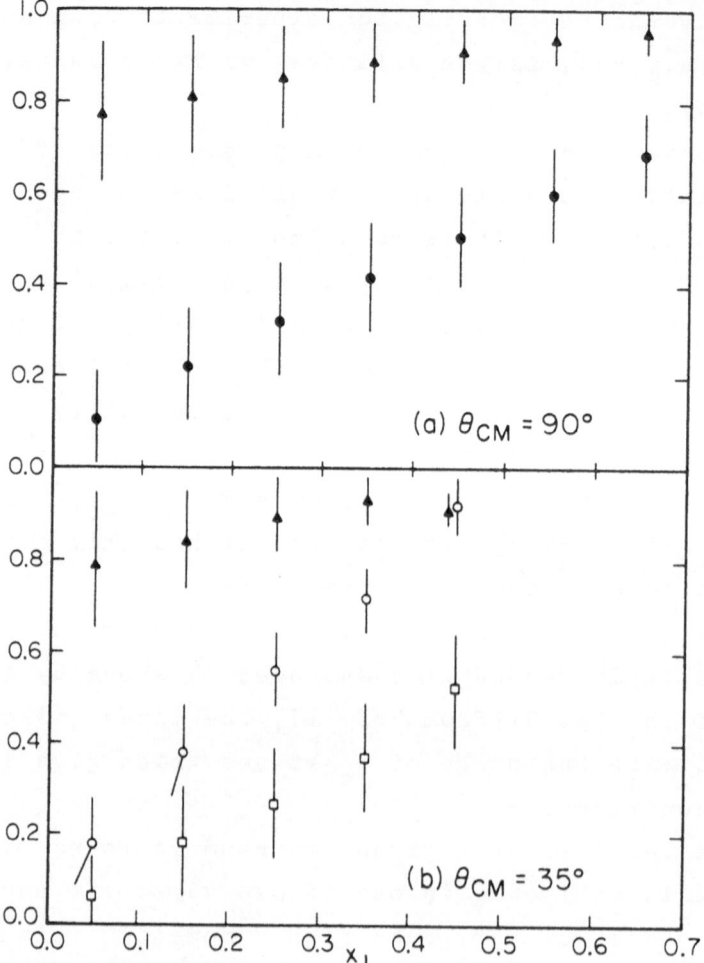

Figure 5: (From Ref. 1.) The mean value of the fraction of the momentum of the parent quark that is carried by the detected hadron z_c, (solid triangles) for pp → π^0 + X versus $x_\perp$ at (a) θ_{cm} = 90° and (b) θ_{cm} = 35°. (Also plotted are x_a (open circles) and x_b (open squares) the fraction of momentum carried by the incoming quarks (x_a = beam, x_b = target). At 90°, $<x_a>$ = $<x_b>$ and are plotted as solid dots.) The error bars are <u>not</u> errors. They represent the standard deviation from the mean, $(<x^2> - <x>^2)^{1/2}$. $x_\perp$ is the transverse momentum of the detected hadron divided by the beam momentum in the center of mass system.

large as it is because we have assumed that the quark
produces hadrons in a range dz which goes as constant
times dz as z approaches 1. This makes it not as difficult as other models (which assume a $(1-z)^p$ dz distribution with p > 0) to get a large z_c particle.

Nevertheless it is rare that a quark at a given
momentum will produce a particle of such high fractional
momentum. For that reason it is characteristic of our
theory that the probability to find a jet of a given
total momentum is a few hundred times larger than the
probability to get a single particle of the same momentum.
This is illustrated in Fig. 6 from Ref. 2. On this figure
we also see an area which indicates the location of this
ratio according to an experiment 260 as was reported by
Fox in his APS talk[2]. Since that talk, work is continuing
in analyzing these jet-triggered experiments and we can
expect experimental information to be very much refined.
I am not allowed to tell you what the present results
are. A gremlin has drawn a point on Fig. 6 but I have
no idea of its significance.

That this result is not trivial and is very model-
dependent can be seen by a comparison with a model that
no one believes, but like a straw man, is easily knocked
down. Suppose collisions were always quark-pion collisions, so we would always get a single pion on one side
and a jet on the other side. The jet to particle ratio
would then be nearer one-to-one than hundreds-to-one.
For this pupose, consider the accompanying figure.

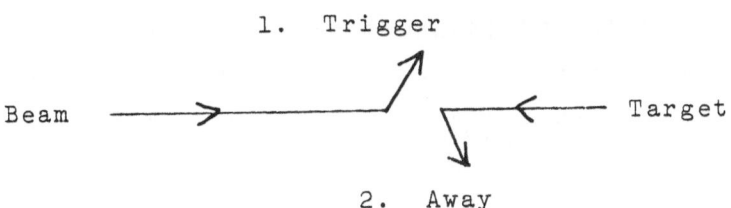

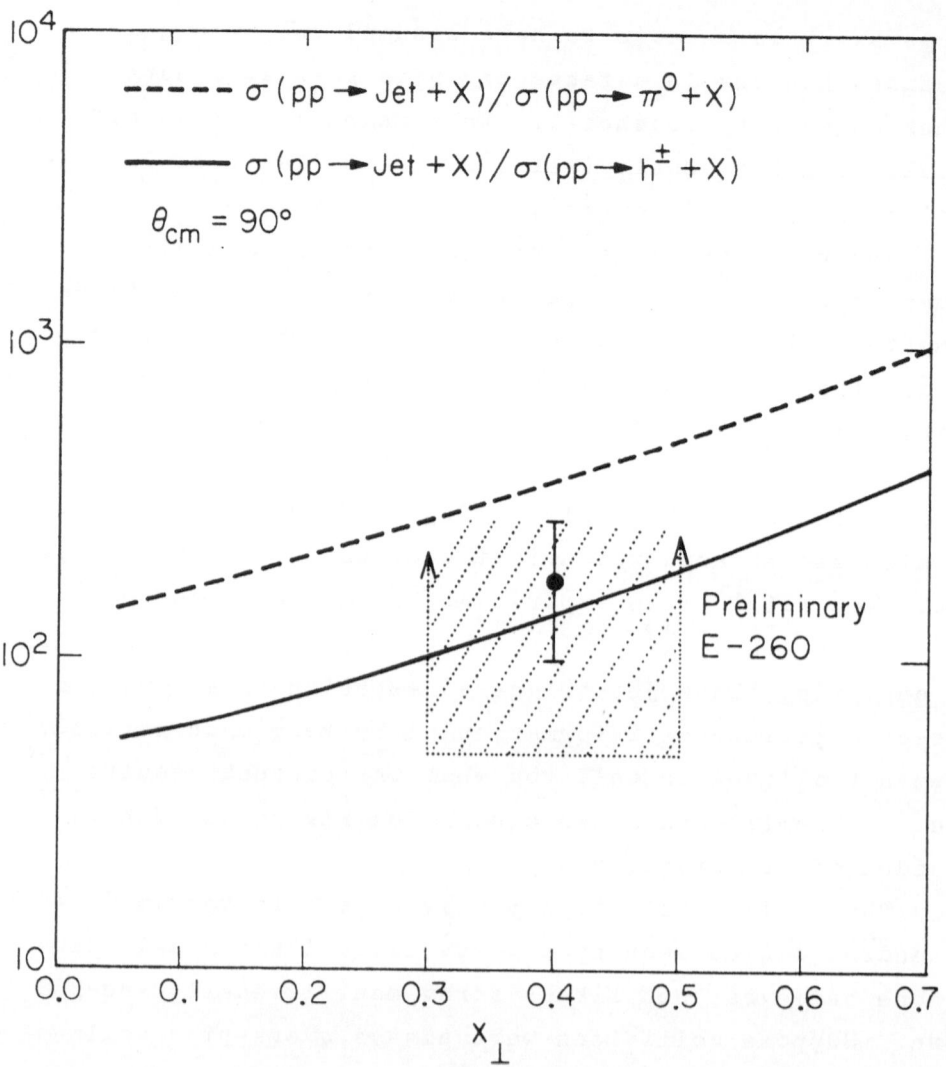

Figure 6: (From Ref. 2.) The ratio of cross-section for producing a jet of particles of total transverse momentum $x_\perp P$ (P is beam momentum in the center of mass system), as compared to the cross-section for producing a single particle with this same transverse momentum. The lines give the ratio for π^0 (dashed) or charged hadrons (solid) expected from the theory of Ref. 1. The dotted area gives the preliminary results of the experiment E260 reported in Ref. 2.

It shows the two particles in the center of mass colliding, the experiment being triggered by particles on one
side of the beam called the "trigger side" and observations made of particles either on the same side or the
opposite side (called the "away side") of the beam. We
begin by discussing correlations for particle triggers,
that is experiments triggered by a single particle on
the "trigger side." First, we will discuss what is seen
on the trigger side. Experiment and theory agree well
with the following results[2]: a) the secondary particles
are very close in angle to the triggering particles.
That is to say, we are getting a jet off to one side.
b) Pairs of particles of unlike charge are more frequent
than those of like charge (for the net charge of the jet,
near that of a quark, should be small), $\pi^+\pi^-/\pi^-\pi^- > 1$.
Theory says, and experiment agrees, that the ratio is
about 3 at 45° C.) The variations of results with angle
and energy are all clear via the expected changes in z_c.
d) The number of jet triggers/particle triggers is of
the order 100 as we have already seen.

      For particle triggers the particles on the "away
side" have the following properties[2]: a) They are found
over a wide distribution of angles. This is expected
because collisions producing a given angle and on the
trigger side come from quarks of various momenta and
therefore are of various angles on the away side. b)
What is observed on the away side is insensitive to the
angle of the trigger. This is expected of the theory
as a kind of numerical coincidence valid for the particular law $1/st^3$ that we have chosen for the quark-quark
collision. c) The charge of the particle on the away
side is virtually independent of the charge of the trigger. This is because the quarks scattered out of one

hadron do not depend much on the character of a quark
from the other hadron which does the scattering. Since
there are more u quarks than d quarks in a proton we
would expect to see more π^+ than π^-. At 90° we expect
that

$$\frac{\pi_1^+ \; \pi_2^+}{\pi_1^+ \; \pi_2^-} = 1.3 \; ,$$

in agreement with experiment. d) Variations of results
with trigger particle is largely understood. For more
details on all these points see the report[2] by Fox at
the Brookhaven APS meeting.

Finally, we compare the away side results for a
jet trigger and a particle trigger. Since a particle
has only a fraction z_c of the momentum of a jet, we would
expect that the result would be exactly the same for a
jet trigger of momentum P_{jet} and a particle trigger of
momentum $P_{part} = z_c \, P_{jet}$ with z_c about .85. Results
are compared in this way on Fig. 7. It is seen that
within the accuracy of these experiments the expectation
is confirmed. Again this is not trivial, for consider
our "straw man" model with a particle on one side and a
jet on the other. A particle trigger on one side sees
a jet on the other, while a jet trigger should produce
but a single particle on the away side - a very different
thing. It appears that, in each event, what comes out
of one side or the other, is qualitatively the same,
and not very different as the above model would have it.

But, we do <u>not</u> agree with experiment in all respects[*].

---

[*] See note added in proof.

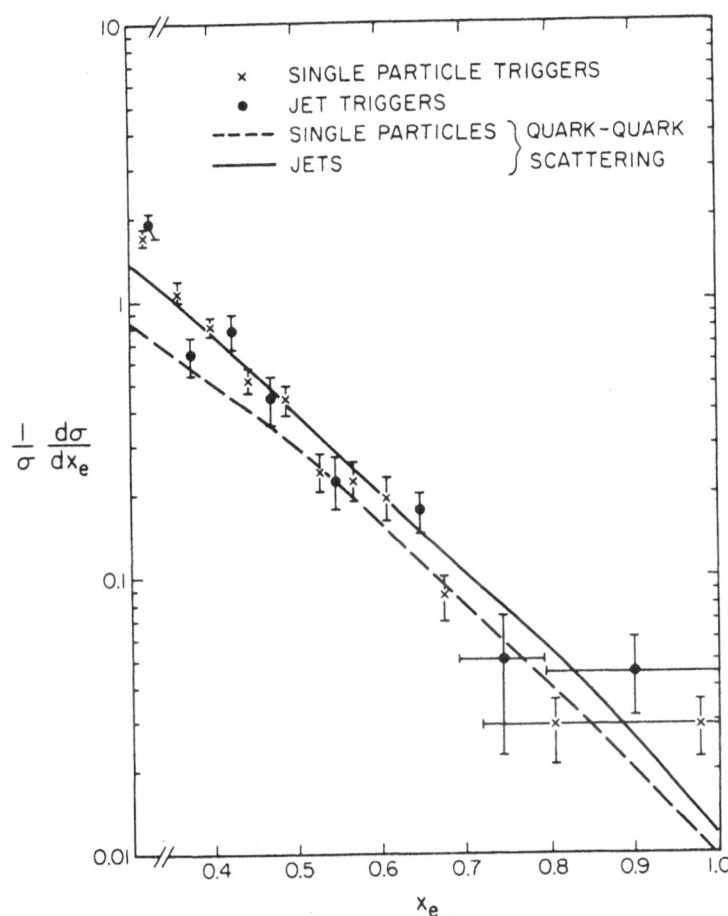

Figure 7: The distribution of hadrons on the side opposite a large $p_\perp$ trigger. We try to plot this data against x_ε, the ratio of transverse momentum of the observed hadron to the transverse momentum of the <u>quark</u> which made the trigger. For a jet trigger we take $p_{\perp jet} = p_{\perp quark}$ but for a single particle trigger $p_{\perp part. trig.} = z_c \, p_{\perp quark}$ with $z_c \approx 0.85$ (see Fig. 5). Thus the data is plotted against $x_\varepsilon = p_{\perp hadron}/p_{\perp jet\ trig.}$ for jet triggers and $x_\varepsilon = p_{\perp hadron} \, z_c / p_{\perp part. trig.}$ for particle triggers. The lines are predictions from the model of Ref. 1.

The absolute number of particles coming out on the away
side is, experimentally, too small! This is a result of
experiment[8] R412 at CERN with pp → π° + h$^{\pm}$ + X. To decrease the statistical error due to detailed binning
define

$$
N^{h^{\pm}} = \int_{0.4}^{\infty} dx_{\varepsilon} \int_{-2.5}^{2.5} dy \int_{-35°}^{35°} d\phi \; \frac{1}{N_{trig}} \frac{dN^{h^{\pm}}}{dx_{\varepsilon} dy d\phi} \quad (1)
$$

as the number of charged hadrons coming out with rapidity
y between -2.5 and 2.5 within 35° in azimuth from the
direction away from the trigger and having a momentum
greater than 0.4 of that of the trigger. x_{ε} is the away
side momentum (in the plane of the beam and trigger)
divided by the trigger momentum, $x_{\varepsilon} = p_x(h^{\pm}$ away$)/p_{\perp}$
($\pi°$ trig.).

The data[8] says that $N^{h^{\pm}} \approx 0.25$, whereas according
to theory it should be .46. But the theory is affected
by a new effect: the transverse momentum that the
original quarks may have in the incoming hadron beam.
If the two quarks should happen to have some small initial
transverse momentum in the direction of the trigger, then
the momentum that must be supplied by the collisions to
make up the total that is observed in the trigger need
not be so great. (See the accompanying diagram.) Since
the cross section rises so rapidly with falling momentum,
the collisions are biased toward those in which the
quarks have initial transverse momentum in the direction
of the trigger, and therefore less momentum on the away
side.

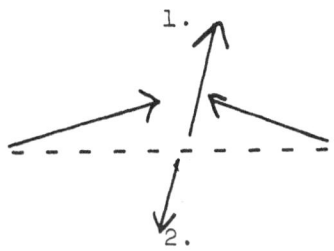

The effect is in the right direction, but too small.
In order to calculate it we shall have to add a new
parameter to our model, the mean transverse momentum
of the quarks inside a proton. There are no experiments
observing only single hadrons that are sensitive to this
parameter, and we have no independent way of determining
it. If we guess the mean transverse momentum of the
quarks in the proton is 330 MeV (like the hadrons in a
hadron collision) independent of their longitudinal
momentum x, and exponentially distributed, our value for
$N^{h^{\pm}}$ is reduced to .41; 500 MeV gives .39. We can get
down to .26 by choosing the large value 700 MeV, but this
is an artifact of our calculation scheme. For such a
high initial transverse momentum, triggers can be generated for which the total momentum observed is practically just the initial transverse momentum (from the tail
of the distribution), only a very soft collision need
occur. Our formula $d\sigma/dt$ makes such collisions very
likely, and 700 MeV permits a number of such soft collisions to contribute. This is not in the spirit of our
original model.

We do not know what to make of the remaining discrepancy. Our calculations should be good, assuming our
model, to ·20%. For completeness, we should mention a
similar experiment, by the CCHK collaboration, even though

it is still unpublished. This experiment was $pp \rightarrow h^{\pm}$
$+ h^{\pm} + X$ with a trigger at 45°. In a similar manner we
calculate

$$
N^{h^{\pm}} = \int_{0.5}^{\infty} dx_\epsilon \int_{-3.0}^{3.0} dy \int_{-90°}^{+90°} d\phi \frac{1}{N_{trig}} \frac{dN^{h^{\pm}}}{dx_\epsilon dy d\phi} ,
$$

(2)

but the range of variables is somewhat different. Data
are given for two different values of the trigger perpendicular momentum. For $p_\perp$ (trig) = 2.1 GeV, they
found $N^{h^{\pm}}$ = .22 where as for $p_\perp$ (trig) = 2.8 GeV they
found $N^{h^{\pm}}$ = .12. Our theory (using 500 MeV for the
initial transverse momentum of the quark), gives $N^{h^{\pm}}$ =
0.26 in either case. In our theory $N^{h^{\pm}}$ can only depend
on $p_\perp$ (trig) via changes in z_c which are too small, in
this case, to have any effect. If this large experimental
dependence of $N^{h^{\pm}}$ with $p_\perp$ (trig) is confirmed in further
study, our theory must be completely abandoned.

The comparison of theory and experiment is illustrated in more detail in the R412 experiment in Fig. 8
and Fig. 9.

The existence of such an initial transverse momentum
can be seen much more directly from another experiment.
If there were no initial transverse momentum, and we
could observe the momenta of the quarks directly, then
the two quark momenta and the beam particles would all
lie in a plane. It is possible to measure the component
(called p_{out}) of the away particles which is perpendicular to the plane of the beam particles and the trigger
particle (in the center of mass system). This is affected
by two things: 1) The transverse momenta that the hadrons
get in a jet so that they do not lie exactly in the plane

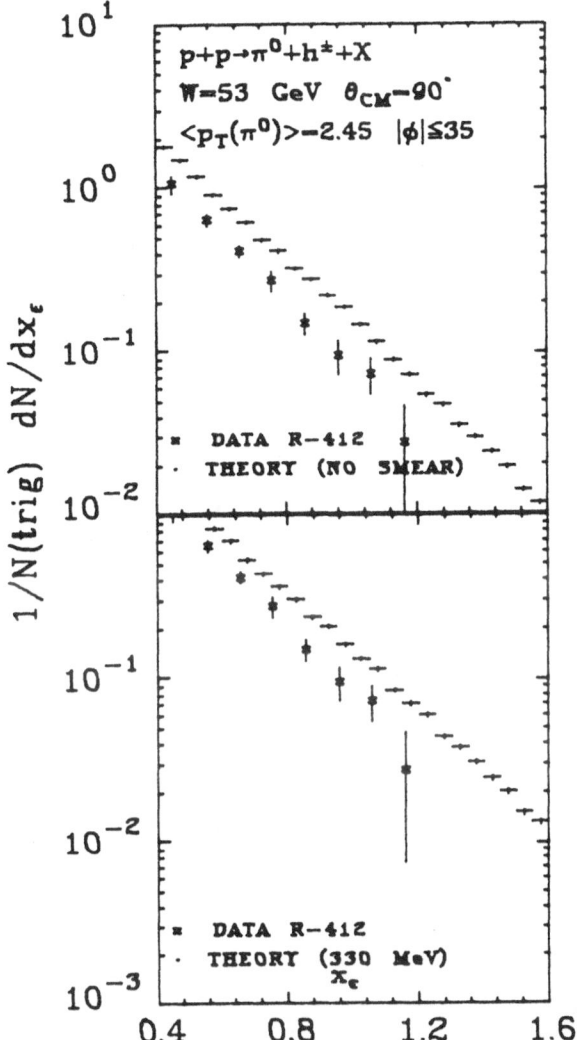

Figure 8: Comparison of theory and data (CERN R-412, Ref. 8) on the away side distribution of hadrons $h^{\pm}$ for the process $pp \rightarrow \pi^0 + h^{\pm} + X$ with $P_{\perp}(\pi^0)$ integrated over the range 2.0 to 4.0 GeV/c and $W = 53$ GeV, $\theta_{cm.trig.} = 90°$. The away hadrons are integrated over a bin with $|Y| \leq 2.5$ and $|\phi| \leq 35°$ and we define $x_{\epsilon} = p_x(h^{\pm})/P_{\perp}(\pi^0)$ (see eq. 1). The theory is calculated with no internal transverse quark momentum (upper) and with mean internal transverse quark momentum of 330 MeV.

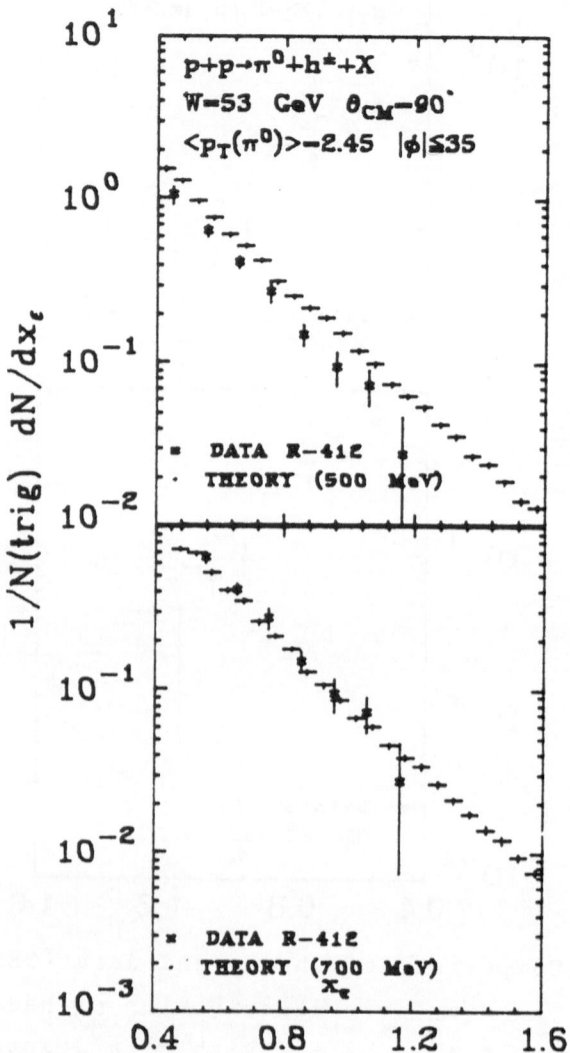

Figure 9: Same as Fig. 8 but the theory is calculated
with mean internal transverse quark momentum of 500 MeV
(upper) and 700 MeV (lower).

of the momentum of the quark producing the jet. We expect this to surely have a mean of 330 MeV and keep it fixed at this value. 2) The initial quarks may have a transverse momentum, we are uncertain of the size of this as we discussed above. The experiment R412 also measured the distribution of this momentum out of the plane, p_{out}. The comparison of experiment[8] with theory is indicated in Fig. 10. Again we see a discrepancy, even with 500 MeV. The theory with 700 MeV agrees with the data very well but we do not believe it is sensible, and consider the discrepancy with the 500 MeV theory as indicating, assuming the experiments are correct, some real incompleteness of our theory. (Recent data on μ-pair production in p-p collisions also suggests initial quarks with wide transverse momentum, if it is really correct to interpret the observed pairs as coming from direct quark-antiquark annihilation.)

At the moment we have further work in progress that we hope to publish soon. We have many more detailed calculations of what we would expect in various correlation experiments, than I have been able to describe here. Secondly, to predict the results of many correlation experiments, particularly on the trigger side, we shall need more detailed understanding of the way that a jet is formed. For that purpose we are making calculations with a very simple theory of the details of the quark-to-hadron cascade. Everything is now set into computer programs so we can calculate what we would expect for many different kinds of experiments, provided they involve only mesons. We enjoy making such calculations before the experiments are done. Only in that way can we gain confidence in, or decide to reject, our model. So if you experimenters tell us what you're going to

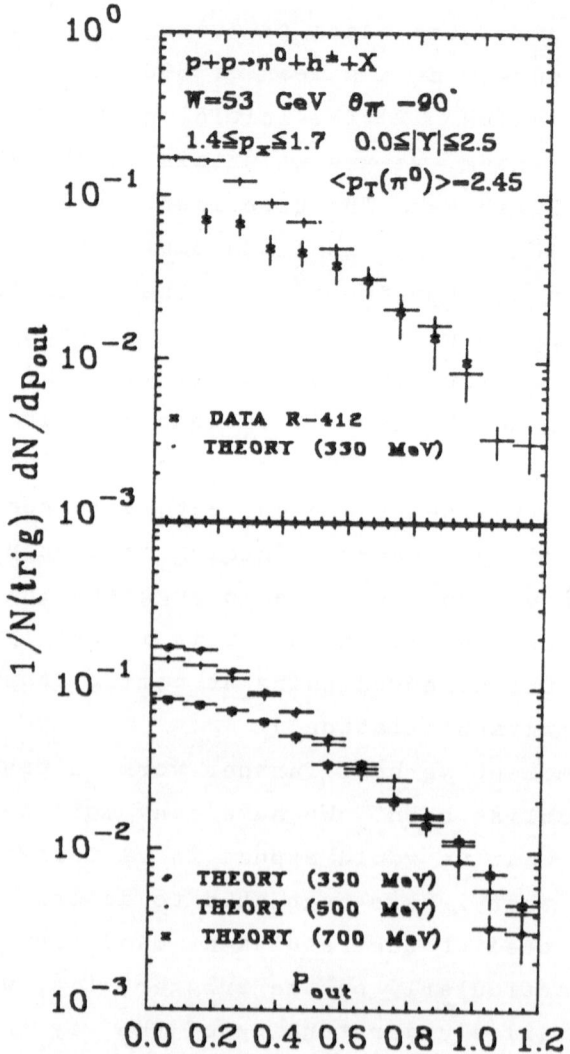

Figure 10: Comparison of theory and data from CERN R-412, Ref. 8 on the P-out distribution of away side hadrons for the process pp → π^0 + h$^\pm$ + X at W = 53 GeV, $\theta_{cm.trig.}$ = 90°, and $p_\perp(\pi^0)$ integrated over a range from 2.0 to 4.0 GeV/c. The theory is calculated using a mean internal transverse quark momentum of 330 MeV and the away hadrons are binned according to 1.4 ≤ p_x ≤ 1.7 GeV/c, 0.0 ≤ |Y| ≤ 2.5. The lower figure shows the dependence of the theory on the choice of the mean transverse quark momentum.

measure, we will be glad to try to make some predictions.

At the present moment we cannot make predictions about everything; we hope in the future to tackle the following problems:

1. We should like to develop a theory of protons and baryons at high $p_\perp$.

2. Single γ's at high $p_\perp$.

3. Experiments with nuclear targets.

4. We should like to unify our theory to that of the main collision at low $p_\perp$.

Finally, of course, we shall have to find out, are we really in trouble from the apparently large transverse momentum of the incident quarks?

This work was done with G. Fox and R. Field. I should like especially to thank R. Field for considerable direct assistance in preparing this talk.

NOTE ADDED IN PROOF

The disagreement with experiment discussed on page 466 and beyond have since been found to be due to an error in the calculational program. All the calculations have now been redone. The theoretical numbers for $N^{h^\pm}$ for the R412 experiment should have been 0.39 (rather than 0.41) for 330 MeV initial transverse momentum, and 0.30 (rather than 0.39) for 500 MeV. The latter is now satisfactorily close to the experimental value ≈ 0.25. For the CCHK experiment the value with 500 MeV should be $N^{h^\pm} = 0.17$ (rather than 0.26).

Figures 8 and 9 must now be changed. The new figures are 8', 9' ($z_p \equiv x_e$).

The P_{out} distributions of figure 10 are also changed and should be as in 10'.

There is no longer any real discrepancy with the choice of 500 MeV for the initial mean transverse momentum and we expect to take this value for future work.

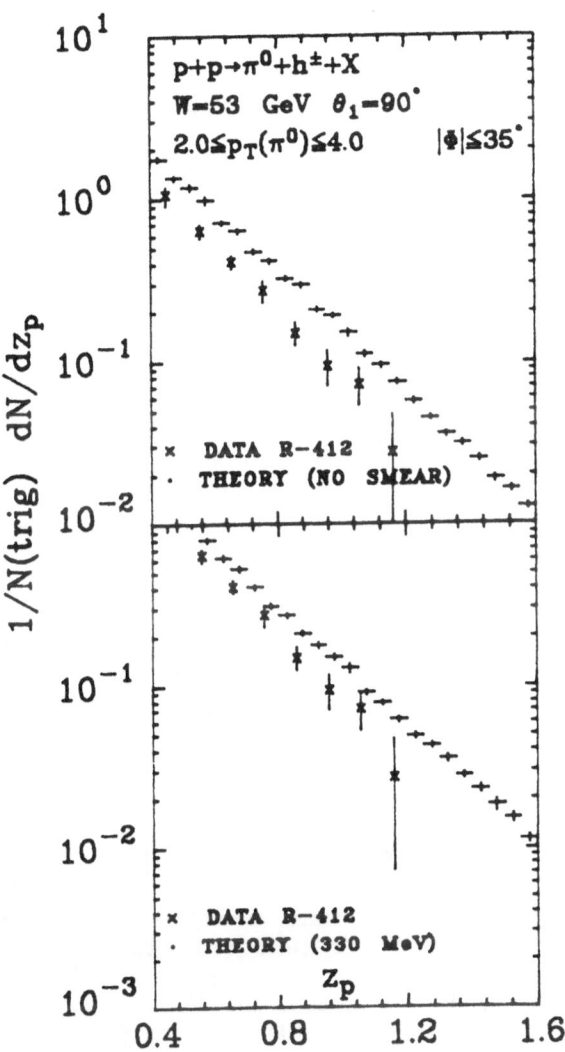

Fig. 8' See NOTE ADDED IN PROOF

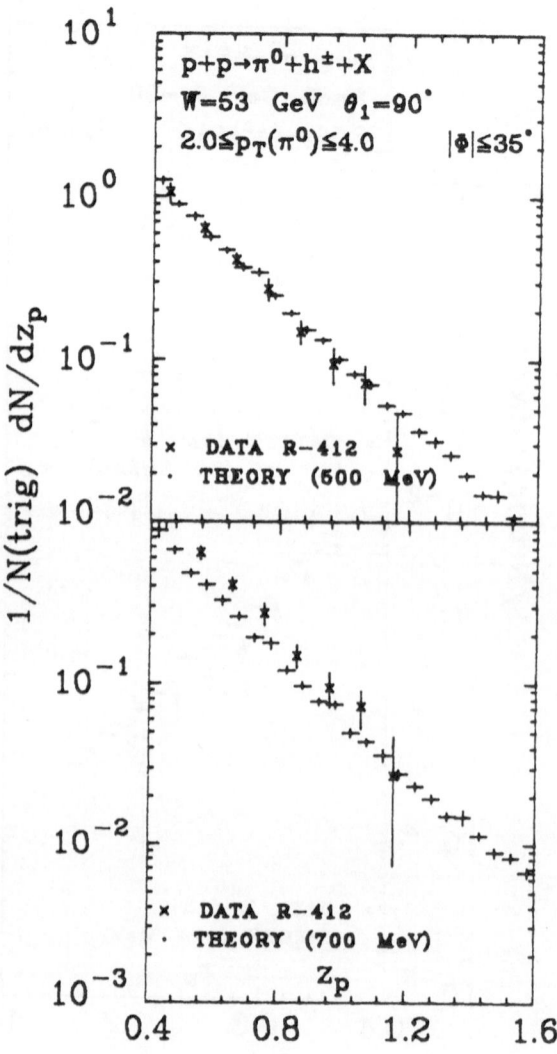

Fig. 9'   See NOTE ADDED IN PROOF

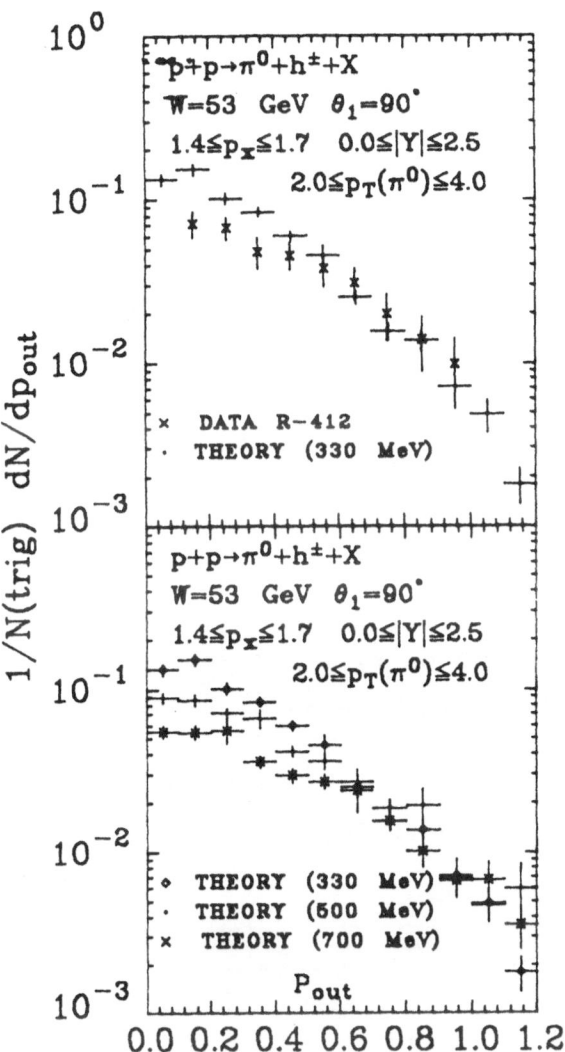

Fig. 10'   See NOTE ADDED IN PROOF

REFERENCES

1. R. Field and R. P. Feynman, CALT-68-565.

2. G. Fox, CALT-68-573.

3. G. Donaldson et al., Phys. Rev. Letters $\underline{36}$, 1110 (1976).

4. R. F. Schwitters, "Hadron Production at SPEAR," International Symposium on Lepton and Photon Interactions, Stanford (1975); J. C. VanderVade "Neutrino-Proton Interactions in the 15-foot Bubble Chamber and Properties of Hadron Jets," IV International Winter Meeting on Fundamental Physics, Salardu, Spain (1976); J. T. Dakin, et al., Phys. Rev. $\underline{D10}$ 1401 (1974).

5. B. Alper et al., Nucl. Phys. $\underline{B100}$, 237 (1975); M. B. Albrow et al., Nucl. Phys. $\underline{B73}$, 40 (1975).

6. M. Della Negra (CCHK collaboration). Private communication and Phys. Rev. Letters $\underline{55B}$, 341 (1975).

7. H. J. Frisch (Chicago-Princeton) Preliminary data; See, for example, H. J. Frisch, Invited Talk, annual meeting Brookhaven National Laboratory, October 1976 (FFL-77-13).

8. P. Darriulat et al., (CERN R412 Experiment), Nucl. Phys. $\underline{B107}$, 429 (1976).

FINAL STATES IN CHARMED PARTICLE DECAYS

Jonathan L. Rosner*

Institute for Advanced Study

Princeton, New Jersey 08540

ABSTRACT

It is shown how weak decays of charmed particles provide information on the isospin of the charm-changing weak interactions, multi-particle production, enhancement of nonleptonic weak interactions, unseen decay modes of known charmed particles and best ways in which to discover new ones, and possible new weak currents and new fermions.

I.   INTRODUCTION

In the period between the discovery of the J/ψ[1,2] and that of charmed[3-5] mesons at 1.87 GeV/c^2,[6-10] a good deal has been learned about the weak decays[11] of heavy particles. This talk - which might be subtitled "An Appreciation of Charm" - is an attempt to show some of

* Work supported in part by ERDA under Contracts No. E(11-1)-1764 and E(11-1)-2220. On leave during 1976-7 from School of Physics and Astronomy, University of Minnesota, Minneapolis, Minnesota 55455

the uses of that information.

We begin (Section II) with a list of the expected properties of charmed particles:[3, 12-14] their masses, and the states to which they decay. Section III presents some results of isospin restrictions on the final states, mentioning both rigorous consequences of the form of H_{weak} and predictions based on statistical models.[16,18-21] These results apply to decays to a __fixed__ number of particles, giving the distribution of charges of these particles.

In Section IV we discuss the branching ratios of charmed particles to final states of __various__ multiplicities. Clues can be taken[13,22] from statistical models, and predictions compared with results on $\bar{N}N$ and e^+-e^- annihilations[26-29] at center-of-mass energies similar to the charmed particle masses.

There remain questions of the applicability of symmetries beyond isospin to charmed particle decays, especially the nonleptonic ones.[13,30-33] These are discussed in Section V. Answers to these questions may shed light on the way in which the nonleptonic decays of strange particles appear to be enhanced.

The results of Secs. III-V are combined into sets of branching ratios for charmed particle decays. Such predictions can help interpret present data (Sec. VI), including those experiments which directly detect particles with lifetimes of 10^{-12} - 10^{-14} sec.,[34-38] and other experiments which observe multi-particle decays of charmed mesons[6-10] or candidates for charmed baryons.[39-42] The estimates of branching ratios also are used to predict new decay modes of known charmed particles, and to suggest the best modes in which to observe predicted particles like $F^{\pm}(2000)$.[13,14,43]

In proceeding from the expected behavior of
charmed particles to the totally unknown world of heavy
leptons and new (charmless) heavy quarks, one may be
able to use many of the results of Secs. III-VI. Some
suggestions are made in Sec. VII. Particular attention
is directed to the chance that some of the new objects
may be present in the final states of charmed-particle
weak decays. Sec. VIII summarizes.

II. CHARMED PARTICLE PROPERTIES

We shall assume the "standard" four quarks are
coupled to left-handed charged currents as shown in
Table I.[3]

Table I. Properties of four quarks

Name	Charge	Strangeness	Charm	Currents	θ = Cabibbo Angle ($\approx 15°$)
u	2/3	0	0	$\begin{pmatrix} u \\ d' \end{pmatrix}_L$	$d' = d\cos\theta + s\sin\theta$
d	-1/3	0	0		
c	2/3	0	1	$\begin{pmatrix} c \\ s' \end{pmatrix}_L$	$s' = -d\sin\theta + s\cos\theta$
s	-1/3	-1	0		

The J/ψ mass was an important factor in setting
the masses expected for charmed particles.[13,14,43,44]
The estimates of Ref. 14 were notably successful for
S-waves. On the basis of the observed charmed particles,
these estimates can be adapted to predict the masses
of the remaining singly-charmed S-wave mesons and
baryons;[43,22] the results are shown in Table II.
(Masses in parentheses denote predictions.)

In Table II only the underlined states are expected to be weakly decaying. Others can decay to the
underlined ones via photon emission, pion emission, or
both.

Our major concern will be with weak decays not involving powers of sin θ. Consequently, the semileptonic decays we shall discuss involve the transitions

$$c \rightarrow s \ell^+ \nu_\ell : \Delta S = \Delta C, \Delta I = 0 \qquad (II.1)$$

while the nonleptonic decays involve

$$c \rightarrow s \, u \, \bar{d}: \quad \Delta S = \Delta C, \Delta I = \Delta I_3 = 1 \quad . \qquad (II.2)$$

All we say will be valid for antiparticles as well, and they will not be discussed separately. The quantum numbers of hadronic systems to which (II.1) and (II.2) give rise have been listed in Refs. 13 and 16.

III. ISOSPIN

As a result of Eqs. (II.1) and (II.2), all particles in Table II decay to final states with definite isospin, with the exception of D° and A° final states, which have a mixture of $I=1/2$ and $I=3/2$. Some restrictions which follow from these total isospin values are noted in Refs. 15-17.

Table II. Observed and predicted charmed particles
(singly charmed S-wave mesons and baryons)

Quark content	Name (Ref.13)	Mass (MeV/c^2)	References	Remarks
$c\bar{u}$ $^1S_0$	$\underline{D}^0$	1867 ± 3	6-10,45	
$^3S_1$	$D*^0$	2005.6 ± 1.5	6-10,45	
$c\bar{d}$ $^1S_0$	$\underline{D}^+$	~1872	6-10,45	Sum of masses:
$^3S_1$	$D*^+$	~2008	6-10,45	3880 ± 5. (Ref.45)
$c\bar{s}$ $^1S_0$	$\underline{F}^+$	(1975)	14, 43	
$^3S_1$	$F*^+$	(~2060)	14, 43	(Not yet observed)
cud $^2S_{1/2}$	$\underline{C}_0{}^+$	2260 ± 10	39-41	$I = 0$
cuu, $^2S_{1/2}$ cud,	$C_1^{++,+,0}$	~2420	39-41	$I=1$. See Refs. 22,
cdd $^4S_{3/2}$	$C*_1^{++,+,0}$	~2480	40,41	43 for interpretation
csu, $^2S_{1/2}$ csd	$\underline{A}^{+,0}$	(2480)		Not yet observed.
csu, $^2S_{1/2}$ csd	$S^{+,0}$	(2570)		Mass predictions are those
$^4S_{3/2}$	$S*^{+,0}$	(2640)		of Ref.14, adjusted upward by 60
css $^2S_{1/2}$	$\underline{T}^0$	(2740)		MeV to fit the
$^4S_{3/2}$	$T*^0$	(2780)		$C_0{}^+$ mass. See Refs.22,43.

A.   Exact relations

The decay $D^o \rightarrow K^- \pi^+$ has been seen.[6-10] Purely because of (II.2), the amplitude for this process is related to that for two others by[16,30,31]

$$A(D^o \rightarrow K^-\pi^+) + \sqrt{2}A(D^o \rightarrow \bar{K}^o\pi^o) = A(D^+ \rightarrow \bar{K}^o\pi^+) . \qquad (III.1)$$

At least one of the two decays involving $\bar{K}^o$ must occur.

The constraint due to (III.1) may be expressed
geometrically: the ratios $\Gamma(D^o \rightarrow K^-\pi^+)/\Gamma(D^o \rightarrow \bar{K}\pi)$ and
$\Gamma(D^+ \rightarrow \bar{K}^o\pi^+)/3\Gamma(D^o \rightarrow \bar{K}\pi)$ lie inside an ellipse.[16] The
first ratio is clearly bounded between 0 and 1. That
the second also lies between 0 and 1 is a special case
of[16, 17]

$$0 \leq \frac{\Gamma(D^+ \rightarrow X^+)}{\Gamma(D^o \rightarrow X^o)} \leq 3 , \qquad (III.2)$$

where (X^+, X^o) denote the totality of charge distributions of a multiparticle final state with (positive,
zero) charge, e.g., $X^+ = (\bar{K} + n\pi)^+$, $K^o = (\bar{K} + n\pi)^o$. The
lower (upper) bound in (III.2) is attained when
$I=3/2$ (1/2) final states are suppressed. While the
$I=1/2$ and $I=3/2$ contributions to $D^o \rightarrow X^o$ do not interfere
in the total rate, they may do so for individual charge
states.

The D^+ and D^o semileptonic decays to $\ell^+\nu_\ell \bar{K}$ +
($\geq$ 0 pions) obey pairwise equalities as a result of
(III.1), e.g.,

$$\Gamma(D^o \rightarrow \ell^+\nu_\ell K^-) = \Gamma(D^+ \rightarrow \ell^+\nu_\ell \bar{K}^o). \quad (III.3)$$

Consequently, the branching ratios for the final states
in (III.3), or for any other pair of isospin-related

final states, obey

$$B(D^o)/B(D^+) = \tau(D^o)/\tau(D^+) , \qquad (III.4)$$

allowing the ratio of lifetimes τ to be measured.[17]

B. Bounds on charge distributions

While $\Gamma(D^o \to K^-\pi^+)/\Gamma(D^o \to \bar{K}\pi)$ is restricted to lie only between 0 and 1 in the absence of further informa-tion, many other such ratios obey less trivial bounds. For charmed-particle decays,[16] these are useful when one can detect only certain charge states, permitting estima-tes for the rates at which others are produced. Similar considerations have been applied to decays of the J/ψ and other $c\bar{c}$ states.[45-47] A sample of such bounds is pre-sented in Table III. These are the lowest-multiplicity ones for which the bounds aren't either (i) 0 and 1, or (ii) calculable with the help of the most elementary Clebsch-Gordan coefficients.

The method employed in Ref. 16 for obtaining the bounds in Table III is quite straightforward and general. It involves the expansion of the decay amplitude in terms of a set of invariant isospin amplitudes. The expansion coefficients, when this amplitude is squared and summed over kinematic configurations, define a symmetric matrix whose maximum and minimum eigenvalues directly yield the required upper and lower bounds. As the algebra is per-formed numerically, we have been able to obtain bounds for up to seven particles in the final state.

C. Bounds on the average number of charged particles

The above methods apply to any quantity linear in branching ratios. One such quantity is the average number $\bar{n}_c$ of charged tracks seen in an experiment. Let us assume that an experiment sees all the charged products of K_S^0 and neutral hyperon decays. Then

$$n_c = n_{pc} + 2 n_V \quad , \qquad (III.5)$$

where n_{pc} is the number of prompt charged tracks and n_V is the number of K_S^0 and neutral hyperons observed to decay to pairs of charged tracks. Fig. 1 shows bounds on $\bar{n}_c$ for D semileptonic and nonleptonic decays.

In Figs. 1a, 1b, the tight bounds for semileptonic decays arise because the I=1/2 final state obeys $n(\pi^0) = [n(\pi^+)+n(\pi^-)]/2$, so 2/3 of the pions must be charged.[15,48] Thus the variations come entirely from the kaons. The lax bounds in Fig. 1c reflect the large number of isospin amplitudes (I = 1/2, 3/2) that can contribute to D^0 nonleptonic decays. If I=3/2 final states were suppressed, these bounds would be tighter. For I=0 final states (arising in some semileptonic decays) one has precise expressions for $\bar{n}_c$, e.g.,

$$\bar{n}_c(C_0^+ \to \ell^+ \nu_\ell \Lambda + n\pi) = 1 + \frac{4}{3} + \frac{2}{3} n \quad , \qquad (III.6)$$

the first term coming from ℓ^+, the second from Λ, and the third from the pions.

D. Simultaneous bounds

In many instances two different charge states of the same multiplicity can be observed, but these still do not exhaust all possible charge states of that multiplicity. Thus, for example, the decays $F^+ \to \bar{K}^0 K^0 \pi^+$, $K^- K^+ \pi^+$ might be observed, but the decay $F^+ \to \bar{K}^0 K^+ \pi^0$ would be missed if neutral pions were not detected. Since the allowed values of the $\bar{K}^0 K^0 \pi^+/\bar{K}K\pi$ and $K^- K^+ \pi^+/\bar{K}K\pi$ ratios must lie in the interior of an ellipse,[16] an improvement on the bound in row 5 of

Table III. Some isospin bounds on final states in charmed-particle decays

Decaying particle	Final state	Relative to:	Lower bound	Upper bound	Statistical value [a]
D^o	$K^- \pi^+ \pi^- \ell^+ \nu_\ell$	$\bar{K}\, 2\pi \ell^+ \nu_\ell$	1/3	2/3	1/2
C_o^+	$\Lambda 2\pi^+ 2\pi^- \ell^+ \nu_\ell$	$\Lambda\, 4\pi \ell^+ \nu_\ell$	1/3	8/15	2/5
D^o	$K^- 2\pi^+ \pi^-$	$\bar{K}\, 3\pi$	0	4/5	4/15
D^+	$K^- 2\pi^+$	$\bar{K}\, 2\pi$	0	4/5	2/5
F^+	$\bar{K}^o K^o \pi^+$ $+ K^- K^+ \pi^+$	$\bar{K}\, K\pi$	1/2	1	3/4
C_o^+	$\Lambda 2\pi^+ \pi^-$ or $\Sigma^- \pi^- \pi^+$	$\Lambda\, 3\pi$ or $\Sigma\, 3\pi$	1/2	4/5	3/5
C_o^+	$K^- p \pi^+$	$\bar{K}\, N\pi$	0	3/4	3/8

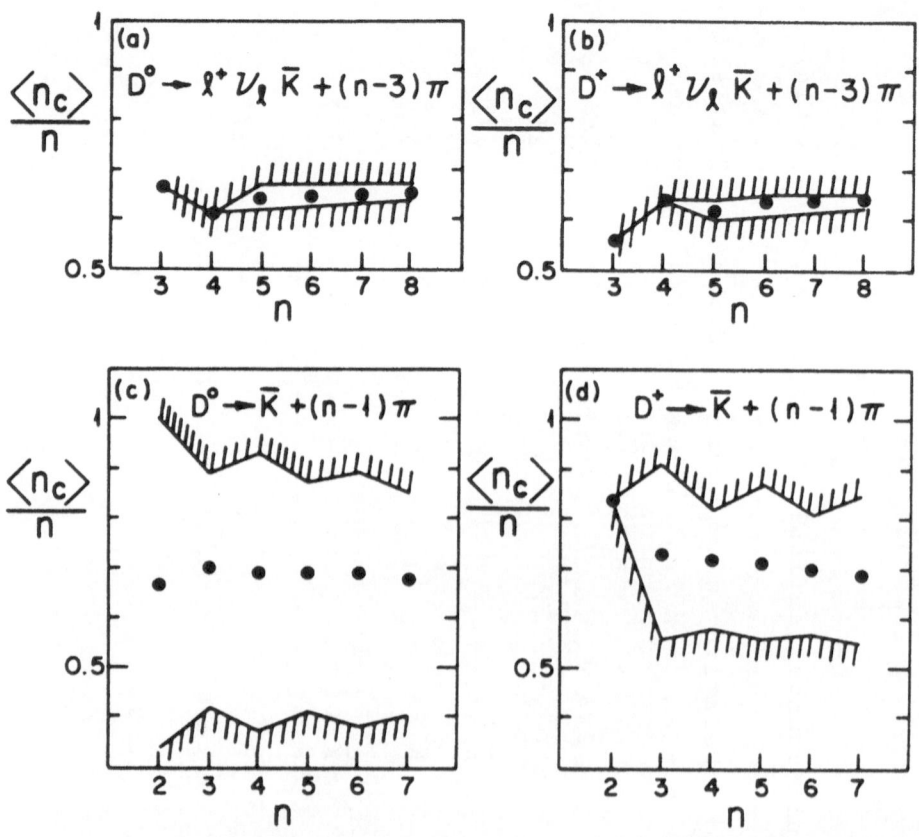

Fig. 1.   Bounds on $\bar{n}_c$ for D decays. $\bar{n}_c$ includes charged
tracks from K_S^o decays.   n denotes total multiplicity
of final state.   Black dots denote predictions of statistical model (Sec. III. E).   (a) D^o semileptonic,
(b) D^+ semileptonic, (c) D^o nonleptonic, and (d) D^+
nonleptonic decays.

Table III is possible. A similar conclusion holds for
any other pair of charge states, with the exception
that the ellipse is replaced by a convex figure containing straight-line and curved segments. As an example, the allowed values of

$$X_{\Sigma 2\pi} \equiv \frac{(C_o^+ \to \Sigma^+ \pi^+ \pi^-)}{(C_o^+ \to \Sigma \, 2\pi)} \qquad (III.7)$$

and

$$Y_{\Sigma 2\pi} \equiv \frac{(C_o^+ \to \Sigma^- 2\pi^+)}{(C_o^+ \to \Sigma 2\pi)} \qquad (III.8)$$

lie within the convex hull of an ellipse defined by
$I_{\pi\pi}$ = even (amplitudes relatively real) and the point
$X_{\Sigma 2\pi}$ = 1/2, $Y_{\Sigma 2\pi}$ = 0, corresponding to $I_{\pi\pi}$ = 1. All
the bounding curves in Ref. 16 appear to be the convex
hulls of ellipses and points, but we do not yet have a
general proof that this should be so.[49] This is because our method[16] allows the direct numerical construction of the bounding curves: we bound

$$X_\Theta \equiv X \cos \Theta + Y \sin \Theta \quad , \qquad (III.9)$$

vary Θ, and make use of the convexity of the bounding
curve to construct the figure using the same eigenvalue
procedure described in Sec. III.B.

The simultaneous-bound method also allows one to
relate D^+ and D^o decays to $\bar{K} + n\pi$,[16] thus generalizing
to all n the ellipse mentioned below Eq. (III.1).
(This ellipse is replaced by a different convex figure
for each n and for each pair of charge states.)

E.   Statistical model

In a multi-particle decay, the number of invariant

isospin amplitudes grows rapidly with the number of
particles.[19]   It was suggested some time ago[18] that the
charge distribution in a multi-particle process be
calculated by letting each of these amplitudes have
equal magnitude, and by assuming that the interference
terms vanish when averaged over kinematic variables.
Some results of applying this postulate to the decays
of charmed particles are shown in Fig. 2.   (The entries
are rational numbers; explicit results may be found in
Ref. 16.)

In Fig. 2 note the statistical suppression of
multi-π^{o} states,[50] and the importance of states with
$n(\pi^{o}) \leq 1$. The neglect of interference between amplitudes may be a poor approximation for low n: for n = 1,
no kinematic averaging can occur, and the statistical
prediction is more a reflection of ignorance than of
anything else.   Another point of caution is that Fig. 2
ignores η, η' production.   (However, we shall estimate
in Sec. V.C that the states with η and η' should com-
prise only about 1/5 of D^{o} nonleptonic decays.

The checking of statistical isospin models is
illustrated in Refs. 21, 27, and 45-47.   From known experimental rates, one uses results such as shown in
Fig. 2 to predict rates for unseen modes.   These predicted rates imply distributions in the number of observed tracks which may be checked with experiment.
Moreover, the branching ratios thus constructed should
add up to 1.   In this manner one finds no clear-cut
difficulty with the statistical model for J/ψ decays.
However, for pionic $\overline{N}N$ annihilations at rest, the model
appears to underestimate the production of neutral
particles.[51]   Whether this shortcoming is due to η and
η', final-state photons, or a fundamental inaccuracy of

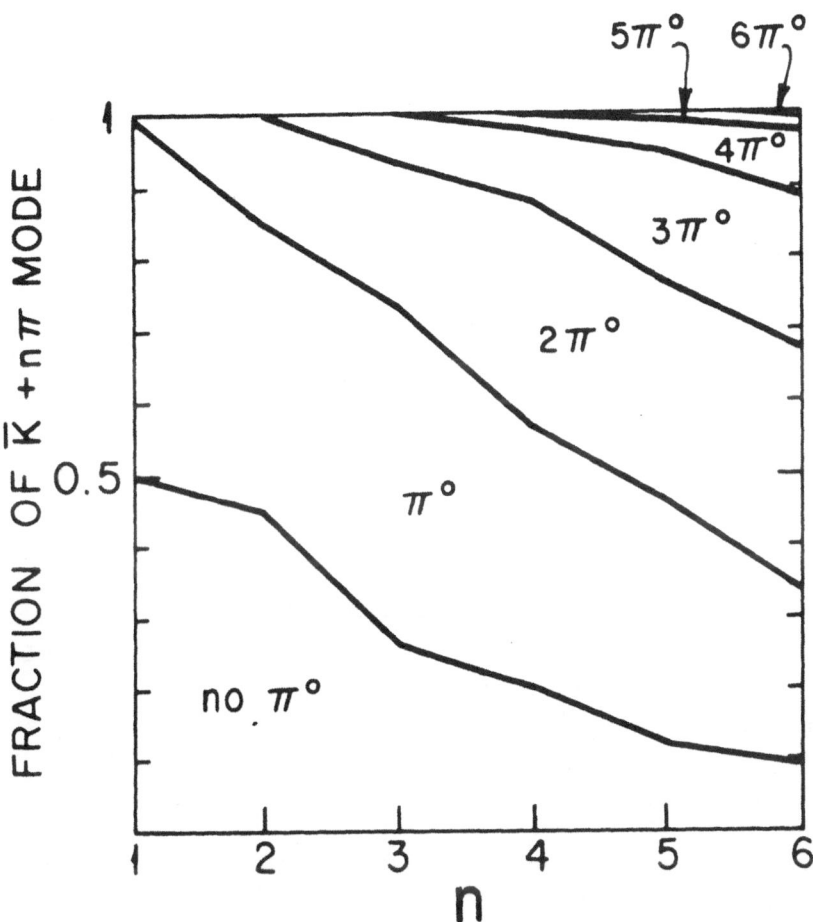

Fig. 2. Fractions of $D \to \bar{K} + n\pi$ decays giving no π^{o}, one π^{o}, etc., on basis of statistical isospin postulate.

the model[27] is unknown. Charmed-particle decays should be very helpful in shedding further light on the question.

IV.   MULTIPLICITIES

When a charmed particle decays, how many particles emerge? What is the shape of the distribution in the number of particles?

A.   Statistical (thermodynamic) model for $\bar{n}$

Imagine the hadronic debris of a charmed-particle decay to be confined to a sphere of radius $R_o \simeq 1$ f at a temperature T.[22, 23] Assume the matter to evolve into $\bar{n}$ pions and the minimum number of additional particles necessary to conserve energy, momentum, and any other quantum numbers. (Hence a 2-body decay corresponds here to n = 0.) Then calculate $\bar{n} \sim V T^3$ and the total energy $\bar{E} \sim V T^4$ available to the $\bar{n}$ particles. (Here $V = 4\pi R_o^3/3$.) Eliminate T; the result is[22]

$$\bar{n} = 0.528(E/E_o)^{3/4}$$

(IV.1)

$$(E_o \equiv \hbar c/R_o = 200 \text{ MeV.})$$

We interpret[22]

$$E = M - M_1 - M_2$$

(IV.2)

for a particle of mass M decaying to particles of mass M_1, M_2, and n additional pions; thus the total multiplicity $\bar{n}_{tot}$ is

$$\bar{n}_{tot} = \bar{n} + 2 \quad .$$

(IV.3)

In this manner one is led to expect values of $\bar{n}_{tot}$

around 4 for nonleptonic D, F, and C_o decays.

Eq. (IV.1) is not expected to hold for values of E so high that the anisotropies seen in e^+ -e^- annihilations ("jet structures"[52]) may occur.[53,54] However, below E = 4 GeV it seems to describe e^+ -e^- annihilations as well as the conventional formula[29]

$$\bar{n}_c = 1.93 + 1.5 \ln E_{c.m.} \qquad (IV.4)$$

if suitably interpreted.[55] Thus Eq. (IV.1) probably can be applied with equal accuracy to weak decays of all particles in Table II.

B. Shapes of Distributions

Two simple multiplicity distributions for $D \rightarrow \bar{K}$ + $n\pi$ whose shapes are specified by $\bar{n}$ alone are shown in Fig. 3. One is a Poisson distribution in n. The other is obtained by assuming that the Lorentz-invariant matrix element for $D \rightarrow \bar{K}\pi + (n+1)\pi$ is a constant times that for $D \rightarrow \bar{K}\pi + n\pi$, and neglecting the masses of the final hadrons. Then one finds,[13] for the probability $P_{\bar{n}}(n)$ of seeing $D \rightarrow \bar{K}\pi + n\pi$ when the average value of n is $\bar{n}$, the relation

$$P_{\bar{n}}(n+1)/P_{\bar{n}}(n) = c^2/[(n+1)(n+2)] , \qquad (IV.5)$$

with c a constant fixed by $\bar{n}$.

The distributions of Fig. 3 give a first impression of the charmed-particle decay multiplicity one might expect.[57] They may be combined with statistical isospin predictions[16, 18-20] to yield branching ratios for any charge state. As the distribution of (IV.5) should be corrected for particle masses,[58, 59] I will not use it further except to note that it is narrower than the

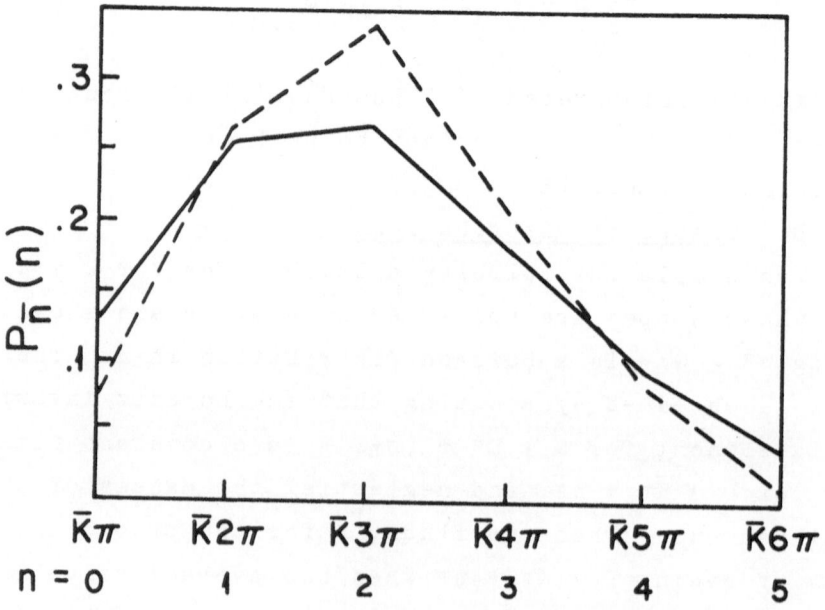

Fig. 3. Two multiplicity distributions for $D \rightarrow \bar{K}\pi + n\pi$, with $\bar{n} = 2.07$ (Ref. 22). Solid line: Poisson distribution in n, $P_{\bar{n}}(n) = \bar{n}^n \, e^{-\bar{n}}/n!$ Dashed line: distribution of Eq. (IV.5). (See Refs. 13, 56.)

Poisson distribution, and gives smaller two-body branching ratios. In this talk I shall present some tables of branching ratios based on Poisson distributions[22] in n, with $\bar{n}$ fixed by (IV.1).

The branching ratios for D nonleptonic decays, relative to all $\bar{K}\pi + n\pi$ decays, are shown in Tables IV and V. The underlined charge states have been observed in Refs. 6-10 (solid) and 41 (dashed).

Table IV. Predicted branching ratios for specific charge states of $D^\circ \rightarrow \bar{K}$ + pions, in percent. Statistical isospin model used.

$n(\pi^{\pm}) =$	0	1	2	3	4	≥ 5	Sum
	$\bar{K}^\circ$	K^-	$\bar{K}^\circ$	K^-	$\bar{K}^\circ$		(Poisson)
K =							
$\bar{K}\pi$	6	<u>6</u>					12.6
$\bar{K}2\pi$	4	10	<u>12</u>				26.1
$\bar{K}3\pi$	2	5	13	<u>7</u>			27.0
$\bar{K}4\pi$	~0	2	6	7	<u>4</u>		18.6
$\bar{K}\geq 5\pi$	~0	1	2	4	5	3	15.7
Sum	12	25	33	18	9	3	100

Table V. Same as Table IV for $D^+ \to \bar{K}$ + pions.

$n(\pi^{\pm})$ =	1	2	3	4	>5	Sum
K =	$\bar{K}^0$	K^-	$\bar{K}^0$	K^-		(Poisson)
$\bar{K}\pi$	13					12.6
$\bar{K}2\pi$	16	<u>10</u>				26.1
$\bar{K}3\pi$	8	9	11			27.0
$\bar{K}4\pi$	2	4	9	4		18.6
$\bar{K}\geq5\pi$	1	1	3	4	3	15.7
Sum	40	24	25	8	3	100

We may construct a Poisson distribution for the decays $D \to \ell^+ \nu_\ell \bar{K}$ + $n\pi$ by starting from the observation that one expects[13,33]

$$\frac{\Gamma(D \to \ell~\nu_\ell \bar{K})}{\displaystyle\sum_{n \geq 0} \Gamma(D \to \ell^+ \nu_\ell \bar{K} + n\pi)} \simeq 0.3 \qquad (IV.6)$$

A much larger number would be ruled out by the softness of the observed direct lepton spectrum in multi-particle events observed at DESY[60] and at SPEAR[45, 61]. A much smaller number would be hard to reconcile with separate estimates[13,33] for three-body and inclusive semileptonic decays.[62]

Eq. (IV.6) may be taken as the first term in a Poisson distribution with $\bar{n} = -\ln(0.3) = 1.2$. The resulting branching ratios are shown in Table VI.

Table VI. Predicted branching ratios for specific charge states of $D \rightarrow \ell^+ \nu_\ell \bar{K}$ + pions, as percent of all such decays. Statistical isospin model used.

$n(\pi\frac{+}{-}) =$	0	1	2	≥ 3	Sum (Poisson)
$D^0: \bar{K} =$	K^-	$\bar{K}^0$	K^-		
$D^+: \bar{K} =$	$\bar{K}^0$	K^-	$\bar{K}^0$		
$\ell^+ \nu_\ell \bar{K}$	30				30
$\ell^+ \nu_\ell \bar{K}\pi$	12	24			36
$\ell^+ \nu_\ell \bar{K}2\pi$	4	7	11		22
$\ell^+ \nu_\ell \bar{K}\geq 3\pi$	~0	2	5	5	12
Sum	46	33	16	5	100

C. Inclusive variables

The branching ratios in Tables IV-VI may be condensed into a form more easily compared with present data. The numbers at the bottom of each column, for example, give distributions of kaons and charged pions. These are, of course, compatible with the bounds of Ref. 15. It is interesting that D^+ nonleptonic decays lead in the model to relatively few neutral pions and charged kaons: $\bar{n}(\pi^0)/\bar{n}(\pi\frac{+}{-}) = 0.5$ (bounds of Ref. 15: 0 to 3/2), and $\bar{n}(K^-)/\bar{n}(\bar{K}^0) = 0.5$ (bounds of Ref. 15: 0 to 4).

If the neutral kaons in D decays are visible as two charged particles 1/3 of the time and invisible the other 2/3 of the time, as occurs in certain experiments,[6-10,41] Tables IV-VI imply the distribution of observed charged particles shown in Table VII.

Table VII.   Predictions of percentage of D decays leading
to n_c observed charged particles, when decays of short-
lived neutral kaons are taken into account.

	n_c = 0	2	4	6	$\bar{n}_c$
D^0 semileptonic	-	69	29	2	2.7
D^0 nonleptonic	9	51	35	5	2.7

	n_c = 1	3	5	7	
D^+ semileptonic	31	59	10		2.6
D^+ nonleptonic	26	55	18	1	2.9

The values of $\bar{n}_c$ in Table VII may be compared with
$\bar{n}_c$ = 3.1 for pionic $\bar{N}N$ annihilations at rest,[27] and
$\bar{n}_c$ = 2.9 obtained by extrapolating the expression
(IV.4) down to $E_{c.m.}$ = 1.87 GeV.

A $D\bar{D}$ pair thus should contribute about 5.4 to
$\bar{n}_c$ (cf. Ref. 63).   The masking of this effect just above
charm threshold has been attributed[63] to the production
of heavy lepton pairs,[64] which contribute a nearly
compensating number of events with low charged-particle
multiplicity.

When discussing F^+ or C_o^+ decays, we cannot identify a single channel which, when summed over pions, is
expected to be dominant.   Thus, $F^+ \rightarrow \bar{K}K$ + pions,
$F^+ \rightarrow$ (η or η') + pions, and $F^+ \rightarrow$ pions all are possible
in principle (though the last may be suppressed; see
Refs. 33 and 65.)   Similarly C_o^+ can decay to many
different types of S = -1 baryonic channels.   Relations
among such decays occur in the next section.

V. SYMMETRIES BEYOND ISOSPIN

A. Semileptonic decays

The isospin method[17] noted in Sec. (III.A) for obtaining $\tau(D^0)/\tau(D^+)$ can be generalized[17] to yield $\tau(F^+)/\tau(D^+)$ if one is willing to trust the SU(3) prediction

$$\frac{\Gamma(F^+ \to \ell^+ \nu_\ell \eta)}{\Gamma(D^+ \to \ell^+ \nu_\ell K^-)} = \frac{2}{3} \quad . \qquad\qquad (V.1)$$

This prediction will be modified slightly by singlet admixture in the η,[66] and by phase space and form factor differences.

As for nonleptonic decays,[32] a number of SU(3) - related channels appear in multi-hadron final states of semileptonic decays.[67] It requires phase-space and form-factor calculations to assess their importance.[58]

Parity violations in semileptonic decays of charmed particles,[68] as in $C_0^+ \to \ell^+ \nu_\ell \Lambda$, can test for right-handed currents involving charmed and strange quarks. The Λ polarization is affected substantially if such currents are present.[69]

B. I-, U-, and V-spin for nonleptonic decays

The weak nonleptonic Hamiltonian (II.2) must transform either as a $\bar 6$ or a 15 of SU(3),[30-33] as shown in Fig. 4.

As Fig. 4 shows, the crucial distinction between the $\bar 6$ and 15 contributions lies in their V-spins. Both representations have the same I-spin and U-spin properties. We shall describe some tests of the suggestions that (i) the $\bar 6$ part may be enhanced,[30-33] and (ii) decays into non-exotic final states may be enhanced.[13] Two-body decays of charmed pseudoscalar mesons provide a clear-cut distinction between these two possibilities.

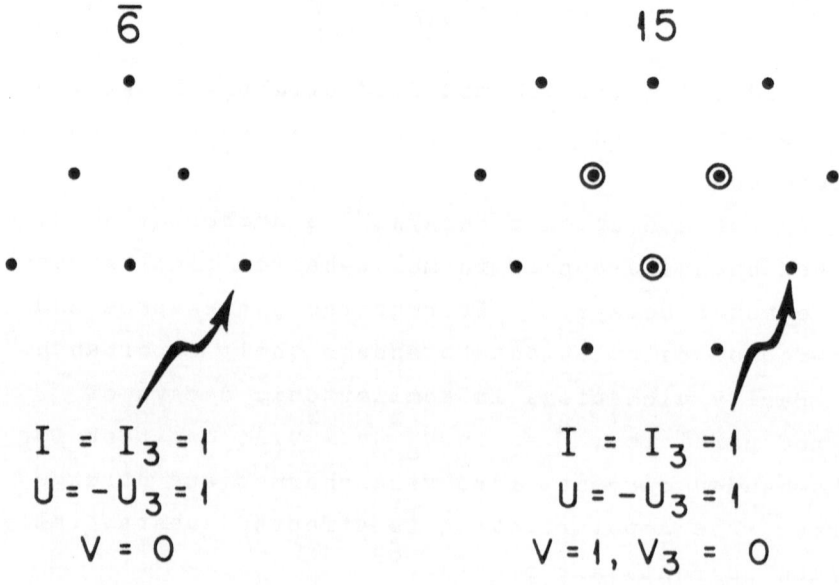

Fig. 4. $\bar{6}$ and 15 representations of SU(3). The part of $H_{weak}^{n.l.}$ giving rise to $c \to s\,u\,\bar{d}$ is shown by an arrow.

(For a related discussion, including estimates of branching ratios, see Ref. 33.)

C. Meson $\rightarrow$ 2 mesons

The I-spin properties of $H_{weak}^{n.l.}(\Delta S = \Delta C)$ lead to the relation (III.1). The U-spin properties can be seen to lead to

$$\Gamma(D^0 \rightarrow \bar{K}^0 \eta) = \tfrac{1}{3} \times 0.9 \times \Gamma(D^0 \rightarrow \bar{K}^0 \pi^0) \quad,^{70} \qquad (V.2)$$

$$A(F^+ \rightarrow \bar{K}^0 K^+) + (3/2)^{1/2} A(F \rightarrow \eta \pi^+) = A(D^+ \rightarrow \bar{K}^0 \pi^+) . \quad (V.3)$$

Note the presence of $A(D^+ \rightarrow \bar{K}^0 \pi^+)$ on the right-hand sides of (III.1) and (V.3). The (spatially symmetric) $\bar{K}^0 \pi^+$ system must be in a 27. It cannot be reached from the D^0 (which belongs to $\bar{3}$) via the $\bar{6}$ piece of $H_{weak}^{n.l.}(\Delta S = \Delta C)$. Thus, one might expect $\Gamma(D^+ \rightarrow \bar{K}^0 \pi^+)$ to be small in comparison with partial widths on the left-hand sides of (III.1) and (V.3) either because of $\bar{6}$ enhancement or because of nonexotic final state enhancement. In either case, if $\Gamma(D^0 \rightarrow \bar{K}^0 \pi^+) = 0$, we have

$$\Gamma(D^0 \rightarrow \bar{K}^0 \pi^0) = \tfrac{1}{2} \Gamma(D^0 \rightarrow K^- \pi^+) \quad, \qquad (V.4)$$

$$\Gamma(F^+ \rightarrow \eta \pi^+) = \tfrac{2}{3} \times 1.06 \times \Gamma(F^+ \rightarrow \bar{K}^0 K^+) . \qquad (V.5)$$

A specific test of the $\bar{6}$ enhancement mechanism is provided by the relation

$$\Gamma(F^+ \rightarrow K^+ \bar{K}^0) = 0.88 \times \Gamma(D^0 \rightarrow K^- \pi^+) \quad,^{70} \qquad (V.6)$$

which may be derived with the $\Delta V = 0$ property of the $\bar{6}$.

The symmetries mentioned above do not predict $\Gamma(D^0 \rightarrow \bar{K}^0 \eta')$. The following remarks are based on the assumption, still open to question, that the nonleptonic D^0 decay proceeds through an octet s $\bar{d}$ state. If one can use a nonet ansatz to relate $\Gamma(D^0 \rightarrow \bar{K}^0 \eta)$ to $\Gamma(D^0 \rightarrow \bar{K}^0 \eta')$, one finds[32,89]

$$\Gamma(D^0 \rightarrow \bar{K}^0 \eta') = 8 \times 0.74 \times \Gamma(D^0 \rightarrow \bar{K}^0 \eta) \ . ^{70} \qquad (V.7)$$
$$(2)$$

The rates for $D^0 \rightarrow 0^- 0^-$ are then in the ratios

$$\Gamma(D^0 \rightarrow K^- \pi^+) \ : \ \Gamma(D^0 \rightarrow \bar{K}^0 \pi^0) \ : \ \Gamma(D^0 \rightarrow \bar{K}^0 \eta) \ : \ \Gamma(D^0 \rightarrow \bar{K}^0 \eta')$$

$$= \quad 1 \quad : \quad 0.50 \quad : \quad 0.15 \quad : \quad 0.88 \qquad (V.8)$$
$$(0.22)$$

Similar methods apply to decays to $0^- 1^-$ and $1^- 1^-$ final states.[30-32,71] In the absence of clear-cut evidence for quasi-two-body decays of this sort, we may instead estimate the multi-body decays of D^0 involving $\bar{K} \eta + n\pi$ and $\bar{K} \eta' + n\pi$ as in Sec. IV, and find

$$\Gamma(D^0 \rightarrow \bar{K}\pi + (\geq 0 \ \text{pions})):\Gamma(D^0 \rightarrow \bar{K}\eta + (\geq 0\pi)):\Gamma(D^0 \rightarrow \bar{K}\eta' + (\geq 0\pi))$$

$$= \quad 0.8 : \quad 0.05 \quad : \quad 0.15 \qquad (V.9)$$
$$(0.9) \quad\quad\quad\quad\quad\quad (0.04)$$

In addition, D^0 should have a branching ratio of a few percent to $K\bar{K}\bar{K} + (\geq 0 \ \text{pions})$.

D. Meson $\rightarrow$ 3 mesons

The best-studied three-body decay of a charmed meson at present is $D^+ \rightarrow K^- 2\pi^+$.[7-10] The three particles appear to be in a spatially symmetric state. On the assumption of $\bar{6}$ dominance, these particles must lie in a $\overline{10}$ of SU(3). Then one expects[32]

$$\Gamma(D^+ \rightarrow K^- 2\pi^+):\Gamma(D^+ \rightarrow \bar{K}^0 \pi^+ \pi^0):D(D^+ \rightarrow \bar{K}^0 \pi^+ \eta):\Gamma(D^+ \rightarrow K^+ 2\bar{K}^0) =$$

$$= \quad 1 \quad : \quad 1/4 \quad : (3/4) \times 0.45 : 1 \times 0.17 . \quad [70] \quad (V.10)$$

The proportion of $D^+ \to \bar{K}^0 \pi^+ \pi^0$ is much smaller than in the statistical isospin model (cf. row 2 of Table V), and in fact is as low as it can be.

E. Baryon → baryon + meson

Another relation implied by $\bar{6}$ enhancement is

$$\sqrt{2} A(C_0^+ \to p\bar{K}^0) + \sqrt{3} A(C_0^+ \to \Lambda \pi^+) + A(C_0^+ \to \Sigma^0 \pi^+) = 0 , \quad (V.11)$$

easily derived from V-spin. Eq. (V.11) implies that $\Lambda \pi^+$ cannot be the only 2-body nonleptonic channel of C_0^+. The maximum of the ratio of $\Lambda \pi^+$ to $(\Lambda \pi^+) + (p\bar{K}^0) + (\Sigma^0 \pi^+) + (\Sigma^+ \pi^0)$ rates, subject to (V.11) and without regard to phase space differences, is $5/11$. From this we are tempted to generalize to multi-body decays: it is unlikely that $C_0^+ \to \Lambda$ + pions accounts for more than half of all C_0^+ nonleptonic decays.

VI. SOME EXPERIMENTAL IMPLICATIONS

A. Short-track detection

The event of Ref. 34 was the first to be interpreted in terms of possible charm production. It involves the production of a pair of shortlived tracks. So, too, does the event of Ref. 35,[37] which can be interpreted as $C_0 \bar{C}_0$ pair production.[22,38] Both these events occurred in cosmic ray exposures. By contrast, at the lower energy of 300 GeV, a sample of 60,000 proton interactions has yielded no candidates for associated charm production, and a corresponding upper limit on the cross section of 1.5 μb/nucleon.[72]

The prospects for charm production in neutrino physics look much brighter: a recent emulsion exposure[36] has obtained one candidate for charm production in only 16 neutrino interactions. This is not far from the expected rate (far above threshold) of $\sin^2 \theta \simeq 1/18$. Since the emulsion stack may contain a total of as many as 200 neutrino interactions, a check of the total rate for short-track production indeed should be possible.

The event found in Ref. 36 consisted of (i) an interaction in the emulsion at a point "A", giving rise to some heavy tracks and five light ones, (ii) the materialization of one of the light tracks into three light ones at a point "B", 182 μm from A, and (iii) an apparent two-body decay ("V^o") of a neutral strange particle, 34 cm down-stream from points A and B. If the V^o is a Λ, it has survived two mean lifetimes, and if it is a K_S^o, it has survived three.

Let us assume that the strange particle comes from the vertex B, as would be expected for the decay of a charmed particle. By arguments of momentum balance,[36] it is possible to rule out certain possibilities like $C_o^+ \rightarrow \Lambda 2\pi^+ \pi^-$ or $D^+ \rightarrow K_S^o 2\pi^+ \pi^-$. The asymmetric decay of the V^o, as well as its lifetime, suggests that it may be a Λ. We have checked that the decays

$$C_o^+ \rightarrow \Lambda 2\pi^+ \pi^- \pi^o \qquad \text{(VI.1a)}$$

$$C_o^+ \rightarrow \Lambda 2\pi^+ \pi^- \eta \qquad \text{(VI.1b)}$$

$$C_o^+ \rightarrow \Lambda \pi^+ \pi^- \ell^+ \nu_\ell \qquad \text{(VI.1c)}$$

all give acceptable kinematic fits,[73] using the momentum measurements of Ref. 36 based on multiple scattering. Eqs. (VI.1) yield a C_o^+ with momentum

$$2.7 \text{ GeV/c} \leq P_{lab} \leq 6.8 \text{ GeV/c} , \qquad (VI.2)$$

and hence lifetime

$$2 \times 10^{-13} \text{ sec.} \leq t \leq 5 \times 10^{-13} \text{ sec.} \qquad (VI.3)$$

The two $C_o(\overline{C_o})$ candidates[22,38] in Ref. 35 live about 7×10^{-13} sec. and 5×10^{-12} sec. if they really are C_o tracks.[38]

The decay $C_o^+ \rightarrow \Lambda 2\pi^+ \pi^-$ already may have been observed.[39-41] By the methods of Secs. III and IV, we have estimated that this decay is 0.16 of all $C_o^+ \rightarrow \Lambda+$ (pions) decays. Interestingly, because of a favorable statistical isospin factor, $C_o^+ \rightarrow \Lambda 2\pi^+ \pi^- \pi^0$ also should be a significant mode: 0.13 of all $C_o^+ \rightarrow \Lambda+$ (pions) decays. Perhaps this is what has been observed in Ref. 36. Unfortunately, it will be impossible to tell from the event in question: there are two solutions for the π^0 momentum, one of which involves a π^0 backward in the laboratory and thus totally undetectable in the present experiment.[73]

It is easier to find charged short-lived particles than neutral ones in emulsions. However, there are several possibilities for charged particles with life-times around 10^{-13} sec.: D^+, F^+, C_o^+, and the heavy lepton L^+. These particles should decay into 1 or 3 charged particles. The D^0 is harder to look for, but among the particles we have discussed it is the only neutral one, and so its signal should be more character-istic. It should decay predominantly to two or four charged particles, according to Table VII. If neutral heavy leptons exist (Sec. VII), they, too, should show up in emulsions.

B.   Other charmed-particle experiments

At $E_{c.m.}$ = 4.028 GeV, SPEAR has measured[9]

$$2B(D^o \to K^-\pi^+) \qquad \sigma(D^o) = 0.52 \pm 0.05 \text{ nb}$$

$$2B(D^o \to \bar{K}^o\pi^+\pi^-) \qquad \sigma(D^o) = 0.80 \pm 0.21 \text{ nb}$$

$$2B(D^o \to K^-2\pi^+\pi^-) \qquad \sigma(D^o) = 0.72 \pm 0.18 \text{ nb}$$

$$2B(D^+ \to K^-2\pi^+) \qquad \sigma(D^+) = 0.27 \pm 0.05 \text{ nb},$$

$$\text{(VI.4)}$$

with additional systematic uncertainties of 50%.[74] The
D^o data are compatible with the predictions of Table IV,
if we regard $\sigma(D^o)$ as unknown. If we estimate $\sigma(D^o)$ in-
depently,[45] we find all the experimental D^o branching
ratios implied by the above numbers are slightly smaller
than the corresponding entries in Table IV.

Note the small value of $B\sigma$ for $D^+ \to K^- 2\pi^+$, in
comparison with the value for $D^o \to \bar{K}^o \pi^+\pi^-$. These are
decays of the same multiplicity, and the statistical
isospin factors[16] are almost the same: 0.4 for the former, and 0.45 for the latter. The disparity may indicate
(i) $\sigma(D^o) > \sigma(D^+)$[43] (cf. the estimate $\sigma(D^o) \simeq \frac{2}{3} \sigma(D)$ in
Ref. 45), (ii) a smaller branching ratio for $D^+ \to \bar{K} +$
pions,[13] (iii) a shortcoming of the model, or (iv) any
combination of the above.

An estimate of the $D^+ \to \bar{K}^o \pi^+$ signal expected on
the basis of the statistical isospin model is possible,
if we use the D^o branching ratios in (VI.4) to obtain a
direct estimate of the relative probabilities of 2-body
and 3-body decays, and apply this estimate to D^+ decays
as well. The resulting prediction, for SPEAR data at
4.028 GeV, is

$$2B(D^+ \to \bar{K}^0 \pi^+) \; \sigma(D^+) = 0.39 \pm 0.13 \text{ nb} . \quad (VI.5)$$

This signal should be detectable. If it is much smaller, the $\bar{K}^0 \pi^+$ mode is undergoing a particular suppression not experienced by the final state $K^- 2\pi^+$. Both are exotic, but the former must belong to a 27 whereas the latter may belong to a $\overline{10}$. Only the latter could be produced if $H_{weak}^{\;n.l.}$ transformed as $\bar{6}$. (See Sec. V.)

The photoproduction experiment of Ref. 41 detects a small $D^0 \to \bar{K}^0 2\pi^+ 2\pi^-$ signal. One expects from Table IV a small branching ratio into this channel: $\approx 4\%$ of $D^0 \to \bar{K}$ + pions. By contrast, according to Table V, the channel $\bar{K}^0 2\pi^+ \pi^-$ should be $\approx 11\%$ of $D^+ \to \bar{K}$ + pions. Its absence could be due to any of the factors (i) - (iv) mentioned above.

C. New modes; new charmed particles

Secs. III-V suggest some interesting new decay modes in which to observe the known charmed particles. Many of these will require the detection of single neutral pions or the imposition of kinematic fits to compensate for single missing neutral particles. Tables IV-VI suggest some prominent modes of D decay. In addition, using a Poisson distribution in n for $C_0^+ \to \Lambda\pi + n\pi$, with $\bar{n}$ set by (IV.1), one finds that[22]

$$C_0^+ \to \begin{matrix} \Lambda\pi^+ \\ \Lambda\pi^+\pi^0 \end{matrix} = \begin{matrix} 17\ \% \\ 30\ \% \end{matrix} \text{ of } C_0^+ \to \Lambda + \text{pions.} \quad (VI.6)$$

Moreover, according to (V.11), if $\bar{6}$ enhancement is valid, we should expect at least one of the modes $C_0^+ \to p\bar{K}^0$, $(\Sigma\pi)^+$ to be visible once $C_0^+ \to \Lambda\pi^+$ is seen.

Many suggestions have been given in Refs. 13 and

43 for observing the F^+ and the remaining particles in Table II. Proceeding from Eq. (IV.1), we would expect

$$F^+ \rightarrow \eta\pi^+ = 12\% \text{ of } F^+ \rightarrow \eta\pi + (\geq 0\pi) \qquad (VI.7)$$

$$F^+ \rightarrow \eta'\pi^+ = 20\% \text{ of } F^+ \rightarrow \eta'\pi + (\geq 0\pi) \qquad (VI.8)$$

$$F^+ \rightarrow \bar{K}^\circ K^+ = 17\% \text{ of } F^+ \rightarrow \bar{K}K + (\geq 0\pi) \qquad (VI.9)$$

Recalling (V.5) and using a nonet ansatz[32] to relate $F^+ \rightarrow \eta\pi^+$ to $F^+ \rightarrow \eta'\pi^+$,[89] we then would find roughly <u>equal</u> rates for $F^+ \rightarrow \eta\pi + (\geq 0\pi)$, $F^+ \rightarrow \eta'\pi + (\geq 0\pi)$, and $F^+ \rightarrow \bar{K}K + (\geq 0\pi)$.[89] Let us assume this to be so, and let us neglect[33,65] the decays $F^+ \rightarrow$ pions. Then we find the F^+ branching ratios listed in Table VIII. The large number of channels available in the nonleptonic decay of F^+ makes it hard to detect in any one of them, though $\bar{K}^\circ K^+$ and $K^- K^+ \pi^+$ look most promising. We shall mention another possibility in the next section.

VII. NEW OBJECTS

It is amusing that just as a quantum number was discovered which restored lepton-hadron symmetry,[3-5] evidence has appeared for new heavy leptons and for possible new quarks which threw the situation off balance again.[45,60,64,75-78]

A. <u>Charm → heavy leptons</u>

If a charged heavy lepton $L^{\pm}$ of mass 1.95 GeV/c^2 [64,60] is coupled via a charged weak current to a massless neutrino ν_L and the F^+ has mass 2 GeV/c^2, the decay $F^+ \rightarrow L^+ \nu_L$ is just barely possible.[79]

Similarly, there is the barest possibility that a <u>neutral</u> heavy lepton could be present in F^+ decay. The process $\mu \rightarrow e\gamma$, whose possible observation has been discussed elsewhere at this conference (refs. 80,75,76)

Table VIII.   Model for F^+ nonleptonic branching ratios, in percent. $F^+ \to$ pions neglected.   Poisson distributions in n.   Statistical isospin model used.

$\eta\pi + n\pi$ ($\bar{n}=2.16$) 33% (44%)[89]			$\eta'\pi + n\pi$ ($\bar{n}=1.63$) 33% (11%)[89]		
$\eta\pi^+$	4	(5)	$\eta'\pi^+$	7	(2)
$\eta\pi^+\pi^0$	8	(12)	$\eta'\pi^+\pi^0$	11	(3)
$\eta2\pi^+\pi^-$	5	(7)	$\eta'2\pi^+\pi^-$	5	(2)
$\eta\pi^+2\pi^0$	4	(5)	$\eta'\pi^+2\pi^0$	3	(1)
$\eta(\geq 4\pi)$	12	(15)	$\eta'(\geq 4\pi)$	7	(3)

$\bar{K}K + n\pi$ ($\bar{n}=1.78$) 33% (44%)[89]					
$\bar{K}^0K^+$	6	(8)	$K^-K^02\pi^+$	1	(2)
$\bar{K}^0K^+\pi^0$	2	(3)	$K^-K^+\pi^+\pi^0$	2	(3)
$K^-K^+\pi^+$	4	(5)	$\bar{K}^0K^0\pi^+\pi^0$	2	(3)
$\bar{K}^0K^0\pi^+$	4	(5)	$\bar{K}^0K^+2\pi^0$	1	(1)
$\bar{K}^0K^+\pi^+\pi^-$	3	(4)	$\bar{K}K(\geq 3\pi)$	8	(12)

is allowed at a level of 10^{-10} or greater[80,75,76] if there exist neutral heavy leptons N_1 and N_2 coupled via right-handed currents:

$$\begin{pmatrix} N_1 \cos \phi + N_2 \sin \phi \\ e^- \end{pmatrix}_R \qquad \begin{pmatrix} - N_1 \sin \phi + N_2 \cos \phi \\ \mu^- \end{pmatrix}_R$$

The present experimental lower limit on the mass of the lightest such lepton N is set only by the absence of the decay $K^+ \to e^+N$.[81]   As long as $m_N < m_F - m_e$, the decay $F^+ \to Ne^+$ can occur, provided the value of ϕ does not prevent it.   If $m_N < m_F - m_\mu$, at least one of the decays $F^+ \to N\mu^+$, Ne^+ must be quite substantial, independent of ϕ, as we now show.

The decay of a pseudoscalar meson of mass M into leptons of mass m_1, m_2 via a V $\pm$ A current involves a Lorentz-invariant matrix element whose square is $M^2(m_1^2 + m_2^2) - (m_1^2 - m_2^2)^2$. For M = 2 GeV/c^2 this factor, multiplied by the appropriate phase space term, is the same to within 5% for F $\rightarrow$ Ne and F $\rightarrow$ Nμ when m_N varies between 0.5 GeV/c^2 (its present lower bound[81]) and 1.75 GeV/c^2. Consequently, we shall neglect both the electron and muon masses in what follows. By an elementary calculation, relating the decays F $\rightarrow$ N (e or μ) to K $\rightarrow$ $\mu\nu$, we find

$$\frac{\Gamma(F \rightarrow Ne) + \Gamma(F \rightarrow N\mu)}{\Gamma(K \rightarrow \mu\nu)} = (f_F/f_K)^2 K_K(m_N) , \qquad (VII.2)$$

with the function $K_K(m_N)$ shown in Fig. 5. Note in (VII.2) the absence of ϕ, the result of summing over e and μ. The ratio of f_F to f_K is unknown. It would be 1 in the SU(4) limit.[82] Since $\Gamma(K \rightarrow \mu\nu) = 5 \times 10^7$ sec^{-1}, one has

$$\Gamma(F \rightarrow Ne) + \Gamma(F \rightarrow N\mu) \geq 2 \times 10^{11} \text{ sec}^{-1}(f_F/f_K)^2$$

$$(1 \text{ GeV/c}^2 \leq m_N \leq 1.3 \text{ GeV/c}^2) \qquad (VII.3)$$

and the sum of these partial widths still is appreciable over most of the remaining allowed kinematic range.[83] Fig. 5 also applies to the decay[79] $F^+ \rightarrow L^+\nu_L$: for $m_L = 1.95$ GeV/c^2, $m_{\nu_L} = 0$, and $m_F = 2$ GeV/c^2, one finds a value (very sensitive to $m_F - m_L$) of

$$\Gamma(F^+ \rightarrow L^+\nu_L) \simeq 3\times10^9 \text{ sec}^{-1}(f_F/f_K)^2 .$$
$$(VII.4)$$

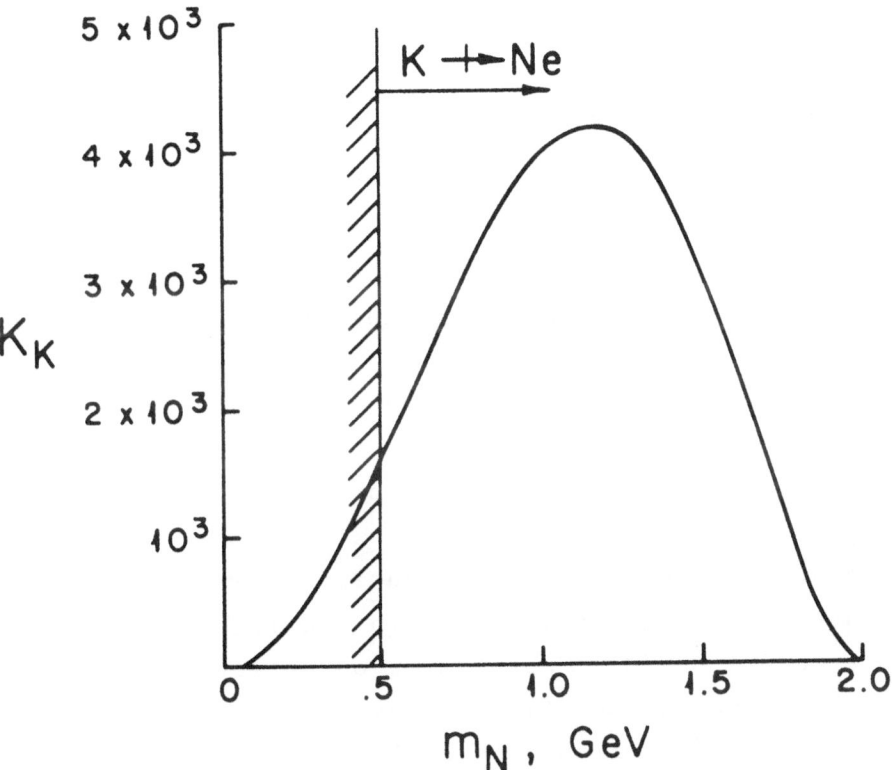

Fig. 5. Kinematic factor describing $F \rightarrow N(e$ or $\mu)$ via Eq. (VII.2).

An estimate of the total decay rate of F^+, taking account of possible nonleptonic enhancement effects,[33] is $\Gamma(F^+) \simeq 8 \times 10^{12}$ sec^{-1} for m_c (the mass of the charmed quark) = 2 GeV/c^2, and less for smaller values of m_c. Thus, for a wide range of values of m_N, the branching ratio

$$\frac{\Gamma(F \to Ne) + \Gamma(F \to N\mu)}{\Gamma(F)} \geq (\text{several } \%) , \qquad (\text{VII.5})$$

as long as $f_F/f_K \geq 1$.

In turn, we have estimated the total decay rate of N and its decay into (e or μ)$^-$ + π^+ via standard methods,[81,33,13] to obtain branching ratios varying from 40% to 10% into the latter channel as m_N varies from 1 to 2 GeV/c^2.

Consequently, the branching ratio for

$$F^+ \to \pi^+ (\text{e or } \mu)^+ (\text{e or } \mu)^- , \qquad (\text{VII.6})$$

if we sum over e and μ, could easily exceed a percent if the heavy lepton N were to lie in the right mass range (m_N around 1 GeV/c^2). Depending on detection capabilities, the channel (VII.6) even could become competitive with those listed in Table VIII as a way of looking for the F^+.

A neutral heavy lepton produced with a branching ratio of several percent in F decays could, in principle, be responsible for anomalous leptons of the same sign as the lepton associated with the incident neutrino in high-energy neutrino reactions.[84,85] An example of such a process is

$$\nu_\mu X \to \mu^- F^+ Y$$

$$N(\mu^+ \text{ or } e^+) \qquad \pi^+$$

$$(\mu^- \text{ or } e^-) + \quad \text{multi-hadrons} \quad . \quad \text{(VII.7)}$$

$$(\mu^+ \text{ or } e^+) + \nu$$

By means of this reaction, a neutral heavy lepton can yield up to four observed leptons in the final state.

Further suggestions for producing neutral heavy leptons have been given.[81,76] We have concentrated on ways associated with charmed-particle decays: if the F^+ can be produced abundantly in any reaction, its decays will help raise the experimental lower limit on the lowest-mass lepton (if any exist) coupled via Eq. (VII.1). (A similar argument could have been given for D^+ decays, but the corresponding rates would have been suppressed by $\tan^2 \theta \simeq 1/18$.)

B. <u>Charged heavy lepton decays</u>

Here we have only a few minor comments, as these decays are well-discussed elsewhere.[45,63,76,81,86]

The statistical isospin model of Sec. III can be applied to those decays involving multi-hadron final states. For example, the pions in $L^- \to \nu_L + (3\pi)^-$ should all be charged 4/5 of the time. Many other such applications are possible, but decay rates into specific channels like $\nu_L \pi^-$, $\nu_L \rho^-$, and $\nu_L A_1^-$, which probably make up the bulk of the hadronic decays, can be estimated directly.[81, 86]

Both charmed particles and heavy charged leptons are sources of massive charged weak currents when they decay. As discussed, these currents can be used to test for new couplings and leptons, as in Eq. (VII.1). Another possibility for the heavy lepton decays is to use

their massive weak currents to test for second-class
currents of hadrons,[87] via final states of definite
G-parity. In this context we note that the decay

$$L^- \rightarrow \nu_L \eta \pi^- \qquad\qquad (VII.8)$$

is absolutely forbidden for first-class currents: only
the vector current can contribute to the production of
a pair of 0^- particles, but the first-class vector
current has G = +, while $\eta\pi$ has G = -.[88]

C. Heavy-quark decays

These are discussed in Refs. 45, 54, 75, and 76.
The statistical considerations of Secs. III and IV are
relevant once the appropriate hadronic quantum numbers
are changed, except that for sufficiently massive quarks,
jet structure of the final state may be appreciable,[54]
and Eq. (IV.1) may not apply. The authors of Ref. 54
have stressed that charmed particles may be important
intermediates in the decays of hadrons containing heavy
quarks. The nonleptonic decays are likely to be affected
seriously by right-handed currents, and one should not
necessarily expect the semileptonic branching ratios of
heavier hadrons to be the same as for charmed ones.

VIII. SUMMARY

The discovery of charmed particles has been a
tremendous source of encouragement in constructing unified gauge theories of the weak, electromagnetic, and
(perhaps) strong and gravitational interactions. This
work is proceeding at an extraordinary rate.

Here, other aspects of charmed particles have been
discussed. Their decays provide information on multi-
body production. This can be helpful in deciding how

hadronic systems evolve, in finding new charmed particles,
and in knowing what to look for in the decays of hadrons
containing new, heavier quarks. Charmed particles also
can provide some insight into the physics of nonleptonic
weak interactions. Their decays (and those of heavy
leptons) provide massive charged weak currents - a
charged version of electron-positron collisions.

From what we have heard from other speakers at
this conference, it looks like there is a whole new
realm of particle physics associated with the mass
scale of 2 GeV/c^2. Charm may be just the first hint
of this physics. The year and a half it took us to discover it, after we knew it should be there, has provided
theorists and experimentalists with some experience which
should be very worthwhile in the near future.

ACKNOWLEDGMENTS

Part of this material originated at Fermilab and
at the Aspen Center for Physics during the summer of
1976, in collaboration with M. Goldhaber, B. Lee, M.
Peshkin, and C. Quigg. I am grateful to them and to
R.M. Barnett, F. Gilman, G. Goldhaber, D. Horn, S.
Nussinov, A. Pais, and S. Treiman for discussions, and
to Dr. Harry Woolf for extending the hospitality of the
Institute for Advanced Study.

REFERENCES

1. J.J. Aubert et al., Phys. Rev. Letters 33, 1404
    (1974).

2. J.-E. Augustin et al., Phys. Rev. Letters 33, 1406
    (1974).

3. S.L. Glashow, J. Iliopoulos, and L. Maiani, Phys.
    Rev. D2, 1285 (1970).

4. The hadronic weak currents of Ref. 3 were first
    proposed by M. Gell-Mann and S.L. Glashow, 1961
    (unpublished); M. Gell-Mann, Phys. Letters 8, 214
    (1964); Z. Maki and Y. Ohnuki, Prog. Theor. Phys.
    32, 144 (1964); Y. Hara, Phys. Rev. 134, B701 (1964);
    and B.J. Bjørken and S.L. Glashow, Phys. Letters 11,
    255 (1964), in which "charm" is named.

5. General quartet schemes for hadrons have been con-
    sidered by the authors of Refs. 3, 4, as well as by
    Y. Katayama et al., Prog. Theor. Phys. 28, 675
    (1962); P. Tarjanne and V.L. Teplitz, Phys. Rev.
    Letters 11, 447 (1963); Z. Maki, Prog. Theor. Phys.
    31, 331, 333 (1964); and D. Amati, H. Bacry, J.
    Nuyts, and J. Prentki, Phys. Letters 11, 190 (1964).

6. G. Goldhaber, F.M. Pierre, et al., Phys. Rev. Letters
    37, 255 (1976).

7. I. Peruzzi, M. Piccolo, G.J. Feldman, H.K. Nguyen,
    J.E. Wiss, et al., Phys. Rev. Letters 37, 569
    (1976).

8. G. Goldhaber, "Study of Charmed Mesons at SPEAR",
    Lawrence Berkeley Laboratory report LBL-5534,
    presented at 1976 SLAC Summer Institute on Particle
    Physics.

9. R. Schwitters, in Proceedings of Division of Particles
    and Fields American Physical Society Divisional
    Meeting, Brookhaven National Laboratory, October,

1976, edited by Ronald F. Peierls, Brookhaven
Natl. Lab. report BNL-50598, February, 1977.

10. J.E. Wiss, G. Goldhaber, et al., Phys. Rev.
Letters $\underline{37}$, 1531 (1976).

11. A description of tests for weak decays of heavy
mesons is given by Benjamin W. Lee, C. Quigg, and
Jonathan L. Rosner, Comments in Nuclear and
Particle Physics, to be published. The most con-
clusive of these tests makes use of the Dalitz-plot
analysis of Charles Zemach, Phys. Rev. $\underline{133}$, B1201
(1964), and has been used in Ref. 10 to infer that
the D is decaying weakly.

12. S.L. Glashow, in Experimental Meson Spectroscopy -
1974, edited by D.A. Garelick, American Institute
of Physics, New York, 1974, p. 387.

13. Mary K. Gaillard, Benjamin W. Lee, and Jonathan L.
Rosner, Rev. Mod. Phys. $\underline{47}$, 277 (1975).

14. A. De Rujula, Howard Georgi, and S.L. Glashow,
Phys. Rev. $\underline{D12}$, 147 (1975).

15. O. Nachtmann and A. Pais, Phys. Letters $\underline{65B}$, 59
(1976).

16. Murray Peshkin and Jonathan L. Rosner, "Isospin
Restrictions on Charge Distributions in Charmed
Particle Decays," Institute for Advanced Study report COO-2220-93, December, 1976, to be published
in Nuclear Physics B.

17. A. Pais and S.B. Treiman, "Charmed Meson Lifetime
Ratios and Production in e^{+} $-e^{-}$ Collisions,
"Rockefeller University report no. COO-2232B-112,
to be published in Phys. Rev. D.

18. E. Fermi, Phys. Rev. $\underline{92}$, 452 (1953); $\underline{93}$, 1434(E)
(1954).

19. A. Pais, Ann. Phys. (N.Y.) $\underline{9}$, 548 (1960), Ibid., $\underline{22}$, 274 (1963).

20. F. Cerulus, Nuovo Cimento (Suppl.) $\underline{15}$, 402 (1960).

21. A. Pais, Phys. Rev. Letters $\underline{32}$, 1081 (1974).

22. Benjamin W. Lee, C. Quigg, and Jonathan L. Rosner, to be published in Phys. Rev. $\underline{D15}$ (1977).

23. E. Fermi, Prog. Theor. Phys. $\underline{5}$, 570 (1950).

24. Some early work in this area is reviewed by A. Jabs, Nucl. Phys. $\underline{B34}$, 177 (1971), and by D.Q. Lamb, in Proceedings of the Colloquium on High Multiplicity Hadronic Interactions, Ecole Polytechnique, 1970, edited by A. Krzywicki et al., p. IV. 89.

25. R. Hagedorn, Nuovo Cimento (Suppl.) $\underline{3}$, 147 (1965); S.C. Frautschi, Phys. Rev. $\underline{D3}$, 2821 (1971); C.J. Hamer and S.C. Frautschi, Phys. Rev. $\underline{D4}$, 2125 (1971).

26. C.J. Hamer, Nuovo Cimento $\underline{12A}$, 162 (1972).

27. S.J. Orfanidis and V. Rittenberg, Nucl. Phys. $\underline{B59}$, 570 (1973).

28. B. Jean-Marie et al., Phys. Rev. Letters $\underline{36}$, 291 (1976).

29. Gary J. Feldman and Martin L. Perl, Phys. Letters $\underline{19C}$, 233 (1975).

30. G. Altarelli, N. Cabibbo, and L. Maiani, Nucl. Phys. $\underline{B88}$, 285 (1975); Phys. Letters $\underline{57B}$, 277 (1975).

31. R.L. Kingsley, S.B. Treiman, F. Wilczek, and A. Zee, Phys. Rev. $\underline{D11}$, 1919 (1975).

32. M.B. Einhorn and C. Quigg, Phys. Rev. $\underline{D12}$, 2015 (1975); Phys. Rev. Letters $\underline{35}$, 1114(C) (1975).

33. J. Ellis, M.K. Gaillard, and D.V. Nanopoulos, Nucl. Phys. $\underline{B100}$, 313 (1975).

34.   K. Niu, E. Mikumo, and Y. Maeda, Prog. Theor.
      Phys. <u>46</u>, 1644 (1971).

35.   H. Sugimoto, Y. Sato, and T. Saito, Prog. Theor.
      Phys. <u>53</u>, 1541(L) (1975), and In <u>Proceedings of</u>
      <u>the 14th International Cosmic Ray Conference</u>,
      Munich, Aug. 15-29, 1975, Max-Planck-Institut,
      1975, paper no. HE5-6, p. 2427.

36.   E.H.S. Burhop et al., Phys. Letters <u>65B</u>, 299
      (1976).

37.   Early emulsion events, including those of Refs. 34
      and 35, are discussed by K. Hoshino et al., in
      Proceedings of the 1975 Cosmic Ray Conference
      (op. cit. Ref. 35), papers no. HE 5-11, p. 2442,
      and HE 5-12, p. 2448, and by G.B. Yodh, in Pro-
      ceedings of the 1975 Cosmic Ray Conference (op.
      cit. Ref. 35), p. 3936. The importance of asso-
      ciated production in reducing background from
      nuclear interactions is stressed by Yodh and in
      Ref. 38.

38.   T.K. Gaisser and F. Halzen, Phys. Rev. <u>D14</u>, 3153
      (1976).

39.   E.G. Cazzoli et al., Phys. Rev. Letters <u>34</u>, 1125
      (1975).

40.   B. Knapp et al., Phys. Rev. Letters <u>37</u>, 882
      (1976).

41.   B. Knapp, in 1976 DPF Proceedings (op. cit. Ref. 9).

42.   S.J. Barish et al., to be published in Phys. Rev.
      <u>D15</u> (1977).

43.   A. De Rujula, Howard Georgi, and S.L. Glashow,
      Phys. Rev. Letters <u>37</u>, 398, 785(C) (1976).

44.   A.W. Hendry and D.B. Lichtenberg, Phys. Rev. <u>D12</u>,
      2756 (1975).

45.   Frederick J. Gilman, these proceedings, and in

1976 DPF Proceedings (op. cit. Ref. 9).

46. Frederick J. Gilman, in High Energy Physics and Nuclear Structure - 1975 (AIP Conference Proceedings No. 26), edited by D.E. Nagle et al., New York, American Institute of Physics, 1975, p. 331.

47. F. Vannucci et al., "Mesonic Decays of the $\psi(3095)$," SLAC and LBL report SLAC-PUB-1862, LBL-5595, December, 1976, submitted to Phys. Rev.

48. M. Peshkin, Phys. Rev. 121, 636 (1961).

49. I thank S. Nussinov for proofs of some special cases.

50. I thank A. Pais for a discussion of this point.

51. Using data quoted in Ref. 27 on up to 7-prong annihilations, I have checked that the sum of all pionic annihilations constructed with the help of tables of Ref. 19 is only about 2/3 of the actual total.

52. Gail Hanson et al., Phys. Rev. Letters 35, 1609 (1975).

53. This possibility has been raised in Ref. 54 for decays of hadrons containing heavier quarks.

54. Robert N. Cahn and Stephen D. Ellis, "How to Look for b-Quarks", Univ. of Michigan report 76.45, January, 1977 (unpublished).

55. Benjamin W. Lee, C. Quigg, and Jonathan L. Rosner, unpublished. In this work one finds a substantial range of overlap between (IV.1) with $E_o = 225$ MeV, $E = E_{c.m.} - 2\,m_\pi$, $\bar{n}_{tot} = n + 2$, and (IV.4), with $\bar{n}_{tot} = \frac{3}{2}\,\bar{n}_c$.

56. I thank C. Quigg for pointing out that Eq. (IV.5) entails $P_{\bar{n}}(n) = c^{2n-1}/[n!\,(n+1)!\,I_1(2c)]$, where I_1 is a modified Bessel function, and $\bar{n} = c\,I_2(2c)/I_1(2c) = c - 3/4 + 3/(32c) + O(1/c^2)$.

57. I thank D. Horn for a discussion of more general
     parametrizations, including the Gaussian distribu-
     tion. A Gaussian would be suitable for describing
     the distribution if its center and width were fixed
     by the data. With present uncertainties in
     branching ratios (see Ref. 9), this is not possible.

58. Benjamin W. Lee, C. Quigg, and Jonathan L. Rosner,
     in progress.

59. The table of branching ratios for $D \to \bar{K}\pi + n\pi$ in
     Ref. 13 contains an additional assumption which is
     not compatible with present data (see Refs. 6-10)
     or with the discussion of Sec. IV. A, namely, the
     fixing of c in Eq. (IV.5), and hence of $\bar{n}$, a priori.

60. H. Meyer, these proceedings; W. Braunschweig et al.,
     Phys. Letters 63B, 471 (1976); J. Burmester et al.,
     Ibid. 64B, 369 (1976).

61. Gary J. Feldman, F. Bulos, D. Lüke, et al., Phys.
     Rev. Letters 38, 117 (1977).

62. For further considerations, see M. Bourquin and
     J.-M. Gaillard, Nucl. Phys. B114, 334 (1976); I.
     Hinchcliffe and C.H. Llewellyn Smith, Nucl. Phys.
     B114, 45 (1976); V. Barger, T. Gottschalk, and R.
     J.N. Phillips, Phys. Letters 64B, 333 (1976), and
     Univ. of Wisconsin report COO-569, August, 1976,
     to be published, and M. Gronau et al., DESY report
     76/62, November, 1976, to be published.

63. See, e.g., Frederick J. Gilman, in Proceedings of
     the 1975 International Symposium on Lepton and
     Photon Interactions et High Energies, Stanford
     University, August 21-27, 1975, edited by W.T.
     Kirk, Stanford Linear Accelerator Center, Stanford,
     Calif., 1975, p. 131, and H. Harari, Ibid., p. 317.

64. M.L. Perl et al., Phys. Rev. Letters 35, 1489 (1975);

Phys. Letters <u>63B</u>, 466 (1976).

65. S. Nussinov, Institute for Advanced Study report COO-2220-85, September, 1976, to be published in Phys. Rev.

66. According to H. Lipkin (these proceedings), this singlet admixture may not have predictable effects.

67. These have been worked out by C. Quigg (private communication).

68. A.J. Buras, Nucl. Phys. <u>B109</u>, 373 (1976).

69. A.J. Buras and John Ellis, Nucl. Phys. <u>B111</u>, 341 (1976).

70. The decimal factor is a phase space correction.

71. S.R. Borchardt and V.S. Mathur, Phys. Rev. Letters <u>36</u>, 1287 (1976).

72. G. Coremans-Bertrand et al., Phys. Letters <u>65B</u>, 480 (1976).

73. I thank C. Quigg for checking the kinematic solutions.

74. The factor of 2 on the left of Eq. (VI.4) comes from our assumption of CP invariance, since only the sum for particles and antiparticles is quoted. Tests of CP invariance in decays of charmed particles are noted by A. Pais and S.B. Treiman, Phys. Rev. <u>D12</u>, 2744 (1975), and Maurice Goldhaber and Jonathan L. Rosner, to be published in Phys. Rev. <u>D15</u> (1977). For an extensive review of a class of models for CP violation see H. Harari, "Beyond Charm", lectures delivered at Les Houches Summer School, August, 1976, Weizmann Institute report WIS-76/54 PH, to be published.

75. H. Fritzsch, these proceedings.

76. R.M. Barnett, these proceedings, and in 1976 DPF Proceedings (op. cit. Ref. 9).

77. F. Gürsey, these proceedings, and in 1976 DPF Proceedings (op. cit. Ref. 9).

78. S.L. Glashow, these proceedings.

79. I. Karliner, Phys. Rev. Letters $\underline{36}$, 759 (C) (1976).

80. T.P. Cheng and Ling-Fong Li, "Nonconservation of Separate μ- and e- Numbers in Gauge Theories with V+A Currents," December, 1976, to be published; T.P. Cheng, these proceedings.

81. J.D. Bjorken and C.H. Llewellyn Smith, Phys. Rev. $\underline{D7}$, 887 (1973).

82. Since $f_K = 1.28\ f_\pi$, these constants might indeed be increasing slightly with mass. The value of f_F may be useful for distinguishing among quark models: see C.H. Llewellyn Smith, Ann. Phys. (N.Y.) $\underline{53}$, 521 (1969). For an extreme view of symmetry-breaking effects, see J. Kandaswamy, J. Schechter, and M. Singer, Phys. Rev. $\underline{D13}$, 3151 (1976).

83. The maximum in Fig. 5 occurs for $m_N = m_F/\sqrt{3} = 1.16$ GeV/c^2.

84. A. Benvenuti et al., Phys. Rev. Letters $\underline{34}$, 419 (1975); Ibid., $\underline{34}$, 597 (1975); Ibid. $\underline{35}$, 1199 (1975); Ibid. $\underline{35}$, 1249 (1975); D. Cline, 1976 DPF Proceedings (op. cit. Ref. 9).

85. B. Barish et al., Calif. Inst. of Technology report CALT 68-567, presented by O. Fackler at 1976 DPF Meeting (op. cit. Ref. 9).

86. Y.S. Tsai, Phys. Rev. $\underline{D4}$, 2821 (1971).

87. A. Pais and S.B. Treiman, Phys. Rev. $\underline{D14}$, 293 (1976).

88. I am indebted to Arthur Halprin for suggesting this rest.

89. Note added: (I am indebted to S. Nussinov for discussions leading to the following remarks):

An alternative version of the nonet ansatz, consistent with the $\bar{6}$ dominance assumption, is the following: if $D^{\circ} = c\bar{u} \rightarrow s\, u\, \bar{d}\, \bar{u} \rightarrow \bar{K}^{\circ} + (\eta \text{ or } \eta')$, the η and η' must be produced through the $u\,\bar{u}$ state, so that one finds the ratio listed in parentheses in Eq. (V.7). This then leads to the alternative conclusions in parentheses in Eqs. (V.8) and (V.9). Similarly if $F^{+} = c\,\bar{s} \rightarrow s\, u\, \bar{d}\, \bar{s} \rightarrow \pi^{+} + (\eta \text{ or } \eta')$, the η and η' must be produced through the $s\,\bar{s}$ state, leading to a ratio of $\Gamma\,(F^{+} \rightarrow \pi^{+}\eta')/\Gamma(F^{+} \rightarrow \pi^{+}\eta)$ which is 1/4 that obtained in Ref. 32. The numbers in parentheses in Table VIII are based on the assumption that the η and η' are produced via the $s\,\bar{s}$ state.

THE WINNER OF THE VECTOR-MODEL LOOK-ALIKE CONTEST*

Howard Georgi[†]

Lyman Laboratory of Physics

Harvard University, Cambridge, Mass. 02138

In the previous talks, Fritzsch and Barnett have discussed the b quark as a possible explanation of the high-y anomaly. The important question is, how can we incorporate the V+A coupling of the b quark to the u quark in a unified model of weak and electromagnetic interactions. The simplest model with the u-b coupling is the vector model,[1] an SU(2) × U(1) model in which both left- and right-handed components of all quarks are in SU(2) doublets. Alas, it is ruled out by neutral current data,[2] which shows significant parity violation in the hadronic neutral current. The standard SU(2) × U(1) model,[3] in which only left-handed quarks are in doublets, agrees well with neutral current data (see Fig. 1), but has no V+A u-b coupling. In the standard model, the high-y anomaly must be attributed to the Q^2

---

*Research supported in part by the National Science Foundation under Grant No. PHY75-20427.

[†]Sloan Foundation Fellow.

dependence of antiquark and strange-quark distributions
in the nucleon induced by QCD. This is probably not
enough to explain data.[4] Furthermore, as we have heard
in Bouchiat's talk, the standard model may be in
trouble with experiments detecting parity violation in
atomic physics.

     In this talk, I will discuss a possible resolution
of the above dilemma in a class of models analyzed by
De Rujula, Glashow and myself[5] and independently (and
differently) by Mohapatra and Sidhu.[6]

     Call the Weinberg-Salam-GIM model, in which only
left-handed quarks display weak SU(2), the standard
model;[3] and call the model in which both left- and
right-handed quarks participate on an equal footing in
weak interactions the vector model.[1] Let us compare
the predictions of these two simple models with experimental data.

     The experiments we are concerned with are of
three kinds: neutral-current neutrino scattering on
electron and nuclear targets; high-energy charged-
current neutrino scattering; and the search for parity-
violating interactions between electrons and nuclei.

     Neutral-current neutrino scattering experiments,
both deep inelastic and elastic are generally in agreement with the standard model and inconsistent with the
vector model. In deep inelastic scattering, the results
are often quoted in terms of the ratios of cross
sections

$$\frac{\sigma_{NC}(\nu d)}{\sigma_{CC}(\nu d)} \quad \text{and} \quad \frac{\sigma_{NC}(\bar{\nu} d)}{\sigma_{CC}(\bar{\nu} d)} \quad ,$$

where NC and CC stand for neutral- or charged-current,
and d indicates the appropriate scattering cross section

off "matter", an equal mixture of neutrons and protons.
Predictions for these ratios for the standard and vector
models according to the naive quark model are shown in
Fig. 1.[7] The standard-model predictions depend upon
the single parameter $\sin^2\theta$, and trace out the nose-like
curve as $\sin^2\theta$ varies from zero to one. The vector
model predictions involve two independent parameters,
the mixing angle and the mass of the Z (which fixes the
overall strength of neutral current effects). But,
because the hadronic neutral current is pure vector,
$\sigma_{NC}(\nu d) = \sigma_{NC}(\bar{\nu}d)$, and the ratio $\{\sigma_{NC}(\nu d)/\sigma_{CC}(\nu d)\}/$
$\{\sigma_{NC}(\bar{\nu}d)/\sigma_{CC}(\bar{\nu}d)\}$ is just the ratio of charged-current
cross section $\sigma(\nu d)/\sigma(\bar{\nu}d) \sim 3$.

These ratios have been measured by several
groups.[2] The experiments involve y cuts to stay away
from $y \simeq 1$, where there are experimental problems. It
is easy to correct for the effect of this cut assuming
scaling and the absence of longitudinal structure
functions. Two corrected experimental points are shown
on Fig. 1 and they are consistent with the standard
model, but several standard deviations away from the
vector model prediction.

A similar result is observed in elastic ν or $\bar{\nu}$
proton scattering. For the vector model to be correct,
the elastic cross sections $\sigma(\nu p \rightarrow \nu p)$ and $\sigma(\bar{\nu}p \rightarrow \bar{\nu}p)$
must be equal. In fact, what is observed is that
$\sigma(\bar{\nu}p \rightarrow \bar{\nu}p)$ is considerably smaller than $\sigma(\nu p \rightarrow \nu p)$,
roughly in accordance with expectations of the standard
model.[2]

In ν_e-e and ν_μ-e neutral-current scattering, the
experimental situation is less clear. Both vector and
standard models are compatible with the data.

The first indication of trouble with the standard

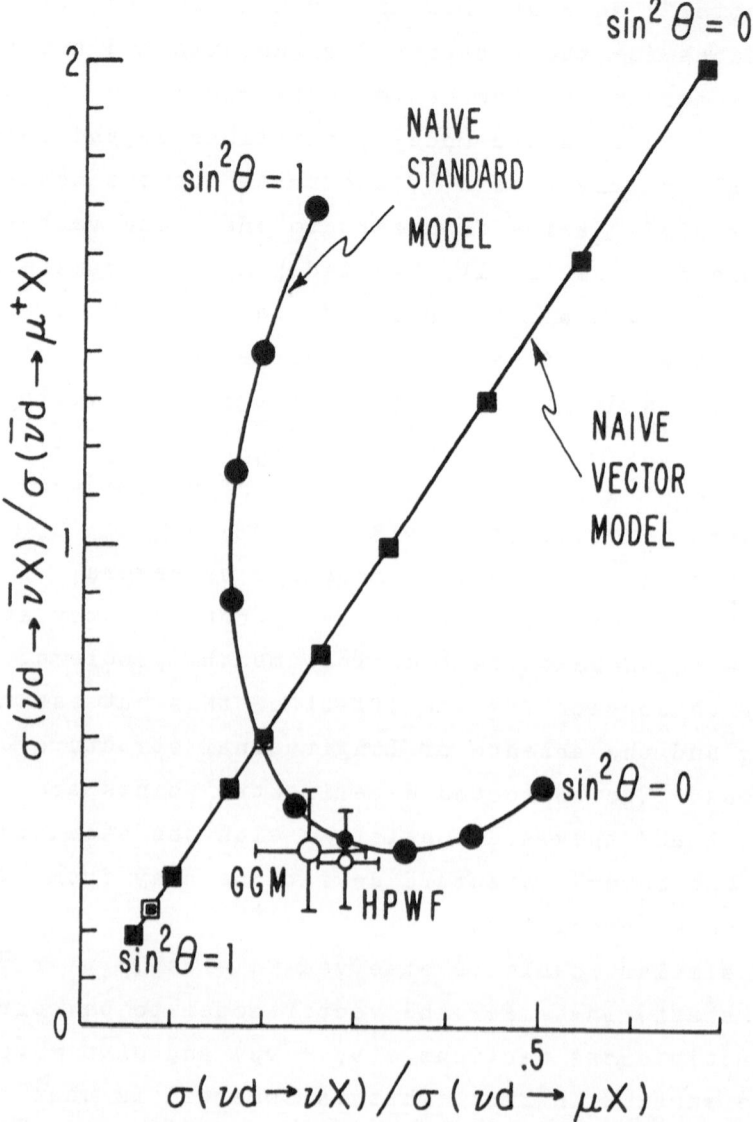

Fig. 1. Ratios of neutral-current to charged-current
neutrino- and antineutrino-induced inclusive cross
sections on "matter" (a target with equal numbers of
protons and neutrons). Naive parton model predictions
are given for standard[3] and vector models.[1] Experimental determinations of these ratios[2] are also
shown.

model comes from the observation of an anomalously large
inclusive $\bar{\nu}$-nucleon charged current scattering at large
y for very energetic $\bar{\nu}$'s, the so-called "high-y anomaly".[8]
At low and moderate energies, it is observed that
$R = \sigma_{CC}(\bar{\nu}d)/\sigma_{CC}(\nu d) \sim 1/3$ and that the y distribution
in ν' scattering is roughly proportional to $(1-y)^2$
while that in ν scattering is roughly independent of y.
All this confirms the picture of a nucleon made up of
quarks with very little antiquark contamination, and the
participation of only left-handed u and d valence quarks
in weak interactions at these energies.

The observed departure of high energy $\bar{\nu}$ scattering
from the expected $(1 - y)^2$ distribution, as well as the
observed significant increase of R suggests that the
charged hadronic weak current has a V + A (right-handed)
component in which some heavy quark couples to one of
the light u or d quarks.[7] This is an effect we have
dubbed fancy. Only at energies well above threshold for
the production of hadrons containing the new heavy quark
do the effects of fancy become significant. Note that
the new quark responsible for the high-y anomaly cannot
be the charmed quark, but must represent a fifth quark
flavor. It must be either a Q = -1/3 quark with right-
handed coupling to u, or a Q = -4/3 quark with right-
handed coupling to d.

It has been suggested, as an alternative to fancy,
that the high-y anomaly is explicable as a renormalization group effect on antiquark distributions and charmed
particle production.[12] It is true that the effective
antiquark and strange quark distribution in a nucleon
increases with q^2, so that the processes $\bar{\nu}d \to \mu^+\bar{u}$ and
$\bar{\nu}s \to \mu^+\bar{c}$ will contribute to a high-y anomaly. But,
detailed analysis suggests that this effect is not large

enough to explain the data.[4]

In the vector model, fancy is naturally incorporated.
There is a right-handed coupling of the u quark to a
heavy b (bottom) quark, and furthermore a right-handed
coupling involving the d quark. At asymptotically large
energies, the vector model predicts identical cross
sections and y-distributions $[1 + (1 - y)^2]$ for both ν
and $\bar{\nu}$ charged deep inelastic scattering. Such models
are in good accord with observed charged current ν and
$\bar{\nu}$ cross sections with a b mass of 4 - 5 GeV.[9]

Figure 2 shows the measured values of R and the
average value of y in $\bar{\nu}$ scattering as a function of
neutrino energy. The dotted line shows the prediction
of the standard model, without the effect of asymptotic
freedom (that is, with exact scaling). The experimental
points show a dramatic increase of both quantities
above the standard prediction at high neutrino energy
-- the high-y anomaly. Corrections due to asymptotic
freedom are included in the solid curve. They improve
the agreement of the standard model with the data, but
the standard model still predicts an energy dependence
very different from that observed. The dashed line shows
the prediction for the vector model with a b mass of 5
GeV. See Ref. 4 for further details.

In toto neutral-current data indicate a large
parity violation in the neutral currents, implying that
left- and right-handed quarks do not participate equally
in weak interactions. On the other hand, high-energy
charged-current data indicate a large right-handed
component of the charged weak current. Neither of the
simplest models, vector nor standard, can accommodate
these seemingly contradictory experimental results. A
more elaborate theory is indicated.

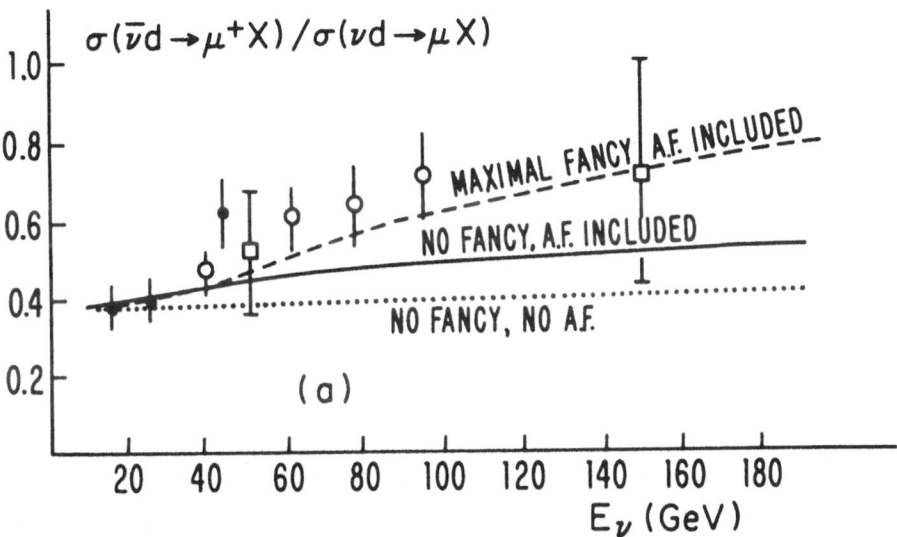

Fig. 2a. Ratio of antineutrino- to neutrino-induced inclusive cross sections on matter, as a function of neutrino energy.

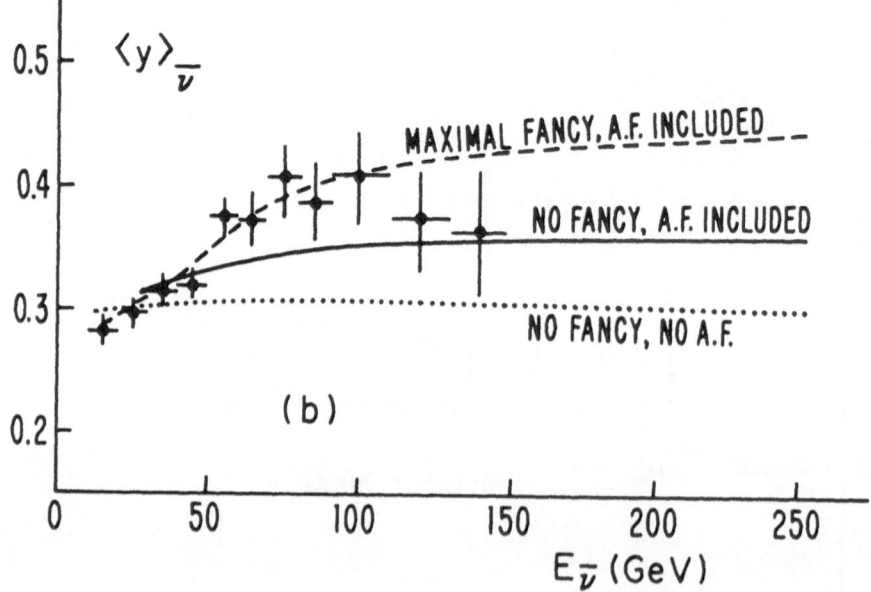

Fig. 2b.   Average value of y in the antineutrino-
induced process, as a function of energy. Predictions
for the standard model with and without asymptotic
freedom corrections are shown.[4]   Also shown is the
prediction of the vector model with m_b = 5 GeV and
asymptotic freedom corrections included.[4]   The vector
model has fancy = 3.   A smaller amount of fancy can be
partially compensated by a smaller b quark mass.   The
round data points are from the HPWF group, the squares
from the Caltech-Fermilab group.[8]

In constructing models to fit the data, we will adopt a set of theoretical prejudices which generalize the GIM mechanism.[10] We will assume that there are no flavor-changing neutral current effects from the gauge couplings in order G or αG.[11] Furthermore, we demand that this situation be obtained "naturally", in a fashion that does not depend on the choice of the parameters of the theory. This assumption is obviously reasonable for the quarks with charge -1/3, where the evidence against strangeness changing neutral currents is very strong. In the charge 2/3 sector, if there were charm changing neutral current effects of order G or αG they would induce a large D^0-$\bar{D}^0$ mixing which would show up in e^+e^- annihilation in the form of doubly strange final states. This effect has not been observed.[12]

Our demand that flavor changing neutral current effects of the gauge couplings be naturally absent is a very strong constraint on model building. Practically, it means that all the left-handed quarks with a given charge must belong to equivalent representations of the gauge group, and similarly for the right-handed quarks. A quark can belong to a different representation only if it carries some new quantum number which is absolutely conserved to prevent mixing with the ordinary quarks. We can ignore this possibility because such isolated quarks would not have a large effect on neutrino scattering and could not help us escape the high-y problem.

If we further limit ourselves to quarks with the standard charges, 2/3 and -1/3, then we are stuck. The only $SU(2) \times U(1)$ models allowed are the standard model or the vector model, both of which seem inconsistent

with data. We must relax some assumption. Barnett
and others[13] have considered a variety of $SU(2) \times U(1)$
models in which flavor changing neutral currents are
not naturally suppressed. In this talk, we explore
another direction. We keep the quarks in doublets and
avoid flavor changing neutral currents, but we change
the structure of the gauge group, replacing $SU(2) \times$
$U(1)$ by $SU(2)_L \times SU(2)_R \times U(1)$. The left-handed quarks
and (almost all) the left-handed leptons are assigned
to doublets under $SU(2)_L$ as follows (ignoring small
mixing angles like the Cabibbo angle),

$$\begin{pmatrix} u \\ d \end{pmatrix}_L \quad \begin{pmatrix} c \\ s \end{pmatrix}_L \quad \begin{pmatrix} t \\ b \end{pmatrix}_L \quad \begin{pmatrix} \nu \\ e^- \end{pmatrix}_L \quad \begin{pmatrix} \nu' \\ \mu^- \end{pmatrix}_L \quad \begin{pmatrix} \nu'' \\ L^- \end{pmatrix}_L .$$

$$\text{(1a)}$$

We have indicated three quark doublets and three lepton
doublets, though of course there could as well be four
or even more. All the members of $SU(2)_L$ doublets are
$SU(2)_R$ singlets. The right-handed quarks and leptons
are assigned to doublets of $SU(2)_R$ as follows,

$$\begin{pmatrix} u \\ b \end{pmatrix}_R \quad \begin{pmatrix} c \\ s \end{pmatrix}_R \quad \begin{pmatrix} t \\ d \end{pmatrix}_R \quad \begin{pmatrix} E^\circ \\ e^- \end{pmatrix}_R \quad \begin{pmatrix} M^\circ \\ \mu^- \end{pmatrix}_R \quad \begin{pmatrix} L^\circ \\ L^- \end{pmatrix}_R ,$$

$$\text{(1b)}$$

or perhaps

$$\begin{pmatrix} u \\ b \end{pmatrix}_R \quad \begin{pmatrix} c \\ d \end{pmatrix}_R \quad \begin{pmatrix} t \\ s \end{pmatrix}_R \quad \begin{pmatrix} E^\circ \\ e^- \end{pmatrix}_R \quad \begin{pmatrix} M^\circ \\ \mu^- \end{pmatrix}_R \quad \begin{pmatrix} L^\circ \\ L^- \end{pmatrix}_R ,$$

$$\text{(1c)}$$

and are, of course, singlets under $SU(2)_L$. Once again, we have suppressed the small Cabibbo angle and any mixing angles from (1b), (1c). It is our hope that an iterative procedure will be found in which all such angles vanish in zeroth order, and are small calculable effects in higher order. The distinction between (1b) and (1c) is experimentally important; for example, (1c) leads to considerably more charm production by incident neutrinos than does (1b). There remain three left-handed fields which are singlets under both $SU(2)_L$ and $SU(2)_R$

$$E_L^O \ , \quad M_L^O \ , \quad L_L^O \ .$$

In addition to the conventional quarks and leptons, this model needs a "top" quark with $Q = 2/3$ and a "bottom" quark with $Q = -1/3$, a heavy charged lepton L^-, and three heavy neutral leptons E^O, M^O, and L^O. Of these, the b quark may have been "seen" in neutrino physics, as mentioned in the discussion of the high-y anomaly. The t quark could produce a small anomaly in ν scattering at very high energy. The L^- (decaying into $e^-\nu\bar{\nu}$ and $\mu^-\nu\bar{\nu}$) may have been produced and observed at SPEAR. Heavy neutral leptons may be very difficult to observe.

There are seven gauge bosons in our model: $\vec{W}_L$ associated with $SU(2)_L$, $\vec{W}_R$ associated with $SU(2)_R$, and X associated with $U(1)$. We shall assume that the Lagrangian is invariant under a parity operation which interchanges $\vec{W}_L$ and $\vec{W}_R$. This ensures that the gauge coupling constants of the two $SU(2)$ factors are identical. Parity must be spontaneously broken by the Higgs meson vacuum expectation values (VEV's). Thus, we can take

the gauge couplings as follows: $\vec{W}_L$ couples to $\frac{e}{\sin\theta}\,\vec{T}_L$
where θ is the weak angle and $\vec{T}_L$ are the weak isospin
generators for the left-handed fields (for a doublet
$\vec{T}_L = 1/2\,\vec{\sigma}$). $\vec{W}_R$ couples to $\frac{e}{\sin\theta}\,\vec{T}_R$ where $\vec{T}_R$ are the
weak isospin generators for the right-handed fields,
and X couples to $\frac{e}{\sqrt{\cos 2\theta}}\,S$ where S is the U(1) generator
normalized so that $T_{3_L} + T_{3_R} + S = Q$, the electric
charge. Notice that the U(1) generator is the same as
in a vector model.

To give arbitrary masses to the quarks and leptons,
and to break the gauge symmetry down to the U(1) of
electromagnetism we need at least three multiplets of
Higgs mesons: a (1,2) multiplet ϕ [transforming like a
singlet under SU(2)$_L$ and a doublet under SU(2)$_R$] with
$\vec{T}_L\phi = 0$, $\vec{T}_R\phi = \frac{1}{2}\vec{\sigma}\phi$, $S\phi = -\frac{1}{2}\phi$ and vacuum expectation
value (VEV)

$$<\phi> = \begin{pmatrix} a \\ 0 \end{pmatrix} \quad ; \tag{2a}$$

and two (2,2)'s, Σ_1 and Σ_2 satisfying $\sigma_2\,\Sigma_i^*\,\sigma_2 = \Sigma_i$
(reality) $\vec{T}_L\,\Sigma_i = \frac{1}{2}\vec{\sigma}\,\Sigma_i$, $\vec{T}_R\,\Sigma_i = -\frac{1}{2}\,\Sigma_i\,\vec{\sigma}$, $S\,\Sigma_i = 0$
with VEV's

$$<\Sigma_1> = \begin{pmatrix} b & 0 \\ 0 & b \end{pmatrix} \tag{2b}$$

$$<\Sigma_2> = \begin{pmatrix} ic & 0 \\ 0 & -ic \end{pmatrix} \quad . \tag{2c}$$

There may be other meson multiplets which do not develop
VEV's (indeed there must be for parity violation to be
spontaneous) but they do not effect the gauge structure

of the theory and we ignore them. There may also be
additional Higgs multiplets which have nonzero VEV's
but do not couple to the fermions. These would compli-
cate the symmetry breaking and we will assume that
there are none. The analog to this in the standard
model is the assumption that there is only a Higgs
doublet, which gives the relation $M_Z = M_W/\cos \theta$. Equation (2) determines the gauge structure of the theory
completely, and we can proceed to compare with experiment.

Our model has twice as many heavy weak intermediaries as conventional models: we need two W's and
two Z's. One might hope that we could find a useful
limit of the model in which one W,Z pair is very heavy
and can be ignored. But it doesn't work out that way.
If a W,Z pair is super-heavy, the model collapses into
one of the $SU(2) \times U(1)$ models we have already discarded.
For example, suppose a >> b,c, then $SU(2)_L \times SU(2)_R$
$\times U(1)$ is strongly broken down to $SU(2)_L \times U(1)$, with
three gauge bosons acquiring a very large mass ($\sim$ e a).
The remaining $SU(2)_L \times U(1)$ is just the gauge group of
the standard model, which is broken down further to
$U(1)$ by b and c. For large a, the effect of the very
heavy intermediaries becomes small and the theory
approaches the standard model (with the constraint
$\sin^2\theta \leq 1/2$ arising from the initial parity invariance).

Similarly, in the limit b >> a,c, the light gauge
bosons are associated with the group $SU(2)_{L+R} \times U(1)$ of
the vector model. So as $b \to \infty$ the theory approaches the
vector model.

This theory, then, interpolates continuously between
the standard and vector models and we must hope to find
physics somewhere in the middle, where all the

intermediate vector bosons have masses of the same
order of magnitude.

The overall scale of the three VEV's is fixed by
the requirement that the ordinary left-handed charged-
current weak interactions have the observed strength,
given by the Fermi constant G. The theory is then
determined by G, the mixing angle θ, and the two angles
α and β, where

$$\sin \alpha = \sqrt{\frac{b^2+c^2}{2a^2+b^2+c^2}} \quad ,$$

(3)

$$\sin \beta = \frac{b^2-c^2}{\sqrt{(b^2+c^2)(2a^2+b^2+c^2)}} \quad ,$$

satisfying $1 \geq \sin \alpha \geq |\sin \beta| \geq 0$. In terms of α
and β, the standard model is obtained in the limit

$$\sin^2\alpha = \sin^2\beta = 0 \ ;$$

the vector model is reached in the limit

$$\sin^2\alpha = \sin^2\beta = 1 \ .$$

In the remainder of this note, we compare this class of
models with existing neutrino experiments. Since we
have three parameters (θ, α, β) at our disposal, it is no
surprise that we can fit the available data tolerably
well. Still, we will find that we can squeeze down on
a narrow allowed range of the parameters in the theory,
and make predictions for results of pending experiments.

To compare our class of models with experiment,
we use neutral current data to fix the three angles α,

β and θ. Keeping the weak mixing angle θ fixed, we determine α and β from the measured ratios σ_{NC}/σ_{CC} for deep inelastic ν_μ and $\bar{\nu}_\mu$ scattering. We use these ratios as our initial input because they are measured with better statistics than other neutral current data. In the analysis of the data, we use the naive quark parton model (which ignores the small antiquark and strange quark contamination in the nucleon). Figure 3 shows the values of $\sin^2\alpha$ and $\sin^2\beta$ obtained from the quoted central values (see Fig. 1) of these ratios in experiments by two groups, Gargamelle (GGM) and Harvard-Pennsylvania-Wisconsin-Fermilab (HPWF).[2] We use two experimental inputs to give an idea of the sensitivity of the parameters (and of the predictions discussed below) to the errors and the apparent time dependence of the data. The parameter $\sin^2\theta$ has been varied to obtain the curves shown in Fig. 3. Since both experimental points lie close to the standard nose in Fig. 1, both curves in Fig. 3 come close to the standard corner, $\sin^2\alpha = \sin^2\beta = 0$ (region A in Figs. 3-7). But at the other extreme, there are allowed values which lie roughly half way between standard and vector (region B in Figs. 3-7).

Having fixed $\sin^2\alpha$ and $\sin^2\beta$ in terms of $\sin^2\theta$ (albeit as double valued functions for some $\sin^2\theta$ values) we proceed to display the predictions of the model for other experimental quantities in terms of $\sin^2\theta$. For example, Fig. 7 shows the predictions for g_A and g_V in νe and $\bar{\nu}$e scattering. The quantities g_A and g_V are defined by writing the effective coupling for elastic $\nu_\mu e^-$ scattering as[7]

$$\frac{G}{\sqrt{2}} \left[\bar{\nu}_\mu \gamma^\alpha (1+\gamma_5)\nu_\mu \; \bar{e}\gamma_\alpha(g_V+g_A \gamma_5)e\right] \; .$$

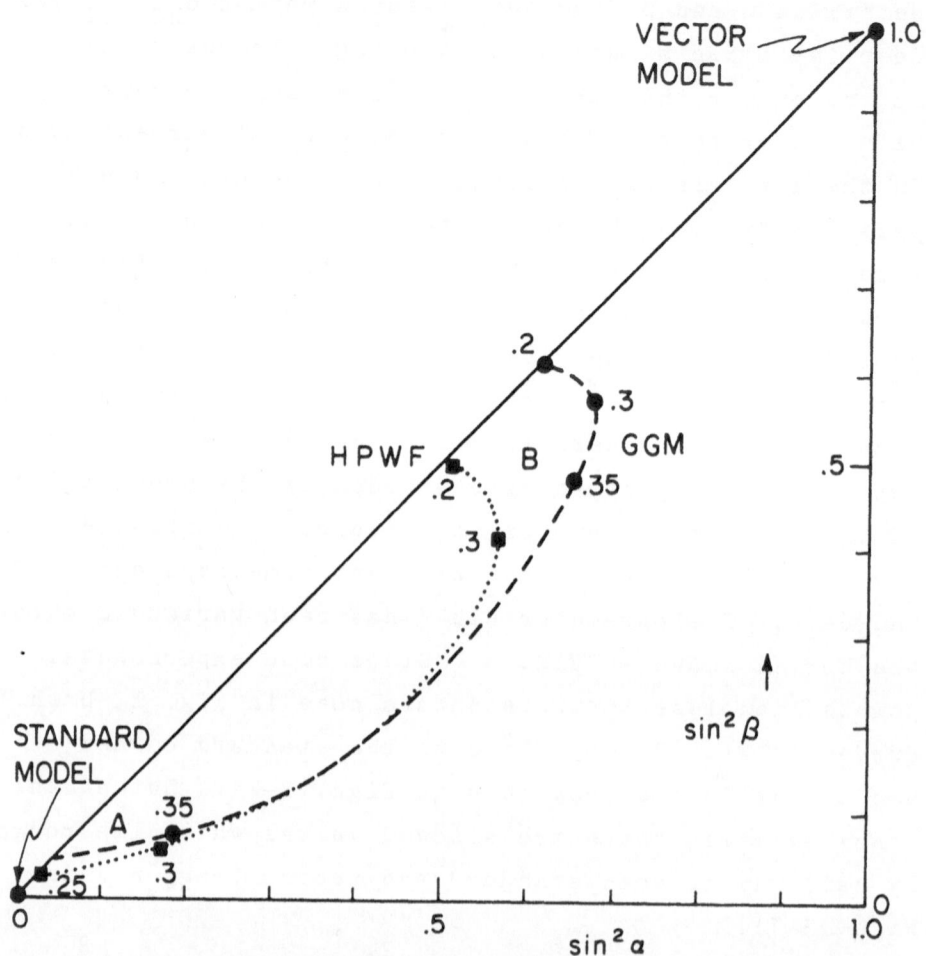

Fig. 3. The triangle is the allowed domain of the
parameters sin α and sin β defined in the text. The
curves are determinations (as functions of $\sin^2\theta$),
from the Gargamelle and HPWF data on inclusive neutral
current cross sections.[2] The numbers on the curves are
values of $\sin^2\theta$. Region A is close to the standard
model limit. Region B, which we favor in our model, is
"half-way" between the standard and the vector models.

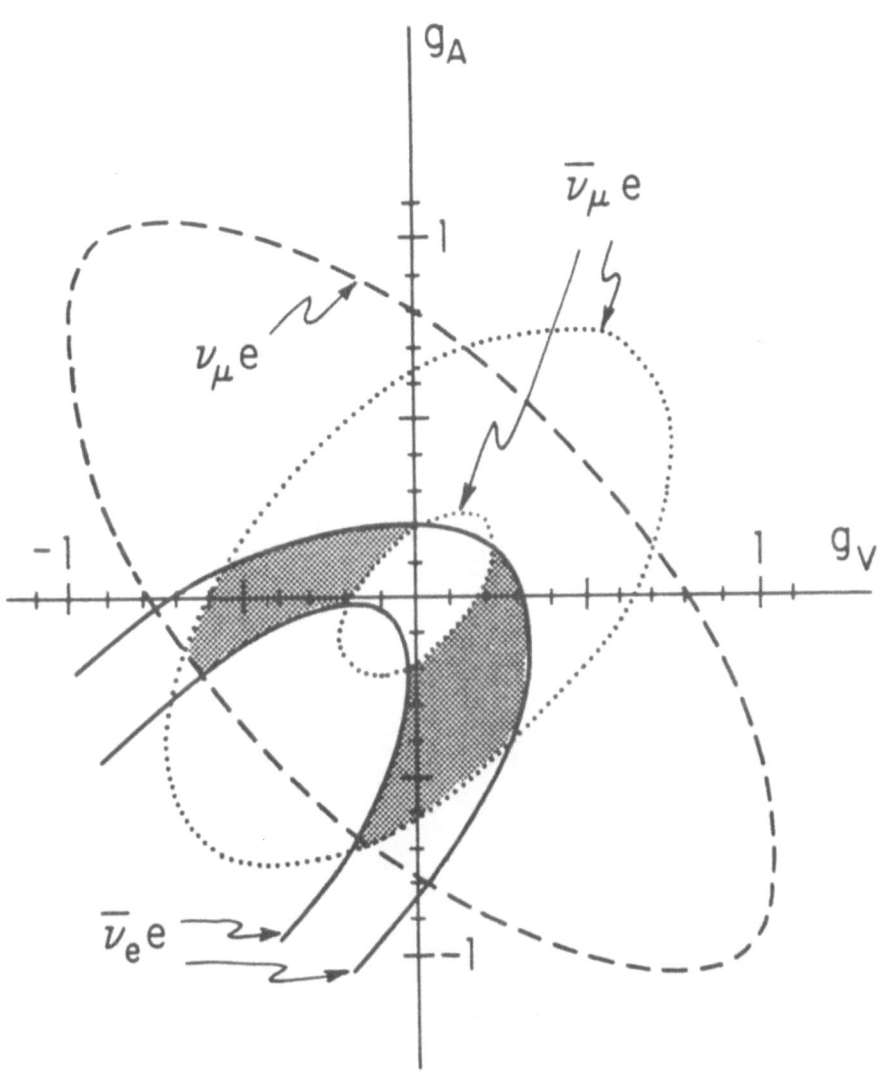

Fig. 4a. 90% confidence level limits on the values of g_A and g_V. The $\nu_\mu e$ and $\bar{\nu}_\mu e$ data are from Ref. 14, the $\bar{\nu}_e e$ data are from Ref. 14. The shaded regions are the combined allowed domains for g_A and g_V.

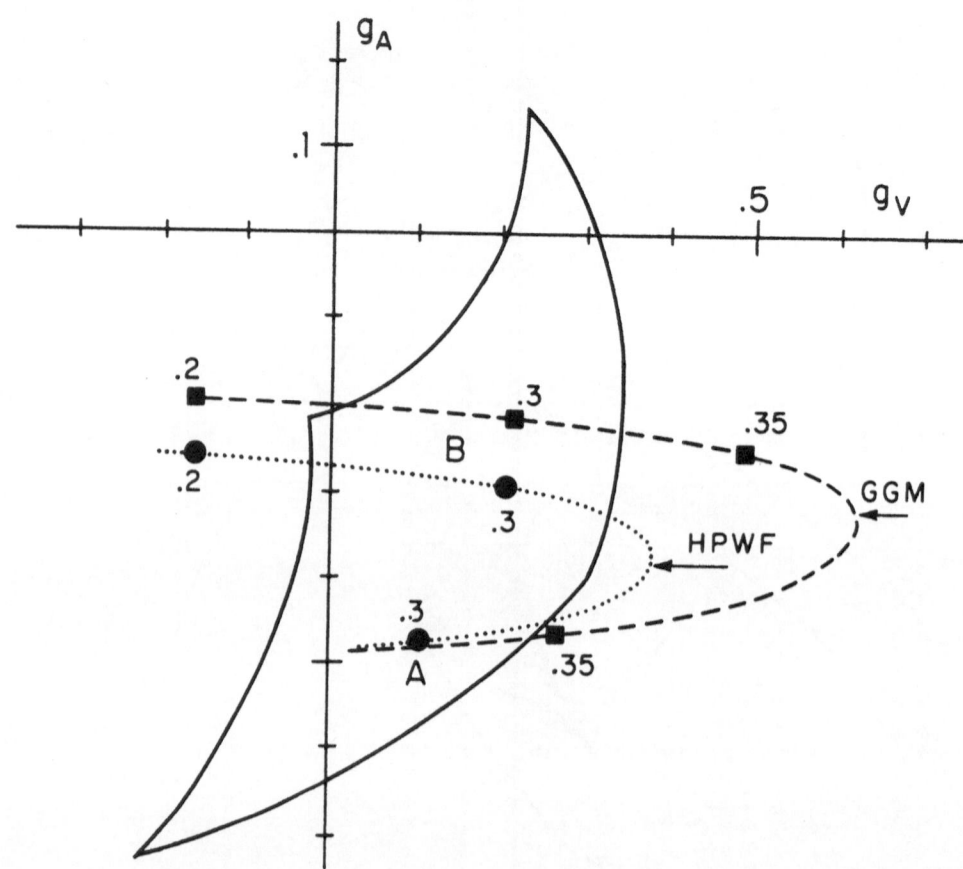

Fig. 4b.   Blow-up of part of Fig. 4a, showing the domain
that is compatible both with experiment and with our
model.   The curves and the regions A and B have the
same meaning as in Fig. 3.

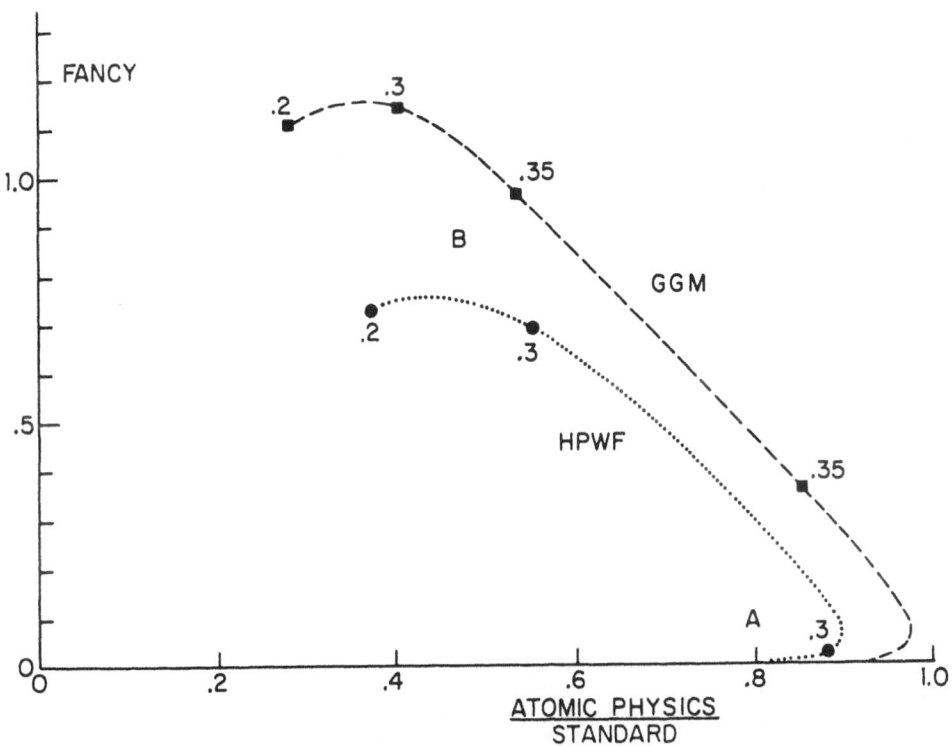

Fig. 5. Plot showing the negative correlation between fancy and the parity-violation signal in "atomic physics experiments" in Bismuth atoms. We normalize the parity violation in atomic physics to the expectation of the standard model with $\sin^2\theta = .33$. The curves and the regions A and B have the same meaning as in Figs. 3 and 4.

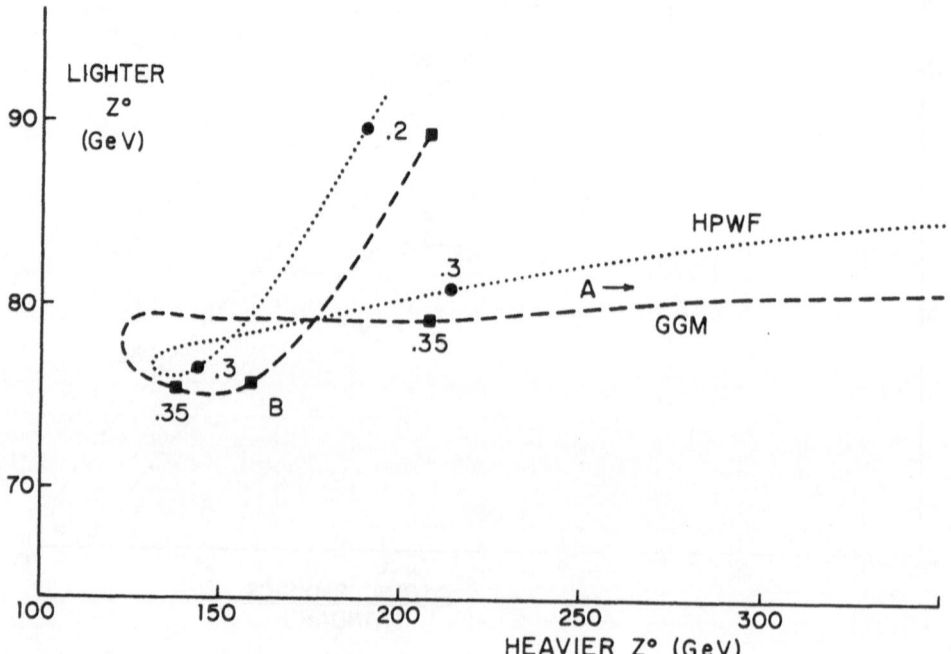

Fig. 6.   Masses of heavy versus light neutral intermediate vector mesons in our model.   The curves and the
regions A and B have the same meaning as in Figs. 3, 4
and 5.

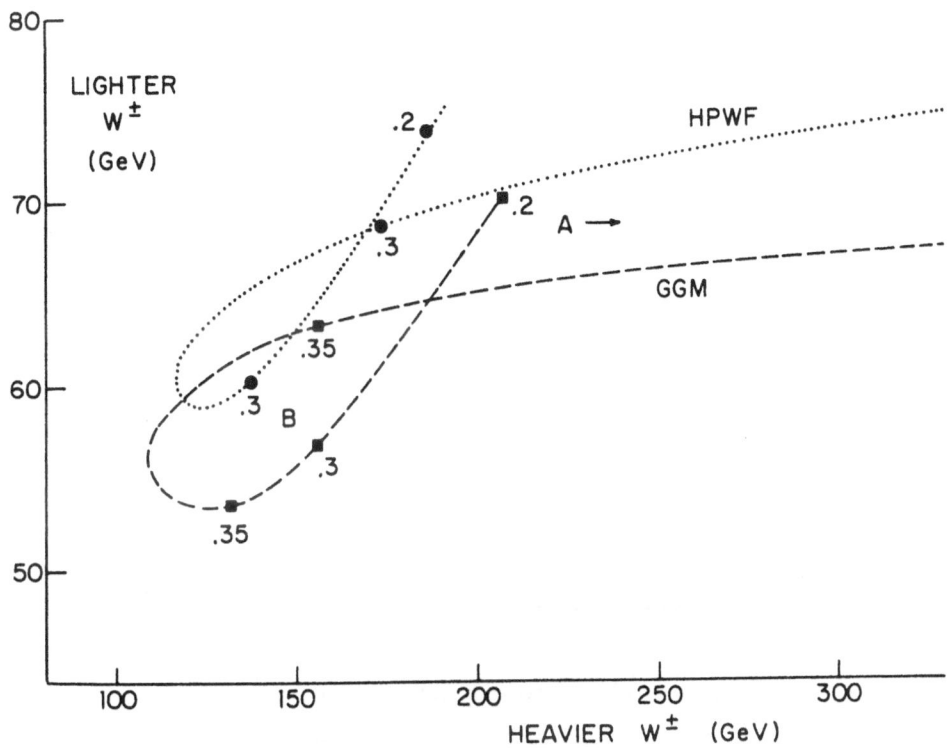

Fig. 7.   Masses of heavy versus light charged intermediate vector mesons in our model.   The curves and
the regions A and B have the same meaning as in Figs. 3,
4, 5 and 6.

The standard model gives $g_A = -1/2$, $g_V = -1/2 + 2\sin^2\theta$. Figure 4a shows the limits on g_A and g_V set by various experiments.[14] Taking the existing neutrino-electron scattering experiments seriously (90% confidence limits), we obtain the two allowed regions shaded in Fig. 4a. Figure 4b is a blow-up of the lower-right allowed region showing the predictions of our class of models. Both HPWF and GGM curves penetrate this allowed region twice. One intersection occurs for $g_A \simeq -0.5$ (region A). This corresponds to region A in Fig. 3, in which the model is very close to the standard model. The other (region B) with $g_A \simeq -0.25$ corresponds to B in Fig. 2, inter-mediate between standard and vector models.

Figures 5 and 6 show the masses of the various intermediate vector bosons in the model. Region A ex-tends out to large values of the heavier $W^\pm$ and Z masses, because this corresponds to the "standard" limit of the theory with $a^2 \gg b^2$, c^2. In region B, the heavier $W^\pm$ and Z are relatively light, only ~ 150 GeV. The lighter $W^\pm$ and Z are substantially less massive than those of the standard model.

Figure 7 shows the amount of asymptotically-produced fancy plotted against the expected parity violation in the experiments of Sandars et al. and Fortson et al.[15] These effects are quantified as follows: Fancy is three times the square of the ratio of the semileptonic couplings $\bar{\nu}_\mu\gamma^\lambda(1+\gamma_5)\mu^-\bar{b}\gamma_\lambda(1-\gamma_5)u$ and $\bar{\nu}_\mu\gamma^\lambda(1+\gamma_5)\mu^-\bar{d}\gamma_\lambda(1+\gamma_5)u$, which is equal to the ratio of b-quark to d-quark production well above b threshold in $\bar{\nu}$-hadron scattering (in the naive quark parton model approximation). In the vector model, fancy is equal to three (see Fig. 2). The relevant source of parity violation in the atomic physics Bismuth experiments is

the neutral current coupling of the vector part of the
hadronic current with the axial-vector part of the
leptonic current. Plotted in Fig. 7 is the quantity

$$\cos^2\beta \ \frac{T_3 - 2 \sin^2 \theta \ Q}{T_3 - .66 \ Q} \quad ,$$

evaluated for the bismuth nucleus. It is the ratio of
the parity violating effect expected in our model to
that predicted by the standard model with $\sin^2\theta$
(standard) = .33.

The important qualitative feature of Fig. 7 is
that fancy and parity violation in atomic physics are
negatively correlated. In region A, there is little or
no fancy and the atomic physics predictions are virtually the same as in the standard model. Region B would
give large fancy production, indeed the maximum consistent with the observed neutral to charged current cross-
section ratios in this model; while parity violation in
atomic physics is less than half that expected in the
standard model.

We favor a choice of parameters in or near region
B because it leads to the natural explanation of the
high-y anomaly as fancy production. We therefore expect parity violation in atomic physics to be slightly
less than half that predicted by the standard model.
If the effect is observed to be much smaller, then our
model would have to be revised. The simplest elaboration consistent with our theoretical prejudices is to
allow Higgs mesons in more complicated representations
of the $SU(2) \times SU(2) \times U(1)$ group. These cannot have
Yukawa couplings to the fermions, but their VEV's can
contribute to the masses of the intermediate vector
bosons. The general (CP conserving) mass matrix

involves 6 parameters, two masses and an angle in the
$W^{\pm}$ sector and the same in the Z sector. Such a model
can obviously be arranged to fit the data, but it is
less attractive than the model we have analyzed in
detail because there is no obvious theoretical need for
the extra Higgs mesons.

We have shown that by enlarging the weak gauge
group to $SU(2)_L \times SU(2)_R \times U(1)$, we can incorporate
both parity-violating neutral-current interactions
(which kill the vector model) and right-handed components
in the charged weak current (which kill the standard
model) and still maintain the natural GIM suppression
of flavor changing interactions. We close by mentioning
some unusual properties of the model and reviewing the
important experimental predictions.

Because of the necessarily rich spectrum of neutral
leptons in this theory it is easy to accommodate violation of muon and electron numbers at a level consistent
with present experimental upper bounds. Indeed, if the
assignment of leptons to doublets involves small angles
as in the quark sector, the decay $\mu \rightarrow e\gamma$ must occur.
The magnitude of this effect depends on the unknown
masses of the heavy neutral leptons, and on their mixing
angles with each other and with neutrinos. The branching
ratio for $\mu \rightarrow e\gamma$ is of order $\alpha(\sin 2\phi)^2 \, (\Delta M_o^2)^2 \, /4\pi M_w^4$,
where ΔM_o^2 is a difference between the squares of masses
of two neutral heavy leptons and ϕ is the mixing angle
between them. With $\sin 2\phi = 1$, $\Delta M_o^2 \approx 10$ GeV^2 and
$M_w \approx 100$ GeV, this branching ratio is $\sim 10^{-9}$. The decay
mode $\mu \rightarrow e e \bar{e}$ should be expected to be suppressed by
an additional factor of α. The loss of independent absolute selection rules for muon number and electron
number should not be upsetting, since there is no

analogous rule involving quark flavors.[16]

The most immediate experimental test of this class of models should come from the searches for parity violation in the electron-nucleon coupling. We predict an effect roughly half as large as that predicted by the standard model. If the effect turns out to be much smaller, the SU(2) × SU(2) × U(1) gauge group may still be the right one but the minimal set of Higgs mesons we used to break the symmetry will have to be enlarged.

Proposals for the direct search for intermediate vector bosons with masses up to a few hundred GeV are presently under discussion. Should these experiments prove feasible, we expect them to be twice as fruitful as expected from SU(2) × U(1) models. We expect two neutral Z's with masses roughly predicted by present data to be in the ranges 75 - 90 GeV and 120 - 200 GeV (Fig. 6, region B). Should the experiments detect charged W's as well, we anticipate the discovery of two of them, with masses 53 - 74 GeV and 110 - 210 GeV (Fig. 7, region B). The existence of two Z's with different couplings will produce a complicated energy dependence of parity violation in e^+e^- annihilation.

Ultimately, the most important prediction of this kind of model may be simply that there are more than four quarks and that at least the b quark is relatively light. If this is true, it should be possible to observe $b\bar{b}$ bound states in e^+e^- annihilation. It may be interesting to try to predict what the spectrum of narrow $b\bar{b}$ resonances will look like.

Eichten and Gottfried[17] have discussed this question in a specific non-relativistic potential model. Their potential is designed to fit the charmonium spectrum and to correspond to theoretical guesses. It is

$$V(r) = -\frac{4}{3} \alpha_s(M^2) \frac{1}{r} + kr \quad,$$

where $\alpha_s(M^2)$ is the running coupling constant of QCD, evaluated at M^2, where M is the quark mass. In their work $\alpha_s(M_{charm}^2) = 0$. The coefficient k of the linear potential is taken to be 0.2 GeV^2. With these parameters, they predict a spectrum of $b\bar{b}$ bound states, ignoring hyperfine splittings. They find that the number of narrow states below the threshold for production of particles with non-zero b-quark number increases with the b mass. They expect three such states for a b mass of $\sim$ 5 GeV. Their spectrum is dominated by the linear part of the potential for such a b mass. For example, the splittings between narrow $b\bar{b}$ states decrease as the b mass increases (in a Coulomb potential, it would increase).

This work raises several questions. The most obvious are: "Is this the right potential?" "What will hyperfine splittings look like?" "Can we describe light quark bound states in the same way with a single potential?" These questions are being studied by a group at Harvard including S.L. Glashow, R. Shankar, W. Celmaster, M. Machecek, A. Yildiz, and myself. Like Eichten and Gottfried, we start with the assumption that the potential is linear at large r and Coulomb-like (up to logarithms) at small r. The question is, what does the transition between the two regions look like? To bracket the possibilities, we have studied two potentials. One, which Shankar calls the Coulinear, is

$$-\frac{4}{3} \alpha_s\left(\frac{1}{r}\right) \cdot \frac{1}{r} \quad,$$

for $r < r_o$ and $kr + V_o$ for $r > r_o$ where V_o and r_o are

chosen so that the potential and its first derivative
are continuous at $r = r_o$ and

$$\alpha_s \left(\frac{1}{r}\right) = \frac{4\pi}{\alpha} \frac{1}{\ln(1/r\Lambda)} \cdot$$

This is as fast a transition as we can imagine from
Coulomb to linear. The Eichten-Gottfried potential, a
sum of Coulomb and linear, is slightly slower. Slower
still would be a sum of Coulomb, linear, and something
intermediate between the two, like $\ln r$, for example

$$V(r) = kr + \delta \ln r$$

$$- \frac{4}{3} \cdot \frac{4\pi}{\alpha} \frac{1}{\ln(1/r\Lambda + 2)} \cdot \frac{1}{r} \cdot$$

$$(4)$$

We have also looked at this potential as an example of
a slow transition between Coulomb and linear.

To describe light quark states, we must include
spin-dependent "hyperfine interactions". For the ρ-π,
K*-K, and D*-D systems, the hyperfine splitting falls
as the mass increases, consistent with an effect proportional to the product of inverse quark masses. But
the ψ/J - X system is peculiar. If the splitting
between ψ and X is really a hyperfine effect, pro-
portional to $\frac{1}{m_1 m_2} |\psi(0)|^2$, then $|\psi(0)|^2$ must have in-
creased enormously from that of the light-quark systems.
A μ^3 dependence on the reduced mass (as in a Coulomb
potential) would be more than enough. This point of
view has some immediate experimental consequences. We
may expect the F*-F splitting to be greater than the
90 MeV or so predicted by ignoring the μ dependence of
$|\psi(0)|^2$.

The Coulinear potential gives a reasonable picture

of s-wave bound state hadrons with $\Lambda = .7$ GeV (determined from e^+e^- annihilation),[18] $k \simeq .3$ GeV^2, $m_c = 1.96$ GeV, $m_{light} = 330$ GeV, $m_s = 650$ GeV. The predicted spectrum of b$\bar{b}$ states looks very much like that obtained by Eichten and Gottfried. It is dominated by the linear term in the potential.

In both the Eichten-Gottfried potential and the Coulinear potential the coefficient k of the linear potential is larger than one would expect from naive comparison with Regge theory. For a classical rotating light-string, one finds $E^2 = 2\pi k J$, where k is the mass-per-unit length of the string. A Regge slope of 1 GeV^{-2} corresponds to $k = .16$ GeV^2. In the potential of Eq. 4, it seems reasonable to demand that k take this value. To get a reasonable ψ'-ψ mass splitting with $k = .16$ GeV^2, we need $\delta \simeq .085$ GeV. We get good values for s-wave bound state masses with $m_c = 1.92$ GeV, $m_{light} = .35$ GeV, $m_s = .625$ GeV. We take $\Lambda = .5$ GeV to agree with electroproduction data.[19] For this potential, the spectrum of b$\bar{b}$ bound states looks quite different from the Coulinear spectrum. The $\ln r$ and Coulomb pieces of the potential play an important role. For example, the splittings between s-wave b$\bar{b}$ states increase with the b mass and the splitting between the two lowest states is much larger than subsequent splittings.

As these extreme examples show, the observation of b$\bar{b}$ states can help to further refine our understanding of the nature of the interactions between quarks. These simple nonrelativistic potential models can be useful quantitative language in which to describe the results of an exciting new generation of experiments.

REFERENCES

1. A. De Rujula, H. Georgi and S.L. Glashow, Phys. Rev. $\underline{D12}$, 3589 (1975); G. Branco, T. Hagiwara and R.N. Mohapatra, CCNY-HEP7518 and COO-223B-84 (1975); E. Golowich and B.R. Holstein, Phys. Rev. Lett. $\underline{35}$, 831 (1975); A. Fernandez-Pacheco, A. Morales, R. Nunez-Lagos, and J. Sanchez-Guillen, GIFT report; H. Fritzsch, M. Gell-Mann, and P. Minkowski, Phys. Lett. $\underline{59B}$, 256 (1975); F. Wilcsek, A. Zee, R.L. Kingsley, and S.B. Treiman, Phys. Rev. $\underline{D12}$, 2768 (1975); S. Pakvasa, W.A. Simmons, and S.F. Tuan, Phys. Rev. Lett. $\underline{35}$, 702 (1975).

2. A. Benvenutti et al., Phys. Rev. Lett. $\underline{37}$, 1039 (1976); J. Blietschau et al., preprint CERN/EP/Phys 76-55; B.C. Barsch, Caltech preprint CALT-68-544; D. Cline et al., Phys. Rev. Lett. $\underline{37}$, 252, 648 (1976); W. Lee et al., Phys. Rev. Lett. $\underline{37}$, 186 (1976); M. Barnett, Harvard preprint; V. Barger and D.V. Nanopoulos, Phys. Lett. $\underline{63B}$, 168 (1976); C.H. Albright et al., preprint, Fermilab-Pub-76/40-TH9; D.P. Sidhu, preprints BNL-21511 and BNL-21468.

3. S. Weinberg, Phys. Rev. Lett. $\underline{19}$, 1264 (1967); A. Salam in Elementary Particle Theory: Relativistic Groups and Analyticity (Nobel Symposium No. 8) edited by N. Svartholm (Almquist and Wiksell, Stockholm 1968) p. 367.

4. M. Barnett, H. Georgi and H.D. Politzer, Phys. Rev. Lett. $\underline{37}$, 1313 (1976); J. Kaplan and F. Martin, Paris preprint, PAR/LPTHE76/18 (1976).

5. A. De Rujula, H. Georgi and S.L. Glashow, preprint, HUTP-77/A002 (1977).

6. R.N. Mohapatra and P.P. Sidhu, preprint, CCNY-HEP-76/14 (1976).

7. A. De Rujula, H. Georgi, S.L. Glashow and H. Quinn, Rev. Mod. Phys. $\underline{46}$, 391 (1974).

8. A. Benvenutti et al., Phys. Rev. Lett. $\underline{36}$, 1478 (1976); ibid. $\underline{37}$, 189 (1976); B.C. Barish, Caltech preprint CALT-68-544.

9. M. Barnett, Phys. Rev. Lett. $\underline{36}$, 1163 (1976).

10. S.L. Glashow, J. Iliopoulos and L. Maiani, Phys. Rev. $\underline{D2}$, 1285 (1970).

11. S.L. Glashow and S. Weinberg, Harvard preprint HUTP-76/A158 (1976).

12. Preliminary data quoted by G. Goldhaber, Loeb Lectures at Harvard University (1976, unpublished).

13. For a review, see M. Barnett, SLAC-PUB 1850.

14. F.J. Hasert et al., Phys. Lett. $\underline{46B}$, 121 (1973) and J. Blietschau et al., preprint CERN/EP/PHYS 76-42 (1976); H. Williams, talk at the APS Meeting at Brookhaven National Laboratory, October 1976; F. Reines et al., Phys. Rev. Lett. $\underline{37}$, 315 (1976).

15. P.E.G. Baird, M.W.S.M. Brimicombe, G.J. Roberts, P.G.H. Sandars, D.C. Soreide, E.N. Fortson, L.L. Lewis, E.G. Lindahl and D.C. Soreide, Letters to Nature $\underline{264}$, 528 (1976); M.A. Bouchiat and C.C. Bouchiat, Phys. Lett. $\underline{48B}$, 111 (1974); M.W.S.M. Brimicombe, C.F. Loving and P.G.H. Sandars, J. Phys. $\underline{B1}$, 237 (1976); E.M. Henley and L. Wilets, Phys. Rev. $\underline{A14}$, 1411 (1976); I. Grand, N.C. Pyper and P.G.H. Sandars (to be published); I.B. Khriplovich, Soviet Phys. JETP (to be published); D.C. Soreide et al., Phys. Rev. Lett. $\underline{36}$, 352 (1976). For a review see I.B. Khriplovich, talk at the XVIII International Conference on High Energy Physics, Tbilisi, USSR, July 1976.

16. See T.P. Cheng's talk for a discussion of $\mu \to e\gamma$ and related questions.

17. E. Eichten and K. Gottfried, Cornell Preprint (1976).

18. R. Shankar, "Determination of the Quark Gluon Coupling Constant", Phys. Rev. D, to be published February 1977.

19. A. De Rujula, H. Georgi, and H.D. Politzer, "Demythification of Electroproduction Local Duality and Precocious Scaling", HUTP-76/A155, Annals of Phys., to be published.

THE ALEXANDER...ZWEIG RULES AND WHAT IS WRONG WITH PSEUDOSCALAR MESONS

Harry J. Lipkin*

Argonne National Laboratory, Argonne, Ill. 60439

Fermi National Accelerator Lab, Batavia, Ill. 60510

The rule forbidding $\phi \rightarrow \rho\pi$ and other stuff has been credited to a number of physicists in various combinations. To avoid arguments about credit, this paper refers to the A...Z rule and allows the reader to insert the names of all desired friends from Alexander[1] to Zweig.[2] Unfortunately the wide distribution of credit has introduced a fuzziness in the definition of the rule. All formulations forbid $\phi \rightarrow \rho\pi$, $f' \rightarrow \pi\pi$ and $J/\psi \rightarrow$ ordinary hadrons. But other predictions are not universal and depend upon whose formulation is used.

The first formulation was Okubo's nonet ansatz[3] for three-meson couplings. Okubo did not draw quark diagrams because the quark had not yet been invented, and did not treat baryons because the quark structure is essential to define the difference between the baryon and meson octets. The SU(3) analog of Okubo's meson ansatz would forbid the coupling of the ϕ and f' from the Σ which has hypercharge zero and occupies the same position in the

---

*Weizmann Institute of Science, Rehovot, Isreal

baryon octet that the ρ and π occupy in the meson octet.[4,5]

The correct extension of Okubo's nonet ansatz to baryons requires at least a mathematical quark description of the baryons as constructed from three fundamental SU(3) triplets. This decouples the ϕ and f' from the nucleon which has <u>strangeness</u> zero. Strangeness, unlike hypercharge, is outside SU(3) and arises naturally only in a quark description where it counts the number of strange quarks. The baryon selection rule arises naturally in the quark-line-diagram descriptions of Zweig[2] and Iizuka,[6] where disconnected diagrams are forbidden. It can also be cast into Okubo's language by representing baryons as third rank tensors and forbidding the appropriate disconnected contractions of these tensors. But in Okubo's terminology a difference still remains between the simple nonet ansatz for mesons and its extension to baryons. The possibility exists that the breaking of the rule may be much stronger for one than for the other, because the breaking mechanisms may be different in the two cases. This is an interesting point which remains to be settled by experiment.

Many possibilities arise in more complicated vertices. The most naive Zweig-Iizuka formulation forbids all disconnected diagrams. The simplest extension of the Okubo ansatz forbids only hairpin diagrams, in which one external particle is disconnected from the rest. Two interesting open questions are whether disconnected diagrams which are not hairpin diagrams might be less forbidden than hairpin diagrams, and whether doubly disconnected diagrams are more forbidden than singly disconnected ones. Interesting cases of these ambiguities arise in simple decays of the J/ψ.

Consider, for example,

$$\psi' \rightarrow \psi + \pi^+ + \pi^- . \qquad (1a)$$

This decay is described by a disconnected diagram which
is not a hairpin diagram. A similar diagram, rotated
by $90°$ but with the same topology describes the crossed
reaction

$$\psi + \pi^+ \rightarrow \psi' + \pi^+ . \qquad (1b)$$

This process is diffractive excitation of the ψ' in
$\psi\pi$ scattering and is allowed by Pomeron exchange. The
same topolgy describes elastic $\psi\pi$ scattering. There is
no reason to forbid the process (1b) and one can question
whether a process related to it by crossing is as
forbidden as a process described by a hairpin diagram
and which is not related to any allowed process by
crossing, e.g.

$$\pi^- + p \rightarrow \phi + n . \qquad (2)$$

Another example is the difference between the
two decays

$$J/\psi \rightarrow \omega\pi\pi , \qquad (3a)$$

$$J/\psi \rightarrow \phi\pi\pi . \qquad (3b)$$

Both processes involve hairpin diagrams and are forbidden
in any formulation. But the decay (3b) is doubly
disconnected, while the decay (3a) is only singly
disconnected. Originally the experimental results seemed
to indicate little difference between the two processes
and that singly and doubly connected diagrams were equally

forbidden. A possible theoretical description of such
a relation was proposed by Okubo[7] as a natural extension
of his ansatz. However, recent experiments[8] suggest that
the doubly forbidden process (3b) is indeed very different
from the singly forbidden one (3a), and that it proceeds
via an intermediate state whose propagator violates the
A...Z rule.

An instructive example of the "forbidden propagator"
mechanism by which the disconnected process (3b) can
take place is seen in the example of the decay of a high
K^{*-} resonance into three kaons via an intermediate
nonstrange resonance M^o

$$K^{*-} \rightarrow K^- M^o \rightarrow K^- K\bar{K} \quad , \tag{4}$$

where M^o is a resonance like the ρ^o, ω, f or A2 which
consists only of nonstrange quarks.

There are two diagrams for this decay with final
transitions from the $u\bar{u}$ and $d\bar{d}$ (or if you prefer $p\bar{p}$ and
$n\bar{n}$)* components of M^o.

---

* Two notations are used commonly for quark flavors, u,
d, s and p, n, λ. For old nuclear physicists like myself, who have resisted the u, d notation because we
could never remember which way was "up" (in nuclear
physics the neutron has isospin up because common stable
nuclei have more neutrons than protons), there is a
simple mnemonic. Simply write the words

up

down.

People who read from left to right naturally call these
u and d. People who read from right to left, as we do
-Footnote continued on the next page-

The $p\bar{p}$ diagram (or $u\bar{u}$) is connected, obeys the A...Z
rule and leads to the final state $K^-K^+K^-$. The $n\bar{n}$
(or $d\bar{d}$) diagram is disconnected, violates the A...Z
rule and leads to the final state $K^-K^0\bar{K}^0$. Thus if the
rule holds,

$$(K^{*-} \rightarrow K^-(u\bar{u}) \rightarrow K^-K^+K^-) \text{ is allowed,} \qquad (5a)$$

$$(K^{*-} \nrightarrow K^-(d\bar{d}) \rightarrow K^-K^0\bar{K}^0) \text{ is forbidden.} \qquad (5b)$$

But if the resonance M^0 has a definite isospin,
the two transitions (5a) and (5b) must be equal by
isospin invariance. Contradictions between the A...Z
rule and isospin invariance are avoided if the nonstrange
meson spectrum consists of degenerate isospin doublets,
like ρ and ω or f and $A2$. In that case the transition
(5a) proceeds via the particular coherent linear
combination of isovector and isoscalar particles which
has the quark constitution $p\bar{p}$ (or $u\bar{u}$). The A...Z rule
is thus intimately related to the existance of the isospin
doublets found in ideally mixed nonets.

---

[*] in the Middle East, naturally call them p and n. This
leads to the correspondence

$$u \longleftrightarrow p$$

$$d \longleftrightarrow n \quad .$$

The invariance of this transformation under 180^0 rotations
is exhibited by turning this page upside down.

In this talk I bounce between both notations because
I am still using old transparencies, in accordance with
the well-known rule: "Old transparencies never die,
they just fade away."

If the M^o in the transitions (5) is not a member
of an isospin doublet, the A...Z rule is inconsistent
with isospin invariance. This is the case if M^o is a
π^o, which has no degenerate isoscalar partner. The π^o
cannot contribute to the reactions (5) but can appear
as an exchanged particle in the analogous two-body
scattering reactions

$$K^+ + K^- \rightarrow K^{*o} + \bar{K}^{*o} \quad , \tag{6a}$$

$$K^+ + K^- \rightarrow K^{*+} + K^{*-} \quad , \tag{6b}$$

$$K^o + K^- \rightarrow K^{*o} + K^{*-} \quad . \tag{6c}$$

The charge exchange reaction (6a) is clearly
allowed by the A...Z rule and can go by pion exchange.
The amplitudes for the pion exchange contribution to the
reactions (6b) and (6c) are uniquely related to the
charge exchange amplitude (6a) by isospin invariance.
But the reaction (6b) is allowed by the A...Z rule and
the reaction (6c) is forbidden when only nonstrange
quark exchange is considered. (The reaction (6c) is
allowed by $s\bar{s}$ (or $\lambda\bar{\lambda}$) exchange but this is irrelevant
to the present argument). The A...Z rule could be
saved from inconsistency with isospin invariance if a
contribution from isoscalar exchange degenerate with
pion exchange cancelled the pion exchange contribution
to the reaction (6c). But no such isoscalar exists.
Thus violations of the A...Z rule might be expected
in processes where pseudoscalar exchange plays a
dominant role.

The above examples are only a few of the puzzles
and paradoxes of the A...Z rule. A consistent theoretical
derivation of the rule should resolve these, but no such

derivation exists. The most promising approach to
such a derivation at present seems to be in the framework of dual resonance models,[9] but there are many
open unanswered puzzles in this approach as well. The
situation may be summed up by the statement that noboby
understands the A...Z rule and don't believe anyone who
claims he does. A comprehensive review of these puzzles
is given in reference 5.

A principal difficulty to be overcome in any
theoretical formulation is that a succession of transitions
all allowed by the A...Z rule can lead to one which is
forbidden. For example, the forbidden couplings $\phi\rho\pi$
and f' $\pi\pi$ can proceed through an intermediate $K\overline{K}$ state
via the following transition amplitudes observed
experimentally and allowed by the A...Z rule.

$$T(\phi \rightarrow K\overline{K}) \neq 0 , \tag{7a}$$

$$T(f' \rightarrow K\overline{K}) \neq 0 , \tag{7b}$$

$$T(K\overline{K} \rightarrow \rho\pi) \neq 0 , \tag{7c}$$

$$T(K\overline{K} \rightarrow \pi\pi) \neq 0 . \tag{7d}$$

The selection rules can thus be broken by the
following allowed higher-order transitions

$$\phi \rightarrow K\overline{K} \rightarrow \rho\pi , \tag{8a}$$

$$f' \rightarrow K\overline{K} \rightarrow \pi\pi . \tag{8b}$$

If the A...Z rule holds only to first order in
strong interactions, much greater violations are expected
than those experimentally observed. Some mechanism for
reducing these violations seems to be present. One
possibility is a cancellation of the violating amplitudes

by other amplitudes, as occurs in the transitions (5)
where the violating amplitudes from isoscalar and
isovector meson states M^o must cancel one another to
preserve the selection rule. This requires a
degeneracy of the intermediate states. For an analogous
cancellation to be effective for the transitions (8)
an additional intermediate state degenerate with the $K\bar{K}$
state is required. But no such state exists in the
spectrum of physical particles. Thus the cancellation
can only be approximate and hold in some higher symmetry
limit where other states are degenerate with the $K\bar{K}$ state.

The kaon plays a crucial ambivalent role in the
A...Z forbidden transitions (8) between a state containing
only strange quarks and a state containing only nonstrange quarks. Since the kaon contains one strange and
one nonstrange quark, it couples equally to strange and
nonstrange systems and can go either way. A kaon pair
state contains one strange and one nonstrange quark-
antiquark pair. It can therefore be created from a
strange pair by the creation of a nonstrange pair or
vice versa. The kaon pair state thus links two kinds
of states between which transitions are forbidden by the
A...Z rule. The quark diagram for the forbidden
transition (8) illustrates the essential features of the
paradox. Viewed as a single topological diagram it is
indeed disconnected and can be separated into two
disconnected hairpin diagrams. But when it is separated
into two individual transitions, each half is connected.
Connecting the two diagrams together results in a
topological disconnected diagram because of the twist in
the quark and antiquark lines in the kaon intermediate
state.[5]

Thus, to save the A...Z rule the connection of allowed

diagrams by a "twisted propagator" must somehow be
forbidden. But a twisted propagator has physical
meaning only if there is additional information in a
kaon pair state to specify "which way it is twisted";
i.e., whether it originally came from a strange or a
nonstrange system. Some memory of the origin of the
pair is necessary to prevent the nonstrange decay of
a pair which originated in a strange system. But a
physical kaon pair state has no such memory. A kaon
pair produced from a nonstrange system is indistinguishable
from a pair produced from a strange system.

In dual resonance models twisted diagrams are
forbidden because of cancellations from contributions
from exchange degenerate Regge trajectories. The
intermediate states in dual resonance quark diagrams
do not represent physical particles but Reggeons, and
these always occur in exchange degenerate pairs with
opposite behavior under charge conjugation. Twisted
diagrams always include cancelling contributions from
exchange degenerate trajectories. This description is
valid, however, only for diagrams where all internal
lines can be interpreted as Reggeons. This clearly does
not hold for the transitions (8) where the intermediate
state can occur as physical particles on their mass
shell, and the corresponding exchange degenerate
trajectories have no such states. This problem has been
considered[9,10] as a possible mechanism for the breaking
of the A...Z rule in the particular case of the reaction
$\pi^- p \rightarrow \phi n$.

We thus see that two types of degeneracies are
required for the particle spectrum in order to avoid the
breaking of the A...Z rule by the propagators of intermediate states. Nonet degeneracy with ideal mixing is

required so that the $u\bar{u}$, $d\bar{d}$ and $s\bar{s}$ states which are not
eigenstates of isospin and SU(3) can be eigenstates of
the mass operator. Exchange degeneracy is required to
avoid breaking of the rule by twisted propagators.
Whenever these degeneracies are not exact, troubles with
the A...Z rule can be expected. The reader is referred
to reference 5 for further consideration of these effects.

The use of the quark model to determine mixing
angles of neutral mesons from experimental data on
neutral meson production processes was first suggested
by G. Alexander.[1] This work also included the first
derivation of the A...Z for four point functions, based
on the Levin-Frankfurt additive quark model[11,12] in
which every hadron transition is assumed to involve
only one active quark with all remaining quarks behaving
as spectators. Hairpin diagrams are naturally forbidden
in this model, since they involve two active quarks in
the same hadron. Thus the prediction that ϕ production
is forbidden in πN reactions was immediately obtained.
Also obtained were a number of other predictions which
have since been shown to be in very good agreement for
vector meson production. These include the prediction
of no exotic t-channel exchanges and some sum rules and
equalities which are listed below.

At the time that the paper of Alexander et al.[1] was
presented for publication, the quark model was ridiculed
by the particle physics establishment and the paper was
rejected by a referee who dismissed the quark model as
nonsense. It was resubmitted for publication and finally
accepted after new experimental data[13] supported
predictions made in the paper before the data was known.
Today papers based on the single-quark-and-spectators
transition are covered by mentioning the magic name of

"Melosh", since no referee will admit that he really doesn't understand what this Melosh transformation[14] is all about. Meanwhile many of the sum rules and equalities of Alexander et al.[1] were rederived[12,15] by people who do not like quarks and give them other names to avoid calling them quark model sum rules which they really are. Reviewing the experimental results over the past decade shows striking agreement with these quark model relations for all processes of vector meson production and strong disagreement with relations for processes of pseudoscalar meson production, particularly for relations involving η' production.[4,13,15,16] We suggest that an appropriate conclusion from these results is that the quark model description indeed holds for these processes, but that something is wrong with the pseudoscalars,[17] particularly the η'.

The relevant sum rules are the charge exchange sum rule (CHEX)

$$\sigma(\pi^- p \to \pi^0 n) + \sigma(\pi^- p \to \eta n) + \sigma(\pi^- p \to \eta' n) =$$

$$\sigma(K^+ n \to K^0 p) + \sigma(K^- p \to \overline{K}^0 n) , \qquad (9a)$$

and the strangeness exchange sum rule (SEX)

$$\sigma(K^- p \to \eta Y) + \sigma(K^- p \to \eta' Y) = \sigma(K^- p \to \pi^0 Y) + \sigma(\pi^- p \to K^0 Y).$$
$$(9b)$$

These sum rules hold for any meson nonet and do not made any assumption about the mixing angle, except for the conventional description of the η and η' as two orthogonal linear combinations of pure SU(3) singlet and octet states defined in terms of a single mixing angle. For the case of ideal mixing, as in the vector mesons

the two sum rules each split into two equalities, CHEX
becomes

$$\sigma(\pi^- p \to \phi n) = 0 \quad , \tag{10a}$$

which is just the A...Z rule, and substituting (10a) into
(9a) gives

$$\sigma(\pi^- p \to \rho^0 n) + \sigma(\pi^- p \to \omega n) = \sigma(K^+ n \to K^{*0} p) + \sigma(K^- p \to K^{*0} n). \tag{10b}$$

With ideal mising SEX becomes

$$\sigma(K^- p \to \omega Y) = \sigma(K^- p \to \rho^0 Y) \quad , \tag{11a}$$

$$\sigma(K^- p \to \phi Y) = \sigma(\pi^- p \to K^{*0} Y) \quad . \tag{11b}$$

The relation (11a) is seen also to be a consequence of
the A...Z rule for the meson vertex. The incident K^-
contains no d or $\bar{d}$ quarks or antiquarks and therfore
the production of a $d\bar{d}$ pair is forbidden. The ρ^0 and
ω are therefore produced via the $u\bar{u}$ component which is
a linear combination of the two with equal weight.

If there is no mixing, which is a rough approximation
for the pseudoscalar mesons, the charge exchange sum
rule simplifies to

$$\sigma(\pi^- p \to \pi^0 n) + 3\sigma(\pi^- p \to \eta_8 n) = \sigma(K^+ n \to K^0 p) + \sigma(K^- p \to \bar{K}^0 n), \tag{12a}$$

$$\sigma(\pi^- p \to \eta_8 n) = 2\sigma(\pi^- p \to \eta_1 n) \quad . \tag{12b}$$

All the vector meson relations (10) and (11) are in
excellent agreement with experiment. However, the
pseudoscalar meson relations (9) and (12b) are in strong
disagreement. The relation (12a) agrees with experiment

if the η is assumed to be pure octet. This suggests
that the conventional picture in which there is small
mixing may be valid for the η, but that something is
wrong with the η', and it is wrong in the direction that
the η' has an inert piece in the wave function which
does not contribute to the sum rules (9) and (12).

We suggest that there is indeed an additional
piece in the η' wave function and that it is a radially
excited configuration. This leads to a re-examination
of the standard mixing folklore and the discovery that
it is completely unjustified.[17] In a formation which
begins with unperturbed singlet and octet states in the
SU(3) symmetry limit, there is no reason to assume that
SU(3) symmetry breaking should admix only the lowest
ground states of the singlet and octet spectra. This
may work for the tensor and vector mesons, where the
entire nonet seems to be degenerate in the SU(3)
symmetry limit and the dominant breaking of nonet
symmetry is by a quark mass term. The degeneracy
suggests the use of degenerate perturbation theory which
diagonalizes the symmetry breaking interaction in the
space of the degenerate unperturbed states. The mass
term has no radial dependence and would not mix ground
state and radially excited wave functions which are
orthogonal and would have a zero overlap integral.

For the pseudoscalars where there is a large
singlet-octet splitting in the SU(3) symmetry limit there
is no reason to use degenerate perturbation theory and
mix only ground state wave functions. Furthermore, the
singlet-octet splitting can only be produced by an
interaction which violates the A...Z rule because it is
not diagonal in the quark basis and mixes $s\bar{s}$ with $u\bar{u}$ and
$d\bar{d}$. The accepted mechanism for such A...Z violation in

the pseudoscalars is annihilation of the quark-antiquark pair into gluons and the creation of another pair. Here there is no reason to restrict the pair creation to the ground state configuration. There is no overlap integral between the two $q\bar{q}$ states, as the intermediate gluon state does not remember which radial configuration it came from. If the annihilation process depends primarily on the value of the $q\bar{q}$ wave function at the origin, then all radially excited configurations couple with equal strength for wave functions from a confining linear potential.

Thus there is considerable reason to suspect that the trouble with the pseudoscalar meson sum rules is in admixture of a radially excited wave function into the η'. One might expect the η to be purer because the SU(3) flavor octet state does not couple to gluons which are singlets and because it is the lowest state, far in mass from the nearest SU(3) singlet radial excitation. The η', on the other hand is sitting in between the ground state and first radially excited octet states and would be expected to mix with both. Note that mixing of the octet ground state and first radially excited octet state by an SU(3)-symmetric potential need not be considered because it is merely a change in the radial wave function. This mixing can be transformed away by choosing a new radial basis (i.e. a slightly different potential) for which the modified ground wave function in the original basis is the exact ground state in the new basis.

With no further theoretical results or suggestions regarding the treatment of the pseudoscalars, let us turn to experiment and see how experimental data on the new particles can provide additional clues and useful

information. The outstanding open question is the
existence of the charmonium ($c\bar{c}$) pseudoscalar, the η_c,
its mass and properties. At the time of this conference
it is seen only at DESY in its $\gamma\gamma$ decay mode.[18] There
is therefore interest in observing the hadronic decay
modes. The $p\bar{p}$ mode, which was first reported and then
faded away, is irrelevant. There is no reason to expect
such a mode to be strong. The $c\bar{c}$ is expected to decay
into normal hadrons by annihilation of the $c\bar{c}$ in an
A...Z violating transition, presumably via gluons, into
a pseudoscalar state of light quarks. One would expect
the coupling of this state to be similar to the coupling
of the known pseudoscalars, the η and η'. The simplest
hadronic decay channels would be $\pi A2$, ηf^o, $K\bar{K}^*$, $\pi\delta$ and
$\eta\varepsilon$.

The difficulty in observing these hadronic states
arises from the presence of neutrals which are not easily
detected and the difficulty of distinguishing photons
from π^o's. Of these decay modes the $\pi A2$ may be the best
for experimental observation. This decay can be
expected to be relatively strong, because the A2-η-π
coupling is known to be strong from the A2 decay. It
may be observable even with unidentified neutrals because
of the peculiar kinematics of the decay

$$\psi \rightarrow \gamma\eta_c \rightarrow \gamma\pi A2 \rightarrow \gamma 4\pi \ . \tag{13}$$

This decay populates a peculiar region of phase space for
four charged pions and a neutral which may be useful
even if the neutral is not identified as a γ. The photon
has a momentum of about 300 MeV, while the pion which
is not in the A2 has a high momentum varying between 1
and 1.2 GeV depending upon the angle between the pion

momentum and the momentum of the photon. This angular
distribution should be isotropic in the center of mass
system of the η_c, since it has spin zero.

The large disparity between the photon and pion
momenta can serve to distinguish between $\gamma\pi^+\pi^+\pi^-\pi^-$ and
$\pi^0\pi^+\pi^+\pi^+\pi^-$ events. The $\pi\pi A2$ final state is required by
isospin invariance to have a symmetrical distribution
in phase space for the two pions, and the probability
that the charged pion has 1 - 1.2 GeV momentum while the
neutral has only 300 MeV can be expected to be small.
Furthermore, exact relations following from isospin
invariance can be used to estimate this background from
data in other regions of phase space and to subtract
the background.

Consider the decay

$$\psi \rightarrow \pi\pi A2 \quad . \tag{14}$$

Isospin invariance requires the two pions to be in a
state of total isospin one, coupled to the isospin one
of the A2 to give an overall isospin of zero. This
isospin coupling gives a unique relation between the
different possible charge states where one pion has a
momentum of 0.3 GeV and the second has a momentum of
1.1 GeV. Thus it is possible to define linear
combinations of branching ratios which must vanish for the
decay (14) and can serve as background subtractions; e.g.

$$\Delta = W(0, +, -) + W(0, -, +) -$$

$$- \frac{1}{2}[W(+, 0, -)+W(-, 0, +)+W(+, -, 0)+W(-, +, 0)] = 0, \tag{15}$$

where $W(q_1, q_2, q_3)$ denotes the branching ratio for a

state with a 300 MeV pion with charge q_1, a 1.1 GeV
pion with charge q_2 and an A2 with charge q_3. The
first two states in eq. (15) appear as background in
the same region of phase space as the desired decay
(15) while the remaining four are in a completely
different region.

Consider an experiment in which four charged pions
are observed and the missing mass indicates an additional
neutral particle which may be either a photon or a π^0.
If all of these events are substituted into the relation
(15) including those where the neutral is a photon as
well as a π^0, the pion contribution must vanish, in view
of the relation (4). For this case the quantity Δ will
measure

$$\Delta = W(\gamma, +, -) + W(\gamma, -, +) - W(+, \gamma, -) - W(-, \gamma, +) \ .$$

$$(16)$$

This is just the difference between the desired decays
(13) and decays in which the final state has the momenta
of the photon and pion interchanged; e.g. a 300 MeV pion
and a 1.1 GeV photon. The background of π^0 events is
completely eliminated by isospin invariance. The
quantity (16) can then be plotted as a function of the
mass of the πA2 system to see if it exhibits a peak at
2.8 GeV. Note that the values of 0.3 and 1.1 GeV for
momenta were picked just for example. In actual practice
the quantity will be defined for a boson of charge q_1
and momentum k_1 and a boson of charge q_2 and momentum k_2
and the boson is either a photon or a pion.

Further information on the properties of the
pseudoscalar mesons can be obtained from the decays of
new particles into channels including the η and η'.

Analysis of the decays involves SU(3) symmetry and the
A...Z rule as well as properties of the pseudoscalars.
But sufficient data are available to enable comprehensive
tests of all these assumptions. We first list some
SU(3) predictions discussed in reference 17. We begin
with those obtained from the assumptions of SU(3)
symmetry and naive mixing without the A...Z rule. Ideal
mixing for the ω and ϕ is assumed, since known deviations
are small. The one equally obtained is the sum rule:

$$2\Gamma(\psi\to\phi\eta)+2\Gamma(\psi\to\phi\eta') = \Gamma(\psi\to\rho^+\pi^-)+\Gamma(\psi\to\omega\eta)+\Gamma(\psi\to\omega\eta'), \quad (17a)$$

where Γ denotes the reduced width without phase space
corrections. In addition the following relations are
obtained,

$$\Gamma(\psi\to\omega\eta)+\Gamma(\psi\to\omega\eta') = \Gamma(\psi\to\rho^+\pi^-)(1+2|A_1/A_8|^2)/3, \quad (17b)$$

$$\Gamma(\psi\to\phi\eta)+\Gamma(\psi\to\phi\eta') = \Gamma(\psi\to\rho^+\pi^-)(2+|A_1/A_8|^2)/3. \quad (17c)$$

These relations give testable inequalities without
additional assumptions, since the quantity $|A_1^2/A_8^2|$
defined in ref. 17 is positive definite.

The validity of SU(3) symmetry for the decay is
tested independently of the nonet mixing assumption by
the relation which does not involve any mixed mesons

$$\Gamma(\psi \to \rho^+\pi^-) = \Gamma(\psi \to K^{*+}K^-) . \quad (18)$$

When the A...Z rule is assumed, Eqs. (2) hold with
$A_1/A_8 = 1$ and the following additional relation is
obtained:

$$\Gamma(\psi \to \omega\eta) = \Gamma(\psi \to \phi\eta') . \quad (19a)$$

This can be combined with the relations (2) to obtain other simple relations

$$\Gamma(\psi \rightarrow \phi\eta) = \Gamma(\psi \rightarrow \omega\eta'), \qquad (19b)$$

$$\Gamma(\psi \rightarrow \omega\eta) + \Gamma(\psi \rightarrow \phi\eta) = \Gamma(\psi \rightarrow \rho^{+}\pi^{-}), \qquad (19c)$$

$$\Gamma(\psi \rightarrow \omega\eta') + \Gamma(\psi \rightarrow \phi\eta') = \Gamma(\psi \rightarrow \rho^{+}\pi^{-}) \qquad (19d)$$

The relations (19b - d) are not linearly independent of the previous relations. However, if some relations disagree with experiment and others agree, these different combinations can furnish clues to determine what has gone wrong. For example, if the A...Z rule holds and naive mixing breaks down for the η, then relation (19c) which involves only the η might agree with experiment while other relations like (17) and (19d) which involve the η might not. If there is an inert piece in the η' wave function, the right hand sides of (19a) and (19b) and the left hand side of (19d) would all be suppressed by the same factor.

We now consider the possibility of SU(3) breaking. A number of SU(3) predictions have been shown to be in qualitative agreement with experiment in the observed branching ration of J/ψ decays. However there is also evidence for appreciable SU(3) symmetry breaking. We introduce here a simple and intuitively attractive symmetry breaking mechanism which proves a consistent description of several very different breaking effects.

We assume that the decay of the J/ψ into a final state containing ordinary hadrons proceeds via an A...Z-rule violating annihilation of a charmed quark-antiquark pair and the creation of a light quark-antiquark

pair. SU(3) symmetry requires the amplitudes for the
production of $u\bar{u}$, $d\bar{d}$ and $s\bar{s}$ pairs to be equal from a
unitary singlet state. We assume that the symmetry
is broken by suppressing the production of strange
quark pairs,

$$\langle u\bar{u}|Z|c\bar{c}\rangle = \langle d\bar{d}|Z|c\bar{c}\rangle = \langle s\bar{s}|Z|c\bar{c}\rangle/(1-\zeta), \quad (20)$$

where Z denotes the operator describing the A...Z-
violating transition and ζ is the parameter specifying
the suppression of strange quark production. The
suppression of strange particle production is a well-
known experimental effect,[19] and it is therefore
reasonable to attribute SU(3) symmetry breaking to a
mechanism which reduces strange quark production. We do
not consider the dynamical origin of the suppression
factor ζ, but rather investigate the implications of
this type of breaking on observed decay branching ratios.

     We consider three SU(3) predictions: 1) The selection
rule forbidding the PP, VV, TT and PT final states,
2) The predicted equality of the K^+K^{*-} and $\rho^+\pi^-$ decay
modes, and 3) The predicted ratio of the $\eta\gamma$ and $\eta'\gamma$
final states, for which the decay into the octet component
η_8 is forbidden by SU(3) if the $c\bar{c}$ pair first emits the
photon and then turns into a light pseudoscalar meson
via the transition (20). The justification of this
assumption is discussed below.

     We first consider the selection rule forbidding the
PP etc. states. This is conveniently parametrized by
examining the ratio of the forbidden K^+K^- and allowed
K^+K^{*-} decay modes. Both transitions involve the
disappearance of a $c\bar{c}$ pair and the creation of a $u\bar{u}$ pair
and an $s\bar{s}$ pair. We assume that the first pair is created
by the transition (20) and consider the possibility of a

similiar additional SU(3) breaking in the second
process. However, the results are qualitatively
unaffected by considering only the breaking given by
eq. (20).

Two diagrams are seen to contribute to the
transitions to the K^+K^- and K^+K^{*-} final states, one in
which the $u\bar{u}$ pair is created first and one in which
the $s\bar{s}$ is created first. In the SU(3) limit these two
contributions are seen to be equal in magnitude and to
have a relative phase depending upon the behavior of
the corresponding octets under charge conjugation.
Thus the two diagrams exactly cancel one another for
the K^+K^- final state and add constructively for the
K^+K^{*-} final state. This gives the selection rule. When
SU(3) is broken by the mechanism (20) the cancellation
no longer holds and the selection rule is violated. This
is expressed quantitatively by the relation

$$\frac{A(K^+K^-)}{A(K^+K^{*-})} = \frac{(1 - \zeta')\langle u\bar{u}|Z|c\bar{c}\rangle - \langle s\bar{s}|Z|c\bar{c}\rangle}{(1 - \zeta')\langle u\bar{u}|Z|c\bar{c}\rangle + \langle s\bar{s}|Z|c\bar{c}\rangle} \cdot \frac{\langle K^+K^-|u\bar{u}\rangle}{\langle K^+K^{*-}|u\bar{u}\rangle} \,,$$

$$\text{(21)}$$

where $A(f)$ denotes the amplitude for the J/ψ decay into
the final state f, $\langle f|u\bar{u}\rangle$ denotes the amplitude for the
decay of the $u\bar{u}$ state into the final state f and ζ' is
a strange quark suppression factor analogous to ζ in
eq. (20) describing SU(3) symmetry breaking in the
transition to the final state, defined by the relation

$$\langle f|u\bar{u}\rangle = \pm\langle f|s\bar{s}\rangle(1 - \zeta') \,, \tag{22}$$

where the phase is - for forbidden processes and + for
allowed processes.

Substituting eq. (20) into eq. (21) then gives

$$|A(K^+K^-)/A(K^+K^{*-})|^2 = [(\zeta - \zeta')/(2 - \zeta - \zeta')]^2 R(K,K^*),$$
$$(23)$$

where $R(K, K^*)$ is a factor of order unity expressing
the ratio of PP and VP widths when both are allowed
by SU(3)

$$R(K, K^*) = |<K^+K^-|u\bar{u}>/<K^+K^{*-}|u\bar{u}>|^2 . \qquad (24a)$$

If symmetry breaking in the second process is neglected
and ζ' is set equal to zero,

$$|A(K^+K^-)/A(K^*K^{*-})|^2 = (\zeta/2 - \zeta)^2 R(K,K^*) \text{if } \zeta' = 0.$$
$$(24b)$$

Eqs. (4) show that the forbidden decay is no
longer zero in the presence of SU(3) symmetry breaking,
but that it still remains very small even if the breaking
is appreciable. Note, for example that $\zeta = 1/2$, which is
appreciable suppression of the strange quark production,
the K^+K^- suppression factor is 1/9 if $\zeta' = 0$. For
$\zeta = 1/4$ the K^+K^- suppression factor is 1/49.

The $\rho\pi$ decay goes via two diagrams analogous to the
$K\bar{K}^*$ decay, but no strange quarks are involved. In both
cases the two contributions add constructively to give

$$|A(K^+K^{*-})/A(\rho^+\pi^-)|^2 = 1 - \left(\frac{\zeta + \zeta'}{2}\right)^2 .$$
$$(25)$$

Here no additional factor analogous to (24b) is necessary
because both final states are in the same SU(3) multiplet
and the ratio is determined by SU(3).

This result is seen to be much more sensitive to
SU(3) breaking than the selection rule (23), because the

breaking is linear in ζ rather than quadratic. For
$\zeta = 1/2$, $\zeta' = 0$ the ratio (25) is 9/16 while for $\zeta = 1/4$,
the ratio (25) is 49/64. Thus a symmetry breaking which
reduces the ratio (25) from the SU(3) predicted value of
unity by a factor of almost 2 keeps the forbidden
transition suppressed by an order of magnitude, while
a smaller breaking which reduces the ratio (25) to 75%
of its predicted value only allows the forbidden
transition to go with a strength of 2% of the allowed
transitions. Note also that this effect is enchanced by
any additional symmetry breaking expressed by setting
$\zeta' \neq 0$. Such breaking reduces the strength of the
forbidden transition (23) and increases the SU(3)
symmetry breaking in the ratio (25).

Radiative decays can proceed via the conventional
unitary octet component of the photon, which couples
to ordinary light quarks or by the unitary singlet
component which couples to charmed quarks. The octet
component contributes to the $\pi^0\gamma$ and $\eta_8\gamma$ decays, while
the singlet component contributes to the $\eta_1\gamma$ decay.
Since the experimentally observed $\pi^0\gamma$ decay is much
weaker than the observed $\eta\gamma$ and $\eta'\gamma$ decays, we neglect
the contribution of the octet component of the photon
and consider only the singlet. This is also consistent
with the picture in which the A...Z-violating transition
(20) is stronger for a pseudoscalar state, where it can
go via a two-gluon intermediate state, than for a vector
state where three gluons are required.

In this approximation the η and η' decays can only
go via the η_1 state in the SU(3) symmetry limit. However,
symmetry breaking via the mechanism (20) can also
introduce a contribution from the η_8 state. Using eq.
(20) we obtain:

$$\frac{A(\eta_8)}{A(\eta_1)} = \frac{\Sigma_q \langle\eta_8|q\bar{q}\rangle\langle q\bar{q}|Z|c\bar{c}\rangle}{\Sigma_q \langle\eta_1|q\bar{q}\rangle\langle q\bar{q}|Z|c\bar{c}\rangle} = \frac{(1/\sqrt{6})(\zeta)\langle u\bar{u}|Z|c\bar{c}\rangle}{(1/\sqrt{3})(3-\zeta)\langle u\bar{u}|Z|c\bar{c}\rangle} .$$

$$(26a)$$

For the physical η and η' states rotated by a mixing angle θ, this becomes

$$\frac{A(\eta\gamma)}{A(\eta'\gamma)} = [\tan\theta + (\zeta/\sqrt{2})(3-\zeta)]/[1 - (\tan\theta)(\zeta)/\sqrt{2}(3-\zeta)].$$

$$(26b)$$

The two terms in the numerator of the right hand side of eq. (26b) express respectively the contributions of the singlet-octet mixing and the SU(3) symmetry breaking of eq. (20). The two effects interfere constructively for positive values of the mixing angle, which corresponds to the reduction in the strange quark composition of the η.

With Isgur's mixing angle,[20] which corresponds to equal amounts of strange and nonstrange quarks in both mesons with opposite phase,

$$\frac{A(\eta\gamma)}{A(\eta'\gamma)} = \frac{\sqrt{2}-1+\zeta}{\sqrt{2}+1-\zeta} = \frac{0.41+\zeta}{2.41-\zeta} . \qquad (26c)$$

For the values of ζ of 1/2 and 1/4 considered above, the values of the ratio (26c) obtained are 1/2.1 and 1/3.2 respectively, which are the right order of magnitude to fit the available data.

Quantitative predictions from eqs. (23 - 26) should not be taken too seriously, because of the ambiguities in the value of ζ' and uncertainties regarding pseudoscalar mixing. However, the qualitative agreement of the strange quark suppression mechanism (20) in determining the order of magnitude of the three symmetry breaking effects is encouraging.

It is a pleasure to thank Hinrich Meyer for discussions of DESY results during the conference which motivated some of the analysis of new particle decays actually carried out during the conference and included in this text. Discussions with Harald Fritzch at the conference are also gratefully acknowledged. Recent work by Fritzsch and Jackson[21] discusses pseudoscalar mesons and SU(3) symmetry breaking from a very different point of view, using QCD-motivated calculations of radiative decays with conventional mixing and arriving at an SU(3) symmetry breaking qualitatively similar to the ad hoc breaking introduced in eq. (20) of this paper. Combining the Fritzsch-Jackson picture with the idea of radial mixing and the applications to strong interaction processes treated in this paper might provide a consistent description of a large body of hadronic phenomena. Unfortunately the Fritzsch-Jackson preprint was received after this work was completed.

REFERENCES

1. G. Alexander, H. J. Lipkin and F. Scheck, Phys. Rev. Lett. 17, 412 (1966).

2. G. Zweig, unpublished, 1964; and in Symmetries in Elementary Particle Physics (Academic Press, New York, 1965) p. 192.

3. S. Okubo, Phys. Letters 5, 165 (1963).

4. H. J. Lipkin, Physics Letters 60B, 371 (1976).

5. H. J. Lipkin, Lectures at the Erice Summer School 1976, FERMILAB Preprint Conf-76/98-THY.

6. J. Iizuka, Supplement to Progress of Theoretical Phys. 37-38, 21 (1966).

7. S. Okubo, Phys. Rev. D13, 1994 (1976) and D14, 108 (1976).

8. F. Gilman, these Proceedings.

9. C. Schmid, D. M. Webber and C. Sorensen, Nucl. Phys. B111, 317 (1976).

10. E. L. Berger and C. Sorensen, Phys. Letters 62B, 303 (1976).

11. E. M. Levin and L. L. Frankfurt, Zh. Eksp. i. Theor. Fiz-Pis'ma Redakt 2, 105 (1965) [JETP lett. 2, 65 (1965)].

12. H. J. Lipkin, Physics Reports 8C, 173 (1973).

13. J. Mott et al., Phys. Rev. Lett. 18, 355 (1967).

14. F. Gilman, M. Kugler and S. Meshkov, Phys. Rev. D9, 715 (1974); Phys. Letters 45B, 481 (1973).

15. H. J. Lipkin, Nucl. Phys. B7, 321 (1968).

16. M. Anguilar-Benitez et al., Phys. Rev. Lett. 28, 574 (1972); Phys. Rev. D6, 29 (1972); V. Barger in Proc. XVII Intern. Conf. on High Energy Physics, London (1974), ed. J. R. Smith, p. I-208.

17. H. J. Lipkin, FERMILAB preprint Pub-76/94-THY, Physics Letters in press.

18.   H. Meyer, these Proceedings.

19.   H. J. Lipkin, Spectroscopy after the New Particles,
      in Proc. Intern. Conf. on High Energy Physics,
      Palermo, Italy, 1975, Section 9.

20.   N. Isgur, Phys. Rev. D13, 122 (1976).

21.   H. Fritzsch and J. D. Jackson, Mixing of Pseudoscalar
      Mesons and M1 Radiative Decays, CERN Preprint
      TH.2264.

A MECHANISM FOR QUARK CONFINEMENT

David J. Gross

Joseph Henry Laboratories

Princeton, New Jersey 08540

I.   INTRODUCTION

I shall describe in this lecture a new approach to
the dynamics of quarks. It is widely believed that the
strong interactions are described by QCD, a nonabelian
(SU(3)) gauge theory of colored quarks and gluons,
permanently confined in color singlet hadronic bound
states. Our knowledge of the nature of the constituents
of hadrons derives from their observed symmetries, the
success of various phenomenological models such as the
quark model and most important from the observed short
distance behavior of hadronic currents. It is this free-
field theory behavior at large momentum (asymptotic
freedom) that singles out nonabelian gauge interactions
as the source of the strong interactions, allows us to
directly observe the quantum numbers of the quarks and
to derive quantative predictions that can test the
validity of the theory.

However, we are still far from possessing a _theory_
of the strong interactions in which we could calculate
the masses, couplings and scattering amplitudes from

first principles. We do not even have a qualitative
understanding of the dynamical mechanism responsible for
quark confinement. To be sure this is an exceedingly
difficult problem. Since quark mass parameters appear
to be unimportant (i.e. we should obtain a good approximation to hadrons which do not contain strange or charmed
quarks by setting the up-down quark mass parameters equal
to zero) the theory contains a single parameter, which
under dimensional transmutation will be replaced by the
hadronic mass scale. Thus we must calculate the properties
of hadrons starting from a strong-coupling relativistic
quantum field theory with no adjustable parameters.

The approach that I shall discuss is based on the
use of a path integral for the calculation of Euclidean
Green's Functions:

$$Z(J) = D\{A_\mu^a\} \exp [-S(A) + F(J)],$$

where

$$S(A) = \frac{1}{4g^2} \int \text{Tr} [F_{\mu\nu} F^{\mu\nu}] d^4x, \qquad (1)$$

$$F_{\mu\nu} = \partial_\mu A_\nu - \partial_\nu A_\mu + i[A_\mu, A_\nu],$$

$$A_\mu = \frac{1}{2} \lambda^a A_\mu^a,$$

and F couples some external source J to gauge invariant
products of fields (I shall suppress the required gauge
fixing and ghost terms in the functional integral). Thus
in a theory with quarks; if we take $F(J) = \int d^4x J^\mu \bar{\psi} \gamma_\mu Q\psi$
then Z(J) generates the connected Euclidean correlation
functions of the electromagnetic current.

For small values of g one is accustomed to calculating Z by expanding the field A_μ^a about $A_\mu^a = 0$ is the absolute minimum of the action $S(A)$. This expansion of Z in powers of g^2 is simply ordinary perturbation theory. Now in a renormalizable quantum field theory one cannot keep the coupling uniformly small; instead renormalization yields an effective coupling $\bar{g}$ which varies as one changes the scale of lengths or momenta. In an asymptotically free theory such as QCD, $\bar{g}$ inevitably grows large at large distances. Polyakov[2] originally suggested that since $\bar{g}$ gets large other field configurations might dominate Z.

Until now attention has focused on the "instanton" solution of the Euclidean Yang-Mills theory discovered by Belevin, Polyaleov, Schwartz and Tyupkin[3] which minimizes the action in the sector with Pontryagin index one. This configuration was understood to be an indication of vacuum tunneling between an infinite number of classically degenerate vacua.[4,5,6] Thus instanton configurations are of fundamental importance no matter how small the coupling is. One must take them into account in order to construct the correct vacuum of QCD, which is a coherent superposition of the classical vacua, labeled by a continuous parameter $|\theta| \leq \pi$.

The main physical implications of this vacuum degeneracy for QCD has been the effect on fermionic symmetries. It was realized that in a θ-vacuum axial baryon number symmetry is broken without generating a Goldstone boson, thereby solving the notorious U(1) problem.[4,5,6] Other consequences of the existence of these θ-vacua are a possible mechanism for CP violation,[4,5,6] and a possible source for the dynamical symmetry breaking of chiral $SU(N)$[6]. However, it is hard to see what direct role

instantons play in confining quarks. Indeed I shall
argue that their net effect on widely separated quarks
is a finite mass renormalization. This argument is
based on the "dilute gas approximation", where one regards $Z(J)$ as the partition function of a dilute gas of
weakly interacting instantons and anti-instantons, which
is only valid for very weak coupling.

This approximation consists of evaluating $Z(J)$ by
saddle point integration about the saddle points consisting of n_+ (n_-) instantons (anti-instantons) located
at x_i^+ (x_i^-) with scale sizes ρ_i^+ (ρ_i^-). After fixing the
position and scales of the instantons, one performs
Gaussian integration about the minimum of $S(A)$, yielding,
for the vacuum functional,

$$Z(0) = \sum_{n_+,n_-} \frac{1}{n_+!n_-!} \int \prod_i \left[C \frac{d^4x_i^\pm d\rho_i^\pm}{\left(\frac{\bar{g}^2}{8\pi^2}\right)^4 (\rho_i^\pm)^5} \right]$$

$$\exp\left[-\frac{8\pi^2}{\bar{g}^2\left(\frac{1}{\mu\rho_i^\pm}\right)} + V(x_i^\pm, \lambda_i^\pm) \right] \qquad (1a)$$

where $\bar{g}$ $(\frac{1}{\mu\rho})$ is the effective coupling evaluated at a
distance ρ in units of the renormalization scale parameter $\frac{1}{\mu}$. This can be regarded as the partition function
of a gas of instantons, where the chemical potential is
given by $\exp(-\beta\mu) = \exp -\frac{8\pi^2}{\bar{g}^2(\frac{1}{\rho\mu})}$, the entropy of position
is $C \frac{d^4x d\rho}{\left(\frac{\bar{g}^2}{8\pi^2}\right)^4 \rho^5}$; and the interaction energy is V. The

The entropy of an instanton is proportional to $\frac{d^4x}{(g^2\rho)^4} \times$

$\frac{d\rho}{\rho}$ since it suffices to change the orientation in group
space by $d\theta \sim g$, to change the scale size by $d\rho = g\rho$
or to shift the position by $g\rho$ to obtain a physically
distinguishable configuration. The value of C (taking
the gauge group to be SU(2) has been calculated by
't Hooft,[4] and equals .26. For sufficiently small $\bar{g}$
higher order radiative corrections (which correspond
to corrections to the Gaussian integral about the saddle
point) are small, the chemical potential is very large
and the instanton interaction energy is short ranged.
Thus in this limit it is a good approximation to regard
the analog gas as dilute.

Once the effective coupling is not small, other
than minimal action field configurations become important; however if the effective coupling is not too
great ($\frac{g^2}{4\pi^2} < 1$) one might still be able to trust semiclassical calculations. I shall discuss in detail a
particular class of configurations which might play a
dominant role in quark confinement. There are configurations which have two important properties:

First they correspond to separated lumps of one-half
topological charge, with an action (interaction energy)
that increases only logarithmically with separation and
an independent entropy of position for each lump proportional to the logarithm of the space time volume.
Thus while the contribution of such a pair separated by
a distance R will be suppressed, for small $\bar{g}$, by

$$R^4 \exp[- (\frac{const.}{\bar{g}^2} \ln R)],$$

(compared to say instantons where the interaction energy
of a pair is less than $\frac{1}{R^4}$ for large R) as $\bar{g}^2$ increases,
configurations with large separation become more proba-
ble. In fact we shall see that $\bar{g}^2$ will be evaluated at
a scale determined by R and will thus increase with
increasing R. Such configurations we call <u>merons</u> (from
the Greek root μερος = fraction), since they are localized lumps of fractional topological charge. If one
approximates the path integral by a sum over such configurations one obtains the partition function for a gas
of merons with attractive logarithmic interactions.
Such a system bears resemblence to the two-dimensional
Coulomb gas. There one expects that at low temperature
the system is in the dielectric phase, composed of a
weakly interacting gas of dipole pairs. However since
both the energy and entropy of charged pair increase
with the logarithm of the separation, at higher tempera-
tures the entropy term in the free energy F = E-TS takes
over and at some critical temperature isolated charges
can appear. The pair will then dissociate and the system
will be in the plasma phase. In our analog gas the role
of the temperature is played by the effective coupling.
We shall find that for small coupling, which is relevent
to the short distance properties of hadrons, the merons
will be bound into instantons, whereas for large coupling
$\left(\text{in fact for } \frac{\bar{g}^2}{4\pi^2} \gtrsim \frac{3}{14}\right)$ merons will dossociate.

    The second important property of the meron configura-
tions is their effect on quark binding. We shall argue
that in a plasma phase merons would confine quarks. Thus
the analog gas corresponding to QCD can be in different
phases depending on the scale of distances being probed.
At short distance merons will be tightly bound and have

no effect on the quarks which are quasifree. For widely
separated quarks they behave as a plasma, and although
the density of merons will be much less than that of
instantons they will control the long range properties
of the system and confine quarks.

I shall first illustrate these ideas with a two
dimensional model. Then I shall argue that instantons
by themselves do not confine quarks, discuss the general
strategy we recommend for dealing with QCD for moderate
coupling, and construct meron configurations. I shall
argue that other configurations are unlikely to be as
important and estimate the critical value of the coupling
at which quarks are confined. Finally I shall suggest
how our program might lead to calculable phenomenological
models of hadrons.

II. A TWO DIMENSIONAL ILLUSTRATION

To illustrate these ideas it is useful to consider
the two dimensional Abelian Higgs model[7]. This model
is simple to analyze. It contains instantons which are
simply the Nielsen-Olesen vortices, whose topological
charge is proportional to the magnetic flux, and whose
size is fixed (proportional to the inverse vector meson
mass M). The theory is superrenormalizable and the Higgs
mesons provide an infrared cutoff so that one can keep
the couplings small at all distances. If one performs
ordinary perturbation theory about the 'vacuum' $A_\mu = 0$
$\phi = \phi_0$, one would conclude that the gauge symmetry is
spontaneously broken, the Higgs phenomena occurs and the
charge is screened. Instantons will restore a discrete
gauge symmetry, wipe out the Higgs phenomena and confine
fractional charges. Integer charges, which can bind to
the charged scalar fields will not be confined.

As an indicator of confinement one may consider, following K. Wilson, the vacuum expectation value

$$C = \frac{<0|\exp\{ie\oint_L A^\mu dx_\mu\}|0>}{<0|0>} ,$$

where L is an Euclidean loop of spatial extent R and time extent T. For large T this proportional to $\exp[-\varepsilon(R)T]$, where $\varepsilon(R)$ is the interaction energy of two fixed sources of charge $\pm e$ separated by a distance R. A linear potential will result if ℓn C is proportional to the area of the loop (RT), whereas if ℓn C is proportional to the circumference (R+T) then this corresponds to a finite mass renormalization.

For weak coupling we may evaluate C in the dilute gas approximation by considering the effect of one instanton. Since a $\oint A_\mu dx^\mu$ equals 2π (or zero) if the instanton is totally inside (outside) the loop, $\varepsilon(R)T$ will be proportional to the volume in which the instanton overlaps the loop, $\frac{1}{M} \cdot (R+T)$, times the density of instantons $M^2 \exp(-S_{cl})$ and thus

$$\varepsilon(R) \underset{R>>\frac{1}{M}}{\sim} M \exp(-S_{cl}). \tag{2}$$

This R-independent interaction energy is simply a finite mass renormalization.

An alternate derivation of this result can be obtained by considering the change in the energy of the vacuum due to the presence of fixed charges $\pm q$ separated by a distance R:

$$\varepsilon_q(R) = - \underset{T\to\infty}{\text{Lim}} C_q/T, \quad C_q = \frac{<\exp[iq\oint A^\mu dx_\mu]>}{<1>} = \frac{I_q}{I_0} .$$

In the dilute gas approximation this is given by
$(V_0 \sim \frac{1}{M^2})$

$$I_q = \sum_{n_+,n_-} \frac{1}{n_+! n_-!} \int \left[\prod_{i=1}^{n_+} \frac{d^2 x_i^+}{V_0} \right] \left[\prod_{j=1}^{n_-} \frac{d^2 x_j^-}{V_0} \right]$$

$$+ i \ (n_+^L - n_-^L) \ \frac{2\pi q}{e} \ ].$$

where we sum over configurations with $n_+ (n_-)$ instantons (anti-instantons) and $n_{\pm}^L$ is the number of such inside the loop L. Thus $\varepsilon(R)$ is simply equal to the change in the energy of the vacuum which occurs if θ is equal to zero outside the loop and $2\pi q/e$ inside the loop.

The vacuum energy is periodic in θ, so that for q/e = integer one only gets a surface contribution to $\varepsilon(R)$ leading to Eq. 2. If however $\frac{q}{e} \neq 1 \pmod 1$, then there is a volume dependence of the vacuum energy and

$$\varepsilon(R)T \sim \exp(-S_{cl})(1 - \cos \frac{2\pi q}{e}) \ \frac{RT}{V_0} \ \cdots$$

Thus fractional charges are confined in the two dimensional Higgs model. The physical reason that instantons are confined in two dimensions is that they are responsible for restoring a discrete gauge invariance to the theory. This then wipes out the Higgs phenomena ($<\phi>=0$), and although the vector meson acquires a mass there then exists a long range Coulomb interaction between charged sources. This will then confine fractional charges, but not integer charged sources which can bind to charged Higgs mesons.

Now note that confinement of integer charges would

also obtain if there existed isolated configurations of
half-integral flux. The insertion of a charged loop
into a plasma of half-fluxons has the effect of forcing
θ to be equal to π inside the loop - thus providing an
energy proportional to RT. Even if the half-fluxons had
an interaction energy (action) which increased logarithmically with the separation of a pair (say $S_{cl}(R) \sim a +$
$\frac{b}{g^2}$ ln R), then for a pair of such half fluxons we would
get

$$\epsilon(R) \sim \int_{|x^+|<R} d^2x^+ \int_{|x^-|>R} d^2x^- \exp[-a-\frac{b}{g^2} \ell n \, R] \sim TR^{3-\frac{b}{g^2}},$$

which would confine once $g^2 \geq \frac{b}{3}$.

Now in point of fact it is impossible to find such
configurations in the two dimensional Higgs model. Lumps
of flux one-half must be connected by a stringlike region
in which the phase of the Higgs field changes from zero
to π. This gives rise to an interaction that increases
linearly with R. Therefore the lumps never separate.
If one were to estimate the contribution to the interaction energy of such configurations it would be proportional to R exp (-R). However, in four dimensional
nonabelian gauge theories I shall show that such configurations do exist, with only logarithmic action and
with equal power to confine quarks.

III. INSTANTONS AND MERONS

I shall now demonstrate that instantons by themselves do not confine quarks, at least as long as the
effective coupling constant remains small enough to

reliably replace the Euclidean functional integral by the partition function of a gas of instantons. In the dilute gas approximation we shall evaluate the ordered loop integral

$$C_L = \frac{<Tr\{T \exp[i\oint_L A^\mu dx_\mu]\}>}{<1>}$$

for a loop in the z-t plane. To be precise let us consider the loop to consist of the spacelike segments: $x = y = t = 0$, $0 \leqslant z \leqslant R$ and $x = y = 0$, $t = T$, $0 \leqslant z \leqslant R$ and the timelike segment: $x = y = z = 0$, $0 \leqslant t \leqslant T$ and $x = y = 0$, $z = R$, $0 \leqslant t \leqslant T$. C_L is gauge invariant but it most easily evaluated in the $A_0 = 0$ gauge where only the space-like segments contribute. Let us now consider the effect on C_L of a single instanton of size $\rho = \frac{1}{\lambda}$ located at $\bar{X}_I$. If the time coordinate $\bar{t}$ is much greater than $T + \rho$ or much smaller than $-\rho$ then $A_\mu(x)$ evaluated at both space-like segments is almost a pure gauge field which has no effect on the loop. Thus the instanton will contribute a nontrivial phase to C_L only if $0 \leqslant \bar{t} \leqslant T$. For large T>>R then $A_z(\vec{x},t)$ can be replaced in both cases by (to derive this simply evaluate the standard instanton solution, gauge transformed to $A_0 = 0$ gauge, for large positive and negative times),

$$A_z(\vec{x}) = U^{-1}(\vec{x} - \vec{x}_I)\partial_z U(\vec{x} - \vec{x}_I),$$

where

$$U(x) = \begin{cases} \exp[i \pi \dfrac{\vec{\tau} \cdot \vec{x}}{\sqrt{x^2 + \rho^2}}], t = T \\ \\ 1 \qquad\qquad t = 0 \end{cases}$$

It, therefore, suffices to consider the ordered path
integral over the segment at t = T. This is simply
equal to $\text{Tr}\{U^{-1}(-\vec{X}_I)\ U[R\hat{Z}-\vec{X}_I]\}$. Since $U(\vec{X})$ is equal to
-1 when $|\vec{X}|>>\rho$ and joins smoothly to 1 as $\vec{X}\rightarrow 0$ we see
that a nontrivial contribution to C_L will arise only
from instantons whose spatial extent overlaps with just
one of the charged sources. [As long as the $\int d\rho$ integral
converges we can show that the effect of instantons that
overlap both sources is to add a term to $\varepsilon(R)$ which falls
like $\frac{1}{R}$ at large distances]. Thus the interaction energy
between two charged sources at distance R is

$$\varepsilon(R)\sim\int^R\frac{d\rho}{\rho^5}\int_{|x_I|\leq\rho}d^3x_I\ \exp\left\{-\left[\frac{8\pi^2}{\bar{g}^2(\frac{\lambda}{\mu})}\right]\right\}\left(\frac{8\pi^2}{\bar{g}^2(\frac{\lambda}{\mu})}\right)^4.$$

This integral converges as $R\rightarrow\infty$ and thus instantons do
not confine quarks. One can, in principle, include the
effects of instanton-anti-instantion interactions.
However, these are rather short range and we do not expect them to alter the conclusion.

For sufficiently large R the effective coupling,
$\frac{\bar{g}^2(\frac{1}{\mu R})}{8\pi^2}$, might grow too large for such a weak coupling
approximation to make sense. However, we believe that
before this occurs other field configurations will become important, and can lead to confinement even for
effective couplings sufficiently small that semiclassical
considerations are still reliable.

The configurations that we believe control the
behavior of the quark loop consist of isolated lumps
of one-half unit of topological charge, merons, whose
positions are arbitrary and whose action only increases

logarithmically with separation. To illustrate the
effect of such configurations on the quark loop we shall
exhibit a singular configuration of this sort. Consider
the ansatz for the gauge field (for SU(2))

$$A_\mu^a = (\varepsilon_{0a\mu\nu} + \tfrac{1}{2} \varepsilon_{abc}\varepsilon_{bc\mu\nu})\partial_\nu \ln\rho(x). \qquad (3)$$

The Yang-Mills equations are then equivalent to $\Box\rho = C\rho^3$
(C is an arbitrary constant). In particular

$$\rho(x) = [(x-x_1)^2(x-x_2)^2]^{-\frac{1}{2}}$$

yields a solution of the field equations everywhere except the singular points x_i. As noted in Ref. 9, this
solution has the property that the topological charge
density is given by

$$Q(x) = \frac{1}{16\pi^2} \text{Tr } [F_{\mu\nu}\tilde{F}_{\mu\nu}] = \tfrac{1}{2}[\delta^{(4)}(x-x_1)+\delta^{(4)}(x-x_2)],$$

corresponding to two point merons. The action of this
solution is infinite, due to the fact that the field is
singular at x_i. However, this singularity can easily
be removed by smearing the topological charge over a
small sphere. It is then easy to see that the action
will be proportional to $\frac{1}{g^2} \ln (x_1-x_2)^2$, for large
separation. These point merons are then analogous to
point one-half fluxons of the type which we discussed
above. If one were to neglect their logarithmic interaction they would confine quarks. This is easily seen
once again by evaluating C_L in the $A_0 = 0$ gauge. An
individual meron will give a nontrivial phase now if it
is located at $0<<t<<T$ and is no further than R from the

loop leading to an energy between fixed quark sources which increases rapidly ($\sim R^3$) with increasing R. Now, of course, the meron interaction cannot be neglected. However, we shall argue below that its coefficient is proportional to $(\bar{g})^{-2}(\frac{1}{\mu R})$, and for large enough R such that $\frac{\bar{g}^2}{8\pi^2} \geq \frac{3}{28}$ the energy will increase, thus leading to confinement for an effective coupling sufficiently small that our semiclassical argument might still be valid.

IV. A SYSTEMATIC APPROACH TO MERONS

Having seen that merons can confine quarks let us now discuss how they would be treated in a systematic way. For small coupling one is used to approximating the Euclidean Functional integral by a saddle point approximation; i.e. by expanding the fields about strict minimum of the classical action. However, in general, the dominant contributions do not arise from strict minima of the action, even in the case of instantons alone, since we must include multiple instanton-anti-instanton configurations which are not solutions of the equations of motion. In an infinite volume system the most important configurations are those for which the action is not too far from a minimum and for which the entropy is large. Such configurations occupy a large volume in function space, thus compensating for the fact that the action is not minimal. Thus if we examine Eq.(1.a) we note that a configuration consisting of n_+ instantons and n_- anti-instantons has an action $= (n_++n_-)\frac{8\pi^2}{g^2}$ (neglecting the short range interactions), but occupies a volume in function space proportional to $\frac{V_0^{(n_++n_-)}}{n_+!n_-!}$

($V_0 =$ volume of space-time).

The dominant contribution to Z thus arises from configurations with $n_+ \sim n_- \sim V_0 \exp[\frac{-8\pi^2}{g^2}]$. In an infinite volume system, therefore, we have a finite density of instantons and anti-instantons. These configurations are treated by introducing constrainst which fix the collective coordinates of the instantons--i.e. their positions, scales and group orientation--then perform-ing a saddle point integration about the minimum of the constrained action and then integrating over the con-straints. An analogous procedure will be used to discuss merons.

The entropy associated with instantons, and with merons, comes from the fact that they are localized lumps of topological charge whose position is abritrary. From the point of view of entropy the important variable is therefore the topological charge density $Q(x)$. We pro-pose to define an effective field theory for $Q(x)$ by integrating over the gauge field and holding $Q(x)$ fixed, and only then integrating over $Q(x)$. The vacuum to vacuum amplitude would be $\int [dQ] e^{-W(Q)}$ where

$$\exp [-W(Q)] = \int \mathcal{D}\{A_\mu^a\} \prod_x \delta \, [Q(x) - \frac{1}{16\pi^2} \, \text{Tr}(F_{\mu\nu} \tilde{F}_{\mu\nu})] \exp[-S(A)],$$

with suitable gauge fixing and ghost terms implied. We can now imagine performing the A_μ^a integral by a straight forward saddle point approximation and then integrating over $Q(x)$ by more sophisticated methods. In first ap-proximation W would be the minimum action for fields that satisfy the constraint and can be calculated by adding to the action a Lagrange multiplier $\frac{1}{4g^2} \, [\lambda(x) \text{tr}(F\tilde{F})]$, solving the modified field equations $D^\mu F_{\mu\nu} + \partial^\mu \lambda F = 0$ and adjusting $\lambda(x)$ to obtain the desired $Q(x)$.

Of course, one would still have to perform the functional integral of $e^{-W(Q)}$. Our purpose here is to suggest a program of studying this approximate field theory for $Q(x)$ and show how it could lead to confinement. The idea is to replace the Q-field theory by an analog model consisting of a gas of instantons and merons. Such a replacement can only be approximate, but our hope is that the model retains those degrees of freedom which are responsible for confinement.

It might be useful to contrast our approach with that of K. Wilson who replaces QCD by a latticized gauge theory. Both approaches attempt to approximate the quantum field theory by a system involving only a finite number of degrees of freedom in a finite volume. Such systems might then be analyzed using methods that have been successful in many-body theory, such as the renormalization group. Our approximation consists of replacing (in a systematic way which in principle can be made as exact as desired) the field theory by a gas of field configurations. Unlike Wilson we do not introduce a space-time lattice, arbitrary cutoffs or parameters; and the nature of the dynamical variables is quite different. In this asymptotically free theory there is no need to introduce an ultraviolet cutoff (as provided by a space-time lattice) since the short distance behavior of the theory is totally under control and in fact calculable. By not introducing such a cutoff we avoid the severe problems that arise in relating the lattice gauge theory to the continuum field theory.

In the lattice gauge theory the role of the temperature is played by the bare coupling, i.e. the effective coupling constant at a distance corresponding to the

lattice spacing. In our analog gas of field configurations the role of the temperature is played by an effective coupling at varying distances. The effective
temperature of the gas will be determined by the distance scale of the correlation function which one cal-
culates. Thus if one discusses the properties of
quarks at short distances 'd' the important field configurations will be, say, instantons of sizes less than
d, and thus the effective temperature will be small. As
one separates the quarks, instantons (and meron pairs) of
larger size will be relevant, thus the effective coupling
= temperature will increase. Thus our analog gas can
appear in different phases depending on the physical
question being asked.

The merons are most easily found by searching for
minima of S subject to the constraint that the integral
of $Q(x)$ over a given volume is equal to $1/2$. This can
be achieved by taking Lagrange multipliers which are
piecewise constant; spherically symmetric configurations
of this type are easy to construct. In the ansatz of
Eq.(3) let $\rho = \rho(\tau = \ln \sqrt{x^2})$, with arbitrary scale r,
and define $\phi(\tau) = -(\frac{\rho'}{\rho}+1)^r$ where prime denotes differentiation with respect to τ. In terms of ϕ the action is
$S = \frac{3\pi^2}{g^2} \int_{-\infty}^{+\infty} d\tau \; [\phi'^2+(\phi^2-1)^2]$ and the topological charge

density is $\frac{6\pi^2}{g^2} \phi'(1-\phi^2)$. The Yang-Mills equations are
then equivalent to a particle in the potential $-(1-\phi^2)^2$
whose velocity can change discontinuously at any τ where
the Lagrange multiplier $\lambda(\tau)$ is discontinuous. The
function $\phi(\tau)$ must be continuous and approach ± 1 at $\tau =$
$\pm\infty$ to yield a finite action. The standard instanton is
the solution which proceeds from -1 at $\tau = -\infty$ to $+1$ at

$\tau = +\infty$: $\phi(\tau) = \tanh \tau$. Now consider a trajectory that leaves $\phi = -1$ at $\tau = -\infty$ and arrives at the stable minimum $\phi = 0$ at $\tau = 0$ at which point its velocity is discontinuously brought to zero. Then at $\tau = \ln \frac{R}{r}$ it is given a velocity kick ($\Delta\phi' = 1$) to make it arrive at $\phi = +1$ at $\tau = \infty$. Such a configuration consists of one-half of an instanton with half a unit of topological charge in the region $x^2 \leq r^2$, another half unit in the region $x^2 \geq R^2$ and zero charge density for $r^2 \leq x^2 \leq R^2$. Thus we have constructed a meron at $x = 0$ and another at $x = \infty$. For this solution we have ($\sigma_\mu = (1, i\vec{\sigma})$):

$$A_\mu^{++} = \sigma^\dagger \cdot \hat{x} \partial_\mu \sigma \cdot \hat{x} \begin{cases} \dfrac{x^2}{x^2 + r^2} & x^2 \leq r^2 \\[2mm] \dfrac{1}{2} & r^2 \leq x^2 \leq R^2 \\[2mm] \dfrac{x^2}{x^2 + R^2} & R^2 \leq x^2 \end{cases} .$$

A more physical configuration is obtained by performing an inversion about a point A_μ, with $A^2 = Rr = D^2$. Taking $(x-a)^\mu \to rR \dfrac{(x-a)^\mu}{(x-a)^2}$ yields a configuration consisting of a meron of size r at the origin and another one of the same size at a distance D away. The topological charge is confined to small spheres, each of size r, containing half a unit. In a similar fashion one can construct configurations consisting of a meron and an antimeron with localized charges $\frac{1}{2}$ and $-\frac{1}{2}$. For such a spherically symmetric configuration we have

$$A_\mu^{+-} = \sigma^\dagger \cdot \hat{x} \partial_\mu \sigma \cdot \hat{x} \begin{cases} \dfrac{x^2}{x^2 + r^2} & x^2 \leq r^2 \\[2mm] \dfrac{1}{2} & r^2 \leq x^2 \leq R^2 \\[2mm] \dfrac{R^2}{x^2 + R^2} & R^2 \leq x^2 \end{cases} .$$

This classical action (unrenormalized) for such a pair of merons is equal to $\frac{8\pi^2}{g_0^2} [1 + \frac{3}{4} \ell n \frac{D}{r}]$. This consists of a term $\frac{4\pi^2}{g_0^2}$ from integrating over each sphere which represents the (bare) chemical potential of a meron and a term $\frac{6\pi^2}{g_0^2} \ell n \frac{D}{r}$ which represents the (bare) interaction energy of a pair of merons of size r separated by a distance D. To determine the entropy of a meron at $x_\mu = \bar{X}_\mu$ as well as the renormalization of g_0 we must perform a Gaussian Functional integral about a pair of merons. For fixed $Q(x)$ there are minima of the action. Examination of the functional integral about a pair of merons indicates that the renormalization length for the chemical potential is r so that $\exp(-\beta\mu) = \exp\left[-\frac{4\pi^2}{\bar{g}^2(\frac{1}{r\mu})}\right]$, which is reasonable since this term comes from integrating over half an instanton to which $\bar{g}$ is evaluated at its scale size. On the other hand the renormalization length for the interaction energy is D, so that the interaction energy of a pair of merons of sizes r and r', separated by $D \gg r, r'$ is

$$\exp(-\beta V) = \exp\left[-\frac{6\pi^2}{\bar{g}^2(\frac{1}{\mu D})} \ell n \frac{D}{\sqrt{rr'}}\right]$$

The entropy of a meron we estimate to be equal to $C \frac{d^4 x dr}{r^5} \left(\frac{8\pi^2}{\bar{g}^2}\right)^4$ where $C \approx 1$ would have to be determined by calculation. It is important to note that the contribution of arbitrarily small merons is suppressed by asymptotic freedom. For an SU(3) gauge group we know that $\frac{\bar{g}^2}{4\pi^2} \left(\frac{1}{\mu r}\right)_{\mu r \not\to 0} - \frac{2}{11} \ell n(\mu r)$ so that the chemical potential vanishes like $r^{11/2}$ as $r \to 0$, which kills the potential divergence $\frac{dr}{r^5}$ in the entropy.

Why do we restrict our attention to meron configurations and not consider isolated lumps of arbitrary fractional charge? It is clear that these would be as effective in producing confinement. The reason is that although one can easily construct such configurations with only logarithmic interaction energy, it appears to be impossible to superimpose them in such a way to obtain an independent entropy of volume for each lump without letting the action grow rapidly with separation.

What is special about merons and instantons is that they are solutions of the equation of motion everywhere (in the case of the instanton) or everywhere except for a small surface (in the case of merons). This is essential if we wish to superimpose merons and instantons and move them about while keeping the action from growing more rapidly than the logarithm of meron separation. For example we can consider the effect of adding a pair of merons separated by a distance R to a dense sea of instantons and anti-instantons. A simple evaluation of the action yields an upper bound on the charge-of the interaction energy of the merons due to the instantons. For a given instanton $\delta S \sim \frac{\ell n\ R}{R^4}$, and since there can be R^4 instanton in between the merons $\delta S \sim \ell n\ R$. On the other hand if we superimpose on the instanton sea a configuration of separated f, 1-f charges, we find that $\delta S \sim R^2$ thus suppressing these configurations.

Thus if we wish to construct configurations which

can be treated as independent objects, they must be
solutions of the equations of motion everywhere but for
finite regions of space-time. If we wish to consider
only isolated lumps of topological charge then we can
prove that only merons and instantons need be considered.
For consider an arbitrary field configuration which has
an isolated topological charge at the origin, by which
we mean that $\mathrm{Tr}(F_{\mu\nu}\tilde{F}_{\mu\nu}) < \frac{1}{r^4}$ for large r, the gauge field,
which is assumed to be a solution of the equations of
motion, must fall as $\frac{1}{r}$ for large r, since $\int d^4x\ \mathrm{Tr}(F\tilde{F}) \neq 0$,
and $\mathrm{Tr}(F\tilde{F})$ is a total divergence. For large r we can use
the spherically symmetric ansatz. In terms of the field
ϕ introduced above:

$$r\ \mathrm{Tr}(F\tilde{F}) \sim \dot{\phi}(\phi^2-1)$$

and

$$D^\mu F_{\mu\nu} = 0 \Rightarrow \ddot{\phi} = \phi(\phi^2-1) \ \ldots,$$

where

$$\dot{\phi} = r\frac{d}{dr}\phi.$$

Thus if the charge is to be localized $\phi(r)$ must approach
a constant ϕ_0 as $r\to\infty$. The equations of motion then require $\phi_0 = \pm 1$ (instanton or vacuum) or $\phi_0 = 0$ (meron).

V.   CONFINEMENT

We shall now discuss how the merons confine quarks
and estimate the critical value of the effective coupling.
If the gas of instantons and merons was sufficiently
dilute one might perform a calculation of C_L as before.

Now we consider the effect on the quark loop of a pair
of merons at x_1 and x_2. A non-trivial phase will obtain
if one meron is inside the loop, the other outside.
Integrating over the volume occupied by the meron outside the loop then yields (dropping irrelevant terms):

$$\epsilon(R) \sim \int\limits_{|\vec{x}|<R} d^3x \ \frac{dr}{r^5} \ \exp[-4\pi^2/\bar{g}^2(\frac{1}{\mu r})] R^{4-6\pi^2/\bar{g}(\frac{1}{\mu R})}.$$

Thus the interaction energy begins to grow with R when
$\frac{1}{8\pi^2}\bar{g}^2(\frac{1}{R\mu}) = \frac{3}{28}$. At this coupling the density of instantons is still reasonably small, i.e. $.52 \ \exp[-(\frac{8\pi^2}{\bar{g}^2})] \times$
$(\frac{8\pi^2}{\bar{g}^2})^4 = .35$. When $\frac{\bar{g}^2}{8\pi^2} = \frac{1}{8}$, $\epsilon(R)$ becomes linear; however, the density of instantons increases to .72. Thus
confinement may be occurring for a relatively weak
coupling. However one must improve on the above argument,
taking into account instantons and meron interactions.

At low temperature = coupling constant, the merons
will be tightly bound and indistinguishable from instanton configurations. To see whether a phase transition would take place one could carry out a mean field
approximation by considering the screening effect of an
instanton or a tightly bound meron-antimeron pair on a
well separated meron pair. In this highly nonlinear gas
the interactions are quite complicated, however in a
crude approximation it appears that the gas is analagous
to a dielectric medium of quadrapolar objects.

Consider superimposing a meron pair of size $D(<<R)$
between a meron pair separated by a distance R

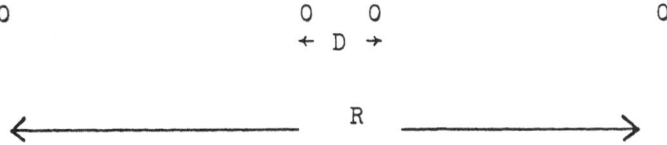

We choose the field which describes the small (large) meron pair to be A_μ(B_μ). Since both A_μ and B_μ are solutions of the equations of motion everywhere except for the surfaces of the merons:

$$\delta S = S(A+B) - S(A) - S(B) \sim \int A^2 B^2 d^4 x.$$

We can choose a gauge where $B_\mu = 0$ at the center of the small meron pair which we choose to be $x_\mu = 0$. Thus for $x \ll R$, $B_\mu \simeq x^\mu F_{\mu\nu}^B(0)$ and

$$\delta S \simeq (F_{\mu\nu}^B(0))^2 \int x^2 A^2 d^4 x \sim D^4 (F_{\mu\nu}^B(0))^2.$$

Thus the effect of the small pair on the large pair is that of a dielectric medium, where the mean field static dielectric constant $\varepsilon(q^2=0)$ is proportional to $\langle D^4 \rangle$. With this analog we can estimate the transition temperature $= \bar{g}^2$ at which a transition to a plasma like phase will occur, i.e. when the pairs will dissociate and the uncorrelated merons can be created. This will occur when

$$\varepsilon(q^2=0) \sim \langle D^4 \rangle \sim \int d^4 D \; D^4 e^{-[6\pi^2/\bar{g}^2(\frac{1}{\mu D})]\ln D}$$

diverges, which occurs at $\dfrac{\bar{g}^2}{8\pi^2} = \dfrac{3}{32}$. (Note that for this

$\bar{g}$ the density of instantons is so small, .17, that this estimate might be meaningful.) Furthermore dimensional reasoning suggests that $\varepsilon(q^2)$ develops a pole, $\frac{1}{q^2}$, when $<D^2>$ diverges, thus leading to exponential screening. This occurs at $\frac{\bar{g}^2}{8\pi^2} = \frac{1}{8}$, which coincides with our previous estimate, based on the quark loop, or when the quark-antiquark interaction energy becomes linear.

VI. CONCLUSIONS AND PROSPECTS

Much work clearly remains in developing all stages of our program. We would like to develop a formalism that would allow us to systematically construct the effective field theory for $Q(x)$. We must understand the precise nature of meron-instanton gas by determining the many-body meron-meron meron-instanton Hamiltonian. Finally we must go beyond arguments that indicate that quarks are confined to actual calculation of the properties of hadrons.

One important feature that has already emerged from the above discussion is the relatively small value of the critical coupling $\bar{g}^2/8\pi^2 = 1/8$, which leads to confinement. This is essentially the value of the effective coupling at a distance corresponding to the size of the hadron. What is remarkable is that such a small coupling, for which semiclassical arguments and low order perturbation theory are still valid, can confine quarks. If the value had turned out to be of order 1 instead of 1/8, our program would have had great difficulties. We also note that $\bar{g}^2/4\pi^2 \approx 1/4$ is consistent with the observed departures from asymptotic freedom.

We are obviously far from being able to calculate hadronic masses (the acid test of a theory of hadrons),

however, one might speculate that our mechanism would
lead to a phenomenological model similar to the MIT bag.
In a system with more than one phase one can have localized regions in the "wrong" phase (e.g. bubbles in a
liquid) at a cost of some energy per unit volume which
may be compensated by other effects. In the presence
of (separated) quarks our Euclidean gas may find it
favorable to be in a dielectric phase (in a cylindircal
region along the "time" axis) at a cost in action of
$BV \cdot T$, where B is some constant energy per unit volume.
The quarks would be almost free particles inside this
region. Since for $\frac{\bar{g}^2}{8\pi^2}$ in the range $\frac{3}{32}$ to $\frac{1}{8}$ the density
of instantons is increasing extremely rapidly one might
expect the transition from the dielectric-asymptotic
freedom phase to the plasma - confining phase to be
rather abrupt, thus producing an effective "bag" with
a sharp boundary. Elaboration of this idea might allow
one to develop a bag whose parameters, range of validity
and dynamics would be calculable.

Finally, we note that our confinement mechanism is
special to nonabelian gauge theories. Not only do we
rely heavily on the asymptotic freedom and infrared
slavery of QCD, but in addition we require meron configurations which satisfy the Euclidean equations of
motion except in small regions and have $F_{\mu\nu} \sim \frac{1}{r^2}$ at
large distances, thus giving rise to long range correlations in $\langle A(x)A(y) \rangle$ which make the quark loop behave as $\exp(-RT)$. Such configurations do not exist in
Abelian theories.

REFERENCES

1. The work reported on in this talk was done in collaboration with C. Callan and R. Dashen (to be published, Phys. Letters).

2. A. M. Polyakov, Phys. Lett. 59B, 821 (1975).

3. A.A. Belavin, A. M. Polyakov, A. S. Schwartz and Yu. S. Tyupkir; Phys. Lett. 59B, 85 (1975).

4. G. 't Hooft, Phys. Rev. Lett. 37, 8 (1976); Phys. Rev. D14, 3432 (1976).

5. R. Jackiw and C. Rebbi, Phys. Rev. Lett. 37, 172 (1976).

6. C. Callan, R. Dashen, and D. Gross, Phys. Lett. 63B, 3341 (1976).

7. This model has been analysed in detail by us. In addition to exhibiting the restoration of gauge symmetry it provides an example of dynamical chiral SU(2) symmetry breaking generated by instantons. C. Callan, R. Dashen and D. Gross, to be published.

8. F. Wilczek, Princeton University preprint (1976); Corrigan and Fairlee, preprint (1976); G. 't Hooft, unpublished.

9. V. de Alfaro, S. Fubini, and G. Furlan, CERN preprint 2232 (1976).

THE STATUS OF ξ-SCALING*

H. David Politzer†

Harvard University

Cambridge, Massachusetts 02138

ABSTRACT

The logic of the "ξ-scaling" analysis of inclusive lepton-hadron scattering is reviewed with the emphasis on clarifying what is assumed and what is predicted. The physics content of several recent papers, which purport to criticize this analysis, in fact confirm its validity and utility. For clarity, I concentrate on the orthodox operator product analysis of electroproduction, local duality and precocious scaling. Other physics discussed includes the successes of QCD in the rate of charm production in muon inelastic scattering and in the energy-momentum sum rule. Gluons carry 30 ± 8% of the proton's energy-momentum at $Q^2 = 3.5$ GeV^2. (Several considerations make this considerably smaller than previous estimates; the largest of these is the inclusion of gluon bremsstrahlung effects.)

---

*Research supported in part by the National Science Foundation under Grant No. PHY75-20427.

†Junior Fellow, Harvard University Society of Fellows.

Can QCD be tested against the available data on inclusive lepton scattering? This question appears absurd given that a major reason for interest in QCD is its explanation of the observed approximate scaling. Yet the question is serious, and its answer is complex. A recent series of papers[1,2,3,4] addresses this issue and concludes that the answer is yes. Certainly, much of the evidence is qualitative and, at best, not even good for two significant figures. And since we still don't know what a proton looks like in QCD, we may yet be barking up the wrong tree.

The Criticisms

I am particularly eager to discuss these issues again in light of certain criticisms that have come up. Several groups[5,6,7,8] have voiced their consternation over particular claims or broad features of the analysis of Refs. 1 and 2. Their papers are often worthwhile: all the mathematics and most of the physics are correct, and much of it is useful. However, the critical conclusions just do not follow from the arguments presented. Reference 3 attempts to meet these criticisms head on, one at a time, systematically. Here, I will sketch the central arguments, clarifying in passing the various questions on which there have been serious misunderstandings. Reference 3 also contains an excellent explanation of ξ scaling and higher twists in the language of the parton model and $P_\perp$ distributions and final state interactions. (For the sake of readers not interested in QCD, I reproduce some of that discussion at the end.)

Summary

To try to avoid any further misunderstanding, I will stick to the orthodox language of the operator product expansion (OPE) and twist. This may be a case of overkill. The same arguments can be presented in simpler, more physical terms. But the OPE has the outstanding virtue of precision. I will concentrate on electro-production off the light quarks in a proton. This case contains all the substantive issues and is the one I've studied in greatest detail.[2]

The following is an outline of this talk: After giving the motivation for the working hypotheses, I sketch the form of electroproduction dictated by the OPE and explain the significance of the variable ξ. I explain the various hypotheses and use the resulting framework to build a phenomenology of precocious scaling and local duality. I then go back and test the hypotheses for consistency. Some other predictions are only briefly reviewed.

Has QCD been verified? No, but it could have been ruled out. Another popular contender is ruled out by the observed behavior at small x and the level of multi-muon production while QCD remains in the running.

I finish with some new work on the energy-momentum sum rule. The significant lesson is not so much that it works but that the proton is rather simpler than pre-viously believed. In particular, gluons carry only 30 ± 8% of the proton's energy-momentum at $Q^2 = 3.5$ GeV^2.

(I add at the end a discussion in parton language taken from Ref. 3.)

<u>Why It Should Work</u>

Precocious scaling for final invariant mass W >2 GeV and local duality[9] show that the simplicity of approxi-mate scaling extends down to $Q^2 \approx 1$ GeV^2. To put it in the most modest terms possible, the purpose of the "ξ-scaling" analysis of Ref. 2 is to see what are the origins and the patterns of this simplicity within the framework

of QCD. We wish to make precise how 1 GeV^2 can be large
compared to the proton's inverse size while not large
compared to its mass. The QCD framework is sufficiently
rigid that if precocious scaling is understandable and
not an accident, then there are many other testable
consequences that must go with it.

The operating procedure will be to assume certain
terms are small on the basis of heuristic arguments and
then use the resulting theory to go back and measure
those terms. Typically we will find that the neglected,
"small" terms are at least an order of magnitude smaller
than the small effects that are kept.

ξ and the OPE

The OPE[10,11] is the starting point for the orthodox
analysis of inclusive lepton scattering.[12,13] Very
schematically

$$<p|JJ|p> = \sum_{n,\alpha} c_{n,\alpha}(q)<p|0^{n,\alpha}|p>$$

$$= \sum_n \vec{c}_n[q;g(Q^2)]e^{\int_{Q_0}^{Q} \gamma \frac{dQ'}{Q'}} <p|\vec{0}^n(Q_0^2)|p>, \qquad (1)$$

where $|p>$ is the target (proton) state of momentum p,
J is the relevant current of momentum q, $0^{n,\alpha}$ are the
local operators of spin n (α labeling all other features),
and $C_{n,\alpha}$ are the operator coefficients in the expansion
of JJ. The virtue of Eq. (1) is that all the q dependence
resides in c's, whose q derivatives are calculable for
large enough q.

A spin projection on Eq. (1) [in the t-channel of
the forward amplitude] gives the following:[14,1]

$$\int_0^1 \xi^n F(\xi, Q^2) d\xi = A_n(Q^2) + \sum_{k=1}^{\infty} B_{n,k}(Q^2) \left[n \frac{M_0^2}{Q^2} \right]^k , \quad (2)$$

where $F(\xi, Q^2)$ is a specific linear combination of structure functions and

$$\xi = x \frac{2}{1 + \sqrt{1 + Q^2/\nu^2}} = x \frac{2}{1 + \sqrt{1 + 4x^2 m^2/Q^2}}$$

$$\approx \frac{x}{1 + x^2 m^2/Q^2} . \quad (3)$$

The significance of the ξ variable is that it is the ξ moments that are related to operators of definite spin. So ξ brings in certain necessary effects proportional to Q^2/ν^2. A_n and $B_{n,k}$ are dimensionless. Their Q^2 dependence is computable (and only logarithmic) for large enough Q^2 if we assume there is only one operator for a given spin and dimension. (I will return to this later.) k is the (twist-2)/2 of the relevant operator where twist is the operator's dimension-minus-spin.[15]

      We are as yet unable to compute $A_n(Q_0^2)$, $B_{n,k}(Q_0^2)$ or M_0^2 at any finite, single Q^2 from first principles. So, in a strict sense Eq. (2) has no predictive content except for $Q^2 = \infty$.

      I factor out the explicit n in the term $[n M_0^2/Q^2]^k$ because the computable coefficient functions do have that n dependence, at least for large n.

      Working Hypotheses

      Within the framework of Eq. (2) either approximate scaling is an accident, i.e. it involves a precise cancellation between terms of all k, or it has a simple

explanation.   To explain scaling using Eq. (2), $A_n(Q^2)$
must be slowly varying in Q^2 and large compared to the
B terms.

The size of the Q^2 variation of $A_n(Q^2)$ is determined by the size of the effective quark-gluon coupling
$g(Q^2)$. $g(Q^2)$ can be related to a mass Λ via

$$\frac{g^2(Q^2)}{4\pi^2} = \frac{a}{\log Q^2/\Lambda^2} + O(g^4), \qquad (4)$$

where $a = 12/27$ for three massless quarks.[1] If precocious scaling is not an accident, then $\Lambda < 1$ GeV.   It
is also true that $\Lambda > 200$ MeV if QCD is to describe
hadrons; were Λ yet smaller, $g^2(Q^2)$ would be smaller,
and hadrons would then be physically larger.

In the QCD phenomenology of electroproduction, Λ is
left as a free parameter to be fit to the data.   If the
data required that Λ fall outside the interval $0.2 < \Lambda <$
1.0 GeV, then QCD would be wrong or, at least, this explanation of precocious scaling would be wrong.

For each n, the twist-2 $A_n(Q^2)$ is the sum of three
terms, each proportional to an unknown matrix element.
To make phenomenological progress, we assume that the
gluon distribution function is negligible compared to the
quark functions for $\xi \gtrsim 1/4$ (or $n \geq 4$).   Under this hypothesis, the three terms reduce to one, with a single
Q^2 dependence.   This assumption can be restated as saying
that there are no valence gluons or virtually no hard
gluons.   Another way to describe it is to note that the
various distribution functions should peak at roughly $\xi =$
$1/N$ where N is the typical number of those quanta present.[16]   So we are assuming that confinement requires lots
of soft gluons.   (I will return to a particular problem

where the glue contribution-- in fact all three operators -- can be determined.)

Regarding the B terms, let me first state our
hypotheses and then explain why I think they're reasonable. We assume

$$\frac{B_{n,k}M_0^2}{A_n} \sim O(\Lambda^2).\qquad(5)$$

With Eq. (5) to estimate which terms contribute for
which Q^2, we use the data to measure $B_{n,1}M_0^2/A_n$ to be
$(375 \pm 25 \text{ MeV})^2$ for $2 \leq n \leq 30$.

Crudely speaking, Eq. (5) says that the ratios of
proton matrix elements of analogous operators of different dimension (but fixed spin) can be estimated by
dimensional analysis, using Λ as the only available mass.
[Note that while this last statement is not renormalization group invariant, Eq. (5) actually is.]

In terms of Feynman graphs (see Fig. 1) the A_n
terms correspond to all possible interactions of the
struck quark with itself [in an expansion in $g^2(Q^2)$],
independent of the spectators. The twist-4 $B_{n,1}$ terms
include graphs where a gluon field strength connects the
struck quark to a spectator. These graphs are proportional to n. Twist-4 also includes four quark operators
in the t channel which measure quark correlations, but
these do not grow like n and in fact are proportional to
$g^2(Q^2)$. The graphs which give the n^k dependence have k
field strengths tying the struck quark to the spectators.
So there is an nM^2/Q^2 for each field strength which connects the struck quark to the others. Thus large numbers
of gluons are unimportant at high Q^2 unless n is also

large. In particular, the above behavior [including Eq. (5)] ensures that the effects of the B terms are fixed in W, the invariant final hadronic mass.[2]

High moments contain the information of the fine detail in ξ of the structure functions. The most interesting fine detail is that shrinking region just below $\xi = 1$ which corresponds to $W \lesssim 2$ GeV. We will see that M_0 is sufficiently small that we can predict the gross features of the region $W \lesssim 2$ GeV using the A_n terms alone. ξ will be of phenomenological utility because there are regions in which Q^2/ν^2 corrections are much bigger the nM_0^2/Q^2.

e-p Data Analysis

To analyze the electroproduction data we ignore nM^2/Q^2 and keep only the A_n terms. Clearly for any M_0^2 and Q^2 there exists a large enough n so that this is a bad approximation. So for each Q^2 we will consider only those moments which satisfy $nM_0^2/Q^2 \ll 1$.

[It is a trivial observation that in a free field theory $M_0 \equiv 0$ and the $B_{n,k}$ terms are absent. (See Fig. 1.) The converse is almost as trivial: if the $B_{n,k}$ terms are identically zero, then the structure functions must be free quark δ functions. This is because the A_n terms alone can not describe how the inelastic threshold approaches the elastic peak in ξ. Nevertheless, saying nM_0^2/Q^2 is small for some n is not equivalent to saying it's small for all n. Furthermore, the n for which it is small grows with Q^2 and is not limited to any finite number like three or four.]

The behavior of the A_n can be summarized by the moment inversion machinery.[13] We can define a smooth function $F^s(\xi,Q^2)$, non-zero for $0 < \xi < 1$ by

$$\int_0^1 \xi^n F^s(\xi,Q^2)d\xi = A_n(Q^2). \qquad (6)$$

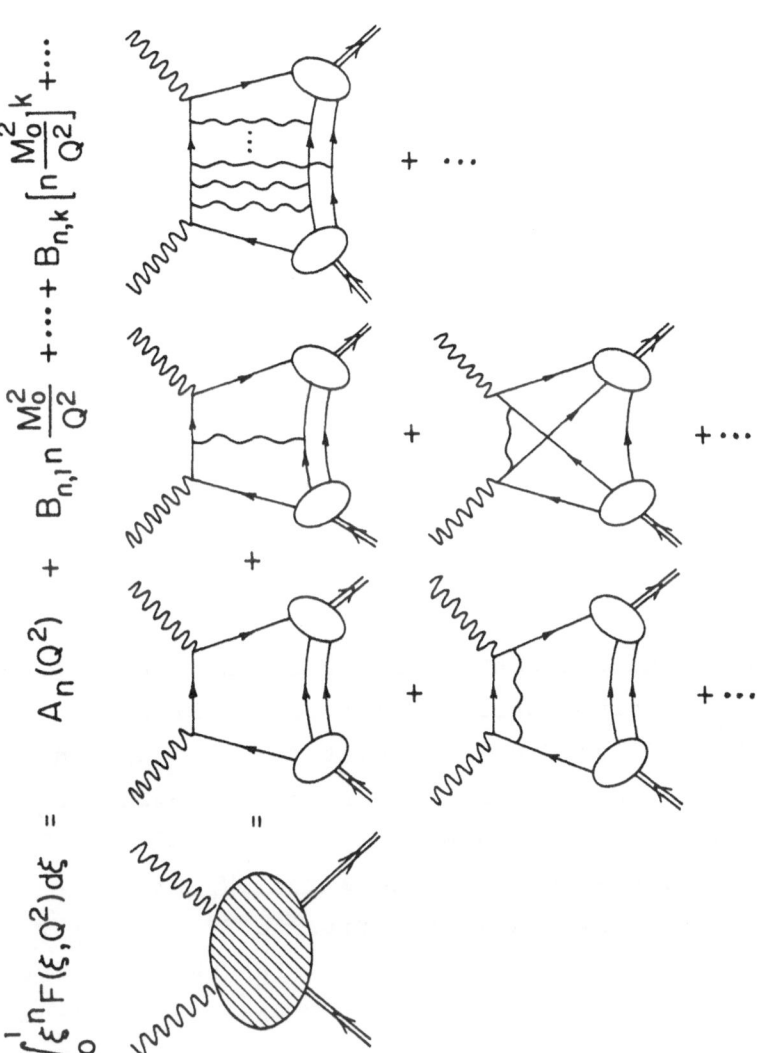

$$\int_0^1 \xi^n F(\xi, Q^2) d\xi \;=\; A_n(Q^2) \;+\; B_{n,1} n \frac{M_o^2}{Q^2} \;+\cdots+\; B_{n,k}\left[n\frac{M_o^2}{Q^2}\right]^k +\cdots$$

Figure 1. The OPE expansion for electroproduction and contributing Feynman graphs. See Eq. (2).

The physical structure function will differ from F^s by the Mellin transform of the B terms, which cannot be positive definite (because its moments grow with n) but is small for $W \gtrsim 2$ GeV [if Eq. (5) is satisfied]. One may regard the inversion machinery as simply a mnemonic for the moments, but I contend it is the simplest way to analyze the data as well as a powerful tool for implementing smoothness assumptions and error estimates. If we deal with data for $W > 2$ GeV we can measure $F^s(\xi, Q^2)$ without interference from the B terms [again, if Eq. (5) is right].

In Fig. 2 I compare the various scaling variables: x, ξ, x' and x_s where

$$x' = \frac{x}{1 + x\frac{m^2}{Q^2}} , \qquad x_s = \frac{x}{1 + x\frac{1.4 \text{ GeV}^2}{Q^2}} . \qquad (7)$$

x_s is constructed so that for $x \gtrsim 0.25$ the SLAC e-p data scales in x_s within experimental errors. From Fig. 2 it is clear that ξ itself will account for only around half of the observed scaling violations for $x > 0.25$. In QCD, the rest must come from Λ.

Figure 3 shows separated data for νW_2 with $W > 2$ GeV.[17] We made three parameter fits to $F^s(\xi, Q_0^2)$ for different values of Λ as indicated. The best $\Lambda = 200$ MeV fit is systematically too flat. The available data favors the best $\Lambda = 500$ MeV fit over the best $\Lambda = 700$ MeV fit.

In Fig. 4 I show $\nu W_2(x, Q^2)$ as predicted by the $\Lambda = 500$ MeV fit for $F^s(\xi, Q^2)$ from $W > 2$ GeV data. The dashed portions correspond to $W \leq 2$ GeV. In that region the

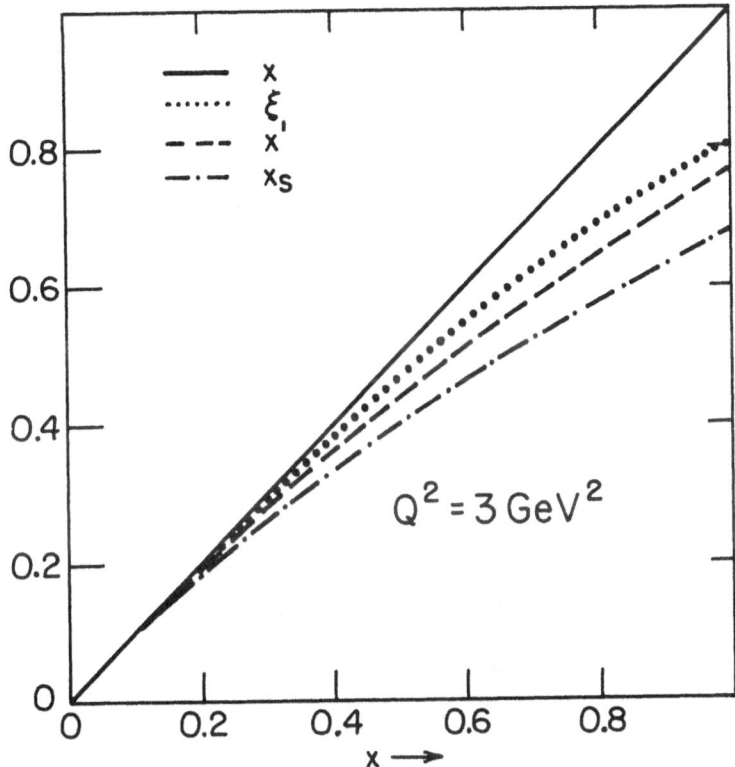

Figure 2. A comparison of various scaling variables, at $Q^2 = 3$ GeV^2. See Eq. (7). x_s is a phenomenological variable which summarizes the observed scaling violations at SLAC.

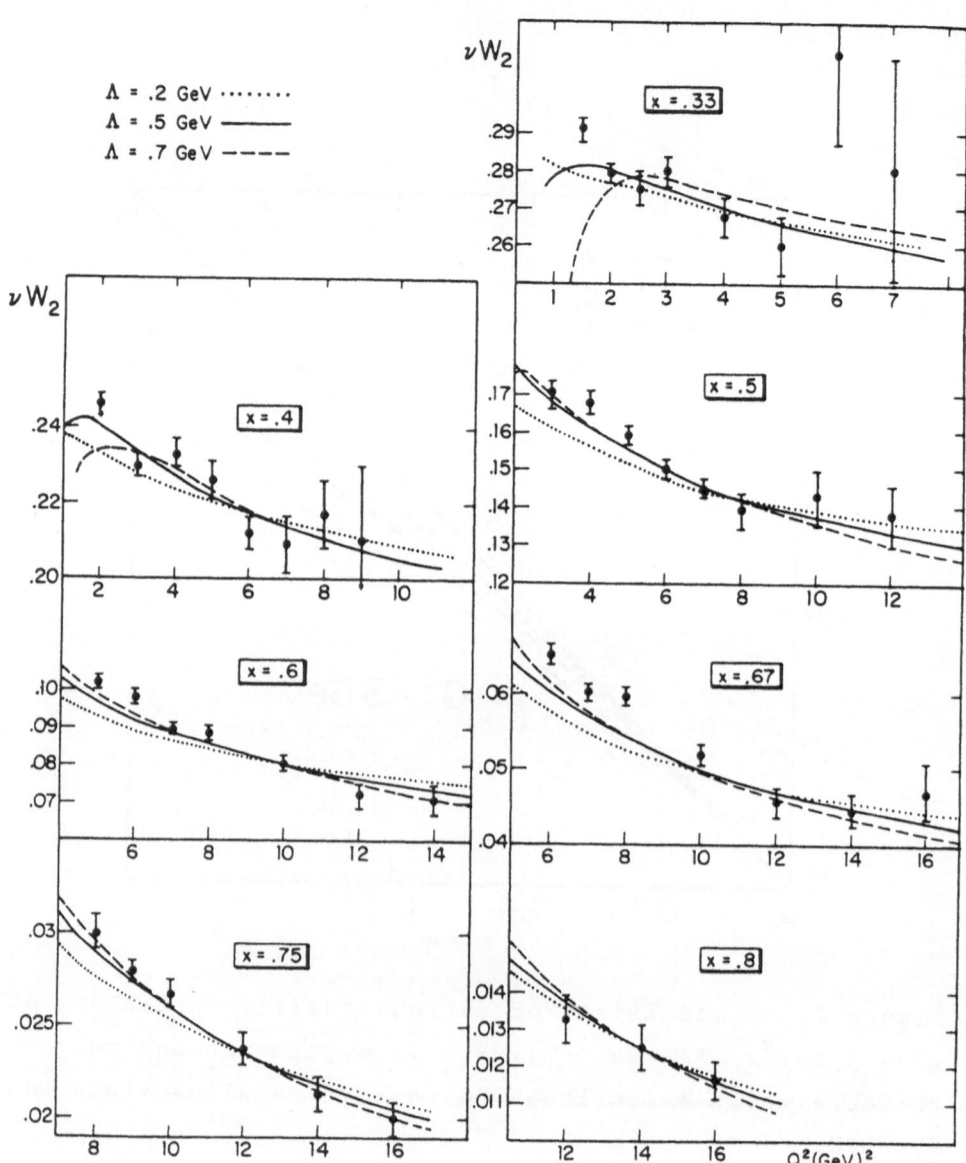

Figure 3. $\nu W_2(x,Q^2)$ versus Q^2 in x bins. The data is
from Ref. 17. The curves are the best three parameter
fits of Ref. 2 for Λ = 200, 500 and 700 MeV.

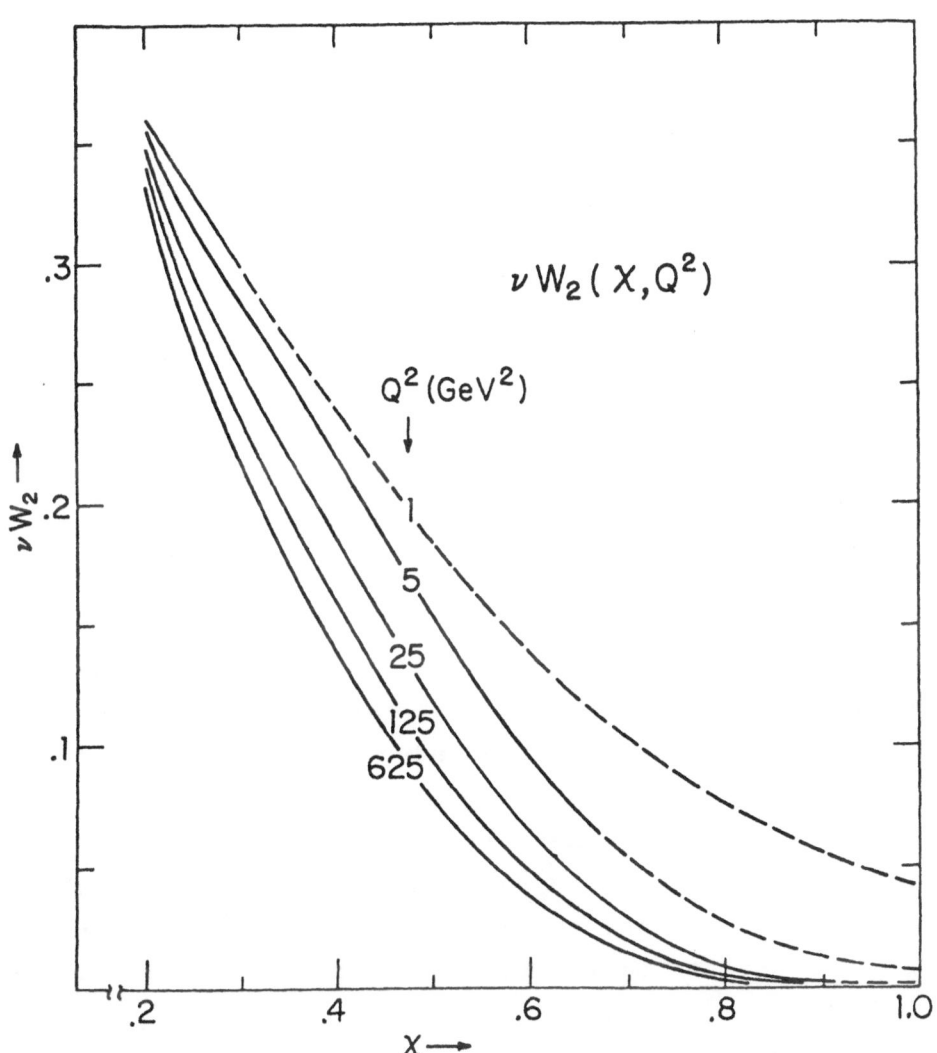

Figure 4. $\nu W_2(x,Q^2)$ versus x for various Q^2 as predicted by the Λ = 500 MeV fit. The dashed portions correspond to W ≤ 2 GeV.

dashed curve is a prediction for νW_2 in the sense of
local duality: the data will differ locally in x but
must agree to $O(\Lambda^2/Q^2)$ when smeared.

Figure 5 shows the comparison of $F^s(\xi,Q^2)$ determined by measurements at W > 2 GeV with the data for
W ≤ 2 GeV. Dig it.

For those who have trouble interpreting Fig. 5
we offer the following more quantitative discussion.
Consider the quantity Δ defined by

$$\frac{n\Delta}{Q^2} \equiv \frac{\int \xi^n [\nu W_2^{\text{experiment}} - \nu W_2^{\text{theory}}] d\xi}{\int \xi^n \, \nu W_2^{\text{experiment}} \, d\xi} \, , \qquad (8)$$

where νW_2^{theory} evolves according to the $A_n(Q^2)$. For n
and Q^2 such that $n\Delta/Q^2$ is small, Δ should be approximately $B_{n,1}M_0^2/A_n$. If we adjust n according to n = $2Q^2$
(in GeV^2), then $n\Delta/Q^2$ is constant within experimental
errors over the entire SLAC region at 0.30. Hence
$B_{n,1}M_0^2/A_n \approx (375 \text{ MeV})^2$, independent of n (at least for
0 < n ≤ 30).

So the n = $2Q^2$ moment is good to 30% over the available Q^2 range. Furthermore, the n = $2Q^2$ moment is about
90% W ≤ 2 GeV. Hence the area of νW_2 in the resonance
region is determined (at least 70% of it) by the twist-
2 operators.

To repeat, ξ is of phenomenological value in this
analysis because Q^2/ν^2 can be bigger than M_0^2/Q^2.

Note also that Δ defined by Eq. (8) measures <u>all</u>
deviations from the twist-2 theory, whether described by
$B_{n,1}$ or anything else.

Some Other Predictions

I will not discuss here the ξ effects for the production of heavy quantum numbers in ν scattering.[1,18]

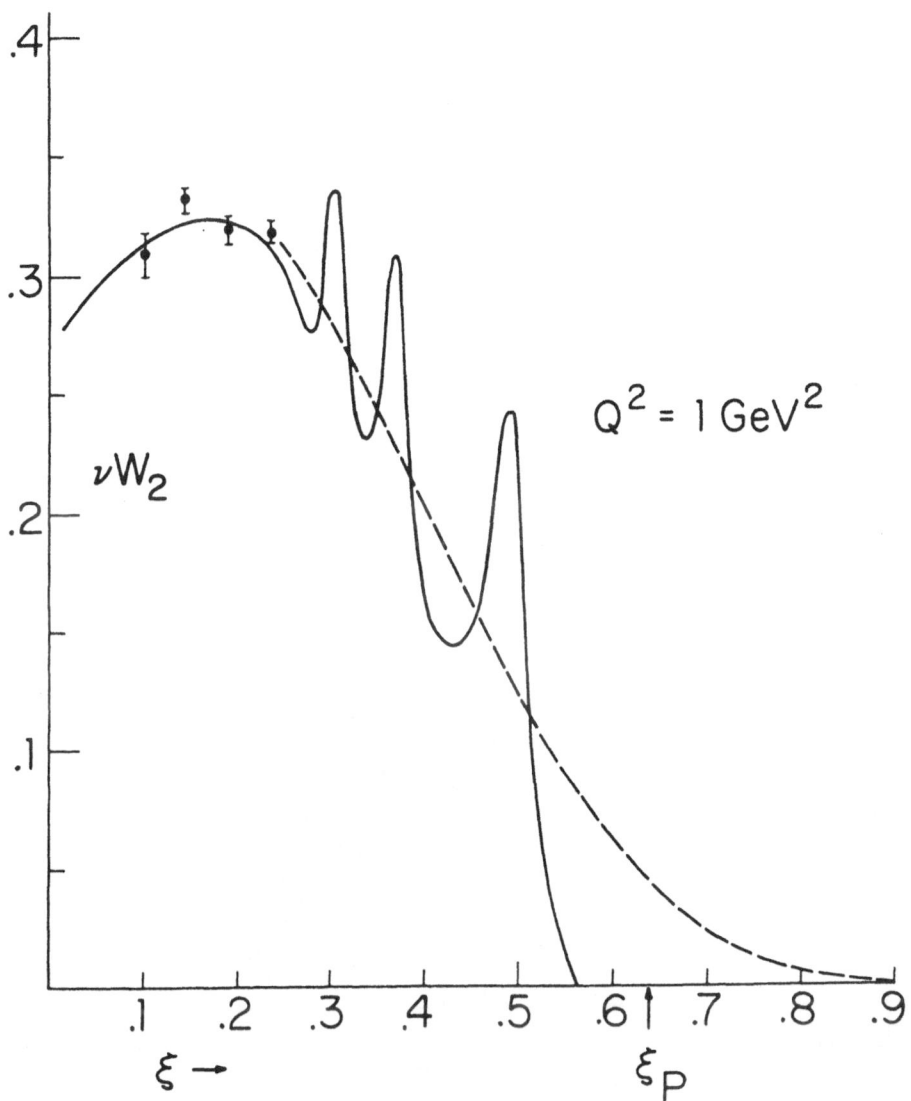

Figure 5a. $\nu W_2(\xi, Q^2)$ versus ξ for various Q^2. The solid line is the data as given in Ref. 17. The dashed line is the $\Lambda = 500$ MeV prediction in the sense of local duality.

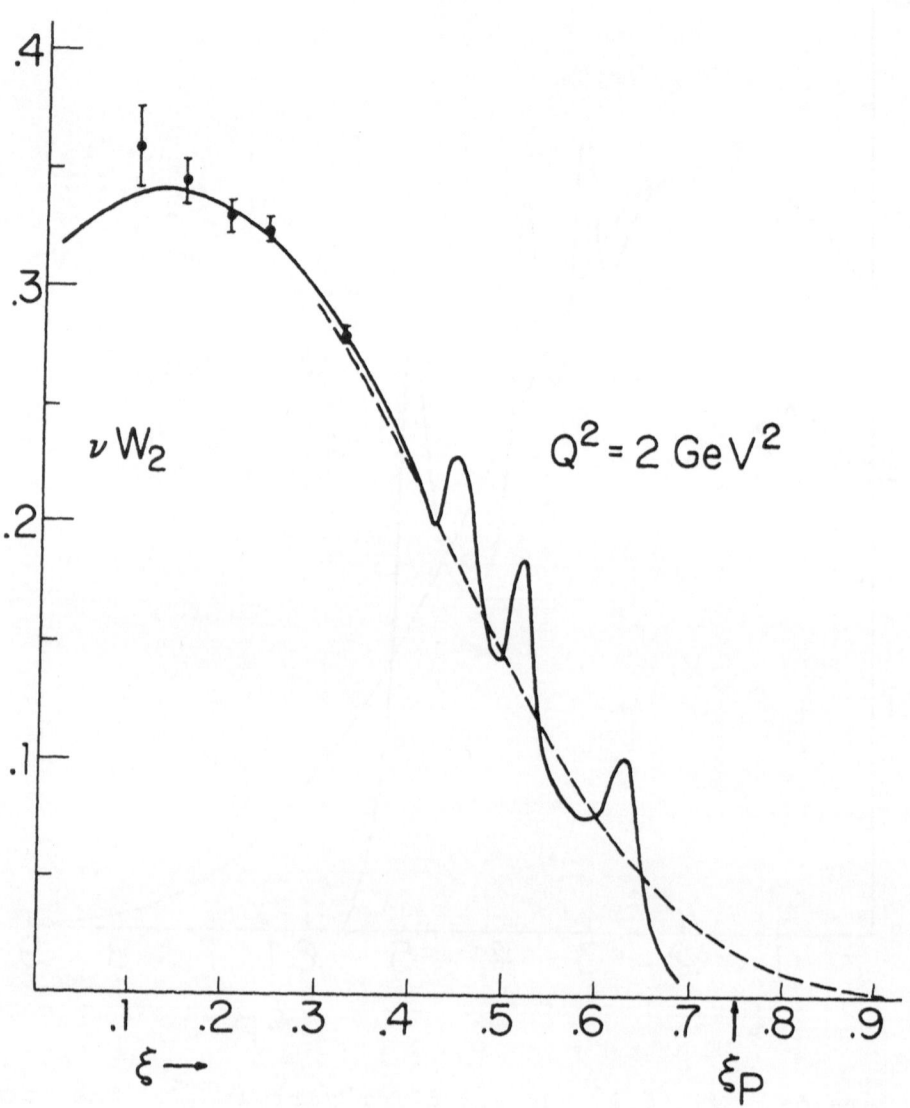

Figure 5b.

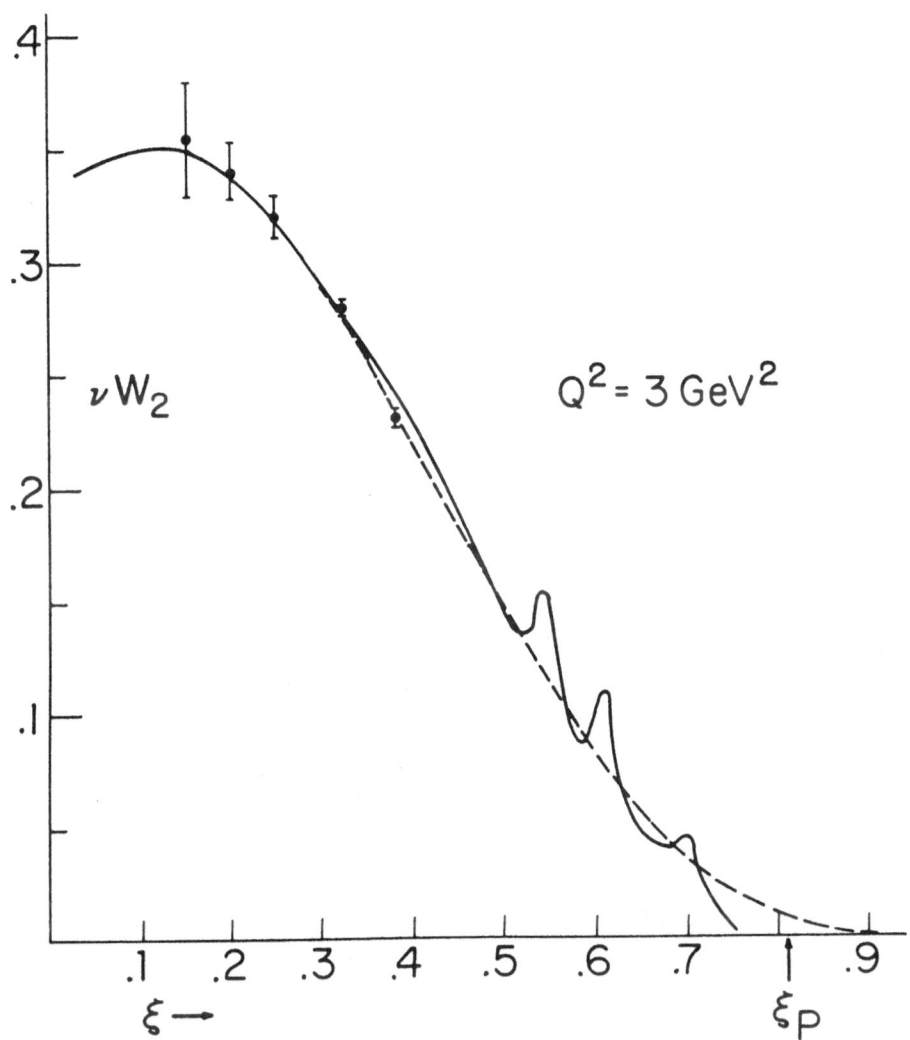

$$\nu W_2 \qquad Q^2 = 3\,GeV^2$$

Figure 5c.

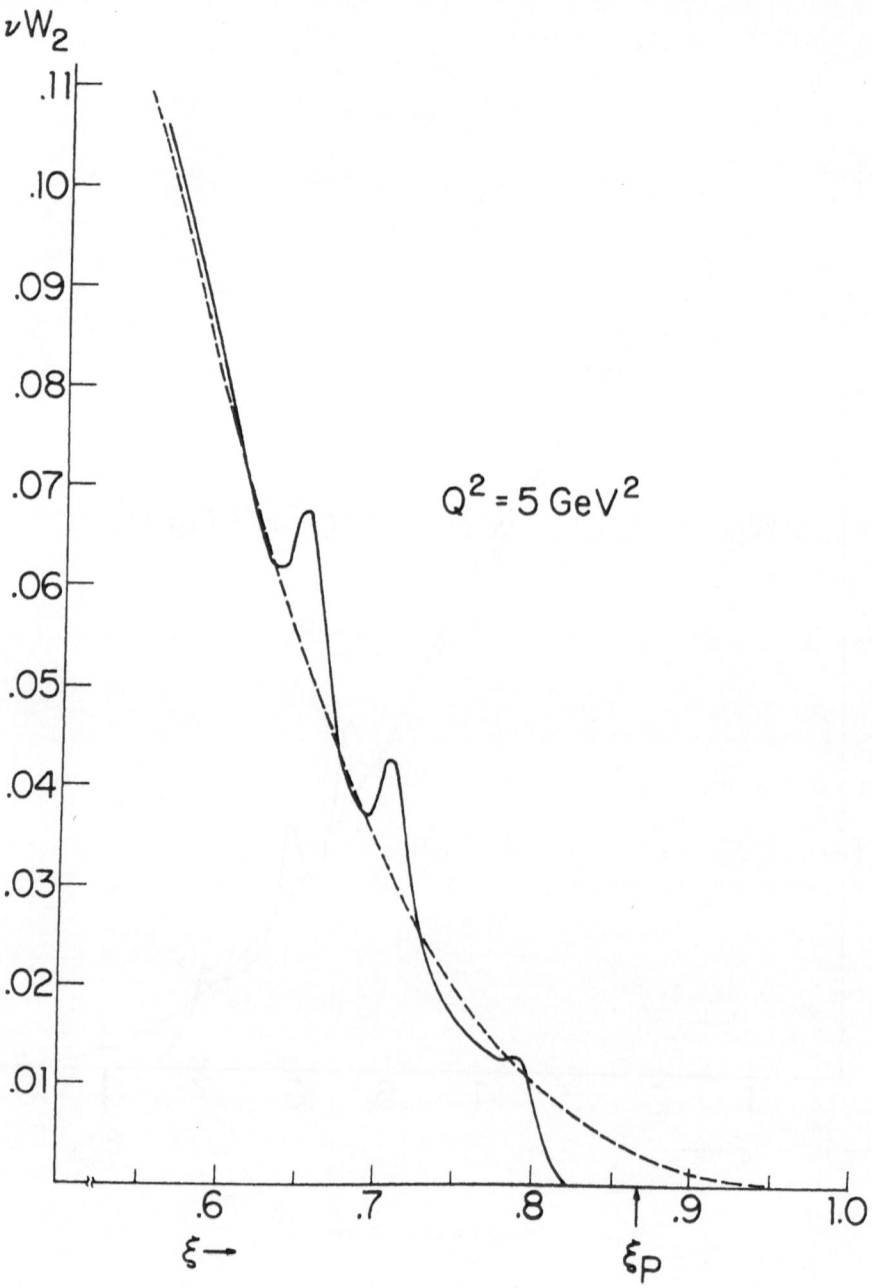

Figure 5d.

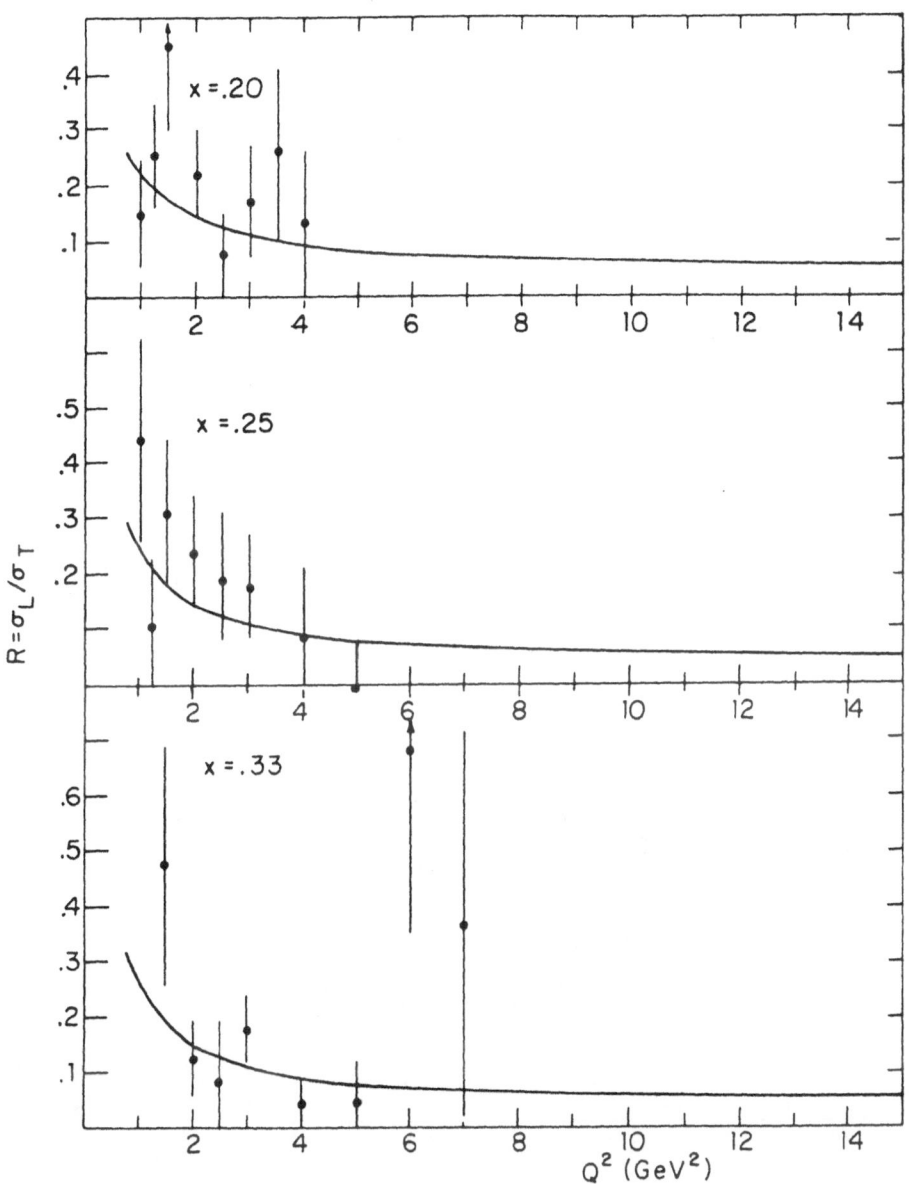

Figure 6a. $R = \sigma_L/\sigma_T$ versus Q^2 for various x.   The data
is from Ref. 17.   The prediction has uncertainties of
order $\alpha_s(Q^2)M_0^2/Q^2$.

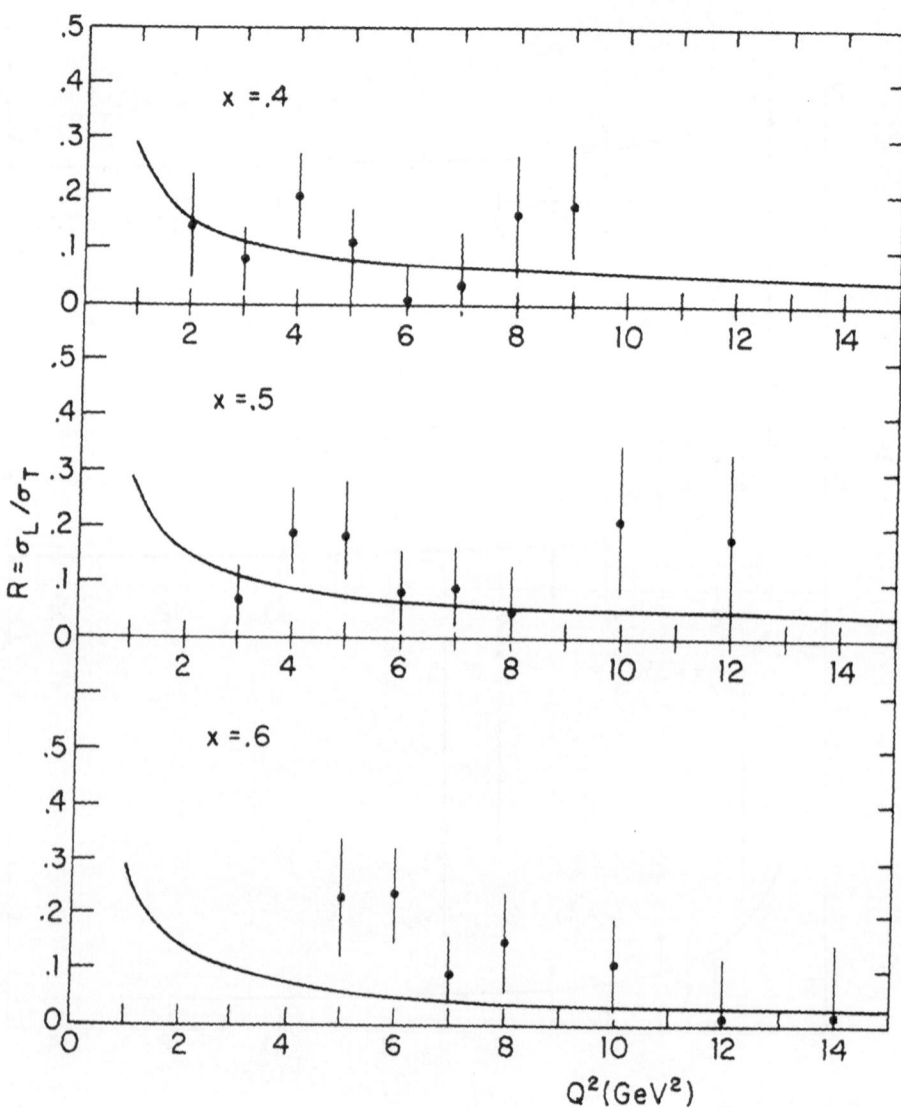

Figure 6b.

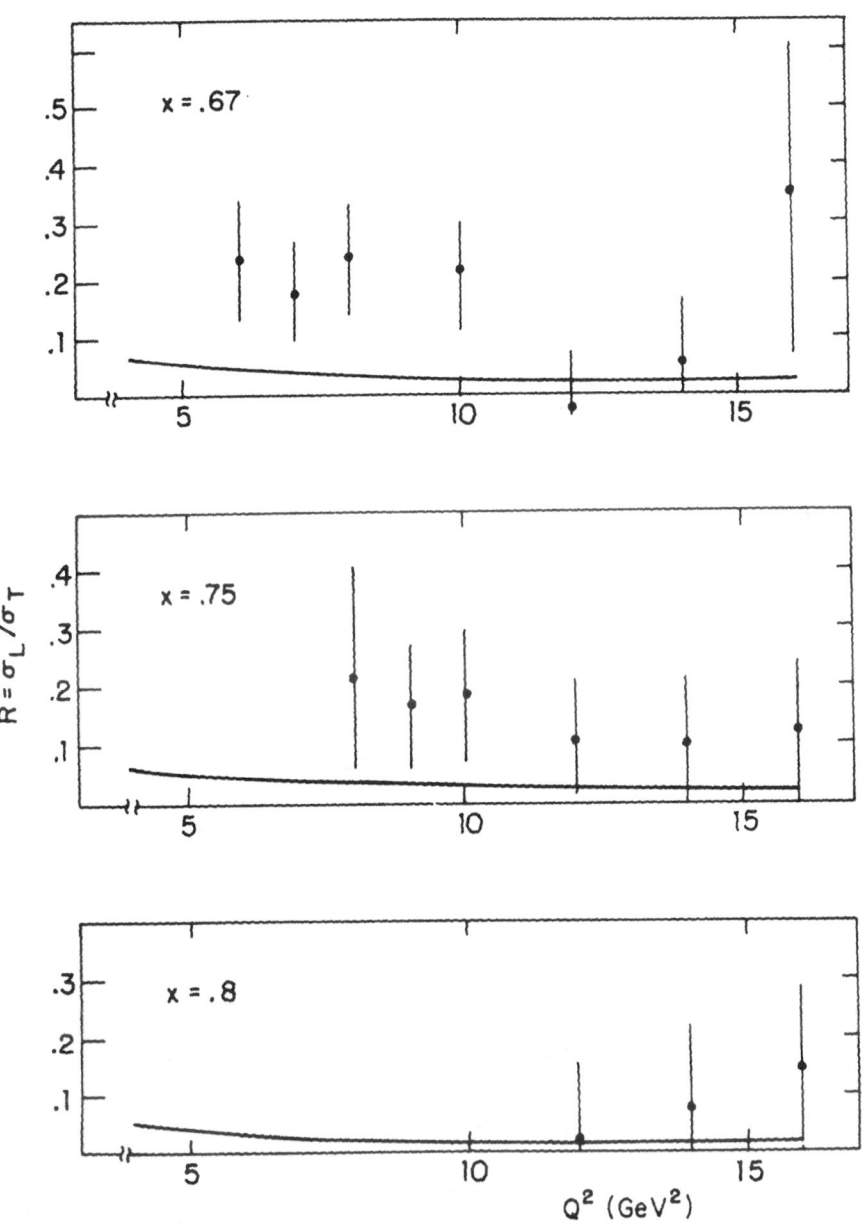

Figure 6c.

That analysis similarly rests on the dimensional analysis estimates of matrix elements. The phenomenological
importance of this analysis is, to put it modestly, in-
estimable: you can't deduce what heavy quarks are being
produced without doing the kinematics right.

A long-standing paradox concerning local duality
and scalar targets has been resolved. Its resolution,
known to many people but not published elsewhere to
my knowledge, rests on the consideration of transition
as well as elastic form factors.[2]

Given a fit to νW_2, QCD makes a fairly definite
prediction for $R = \sigma_L/\sigma_T$. We have calculated the term
in R of order m^2/Q^2 and the term of order $g^2(Q^2)$. There
is no term of order M_0^2/Q^2, so the first missing piece
in our prediction is order $g^2(Q^2)M_0^2/Q^2$. Figure 6 shows
that prediction.[2] A possible interpretation of the g^2
(Q^2) piece (which comes from the graph in Fig. 7) is that
gluon bremsstrahlung gives the quarks a $\langle p_\perp^2 \rangle \sim Q^2/\log Q^2$
independent of the "primordial" $p_\perp$ distribution, which is
probably $\langle p_\perp^2 \rangle \sim M_0^2$.

Does This Prove QCD?

No.

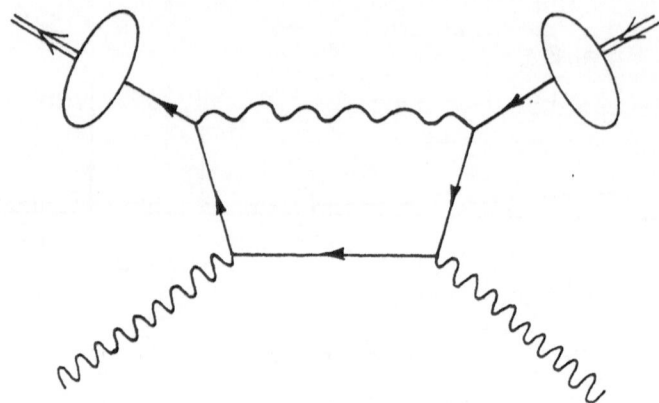

Figure 7.   Feynman graph for the $O(\alpha_s)$ contribution to
σ_L/σ_T.

Small x

The qualitative prediction of QCD is shown in Fig. 8. But a quantitative theory of the rise at small x requires knowledge of the gluon distribution. This is because both graphs of Fig. 9 are the same order in $g^2(Q^2)$. Geoffrey Fox has been able to fit the available data using a plausible gluon distribution.[19] But I wish to make some quantitative statements in which the effects of the gluons are calculable.

First, allow me to knock over a straw man. Some people have suggested that while the qualitative behavior of Fig. 8. appears to be true it does not confirm asymptotic freedom. They suggest instead that what is being seen is scaling in x' or x_s which "explains" the decrease for $x \gtrsim 0.2$; the increase at small x is then supposed to be due to the production of new quantum numbers out of the "sea". If this is the case, then at $Q^2 \approx 10$ GeV^2 14% of the cross section must contain new quantum numbers. QCD predicts about $2 \pm 1\%$ charm and no observable amount of stuff made of possible 4 GeV or heavier quarks.[1,20] From preliminary reports of di- and tri-muons in μ scattering, which suggest about 1% charmed events for $\langle Q^2 \rangle \sim 10$ GeV^2,[21] I think QCD is in better shape.

Energy-Momentum Sum Rule

One of the three spin-2, twist-2 operators is the energy-momentum tensor. Its proton matrix element can be measured with a mass spectrometer. The quark contribution to the energy-momentum tensor can be determined from νW_2 at a single Q_0^2 using e, μ and ν projectiles. So the gluon contribution is the difference. Having measured all n = 2 twist-2 matrix elements at a single Q_0^2, we can predict what will happen for any other Q^2 (large).[4]

The integral of reasonably direct theoretical

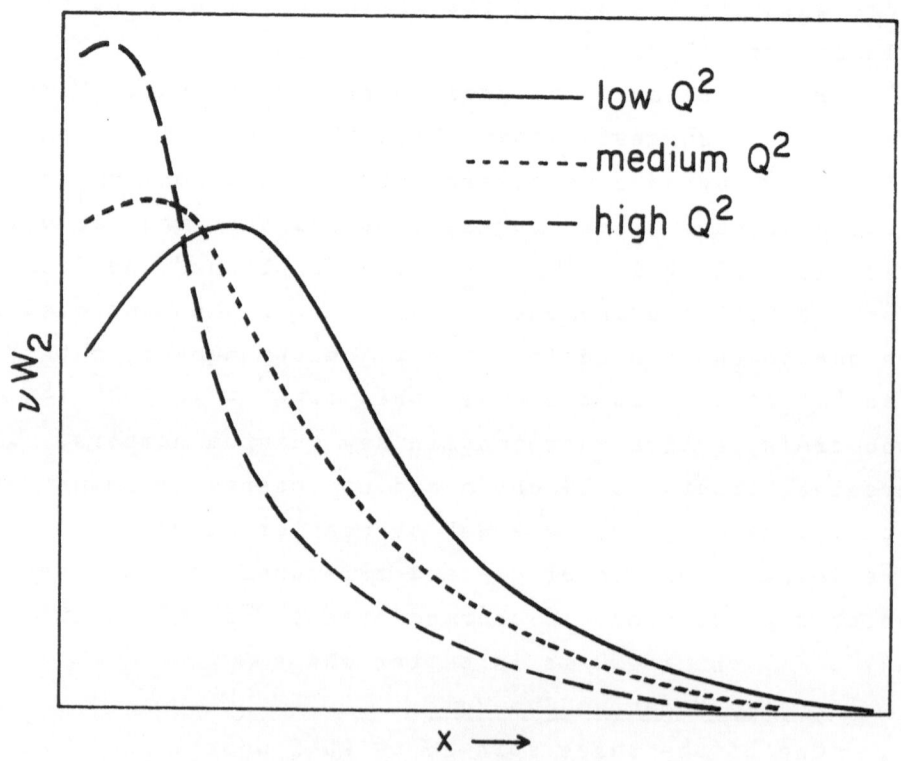

Figure 8.   The qualitative prediction for scaling violations of νW_2 in QCD.

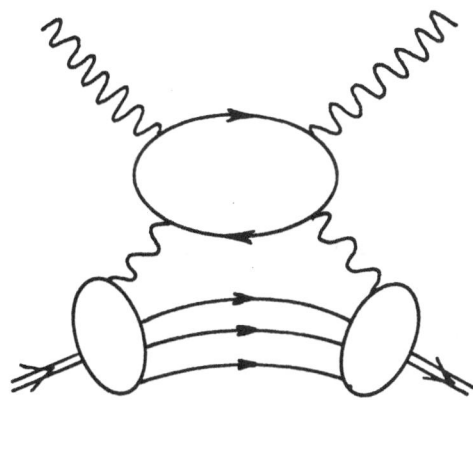

(a)

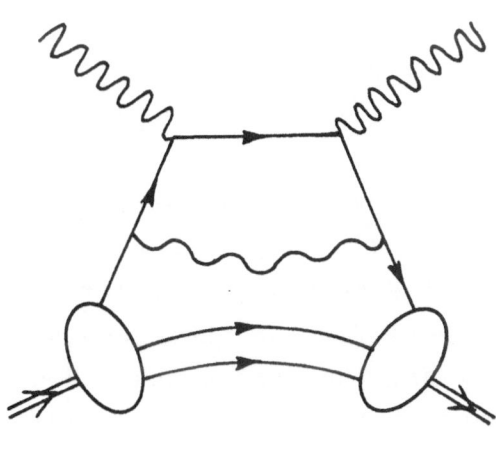

(b)

Figure 9. (a) The graph which gives scaling violations
in scattering off gluons. (b) The graph which gives scaling violations in scattering off quarks.

importance is

$$\int_0^1 (1+4x^2\,\frac{m^2}{Q^2})^{3/2}\,\frac{\xi^2}{x^2}\,\nu W_2(\xi,Q^2)d\xi. \qquad (9)$$

The integral from 0 to 0.25 I deduced from the μ-p data[22] exhibited in Fig. 10. The integral from 0.25 to 1 I took from our Λ = 500 MeV fit to SLAC e-p data. The two regions make roughly equal contributions.

There are three terms in the Q^2 dependence:

$$\int_0^1 F_2 d\xi = \sum_i Q_i^2(Q_i+\bar{Q}_i)\left[\frac{\log Q^2/\Lambda^2}{\log Q^2/\Lambda^2}\right]^{-d_1}[1-a_1 g^2(Q^2)]$$

$$-\text{Glue}\sum_i Q_i^2\left[\frac{\log Q^2/\Lambda^2}{\log Q^2/\Lambda^2}\right]^{-d_2} a_2 g^2(Q^2) \qquad (10)$$

$$+\Delta(Q_i)\left[\frac{\log Q^2/\Lambda^2}{\log Q^2/\Lambda^2}\right]^{-d_3}[1-a_3 g^2(Q^2)].$$

To determine the relative coefficients we need not only $F_2(\xi,Q_0^2)$ and the mass of the proton but also a way to determine the relative abundances of different types of quarks. In a standard fashion these can be deduced from ν and $\bar{\nu}$ scattering and the proton-neutron difference.[23] Allowing ± 100% uncertainties in these estimates for anti-quarks and the u-d difference, I find the gluons carry 30 ± 8% of the total energy-momentum as measured by Q^2 = 3.5 GeV^2. At Q^2 = 1 GeV^2, this is only 22 ± 9%. Note that the a_i terms in Eq. (10), which reflect gluon brems-strahlung and quark pair creation, have a significant effect on this determination. The consequent prediction

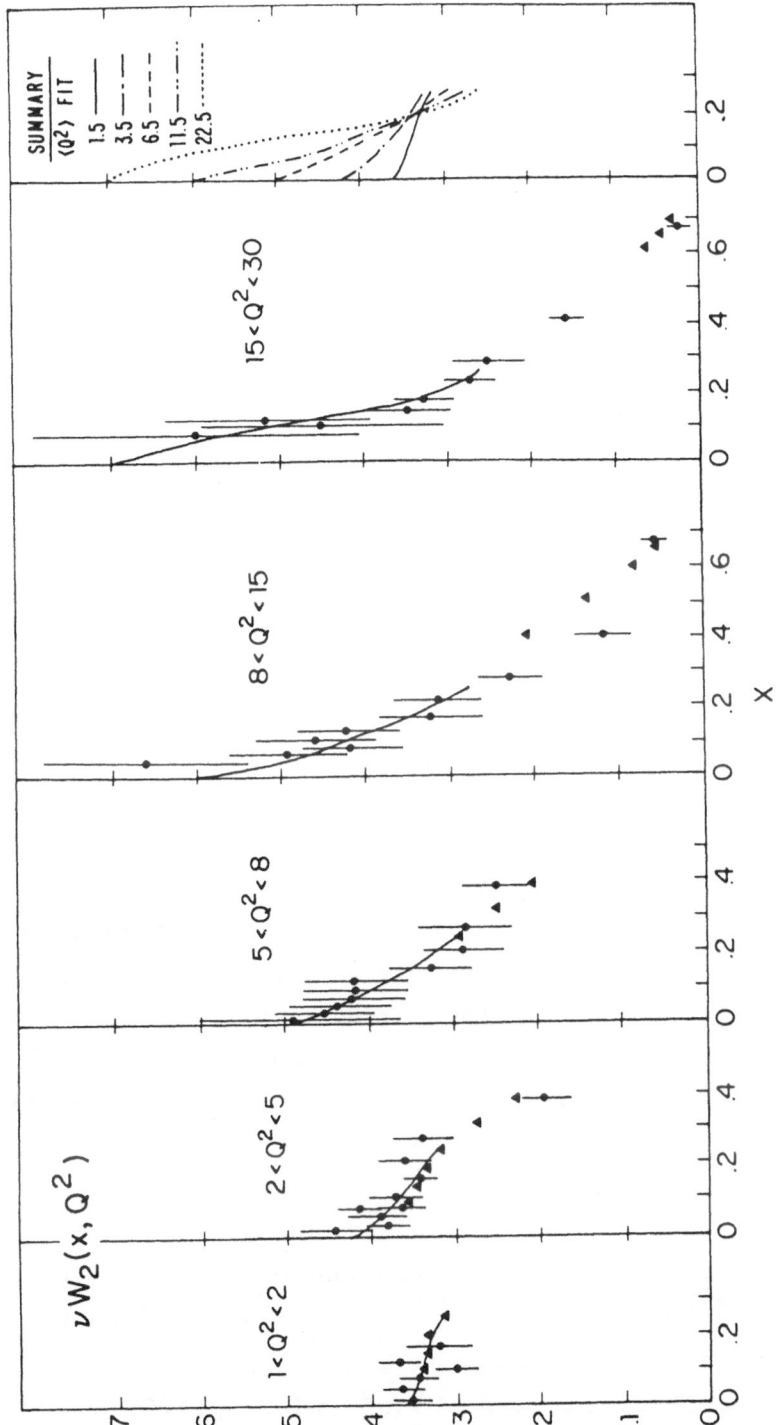

Figure 10. $\nu W_2(x, Q^2)$ versus x in Q^2 bins from Ref. 22. The solid lines are hand-drawn fits for the purpose of x integration.

for the Q^2 - evolution of the energy-momentum sum rule is compared to the data in Fig. 11.

Knowing the area of the glue distribution and assuming that it is confined to $\xi < 0.25$, I computed the glue contribution to σ_L/σ_T. $\bar{R}$ defined by

$$R = \frac{\int_0^{.25} W_L d\xi}{\int_0^{.25} W_T d\xi} \tag{11}$$

is shown in Fig. 12.

The Neutron

It would be a nice test of these ideas if someone carried out this whole analysis for the neutron. There the scaling violations are known to be weaker. That is in fact what the theory predicts, but it should be worked out in detail to show that the same Λ and M_0 describe the neutron.

Of course, it would be really nice if someone would calculate νW_2 from scratch.

ξ-Scaling in Parton Language

Apparently paradoxical conculsions may be reached in a parton model description of our analysis. Scaling in the ξ-variable can be obtained in a parton approach that keeps terms of order m^2/Q^2. This is done with the usual assumptions (the impulse approximation, on-shell quarks with zero mass), but a new prescription is necessary to modify the statement that $p_\mu(\text{quark}) = xp_\mu(\text{target})$, which is inconsistent if $m_p \neq 0$. The correct field-theoretic result is obtained with the prescription p^+ $(\text{quark}) = \xi p^+(\text{target})$, where $p^\pm = (p_0 \pm p_3)$ are "light-cone variables". This leads to ξ-scaling, with $\xi = -q^2/p_+q_-$, provided that effects of order $\langle p_T^2 \rangle /Q^2$ are

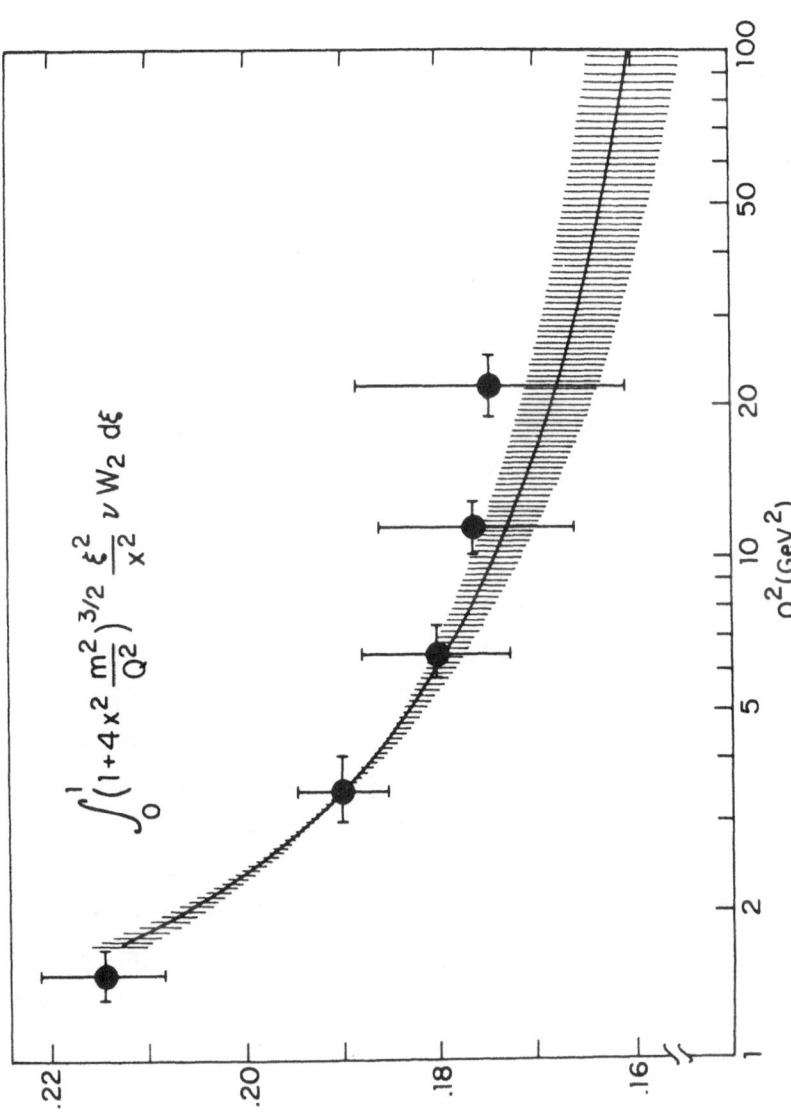

Figure 11. The energy-momentum sum rule versus Q^2. The prediction is normalized at $Q^2 = 3.5$ GeV^2; the shaded region reflects the uncertainty in the glue fraction.

neglected.   Here $\vec{p}_T$ is the transverse parton momentum
in the collinear photon-target frame.   We expect the
quantity $<p_T^2>$ to be related to the mean square radius
of the target, $<p_T^2> \approx 3^{2/3} <r_p^2>^{-1} = (400 \text{ MeV})^2$ for a
proton.

Paradoxes arise as follows: exact ξ-scaling is
obtained in the approximation $<p_T^2>/Q^2 = 0$.   If $<p_T^2> \equiv$
0, the quarks cannot be confined in the transverse direction.   Consequently, they are not confined in any
direction, i.e. they are free.   But if the proton is
simply a system of three, free quarks (at rest in the
proton rest frame), then $\nu W_2 \propto \delta(x-1/3)$.   Hence it is
internally inconsistent to assume that νW_2 is anything
but a δ-function once we have assumed $<p_T^2> = 0$.   To
resolve this paradox we must ask what are the conse-
quences of a small $<p_T^2> \sim (400 \text{ MeV})^2$.   While $<p_T^2>/Q^2$
may be tiny at some large Q^2, the fact of confinement as
evidenced by $p_T^2 \neq 0$ may drastically alter the appearance
of νW_2, no matter how large Q^2 may be.   As a qualitative
but specific example consider the following "bag" calculation.   Jaffe has computed what νW_2 would look like for
a proton made of three noninteracting massless quarks
confined to a sphere of radius R = 1 fermi.[16]   The δ-
function νW_2 typical of free quarks is broadened by
confinement.   In Jaffe's approximation, νW_2 is Q^2-in-
dependent and has a shape quite like the observed structure function (see Fig. 13).   In the same model the momentum per quark is 2.02/R ~ 400 MeV.   Thus there will
be a range in Q^2 for which $<p_T^2>/Q^2$ scaling violation
effects are small compared to m_{proton}^2/Q^2.   However, the
effects of nonvanishing $<p_T^2>$ on the quark distribution
functions are not small.

In the OPE analysis, the twist expansion gives a

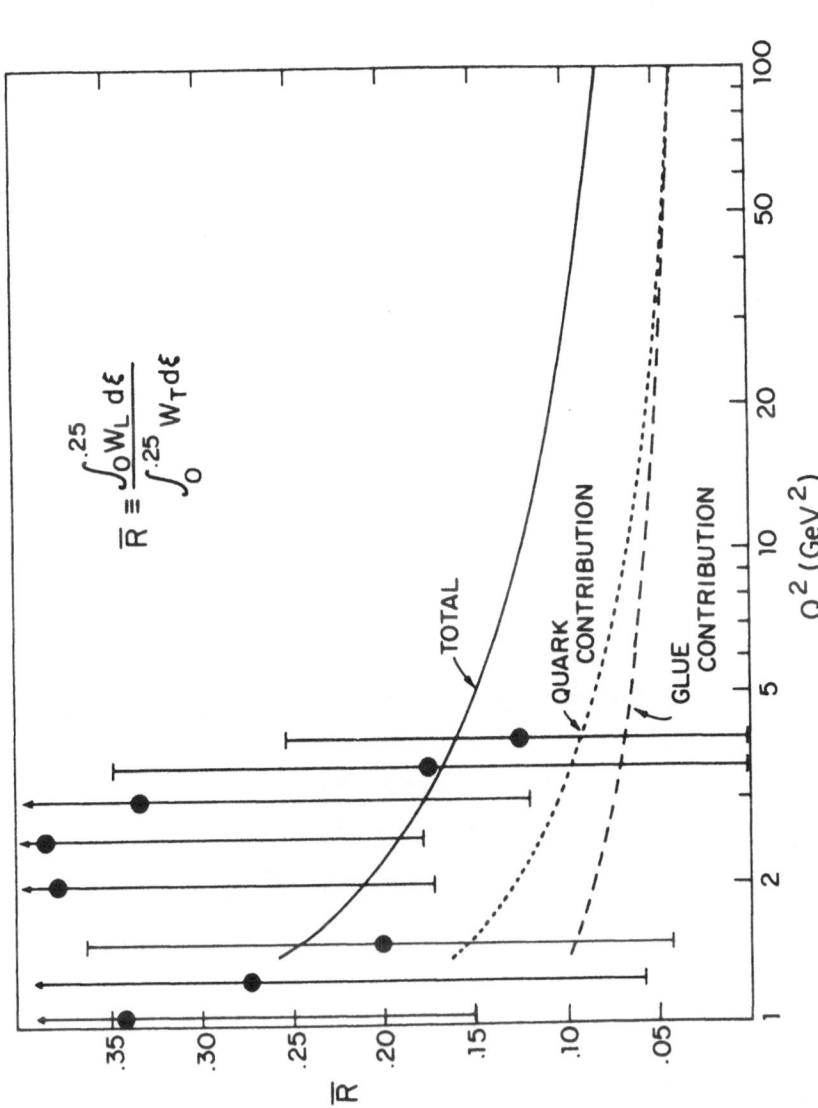

Figure 12. $\bar{R}$ versus Q^2 for $\Lambda = 500$ MeV and 30% glue at $Q^2 = 3.5$ GeV^2. The data is from Ref. 17.

description of ξ-moments of structure functions in
powers of M_0^2/Q^2 with twist-2n corresponding to the $(M_0^2/Q^2)^{2n-2}$ term. The leading (twist-2) contribution corresponds to the parton model impulse approximation.
The log Q^2/Λ^2 interaction corrections to twist-2 correspond to vertex corrections and gluon bremsstrahlung
in the photon-struck quark interaction (see Fig. 1).
In what follows we develop an intuitive understanding of
the phenomena related to twist greater than two (see
Fig. 1). Particular attention is given to twist-4 which
we were able to study phenomenologically.[2]

   There are wavefunction (or initial state) effects
and final state effects contained in the terms of twist-4
and higher. To get a feel for the initial state effects,
consider the following parton discussion. The quarks will
in general have a $p_T \neq 0$ and be off-shell by some amount
proportional to $\langle p_T^2 \rangle$. Let their distribution be described
by a function $f_i(z, p_T^2)$, where z is the momentum fraction
carried by the quark in the +, light-cone direction (as
in the leading $p_T = 0$ analysis). Repeat the parton model
impulse approximation calculation to get the structure
function as an incoherent sum over the constituent partons. Integrate over z and p_T to obtain

$$F \propto \sum_i Q_i^2 \int dz \, dp_T^2 f_i(z, p_T^2) \delta(z - \xi_{eff}) = \int dp_T^2 f(\xi_{eff}, p_T^2),$$

(12a)

where

$$\xi_{eff} = \xi \left[1 + \frac{p_T^2}{Q^2} + 0\left(\frac{p_T^4}{Q^4}\right) \right].$$ (12b)

Expand in powers of p_T^2/Q^2 to find

$$F \propto \int dp_T^2 f(\xi, p_T^2) + \int dp_T^2 \frac{p_T^2}{Q^2} \xi \frac{\partial}{\partial \xi} f(\xi, p_T^2) + 0\left(\frac{\langle p_T^4 \rangle}{Q^4}\right).$$ (13)

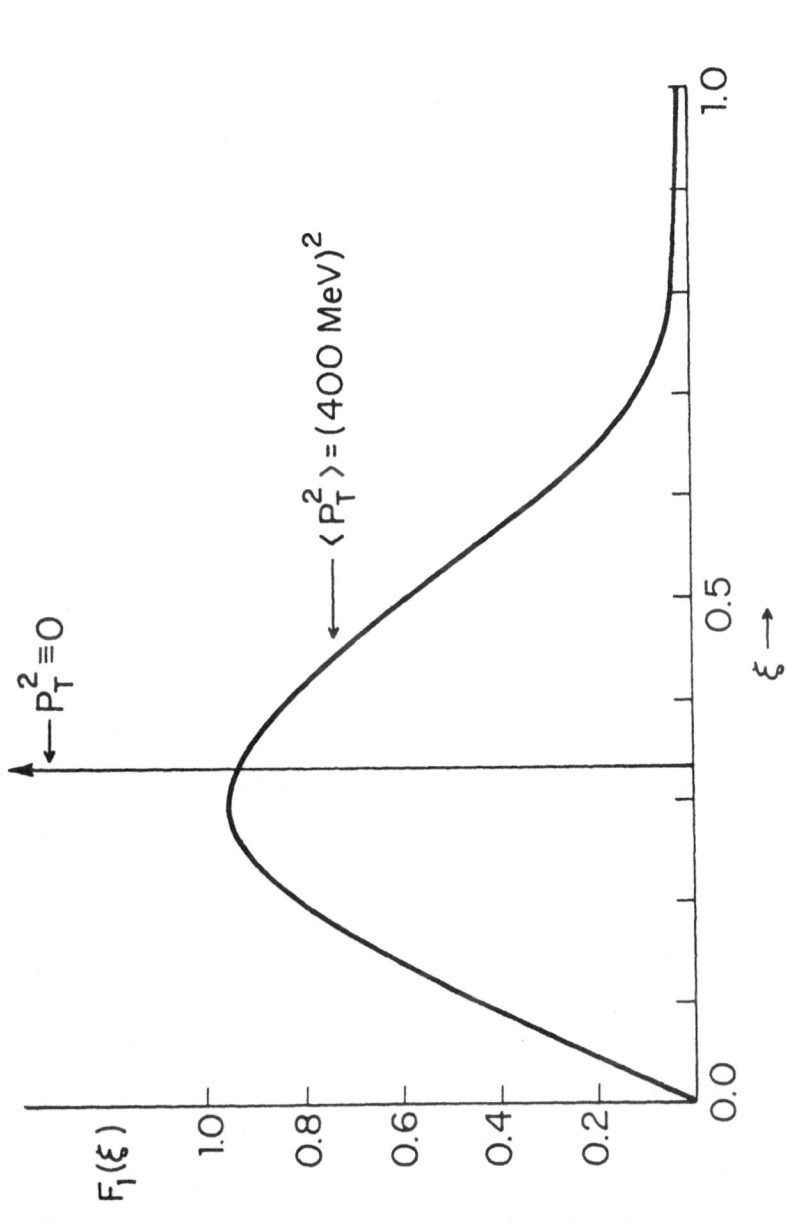

Figure 13. νW_2 versus x in a bag model with radius 1 fermi [or $\langle p_\perp^2 \rangle = (400$ MeV$)^2$] and with infinite radius [or $\langle p_\perp^2 \rangle = 0$].

Compute the ξ moments of the structure function F:

$$\int_0^1 d\xi \; \xi^n F(\xi,Q^2) = \int_0^1 d\xi \; \xi^n \int dp_T^2 f(\xi,p_T^2)$$

$$- (n+1) \int_0^1 d\xi \; \xi^n \int dp_T^2 \; \frac{p_T^2}{Q^2} \; f(\xi,p_T^2) + 0 \left(\frac{\langle p_T^2 \rangle^2}{Q^4} \right) \; , (14)$$

where to obtain the second term we have integrated by parts and discarded surface terms. If

$$\frac{\int_0^1 d\xi \; \xi^n \int dp_T^2 \; \frac{p_T^2}{Q^2} \; f(\xi,p_T^2)}{\int_0^1 d\xi \; \xi^n \int dp_T^2 f(\xi,p_T^2)} \; \simeq \; \frac{\langle p_T^2 \rangle}{Q^2} \; , \qquad (15)$$

independent of n, Eq. (14) becomes

$$\int_0^1 d\xi \; \xi^n F(\xi,Q^2) = A_n \left[1 - (n+1) \; \frac{\langle p_T^2 \rangle}{Q^2} + \ldots \right] \; . \qquad (16)$$

The effect of a nonzero $\langle p_T^2 \rangle$ is of the twist-four form, but lo! it even has the n-dependence obtained from QCD.

Keeping successively higher powers of p_T^2 in the derivation leading to Eq. (16) will introduce powers of $1/Q^2$ times higher moments of the p_T distribution. These moments of the p_T distribution as a function of ξ are contained in the matrix elements of successively higher twists (just as twist-two operators give the ξ distribution times p_T^0.)

Higher twists (>2) also include final state interactions (with high twist the initial-final distinction

breaks down). The most significant final state effect
is due to the exchange of gluons between the struck and
spectator quarks. This is an attractive interaction because the final state has the color quantum numbers of
the target. We know this attraction has a particularly
striking, nonscaling effect for small W, i.e. $1 \lesssim W \lesssim 2$
GeV (at high W the colors are presumably shielded by
pair production). An attractive final state interaction
increases the cross section; so moments of the structure
functions will have the form

$$A_n \left(1 + n \frac{\mu^2}{Q^2} + \ldots \right), \qquad (17)$$

where the + sign is dictated by the attraction, and the
form $n\mu^2/Q^2$ is determined by the phenomenon being fixed
in W; μ is some small mass, corresponding to $1 \lesssim W \lesssim 2$
GeV.

The OPE analysis suggests

$$A_n \left(1 + \frac{B_{n,1}}{A_n} n \frac{M_0^2}{Q^2} + \ldots \right). \qquad (18)$$

A *priori*, we expect the parton parameters $\langle p_T^2 \rangle$, μ^2 and
the QCD parameters $B_{n,1} M_0^2 / A_n$ to all be of order Λ^2, where
Λ sets the coupling strength and was determined to be
500 ± 200 MeV from the logarithmic scaling violations in
electroproduction. Indeed, we found that $B_{n,1} M_0^2 / A_n =$
(350 - 400 MeV)^2, independent of n over the range $0 \leq n$
≤ 30. The observed + sign in Eq. (18) suggests that final
state interactions are in fact more important than the
initial $p_T \neq 0$ in determining $1/Q^2$ effects.

REFERENCES

1. H. Georgi and H. D. Politzer, Phys. Rev. $\underline{D14}$, 1829 (1976) and Phys. Rev. Lett. $\underline{36}$, 1281 (1976).

2. A. De Rújula, H. Georgi and H. D. Politzer, Ann. Phys. (to be published) and Phys. Lett. $\underline{64B}$, 428 (1976).

3. A. De Rújula, H. Georgi and H. D. Politzer, "Trouble with ξ-Scaling?", Harvard preprint HUTP-76/A179 (12/76).

4. H. D. Politzer, "μ-P Scattering and the Glue Fraction of the Proton", Harvard preprint HUTP-77/A001 (1/77).

5. V. Baluni and E. Eichten, Institute for Advanced Study Report No. COO-2220-76 (1976).

6. R. K. Ellis, G. Parisi and R. Petronzio, "Mass Dependent Corrections to the Bjorken Scaling Law", University of Rome preprint (1976).

7. R. Barbieri, J. Ellis, M. K. Gaillard and G. G. Ross, Phys. Lett. $\underline{64B}$, 171 (1976) and CERN preprint TH2223-CERN (1976).

8. D. J. Gross, S. B. Treiman and F. A. Wilczek, "Mass Corrections in Deep Inelastic Scattering", Princeton preprint (1976).

9. E. D. Bloom and F. J. Gilman, Phys. Rev. $\underline{D4}$, 2901 (1971).

10. K. Wilson, Phys. Rev. $\underline{179}$, 1499 (1969).

11. R. Brandt and G. Preparata, Nucl. Phys. $\underline{B27}$, 541 (1971).

12. N. Christ, B. Hasslacher and A. Mueller, Phys. Rev. $\underline{D6}$, 3543 (1972).

13. H. D. Politzer, Phys. Reports $\underline{14C}$, 129 (1974).

14. O. Nachtmann, Nucl. Phys. $\underline{B63}$, 237 (1973).

15. D. J. Gross and S. B. Treiman, Phys. Rev. D4,1059 (1971).

16. R. Jaffe, Phys. Rev. D11, 1953 (1975).

17. E. Riordan *et al.*, SLAC-PUB 1634 (8/75).

18. See R. M. Barnett's and H. Georgi's contributions to this conference.

19. G. C. Fox, unpublished.

20. E. Witten, Nucl. Phys. B104, 445 (1976).

21. K. W. Chen in *Proceedings of the International Conference on New Particles with New Quantum Numbers* (University of Wisconsin, Madison, 1976).

22. H. Anderson *et al.*, "Measurement of the Proton Structure Function", submitted to the conference proceedings, Tbilisi, 1976.

23. D. H. Perkins, in *XVI International Conference on High Energy Physics - Chicago 1972* (NAL, Batavia, 1972).

MUON NUMBER NONCONSERVATION IN GAUGE THEORIES

Ta-Pei Cheng[*] (Presented by Ta-Pei Cheng)

University of Missouri-St. Louis

St. Louis, MO   63121

and

Ling-Fong Li[**]

Carnegie-Mellon University

Pittsburgh, PA   15213

Recent rumors of possible observation at SIN (the Swiss Institute of Nuclear Research) of the famous $\mu \to e\gamma$ process has aroused a great deal of excitement. But the experimentalists themselves have so far not made any public statements; presupposition on our part about their actual result will of course be totally inappropriate. However, regardless of the ultimate outcome of this particular experiment the question of separate conservation of muon and electron number is an interesting and important problem to investigate, especially in the context

[*] Work supported by the National Science Foundation.

[**] Work supported by the Energy Research and Development Administration.

of unified gauge theories of weak and electromagnetic
interactions.

In this talk we shall describe the result of our
work[1] which was completed over a month ago, before we
were made aware of the news of possible detection of
this process. We shall also briefly discuss some of
the very interesting papers that have since been completed by authors at Stanford[2,3], Princeton,[4] and
Rockefeller[22]. All are $SU_2 \times U_1$ gauge theories that
yield a small rate for $\mu \to e\gamma$ without introducing arbitrarily small parameters. Of course theories free of
these restrictions can also be constructed; in fact,
one can anticipate many papers of such nature in the
near future.[5]

A. THEORIES WITH HEAVY NEUTRAL LEPTONS

The observation made by us in Ref. 1 is a very
simple one: In gauge theories, eigenstates of weak
interactions being generally not eigenstates of the
mass matrix, states with the same charge and helicity
will mix. Thus one should not expect muon and electron
number conservation; just as we have a nonvanishing
Cabibbo angle and strangeness is not conserved in weak
interactions. The apparent conservation of muon number
in the standard V-A theory should be interpreted as re-
flecting the fact that neutrino masses (if not identically zero) are almost degenerate when viewed on the
normal mass scale. In theories containing V+A currents,
the right handed muon and electron are expected to couple to intermixing heavy leptons in the GeV range. In
such theories muon number violation effects will be dra-
matically larger. In fact, we found that a reasonable

lepton mass difference would lead us to predict branching ratios for $\mu \to e\gamma$ and $K_L \to e\mu$ comparable to their present experimental limits.

First we will illustrate our point about the relationship between muon number violation effects and the lepton mass spectrum by considering the standard V-A theory of Weinberg and Salam[6] except that we will assume that neutrinos are not massless.

$$\begin{pmatrix} \nu_e \\ e \end{pmatrix}_L \quad \begin{pmatrix} \nu_\mu \\ \mu \end{pmatrix}_L \quad e_R, \ \mu_R, \ \nu_{eR}, \ \nu_{\mu R} \ . \qquad (1)$$

Mass eigenstates being generally not eigenstates of the weak interactions would lead us to expect ν_e, ν_μ to be orthogonal linear combinations of neutrino mass eigenstates ν and ν':[7]

$$\nu_e = \cos\theta \ \nu + \sin\theta \ \nu' \quad ,$$

$$\nu_\mu = -\sin\theta \ \nu + \cos\theta \ \nu' \quad . \qquad (2)$$

A priori there is no reason to expect the mixing angle θ to be particularly small. Even so, effects that violate muon number conservation will still be extremely small, because these effects must also be proportional to the neutrino mass differences $m(\nu) - m(\nu')$: In the limit of $m(\nu) = m(\nu')$, the theory is invariant under a rotation in the ν-ν' plane and we can set $\theta = 0$, leading to a muon number conserving gauge interaction.

$\mu \to e\gamma$

The most general form for the $\mu \to e\gamma$ on-shell amplitude is of the form:

$$T_{\mu e\gamma} = \bar{U}_e(p) \, (a+b\gamma_5) \, i\sigma_{\lambda\nu}q^{\nu}U_{\mu}(p')\varepsilon^{\lambda}, \qquad (3)$$

where $q = p-p'$ ($q^2=0$) and ε^{λ} are the photon momentum and polarization vectors respectively. The decay rate is

$$\Gamma_{\mu e\gamma} = \frac{m_{\mu}^3}{8\pi} \, (|a|^2 + |b|^2). \qquad (4)$$

In this theory of Eqs. (1) and (2), the $\mu \rightarrow e\gamma$ process proceeds via the diagrams shown in Fig. 1.

For each neutrino intermediate state, the leading contribution to amplitudes a and b in Eq. (3) is $O(eG_F m_{\mu})$ corresponding to a branching ratio of $\Gamma_{\mu e\gamma}/\Gamma_{\mu} = O(\frac{\alpha}{\pi})$. This is just the famous result known since the late 1950's that in a weak interaction theory with intermediate vector bosons and one neutrino the $\mu \rightarrow e\gamma$ rate would be much too large[8]. This naturally led to the two neutrino hypothesis.

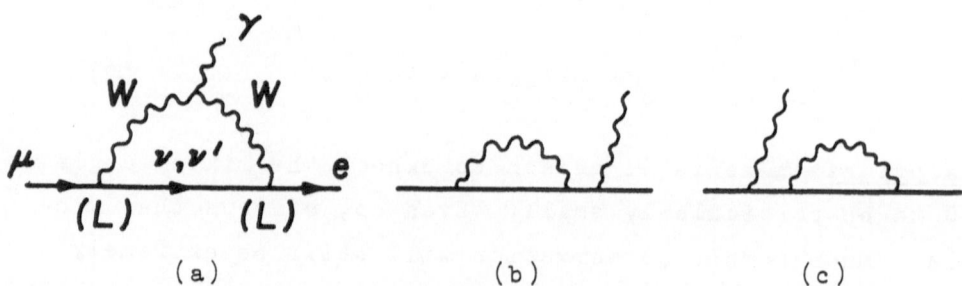

(a)   (b)   (c)

Fig. 1. One loop diagram for the $\mu \rightarrow e\gamma$ decay mediated by intermixing neutrinos. (L) stands for the V-A coupling. One need not explicitly calculate diagrams (b) and (c) since they contribute only to the $\bar{U}_e\gamma_{\lambda}U_{\mu}\varepsilon^{\lambda}$ amplitude which, after summing all graphs, vanishes because of the requirement of electromagnetic current conservation.

Here, because these leading left-left (LL) amplitudes are independent of the masses of the neutrinos ν and ν', and the coupling structures are such that they precisely cancel[9]. Putting it in a language which we shall use later: The leading (LL) amplitude is a "zero-mass-insertion" diagram[10]. If the intermediate fermion states are labeled by the weak eigenstates ν_e or ν_μ, then this diagram clearly must vanish since neither ν_e nor ν_μ couple to μ and e both.

Our task is then to calculate the next leading term which for an (LL) amplitude is a "two-mass-insertion diagram". (Fig. 2)

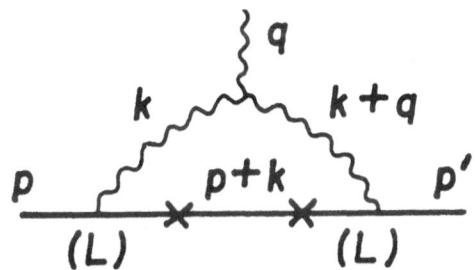

Fig. 2. "Two-mass-insertion" diagram for a (LL) amplitude.

In terms of weak eigenstates the two mass-insertions are: $m(\bar{\nu}_{e_L} \nu_{e_R}) m(\bar{\nu}_{e_R} \nu_{\mu_L})$ [or $m(\bar{\nu}_{e_L} \nu_{\mu_R}) m(\bar{\nu}_{\mu_R} \nu_{\mu_L})$]. The off-diagonal mass term $m(\bar{\nu}_e \nu_\mu)$ cannot be zero in this theory because the mixing angle θ is assumed to be non-zero. In terms of mass eigenstates, the two-mass-insertion corresponds to:

$$\gamma_\mu (1-\gamma_5) [\frac{1}{\not{p}+\not{k}-m} - \frac{1}{\not{p}+\not{k}-m'}] \gamma_\nu (1-\gamma_5)$$

$$\simeq (m^2 - m'^2) \gamma_\mu (1-\gamma_5) \frac{1}{(\not{p}+\not{k})(\not{p}+\not{k})(\not{p}+\not{k})} \gamma_\nu (1-\gamma_5) .$$

(5)

The resulting amplitudes a and b, to leading order in $m^2(\nu) - m^2(\nu') \equiv \Delta m_\nu^2$, are

$$a = b = e\frac{g^2}{8M_W^2} \frac{m_\mu}{32\pi^2} \delta_\nu \quad , \tag{6}$$

with

$$\delta_\nu = \cos\theta\sin\theta \; \Delta m_\nu^2/M_W^2 \quad . \tag{7}$$

From Eq. (4), $g^2/8M_W^2 = G_F/\sqrt{2}$ and $\Gamma_\mu = G_F^2 m_\mu^5/192\pi^3$ being the rate for $\mu \to e\nu\nu$, we obtain a branching ratio of

$$\Gamma_{\mu e\gamma}/\Gamma_\mu = \frac{3\alpha}{32\pi} \delta_\nu^2 \quad . \tag{8}$$

Now δ_ν is an extremely small factor, not necessarily because θ can be small but because $\Delta m_\nu^2/M_W^2$ is probably less than 10^{-20}. This would lead to a branching ratio which for all practical intent and purpose is zero.

One may ask the converse question: given the experimental bound[11] for $\Gamma_{\mu e\gamma}/\Gamma_\mu$ to be 2×10^{-8}, what upper bound of neutrino mass difference do we get? It turns out to be $(\Delta m_\nu^2)^{1/2} < 7$ GeV.[12] While this does not provide us with any useful information about the neutrino mass spectrum, it did occur to us that in a theory with V+A currents the corresponding right-right (RR) diagrams would involve a pair of "heavy neutrinos". Then a reasonable mass difference for the heavy leptons in the GeV range would lead us to expect a rate for $\mu \to e\gamma$ not significantly smaller than the present experiment limit.

For definiteness we propose to consider the following model involving a pair of heavy neutral leptons:

$$\begin{pmatrix} \nu_e \\ e \end{pmatrix}_L, \quad \begin{pmatrix} \nu_\mu \\ \mu \end{pmatrix}_L, \quad \begin{pmatrix} N_e \\ e \end{pmatrix}_R, \quad \begin{pmatrix} N_\mu \\ \mu \end{pmatrix}_R \quad \ldots, \quad (9)$$

where

$$N_e = \cos\phi \; N + \sin\phi \; N' \quad,$$

$$\hspace{6cm} (10)$$

$$N_\mu = -\sin\phi \; N + \cos\phi \; N' \quad,$$

where N and N' are mass eigenstates with masses m and m'. The (RR) amplitude for Eq. (3) is then

$$a^{RR} = -b^{RR} = e\frac{g^2}{8M_W^2} \frac{m_\mu}{32\pi^2} \; \delta_N \quad, \hspace{2cm} (11)$$

with $\delta_N = \cos\phi\sin\phi \; \Delta m_N^2/M_W^2$ and $\Delta m_N^2 = m^2 - m'^2$.

Before proceeding to consider other implications of such a theory, let us first discuss the following important question: is such a theory "natural"? Since N and N' cannot be Majorana particles[13], one should then ask how do the left handed N, N' mix with the neutrinos. Related to this mixing, there are two constraints any physically sensible theory must satisfy:

(i) Universality: The requirement to satisfy the present experimental limit is not very stringent. For the general form of

$$\begin{pmatrix} (1-\theta^2/2)\nu + \theta N \\ \ell \end{pmatrix}_L \hspace{3cm} (12)$$

θ can be as large as 0.1.

(ii) Left-right (LR) amplitude for μ→eγ: If the
heavy neutral leptons N and N' mixes with neutrinos,
then there will be a set of (LR) diagrams similar to
Fig. 1 which can contribute to the decay μ→eγ. But the
leading term is of the one-mass-insertion type, which
will, at first sight, lead to a castastrophically large
branching ratio:

$$\sim \frac{\alpha}{\pi} \left(\frac{\sin\phi \ f'm' + \cos\phi \ fm}{m_\mu} \right)^2, \qquad (13)$$

where f, f' are the couplings between left handed muon
and N, N' respectively. A viable theory must have f and
f' such that the expression in Eq. (13) is (if not zero)
vanishingly small.

These problems concerning naturality led Wilczek
and Zee[4] to propose an alternative set of theories involving doubly charged heavy leptons (which we shall
discuss briefly at the end of this talk). That these
difficulties have a natural resolution within the context of our original heavy neutrino schemes were noted
by Bjorken, Lane and Weinberg[3]. These authors made the
following interesting observation about the theory in
Eq. (9). If the charged lepton mass arises from gauge
invariant bare terms and/or the vacuum expectation values
of singlet Higgs scalars, then the mixings between the
left handed neutrinos and heavy leptons will be uniquely
determined. These mixings are such that they automatically satisfy the two constraints discussed above
[universality and μ→eγ (LR) amplitudes].

With a right handed mixing given as

$$\begin{pmatrix} \cos\phi \ N \ + \ \sin\phi \ N' \\ e \end{pmatrix}_R \quad \begin{pmatrix} -\sin\phi \ N \ + \ \cos\phi \ N' \\ \mu \end{pmatrix}_R \quad (14)$$

their result can then be obtained after a straightforward exercise in diagonalizing a 2×4 mass matrix. It reads as follows:[14]

$$\begin{pmatrix} \alpha\nu + \beta\nu' \ + \ \dfrac{m_e}{m} \cos\phi \ N \ + \ \dfrac{m_e}{m'} \sin\phi \ N' \\ e \end{pmatrix}_L$$

$$\begin{pmatrix} \gamma\nu + \delta\nu' \ - \ \dfrac{m_\mu}{m} \sin\phi \ N \ + \ \dfrac{m_\mu}{m'} \cos\phi \ N' \\ \mu \end{pmatrix}_L \quad (15)$$

plus two orthogonal left handed singlets.

It is clear that the universality constraint can be satisfied easily and the expression in Eq. (13) is identically zero. The origin of this rather remarkable cancellation of the one-mass-insertion diagram may be understood in a number of ways. The following argument is perhaps the most transparent.

The fermion content of this theory is

$$\begin{pmatrix} n_1 \\ e \end{pmatrix}_L \quad \begin{pmatrix} n_2 \\ \mu \end{pmatrix}_L \quad n_{3_L}, \ n_{4_L}, \quad \begin{pmatrix} n_1 \\ e \end{pmatrix}_R \quad \begin{pmatrix} n_2 \\ \mu \end{pmatrix}_R . \quad (16)$$

The triplet Higgs scalars are assumed to be absent, the mass matrix for neutral fermions (before diagonalization) takes on the form of

$$m(n_{iL}n_{jR}) = \begin{pmatrix} m_{11} & 0 \\ 0 & m_{22} \\ m_{31} & m_{32} \\ m_{41} & m_{42} \end{pmatrix} . \qquad (17)$$

In particular we note that $m_{11} = m_e$ and $m_{22} = m_\mu$ and
$m_{12} = m_{21} = 0$ since whatever diagonalizes the mass matrix
for the charged leptons automatically diagonalizes the
(1,2) sector of the neutral mass matrix as well. The
one-mass-insertion diagram clearly vanishes since in
terms of these weak eigenstates it must involve a m_{21}
term as shown in Fig. 3.

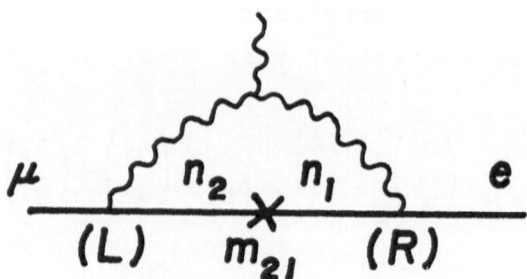

Fig. 3.   One-mass-insertion diagram for $\mu \to e\gamma$.

We have calculated the next leading term, which corresponds to a three-mass-insertion diagram, and obtain

$$a^{LR} = -b^{LR} = -6e \; \frac{g^2}{8M_W^2} \; \frac{m_\mu}{32\pi^2} \; \delta_N \; . \qquad (18)$$

Independent calculation by Bjorken et al.[3] agrees with
this result. This, when combined with the amplitude in
Eq. (11), leads to a rate for $\mu \to e\gamma$ as

$$\Gamma_{\mu e\gamma} = \frac{m_\mu^3}{8\pi} \; (|a^{RR} + a^{LR}|^2 + |b^{RR} + b^{LR}|^2)$$

$$= \frac{75}{32} \; \frac{\alpha}{\pi} \; \delta_N^2 \; \Gamma_\mu \; . \qquad (19)$$

For $(\cos\phi\sin\phi \; \Delta m_N^2) = 1 \; GeV^2$ and $M_W = 50 \; GeV$ we would have
a branching ratio $\Gamma_{\mu e\gamma}/\Gamma_\mu \approx 10^{-9}$.

To show that our one loop calculation[15] is (nonabelian) gauge invariant, we display in Table 1 the detailed results in the general R_ξ gauge[16].

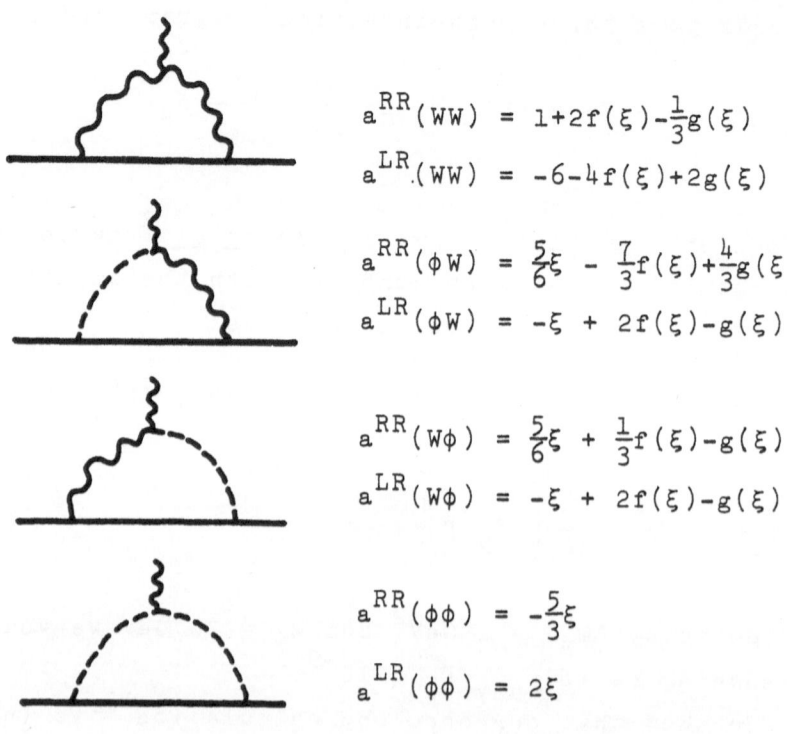

$$a^{RR}(WW) = 1+2f(\xi)-\frac{1}{3}g(\xi)$$

$$a^{LR}(WW) = -6-4f(\xi)+2g(\xi)$$

$$a^{RR}(\phi W) = \frac{5}{6}\xi - \frac{7}{3}f(\xi)+\frac{4}{3}g(\xi)$$

$$a^{LR}(\phi W) = -\xi + 2f(\xi)-g(\xi)$$

$$a^{RR}(W\phi) = \frac{5}{6}\xi + \frac{1}{3}f(\xi)-g(\xi)$$

$$a^{LR}(W\phi) = -\xi + 2f(\xi)-g(\xi)$$

$$a^{RR}(\phi\phi) = -\frac{5}{3}\xi$$

$$a^{LR}(\phi\phi) = 2\xi$$

Table 1.   Detailed result for the $\mu \to e\gamma$ amplitudes in R_ξ gauge given in units of $e\dfrac{g^2}{8M_W^2}\dfrac{m_\mu}{32\pi^2}\delta_N$.   W and ϕ stand for intermediate vector boson and would-be-Goldstone boson respectively. $f(\xi) = \dfrac{\xi}{\xi-1}(1-\dfrac{\ln\xi}{\xi-1})$ and $g(\xi) = \dfrac{\xi\ln\xi}{\xi-1}$.

$K_L \rightarrow e\mu$, $\mu \rightarrow 3e$, etc.

    We shall mention very briefly the implications of the model in Eqs. (9) and (10) for other μ number violation processes.

    First consider $K_L \rightarrow e\mu$. The free quark diagrams for this process $sd \rightarrow \mu e$ are displayed in Fig. 4.

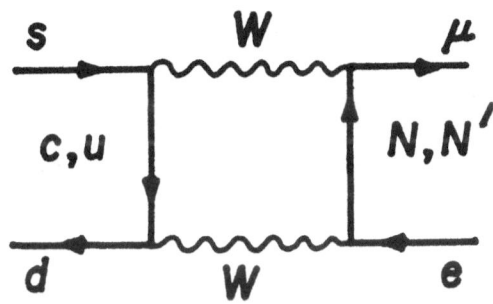

Fig. 4. Free quark diagram which contributes to $K_L \rightarrow e\mu$.

The amplitude for this diagram can be easily calculated

$$T_{sd \rightarrow \mu e} = \frac{g^4}{16\pi^2 M_W^2} \; \delta \, [\bar{d}\gamma^\alpha \tfrac{1}{2}(1-\gamma_5)s] \, [\bar{\mu}\gamma_\alpha \tfrac{1}{2}(1+\gamma_5)e] ,$$

where

$$\delta = (\frac{\sin\theta_c \cos\theta_c \, \Delta m_q^2}{M_W^2})(\frac{\sin\phi\cos\phi \, \Delta m_N^2}{M^2}) .$$

The suppression factor clearly shows that in this particular process both the familiar GIM and the new
leptonic GIM mechanisms are involved[9]. But the important feature to note is that the denominator in δ
is not M_W^{-4} but $M_W^{-2}M^{-2}$ where M^{-2} is some rather complicated combination of fermion mass factors.[17] Thus in
this case here, unlike the situation in $\mu \to e\gamma$, the cancellation between the N and N' propagators does not
introduce further severe suppression factors. We conclude that $K_L \to e\mu$ is not strongly suppressed when compared
to $K_L \to \mu\mu$. We would guess that its branching ratio is in
the 10^{-10} range.

Because the denominator in the suppression factor
δ for $K_L \to e\mu$ is not simply powers of M_W, we will not be
able to make any precise prediction for the $K_L \to e\mu$ even
if $\mu \to e\gamma$ rate was given (and if we know how to take strong
interactions into account). But this would not be the
situation for the pure leptonic process $\mu \to 3e$. Here the
suppression factor should be the same, up to logs, as
in Eq. (11). This theory should give an almost param-
eter-free prediction for the ratio of $\Gamma_{\mu \to 3e}/\Gamma_{\mu \to e\gamma}$.[18]
The calculation involves diagrams shown in Fig. 5.

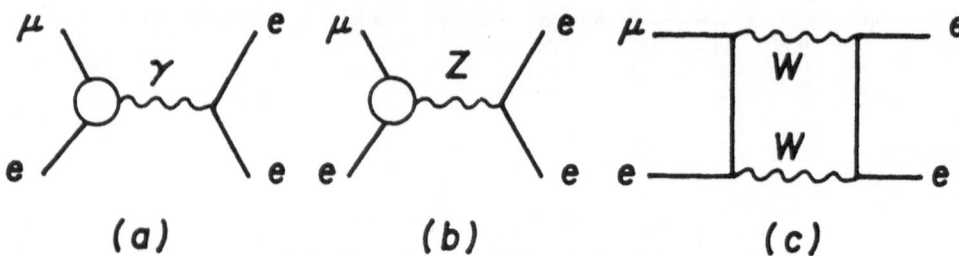

Fig. 5.   Three classes of diagrams for $\mu \to 3e$.

Summary

(1) In gauge theories, weak eigenstates generally are not mass eigenstates; we should not expect separate conservation of muon and electron number. This situation is the same for strangeness and any other quark and lepton flavor numbers. However, in the standard V-A Weinberg-Salam theory degeneracy of neutrino masses gives rise to muon number conservation. Thus symmetry breaking pattern for muon and electron number conservation is directly correlated to some particular features of the fermion mass matrix. This provides the possibility of viewing these conservation laws as having a dynamical origin. This is a view favored by many of us for all the symmetries observed in nature.

(2) In gauge theories with V+A currents and heavy leptons, muon number violation effects should be proportional to heavy lepton mass differences. We expect the rates for $\mu \to e\gamma$ and $K_L \to e\mu$ to be comparable to their present experimental limits.

(3) In constructing theories for leptons beyond the standard Weinberg-Salam scheme, one should be mindful of the need for a "leptonic GIM mechanism". The analogy between $K_L \to \mu\mu$ and $\mu \to 3e$ is particularly close as shown by the diagrams in Fig. 6. Thus conditions for "natural GIM" will be equally applicable for the leptonic situations.[19]

(4) We have proposed a simple theory incorporating all the features discussed above.

$$\begin{pmatrix} \nu_e \\ e \end{pmatrix}_L, \quad \begin{pmatrix} \nu_\mu \\ \mu \end{pmatrix}_L, \quad \begin{pmatrix} N(\phi) \\ e \end{pmatrix}_R, \quad \begin{pmatrix} N'(\phi) \\ \mu \end{pmatrix}_R \cdots$$

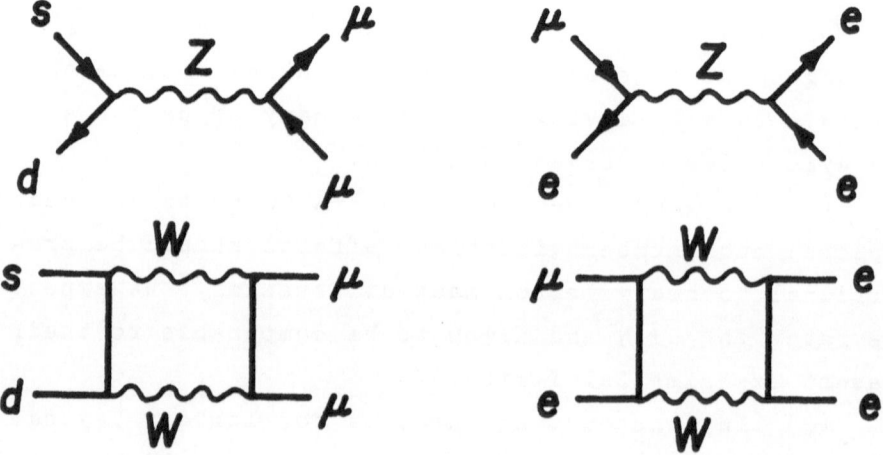

Fig. 6. Direct and induced flavor-changing neutral
current effects.

We have placed the right handed electron and muon in
weak isodoublets. This is mainly out of aesthetic reasons, but it also has the implication that the electronic
neutral current is purely vector. Parity violation in
high Z atoms should be small.[20] With the Bjorken-Lane-
Weinberg observation, this model can indeed be "natural".

(5) Such theories will have numerous experimental
predictions. We have discussed the rates for $\mu \to e\gamma$. If
these decays are in fact observed, then the next step
should be the measurement of the asymmetry parameter

$\alpha = \dfrac{2\text{Re}(a^*b)}{|a|^2 + |b|^2}$. We predict it to be -1 corresponding

to an outgoing right handed electron. Our approach
requires the existence of at least two heavy neutrinos
in the GeV range. They will be difficult to detect but
probably not hopeless. A detailed discussion of searching for N's may be found in the paper by Bjorken and
Llewellyn Smith, and the report by Barnett to this conference.[21]

(6) The interactions we have discussed are of great
interest because they are leptonic higher order weak processes. However, the predictions involve parameters such
as the mixing angle ϕ and heavy lepton masses, which are
not likely to be determined in the near future. The ratio
$\Gamma(\mu \to 3e)/\Gamma(\mu \to e\gamma)$ may turn out to be the cleanest test of
our theory.

Now we shall describe very briefly some alternative
schemes for muon number nonconservation which also do not
arbitrarily restrict parameters in the theory to some
particular set of values.

B. Theories with Higgs Scalars

In the discussion of most weak interaction processes
we can generally ignore the physical Higgs scalars since

their couplings to the fermions are expected to be very weak. However, in rare and ultrarare decay processes they may very well be important. Bjorken and Weinberg[2] pointed out that in the standard Weinberg-Salam theory, if there is more than one independent doublet of Higgs and if one allows the most general Yukawa couplings, one would naturally have a $\bar{\mu}e\phi$ coupling. Assuming that all Yukawa couplings are of the same order, then this coupling is $O(g\frac{m_\mu}{M})$. The extreme smallness of this coupling has the following interesting implication. The diagrams with the least number of Higgs-fermion vertices are the most important ones. They turn out to be two loop diagrams shown in Fig. 7. For a reasonable range of Higgson masses, such diagrams are estimated to yield a $(\frac{\alpha}{\pi})^3$ branching ratio for $\Gamma_{\mu e\gamma}/\Gamma_\mu$.

C. Theories with doubly charged heavy leptons

Wilczek and Zee[4] proposed a class of theories where the muon and electron are coupled to intermixing doubly charged heavy leptons. The decay $\mu \to e\gamma$ then proceeds via diagrams shown in Fig. 8, which yield basically the same branching ratio as the neutral lepton model shown in Eq. (19)[22].

These authors discussed two examples of such theory:

$$\text{I} \qquad \begin{pmatrix} \nu_e \\ e \\ h(\theta) \end{pmatrix}_L \qquad \begin{pmatrix} \nu_\mu \\ \mu \\ k(\theta) \end{pmatrix}_L \qquad \cdots$$

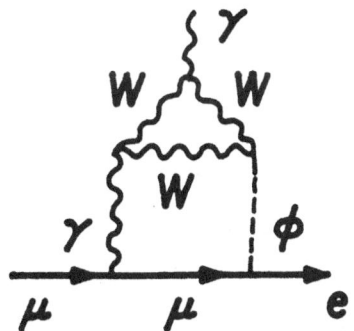

Fig. 7. Two-loop diagrams for μ→eγ with an off-diagonal
Higgs coupling.

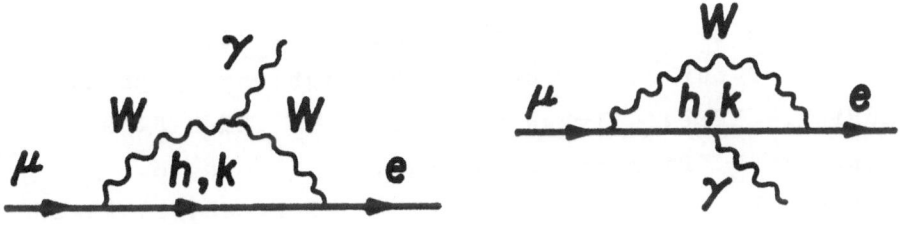

Fig. 8. μ→eγ in theories with doubly charged leptons.

and

$$
\text{II} \quad \begin{pmatrix} \nu_e \\ e \end{pmatrix}_L \begin{pmatrix} \nu_\mu \\ \mu \end{pmatrix}_L \begin{pmatrix} e \\ h(\theta) \end{pmatrix}_R \begin{pmatrix} \mu \\ k(\theta) \end{pmatrix}_R \cdots \quad .
$$

These models have the virtue of being automatically natural. The experimental signatures of doubly charged leptons will of course be striking. Also, model (I) would predict a whole spectrum of quarks and model (II) would predict parity violation in atomic physics twice as large as the Weinberg-Salam theory. Another interesting feature is that, because the μe charge radius of such theory is huge, the rate for $\mu \to 3e$ is predicted to be large

$$
\Gamma(\mu \to 3e)/\Gamma(\mu \to e\gamma) = \frac{32}{225} \left(\frac{\alpha}{\pi}\right)\left(\frac{M_W}{m}\right)^4 ,
$$

which is less than 15 for heavy lepton mass m larger than 4 GeV. Independently, Marciano and Sanda[22] have also done calculations which are similar to Ref. 4 but with general coupling matrices.

We have benefitted from stimulating discussions with many of our colleagues in the high energy physics community. Especially we would like to thank R. M. Barnett, J. D. Bjorken, H. Georgi, F. Gilman, M. Gell-Mann, K. Lane, P. Minkowski, S. Pakvasa, P. Ramond, S. P. Rosen, S. Treiman, H. S. Tsao, S. Weinberg, F. Wilczek, L. Wolfenstein and A. Zee.

REFERENCES

1. T. P. Cheng and L. F. Li, to be published in the Phys. Rev. Letters.

2. J. D. Bjorken and S. Weinberg, submitted to the Phys. Rev. Letters.

3. J. D. Bjorken, K. Lane and S. Weinberg, in preparation.

4. F. Wilczek and A. Zee, submitted to the Phys. Rev. Letters.

5. For a related discussion of the weak interaction models and the $\mu \to e\gamma$ process, see the report by H. Fritzsch to this conference.

6. S. Weinberg, Phys. Rev. Letters 19, 1264 (1967); A. Salam, Elementary Particle Theory, edited by N. Svartholm, (Almquist, Stockholm 1968).

7. Such an exercise has been considered before by S. Eliezer and D. A. Ross, Phys. Rev. D 10, 3080 (1974). However their $\mu \to e\gamma$ calculation has a number of problems: among others it does not satisfy the requirement of (electromagnetic) current conservation and the result has incorrect mass dependences. See also, S. M. Bilenky and B. Pontecorvo, Phys. Letters 61B, 248 (1976).

8. G. Feinberg, Phys. Rev. 110, 1482 (1958).

9. This cancellation only requires that the electron and muon couple to independent fields. This is reminiscent of the cancellation needed to suppress αG_F amplitudes for strangeness changing neutral current effects first discussed by S. L. Glashow, J. Iliopoulos, and L. Maiani, Phys. Rev. D 2, 1285 (1970). We shall call it "leptonic GIM mechanism".

10. By "mass-insertion" we mean either the expansion of the fermion propagator or perturbation of the symmetric theory by mass terms: $m_{ij}\bar{\psi}_i\psi_j$.

11. S. Parker, H. L. Anderson and C. Rey, Phys. Rev. 133B, 768 (1964).

12. Similar conclusion was reached by A. K. Mann and H. Primakoff, University of Pennsylvania preprint (1976).

13. A. Halprin, P. Minkowski, H. Primakoff, and S. P. Rosen, Phys. Rev. D 13, 2567 (1976); T. P. Cheng, Univ. of Missouri-St. Louis report, 1975 (unpublished), and Phys. Rev. D 14, 1367 (1976).

14. In this theory the electron and muon neutrios are two non-orthogonal combinations of the mass eigenstates ν and ν': $\nu_e = \alpha\nu + \beta\nu'$ and $\nu_\mu = \gamma\nu + \delta\nu'$, with $(\nu_e \cdot \nu_\mu) = \alpha\gamma + \beta\delta = \cos\phi\sin\phi\, m_e m_\mu (m^2 - m'^2)/m^2 m'^2$. Thus $\nu_\mu + p \rightarrow e^+ + n$ can take place with a very small probability. This phenomenon is distinct from the question of "neutrino oscillations" as discussed by B. Pontecorvo, Zh. Eksp. Teor. Fix. 53, 1717 (1967) [Soviet Phys. JETP 26, 934 (1968)].

15. Since our one loop result involves a small factor δ_N, multiloop contributions (especially those involving virtual photons) may not be small. Preliminary calculations indicate that they are indeed negligible. In this connection we can use the theorem of G. Feinberg, P. Kabir and S. Weinberg [Phys. Rev. Letters 3, 527 (1959)] to deduce that certain two loop diagrams which are individually large will be cancelled when all diagrams of the same order are summed.

16. K. Fujikawa, B. W. Lee and A. I. Sanda, Phys. Rev. D 6, 2923 (1972); Y. P. Yao, Phys. Rev. D7, 1647 (1973).

17. T. P. Cheng and L. F. Li, in preparation. Details of all our calculations on $\mu \to e\gamma$, $K \to e\mu$, $\mu \to 3e$ and $\mu N \to eN$, etc. will be given in this paper.

18. S. B. Treiman, F. Wilczek and A. Zee (in preparation), and Ref. 17; A. Pais, talk at the Symposium "Five Decades of Weak Interactions", City College of New York, Jan. 21-22, 1977.

19. See for example, S. L. Glashow and S. Weinberg to be published in the Phys. Rev.

20. P. E. G. Baird et al. and E. N. Fortson et al. Nature 284, 528 (1976); C. Bouchiat's talk at this conference.

21. J. D. Bjorken and C. H. Llewellyn Smith, Phys. Rev. D 7, 887 (1973); R. M. Barnett's talk at this conference.

22. W. J. Marciano and A. I. Sanda (unpublished).

SEMICLASSICAL PHYSICS AND CONFINEMENT[*]

John M. Cornwall

Department of Physics

University of California, Los Angeles, CA. 90024

I.   INTRODUCTION

Let us first consider the status of continuum quantum chromodynamics (QCD) as it was presented here last year.[1,2] On one hand, George Tiktopoulos and I found evidence from perturbation theory for exponentiation of infrared singularities of various processes involving colored external particles.[1] This exponentiation suggested the vanishing of soft-gluon-emission amplitudes in the limit of zero gluon mass, and led to the appearance of factors of g^{-1} (g is the coupling constant) in certain integrals related to total cross-sections. Clearly these factors could not be recovered in perturbation theory. There were no real-gluon singularities, so the virtual singularities could act to confine quarks. On the other hand, calculations of various color-blind cross-sections order by order in g showed the usual Bloch-Nordsieck cancellation between

*Work supported in part by the National Science Foundation.

real and virtual infrared singularities.[2] Both of these
alternatives were examined in terms of the bare coupling
constant (or a coupling constant renormalized off-shell).

Although it is amusing that, even at this level,
QCD may have a nonperturbative phase, this is not the
essence of the confinement problem. Two central issues
must be attacked: (1) What is the infrared-singular
behavior of the renormalized coupling constant, or more
cogently, what is the behavior of the renormalization-
group-invariant effective charge $\bar{g}(k)$ for small k; and
(2) what is the role of classical solutions of QCD in
confinement, or more cogently, the role of semiclassical
solutions in which quantum corrections to the classical
field equations are kept? These issues are closely
related, needless to say.

I will describe here not a headon attack on these
issues, but a flanking maneuver. This is excusable,
because we are only beginning to understand how to study
the real problems. The first step is to see whether the
sum of leading logarithms has any features which could
conceivably be true in the full theory; the second step
is to construct a phenomenology based on these features.
The final step, of course, would be to show what has
been done so far is not just phenomenology but an unfinished theory. One shows this by finishing the theory.

This year considerable effort has been spent on
looking at leading logarithms in perturbation theory,
including coupling-constant renormlization effects.
(However, more of these works consider the possibility
that classical or semiclassical solutions play any role).
The result is that, considering the only leading logarithms,
all infrared singularities of processes involving quarks
are either of the QED type (e.g. ladder graphs) or can

be subsumed in $\bar{g}(k)$, where k is a small momentum defined by the process itself. In some recent works[3,4,5] it is made explicit that $\bar{g}$ is evaluated at some physical momentum defined by the process; in others,[6,7,8] this point is not made, but it is shown that all non-QED infrared singularities appear in a renormalized coupling constant equivalent to $\bar{g}(\lambda)$, where λ is an infrared cutoff. The cutoff is necessary in these latter works because mass-shell renormalization is attempted. These works tell us nothing about $\bar{g}(k)$ that was not already known, so that any phenomenology must be based on sheer guesswork for $\bar{g}$.

The results of Ref. 4--described in Section II-- have a simple and elegant structure in space-time, resembling closely the eikonal formalism of QED. Green's functions are expressed as path integrals over a classical quark action-at-a-distance action, which however is modified by quantum effects as expressed in $\bar{g}(k)$. By choosing $\bar{g}(k)$ to be sufficiently singular (say, $\bar{g}^2 \sim m^2/k^2$ at small k which corresponds to a linearly rising Schroedinger potential) we construct a phenomenology of confinement which deals fully with relativistic effects such as retardation. Belief in this machinery is augmented by the observation that it works for two other confined theories - d = 3 QED, and d = 2 QCD· in the light-cone gauge - without having to make any guesses for $\bar{g}(k)$.

At this stage, no account has been taken of possible classical or semiclassical gauge-field degrees of freedom, such as would be associated with strings or bags. A linearly-rising potential corresponds at best to a frozen string, which may of course be a useful limiting case (note that for d = 2 QCD, strings are necessarily frozen). In Section III I describe an unabashedly phenomenological approach to the question of classical gauge-field solutions <u>as modified by quantum effects</u>. The result is that one may find <u>confined</u> analogs of classical solutions such as monopoles[9] and instantons.[10] The confined solutions can be interpreted as strings[11] or glueballs (a glueball is an empty bag[12]). The glueballs turn out to be rather heavy, perhaps several GeV; their decay into hadrons might be retarded by Zweig-rule effects.

Evidently we wish to replace the frozen string of Section II with the dynamical strings and bags discussed in Section III. I will describe elsewhere the modifications to the formalism of Section II which ensue. Finally, since the action associated with classical or semiclassical solutions is $\sim g^{-2}$, there is hope of using these solutions as a non-perturbative starting point for the calculation of $\bar{g}^2$; indeed, this is a powerful motive for introducing semiclassical solutions in the first place. Work is in progress on this final point.

II.   THE FROZEN STRING

The leading-logarithm results of Refs. 3,4, can be described by a classical action-at-a-distance theory of quarks, just as in the eikonal formula for QED. But the "potential" in k-space is not g^2/k^2, but $\bar{g}^2(k^2)/k^2$ where $\bar{g}$ is the effective charge. (In QED there is no difference for small k). Nothing is known about the non-leading logarithms; the possibility is open that they may cancel (except for those which go into $\bar{g}$), as in QED.

Consider the Green's function

$$\Gamma(x,y;z) = <0|T(J(z)\ \psi(x)\ \bar{\psi}(y))|0> \qquad (1)$$

where $J(z)$ is a nearly local, color-singlet source for a quark q and an antiquark $\bar{q}$, and the Fourier transform of Γ:

$$\Gamma_p(x-y) = \int dz \; e^{ipz} \; \Gamma(x,y;z) \quad . \tag{2}$$

The infrared singularities of Γ_p are revealed as $x-y$ becomes very large compared to the quark Compton wave length M^{-1}. For example, in the space-like limit $x_o = y_o$; $|\underset{\sim}{x}-\underset{\sim}{y}| = r \to \infty$ Γ_p will go like e^{ipr}/r (just like a Schroedinger wave function) if quarks can be realized as asymptotic free particles, but if there is confinement, Γ_p will vanish exponentially in r. Of greater practial interest is the timelike limit $t \equiv x_o - y_o \to \infty$; $|\underset{\sim}{x}-\underset{\sim}{y}|$ fixed, which reveals the $q\bar{q}$ bound-state spectrum as oscillations like $e^{ip_o t}$.

There is not space here to record the details of calculating the infrared-singular part of (1); these are given in Ref. 4. The main approximation is to replace the quark Dirac matrix γ_μ by the quark proper velocity $\dot{x}_\mu$ and the main technical tool is an approximate solution of the Ward identities for QCD.

The leading-logarithm version of Γ is:[13]

$$\Gamma(x,y;0)=S(x) \; \bar{S}(y) \int_0^\infty ds \int_0^\infty ds' \int (d \text{ path}) \; e^{iA_{c\ell}(s,x;s',y)} \tag{3}$$

where $S(\bar{S})$ is the full quark (antiquark) propagator, and $A_{c\ell}$ is a classical action constructed from a special sort of gluon propagator:

$$A_{c\ell} = - \frac{M}{2} \int_0^s d\tau(\dot{x}^2 + 1) - \frac{M}{2} \int_0^{s'} d\tau(\dot{y}^2 + 1)$$

$$+ C_F \int_0^s d\tau \int_0^{s'} d\tau' \; \dot{x}_\mu \dot{y}_\nu \Delta^{\mu\nu}(x-y) \tag{4}$$

where C_F (=4/3 for QCD) is the quark Casimir eigenvalue.

In the Feynman gauge,

$$\Delta_{\mu\nu}(x-y) = - \frac{g_{\mu\nu}}{(2\pi)^4} \int dk \frac{\bar{g}^2(k)}{k^2+i\varepsilon} e^{-ik(x-y)} . \qquad (5)$$

(Naturally, in order to find real bound-state energies using (4) it is necessary to use a _real_ $\Delta_{\mu\nu}(x-y)$ in what follows). In (5), $\bar{g}$ is the leading-logarithm approximation to the effective charge. In (2), the path integral is over all q and $\bar{q}$ classical paths which lead from the origin to x or to y. The quark propagator S can also be expressed as a proper-time and path integral over the exponential of a classical action referring to only one quark and its self-interactions through $\Delta_{\mu\nu}$.

Of course, there is every reason _not_ to believe in the physical significance of (3)-(5) insofar as they merely express the sum of leading logarithms. But the transparent simplicity of these formulas (in which all quark propagator and vertex radiative corrections cancel except for the factors $S\bar{S}$) give me hope that they can serve as a basis for a very effective phenomenology of confinement, which goes well beyond analogous phenomenological approaches that use either potential theory or some simple Bethe-Salpeter equation. The idea, of course, is to put in something for $\bar{g}^2(k)$ in (4) which reproduces the sort of action-at-adistance potential that one guesses will confine. The usual folk-lore is $\bar{g}^2 \simeq -m^2/k^2$, which yields a linear Schroedinger potential $V(r) \simeq (8\pi)^{-1} m^2 r$. This does not, of course, account for the string or bag degrees of freedom, but it may still be useful as a description of a frozen string.

For d = 4 QED, it is quite adequate to approximate

(2) by dropping the path integration and substituting straight-line paths in $A_{c\ell}$, but this is clearly wrong for a confining theory. The orbits which minimize the classical action are more nearly circles than straight lines. It is a forbidding task to solve the classical equations following from $A_{c\ell}$;[14] an easier starting point is a variational principle. The first point to note is that the equations of motion following from (3) show that $\dot{x}^2$ and $\dot{y}^2$ are constants; it is necessary to choose these constants to be unity (that is τ and τ' in (3) are proper times). This is achieved by requiring[15]

$$\frac{\partial}{\partial s} A_{c\ell}(s,x;s',y) = 0 = \frac{\partial}{\partial s'} A_{c\ell} . \qquad (6)$$

Secondly, when the equations of motion are satisfied, it is easy to show that

$$\frac{\partial A_{c\ell}}{\partial x^\mu} (s,x;s',y) = -p_\mu^{(x)} \qquad (7)$$

where $p_\mu^{(x)}$ is the canonical momentum of the particle labeled x. A similar equation holds for $p_\mu^{(y)}$. All that is necessary is to parametrize the orbits $x_\mu(\tau), y_\mu(\tau')$, compute the trial action and trial momenta $p_\mu^{(x,y)}$, and fix the variational parameters by requiring that (6) and (7) hold. Having done this, we compute the WKB energy by requiring $\int \underline{p} \cdot d\underline{q} = 2\pi N$; quantum corrections are straightforwardly calculated in principle (if not in practice). I have calculated these WKB energies using circular orbits for d = 4 QED, d = 3 QED, and d = 2 QCD,[16] and am in the process of studying various cases for $\bar{g}^2(k)$ in d = 4 QCD. It turns out that d = 4 QED (relativistic Coulomb case) has an exact solution to the

equations of motion following from (3), with circular
orbits.[17] For d = 2 QCD, the WKB results are those of
t'Hooft[18] for large N (as they must be) with masses
growing like $N^{1/2}$; for d = 3 QED, masses grow like
$\ln N$. Details of the method and results will be
published elsewhere.

Let us return briefly to the quark propagator,
which appears as a factor in (2). It turns out[4] that
for a confining theory S(p) is an entire function
(in the Feynman gauge, at least) with no poles in
momentum space. If $\bar{g}^2 \sim -m^2/k^2$, the propagator is
essentially (α is an inessential constant)

$$S(p) \sim -i \int_0^\infty ds \, e^{is(\not{p}-M)-\alpha s^2} \qquad (8)$$

which is very well-behaved at $\not{p}$ = M. The same thing
ought to happen for the gluon propagator, too, and I
will make heuristic use of this in Section III.

III. SEMICLASSICAL CONFINED SOLUTIONS

Although it is not standard, I use the word
"semiclassical" to refer to solutions of the classical
field equations, as modified by quantum effects
(to all orders). We are all familiar with the explosion
of interest in solutions to the purely classical gauge
field equations (e.g., Refs. 9, 10). But it is open to
serious question whether these solutions are even
qualitatively correct for a confined theory, which has
a large (i.e., infrared-singular) coupling constant.
The point is that for a gauge theory an expansion in $\hbar$
is equivalent to an expansion in g^2, the true expansion
parameter being $g^2\hbar$.

To put the point another way: classical solutions
to the pure gauge-field equations either have no scale

of length, or they have an arbitrary scale (instantons).
Yet as soon as quantum effects are included, there is
a scale: the renormalization-group-invariant mass
scale m, given by

$$m = \mu \, \exp \int_{g(\mu)}^{g(m)} \frac{dg'}{\beta(g')} \qquad (9)$$

where μ is an arbitrary renormalization mass and $g(\mu)$
the coupling constant renormalized at that mass.

The question is: does the scale m enter the
quantum corrections to the classical field equations
in an essential way? The answer must be that it does,
that it confines static solutions, just as it does their
pure quantum counterparts, so they are large only for
distances $\lesssim m^{-1}$. (Instantons are probably not confined).

The classical gauge-field equations with quantum
corrections follow from the expression ($c, \bar{c}$ are ghost
fields)

$$0 = \int (dA_\mu \, dc \, d\bar{c}) \, \frac{\delta}{\delta A_\nu(x)} e^{i \int L \, + \, L_{gauge} \, + \, L_{ghost}} \qquad (10)$$

and the decomposition of A_μ^a into a classical field $\hat{A}_\mu^a$
and a quantum field B_μ^a :

$$A_\mu^a(x) = \hat{A}_\mu^a(x) + B_\mu^a(x) \; ; \quad <B_\mu^a(x)> = 0. \qquad (11)$$

For present purposes, it is only necessary to know that
quantum corrections to the classical equations

$$(\partial_\mu + g\hat{A}_\mu \times) \, \hat{G}^{\mu\nu} = 0 \qquad (12)$$

involve such objects as $<B_\alpha(x) \times B_\beta(x)>$, the antisymmetric part of the gluon propagator. This in turn

is a complicated functional of $\hat{A}$, vanishing at $\hat{A} = 0$.
The part which is a <u>local</u> function of $\hat{A}$ depends on the
behavior of certain B-field Green's functions at
vanishing momentum. Perturbation theory suggests that
these Green's functions are singular at zero momentum
(like the free propagator) but, if confinement works,
these Green's functions will in fact be finite (that
is, just like the quark propagator (8), they will be
entire functions). This heuristic reasoning indicates
that one important aspect of the quantum corrections to
(12) may simply be a mass term, with the scale set by
the mass m in (9).

The knee-jerk response to this is that you can't
add a symmetric mass term to the classical field equations
and maintain gauge invariance, but this is wrong. You
can add a mass term, provided that you also add a set
of scalar fields ϕ^a which transform in a special way
under the gauge group, and which can be solved for in
terms of the $\hat{A}^a_\mu$ with the help of the ϕ field equation.[19]
This is just a realization at the classical level of the
Schwinger mechanism, which is <u>not</u> operative at the
quantum level (the B-field Green's functions are entire,
and have no poles at $k^2 = m^2$).

As in Ref. 19, consider the (classical) Lagrangian

$$L_{c\ell} = -\frac{1}{4} \hat{G}^{a}_{\mu\nu}{}^2 - \frac{m^2}{2} [\hat{A}^a_\mu - \frac{1}{g} (\beta^T \partial_\mu \phi)^a]^2 \qquad (13)$$

where

$$\beta(\bar{\Phi}) = \frac{1}{\bar{\Phi}} (e^{\bar{\Phi}} - 1), \quad \bar{\Phi} = \Sigma T^a \phi^a \qquad (14)$$

and the T^a are the real, antisymmetric matrices of the
adjoint representation (i.e., structure constants).
$L_{c\ell}$ is gauge-invariant, provided that the transformation

law for the ϕ^a is

$$e^{\bar{\phi}'} = e^{\bar{\phi}} e^{-\bar{\theta}} \tag{15}$$

for transformation by the group element $e^{\bar{\theta}}$. The field equations are

$$(\partial_\mu + g\hat{A}_\mu \times) \hat{G}^{\mu\nu} + m^2(\hat{A}^\nu - \frac{1}{g}\beta^T\partial^\nu\phi) = 0. \tag{16}$$

Gauge invariance requires that

$$(\partial_\nu + g\hat{A}_\nu \times)(\hat{A}^\nu - \frac{1}{g}\beta^T\partial^\nu\phi) = 0 \tag{17}$$

which happens to be the ϕ field equation, so (15)-(17) are mutually consistent. Actually, the ϕ^a are not independent degrees of freedom, since they can be gauged away; they are analogous to the $k_\mu k_\nu$ part of the gluon self-energy which accompanies the $g_{\mu\nu}$ part to insure current conservation.

Let us now use (16) and (17) to create a static glueball, or empty bag: a confined, finite-energy solution to (16) and (17). For simplicity, only the group SO(3) will be considered for the Yang-Mills symmetry. Begin with the Wu-Yang-'t Hooft-Polyakov gauge choice

$$\hat{A}_i^a = \varepsilon_{iak}\hat{r}_k \frac{G(r)}{gr}; \quad \hat{A}_o^a = 0 ; \tag{18}$$

$$\phi^a = \hat{r}^a \phi(r),$$

(it is straightforward to show that $\hat{A}_o^a$ must vanish from (16)). Eq. (17) for ϕ is

$$\frac{\sin\phi}{\phi} \{\phi'' + \frac{2\phi'}{r} - \frac{2\phi}{r^2} + (\phi')^2[\text{ctn}\phi - \frac{1}{\phi}] + G\frac{\phi}{r^2}\} = 0 \tag{19}$$

and have the solutions $\phi = N\pi$, $N = 0,1,2,\ldots$. For even N, Eq. (16) is

$$G'' = m^2 G + \frac{1}{r^2} G(G+1)(G+2) \qquad (20)$$

while for odd N the mass term becomes $m^2(G+2)$. Therefore, shifting ϕ from an even to an odd mutiple is equivalent, in (20), to the gauge transformation $G \rightarrow -G-2$; (20) is invariant when both substitutions are made at the same time.

Eq. (20) has been solved numerically; to within the accuracy of the computation a key number which emerges is an integer which suggests the existence of a simple, exact solution. The solution behaves like

$$G \sim -2e^{-mr} \qquad r \rightarrow \infty$$
$$G \sim -2 + 0(r^2 \ell n\ r) \quad r \rightarrow 0 \qquad (21)$$

(of course, the -2 in the $r \sim 0$ regime is exact; the other 2 is approximate). In the absence of the mass term, $G = -2$ is an exact, pure gauge solution with vanishing fields and energy. With the mass term, the energy is

$$H = \frac{4\pi}{g^2} \int_0^\infty dr \{G'^2 + m^2 G^2 + \frac{1}{2r^2} G^2(G+2)^2\} = 5.3(\frac{4\pi}{g^2})\ m. \qquad (22)$$

It is finite because G approaches a pure gauge solution at small distances.

Similar confined solutions can be found for infinitely long cylindrical strings; their energy per unit length is finite and scales as $m^2(4\pi/g^2)$ which can roughly be identified with the universal inverse

Regge slope or the coefficient of r in the charmonium potential. This means that $m^2(4\pi/g^2) = 0(1 \text{ GeV}^2)$ and that the glueball energy is of order GeV $\times (\sqrt{4\pi}/g)$. This is rather large, possibly up to 10 GeV, if we identify g in (22) with the gluon coupling constant renormalized at $\sim$ GeV energies.

So there is a heavy glueball and a string which can be used to join quarks, justifying the phenome- nological choice of $\bar{g}^2(k)$ discussed in connection with the frozen string. Of course it is now necessary to go back and incorporate the dynamics of the semiclassical strings in the interactions of quarks. This is not particularly difficult in principle, but I will not discuss it here. It will certainly be a messy practical problem, just as the MIT bag[12] is. The final step which would justify calling all of this an unfinished theory, rather than mere phenomenology, is to use the semi- classical solutions as a starting-point for calculating $\bar{g}^2(k)$.

ACKNOWLEDGEMENTS

Thanks go first to George Tiktopoulos for innumerable challenging discussions during his stay at UCLA. I have also enjoyed stimulating conversations with Peter Carruthers and Fred Zachariasen, as well as with Henry Abarbanel.

REFERENCES

1. John M. Cornwall and G. Tiktopoulos, in <u>New</u>
    <u>Pathways in High Energy Physics II</u>, edited by
    A. Perlmutter, Plenum Press, N.Y., 1976, p. 213.

2. Thomas Appelquist, ibid., p. 199.

3. John M. Cornwall, UCLA Report 76/TEP/10, June 1976
    (talk given at the US-Japan Seminar on Geometric
    Models of the Elementary Particles, Osaka).

4. John M. Cornwall and G. Tiktopoulos, UCLA preprint
    76/TEP/22, November 1976.

5. J. Frenkel, R. Meuldermans, I. Mohammad, and J. C.
    Taylor, Phys. Lett. <u>64B</u>, 211, (1976), and Oxford
    preprint 67/76;
    J. Frenkel, Phys. Lett. <u>65B</u>, 383 (1976);
    T. Kinoshita and A. Ukawa (unpublished).

6. C. P. Korthals Altes and E. de Rafael, Phys. Lett.
    <u>62B</u>, 320 (1976).

7. P. Cvitanović, CERN preprint TH. 2231-CERN,
    September 1976.

8. Akio Sugamoto, University of Tokyo preprint UT-721,
    September, 1976.

9. G. 't Hooft, Nucl. Phys. <u>B79</u>, 276 (1974);
    A. Polyakov, JETP Letters <u>20</u>, 194 (1974).

10. A. Belavin, A. Polyakov, A. Schwartz, and Y.
    Tyupkin, Phys. Lett. <u>59B</u>, 85 (1975); G. 't Hooft,
    Phys. Rev. Letters <u>37</u>, 8 (1976). Note that con-
    fined instantons necessarily have zero topological
    charge. Actually, it is more reasonable that
    instantons oscillate, rather than decrease ex-
    ponentially, because they are solutions in
    Euclidean space.

11. H. B. Nielsen and P. Olesen, Nucl. Phys. <u>B61</u>, 45
    (1973).

12. A. Chodos, R. L. Jaffe, K. Johnson, C. B. Thorn, and V. Weisskopf, Phys. Rev. D9, 3471 (1974).
    A. Chodos, R.L. Jaffe, K. Johnson, and C.B. Thorn, Phys. Rev. D10, 2599 (1974).

13. Some of the non-leading logarithms can be summed up by a path-ordering process described in Ref. 4. Strictly speaking, the sum of leading Feynman graphs in (2) corresponds to approximating the path integral by substituting straight-line paths in $A_{c\ell}$.

14. But it is not a chimerical or impossible task, in spite of the sentiments expressed a dozen or more years ago about the inconsistency of action-at-a-distance theories with canonical Poincaré invariance.

15. For a single free particle going from 0 to x, the first term on the RHS of (3) leads to

$$A_{c\ell} = - \frac{Ms}{2} - \frac{Mx^2}{2s} \; ; \; \partial A_{c\ell}/\partial s = 0 \text{ implies } s=\sqrt{x^2}.$$

16. For d = 2 QCD this was done with my long-time collaborator, George Tiktopoulos.

17. Later I discovered that this solution is due to A. Schild, Phys. Rev. 131, 2762 (1963).

18. G. 't Hooft, Nucl. Phys. B75, 461 (1974).

19. John M. Cornwall, Phys. Rev. D10, 500 (1974).

SOME OBSERVATIONS ON QUANTUM CHROMODYNAMICS

G. 't Hooft*

Stanford Linear Accelerator Center

Stanford, California

and

Institute for Theoretical Physics

University of Utrecht, Netherlands

1.   INTRODUCTION

It need not be repeated here why the theory referred to as "quantum chromodynamics" is considered to
be the best candidate for a successful theory for the
strong interactions.   The Lagrangean is written as

$$L(A,\bar{\psi},\psi) = -\frac{1}{4} G^a_{\mu\nu} G^a_{\mu\nu} - \sum_f \bar{\psi}_f(\gamma D + m_f)\psi_f, \quad (1.1)$$

where the gauge group is SU(3):

$$G^a_{\mu\nu} = \partial_\mu A^a_\nu - \partial_\nu A^a_\mu + g f_{abc} A^b_\mu A^c_\nu ,$$

$$a,b,c = 1,\ldots,8, \quad\quad (1.2)$$

---

*Present address: Institute for Theoretical Physics,
University of Utrecht, Princetonplein 5, Utrecht, The
Netherlands.

and the quark fields ψ_f come in triplets, for which

$$D_\mu \psi_f = \partial_\mu \psi_f - \frac{1}{2} ig \; \lambda^a \; A_\mu^a \psi_f \; ,$$

$$[\lambda^a, \lambda^b] = 2i \; f_{abc} \; \lambda^c \; . \tag{1.3}$$

The free parameters of the theory are (seem to be) the masses m_f and a dimensionless coupling constant g.

In spite of the simplicity of the basic Lagrangean the proposed system is the most complicated field theory particle physicists ever dealt with. The reason is that there does not exist a simple free field theory that even remotely describes the physical particle spectrum and therefore could be used as the first of a successive series of approximation. In other quantum field theories such as quantum electrodynamics and the weak interactions the perturbation series in terms of the coupling constant is successful because the coupling is small enough to guarantee rapid convergence.

In quantum chromodynamics the coupling is large, and in a certain sense even infinite (the quarks cannot come free). Only in the far Euclidean region for the momenta involved is the effective coupling small and the perturbation expansion with respect to it can be used as an asymptotic expansion (the radius of convergence is zero). It is therefore understandable that most particle theorists seek other ways to approximate the infinity of field variables by some simpler model. Those approximations, however useful they may be, do not answer the question whether the given Lagrangean really defines a unique theory. It is easy to argue (as we will do in the following chapters) that the formal series in the coupling constant g diverges badly for

all values of g. Due to the renormalization group the
series has a direct physical interpretation as an
asymptotic expansion for very large (Euclidean) momenta.
Although the expansion diverges, does it perhaps in
combination with physical requirements such as unitarity and causality define a theory uniquely? Can we in
principle replace the divergent series by a convergent
one, no matter how complicated? It is this question
that we shall investigate.

In section 2 we show that the most essential renormalization factors in our theory are determined by
the one- and two loop Feynman graphs only, due to
asymptotic freedom, and are therefore known. This
implies that we can give a simple definition of the
theory and its physical parameters in such a way that
all physical quantities remain finite at all orders of
perturbation theory. It is tempting, then, to assume
that this defines also the nonperturbative theory, but
that is not true (except possibly when one of the quark
masses is put equal to zero). There is known to be yet
another free parameter in the form of an angle θ that
has to be defined as well. Its effects (arising from
the so called instantons) simply do not show up in any
finite order of the perturbation expansion. Are there
no other such parameters θ', θ'', ...? Perhaps an infinity of them? As yet we were unable to settle that
question.

In the next chapters we assume that a theory with
physically acceptable properties is defined this way
and consider it as a function of the renormalized coup-
ling constant g^R and fixed value(s) for θ (θ', θ'',...).
In the massless case ($m_f = o$) we find the complete
analytic structure in the complex $(g^R)^2$ plane (fig. 2).

In view of the analytic structure that we find we consider it unlikely that any attempts to solve QCD using Padé approximants will be successful; they will certainly diverge sooner or later. Much more interesting is the claim that perhaps Borel resummation can be applied[1]. We show how the instantons enter in the Borel resummation expression. They are an interesting complication but do not seem to be the most essential obstable in formulating convergent expressions. Much harder we think is the problem of justifying the interchange of this resummation procedure with the limit where renormalization and infrared cutoffs Λ go to infinity. And there is another complication that might make the Borel approach essentially worthless here: we show that the Borel integral diverges for all values of the coupling constant. This is again a consequence of the bad analytic structure for complex g^R. We conclude that there is still no convergent resummation procedure for QCD, the same conclusion as in ref. 2) but on different grounds.

Finally we observe some infrared divergencies of QCD in less than four dimensions (independent of the masses of the quarks). In four dimensions the relevant diagrams (fig. 4) are not infinite but may become so large that the Borel summability properties of the theory are endangered.

2. DEFINITION OF THE THEORY

The Lagrangean (1.1) defines the unrenormalized perturbation expansion in the coupling constant g. A convenient way to define a (perturbatively) finite expansion parameter g_D is through dimensional renormalization. The subscript D stands for dimensionally renormalized. At $4-\varepsilon$ dimensions, with ε nonrational, the integrations in momentum space can be defined unambiguously. The S-matrix in the limit $\varepsilon \to o$ is finite if g (the bare coupling constant occurring in (1.1)) is taken to be a function of g_D and ε (see ref. 3)

$$g^2 = \mu^\varepsilon \, [g_D^2 + \frac{g_D^4 \, b_1(g_D^2)}{\varepsilon} + \frac{g_D^6 \, b_2(g_D^2)}{\varepsilon^2} + \ldots], (2.1)$$

and similarly the bare mass m:

$$m_f = \mu m_{fD} \, [1 + \frac{g_D^2 \, a_1(g_D^2)}{\varepsilon} + \frac{g_D^4 \, a_2(g_D^2)}{\varepsilon^2} + \ldots].$$
$$(2.2)$$

Here, the quantities μ^ε and μ are there to show that in $4-\varepsilon$ dimensions g^2 has dimension ε and m_f has dimension one, as can easily be read off from the Lagrangean. The functions $a_i(g_D^2)$ and $b_i(g_D^2)$ are uniquely determined[3] if we require only poles in $1/\varepsilon$ in (2.1) and (2.2), but the definition of g_D^2 depends on the choice of the parameter μ.

At this point we emphasize that field theory in a noninteger number of dimension has never been defined beyond the perturbation expansion. Therefore the above defined parameters g_D^2 and m_D may not exist beyond perturbation theory. We prefer to construct other parameters whose existence follows undisputably from the assumption that the theory is approached by the asymptotic expansion in g in the deep Euclidean region.

First we go back to the perturbation series for g_D^2 and m_D. The S-matrix should not depend on the choice of μ (invariance under the renormalization group), but the above definitions of g_D and m_D do, so we write $g_D(\mu)$ and $m_D(\mu)$. Invariance of the S-matrix under changes of μ is guaranteed if $g_D(\mu)$ and $m_D(\mu)$ satisfy the equations[3]

$$\frac{\mu dg_D^2}{d\mu} \equiv \beta^D(g_D^2) = (1 - g_D^2 \frac{d}{dg_D^2}) \ g_D^4 \ b_1(g_D^2) \ , \qquad (2.3)$$

and

$$\frac{\mu dm_D}{d\mu} \equiv m_D(-1+\alpha^D(g_D^2)) =$$

$$= -m_D -m_D \ g_D^2 \frac{d}{dg_D^2} \ (g_D^2 \ a_1(g_D^2)) \ . \qquad (2.4)$$

The lowest order terms of α and β for our theory are known[4]:

$$\beta^D(g_D^2) = \beta_1 \ g_D^4 + \beta_2 \ g_D^6 + \beta_3^D \ g_D^8 + \cdots \ ,$$

$$\alpha^D(g_D^2) = \alpha_1 \ g_D^2 + \alpha_2^D(g_D^4) + \cdots \ ,$$

with

$$\beta_1 = - \frac{1}{8\pi^2} \ (11 - \frac{2}{3} N_f) \ ,$$

$$\beta_2 = \frac{1}{(8\pi^2)^2} \ (\frac{19}{3} N_f - 51) \ ,$$

$$\alpha_1 = - 1/2\pi^2 \ . \qquad (2.5)$$

We limit ourselves to $N_f \leqslant 16$, so that $\beta_1 < 0$.

The importance of eqs. (2.3) and (2.4) is that we can let our choice of μ depend on the problem considered. For any process that is free of infrared divergencies in the massless limit the essential expansion parameter is

$$g_D^2(\mu) \ \log \frac{k^2}{\mu^2} \ , \qquad (2.6)$$

where k^2 stands for the typical external momenta. This is smallest if μ^2 is chosen to be of the same order as k^2. If β_1 is negative (as is the case here) then $g_D(\mu) \to o$ if $\mu \to \infty$. Thus, if the external momenta k^2 go to infinity then the expansion parameter (2.6) goes to zero and we have a rapidly converging series. This phenomenon is called asymptotic freedom.

Numerous authors discuss the consequences of a non-trivial zero of the function β. We stress that with our definition the presence or absence of zero's is irrelevant. We say this because g_D is not directly connected with a physically measurable quantity such as a gauge-invariant Green's function. Suppose there were a zero at g_D^o. Any redefinition of the form

$$\frac{1}{g_x^2} = \frac{1}{g_D^2} - \frac{1}{(g_D^o)^2} \tag{2.7}$$

would remove the zero from the β function for the new g_x^2. The $\theta(g^4)$ corrections to the renormalized coupling constants are usually defined in a rather arbitrary fashion and the "correction" (2.7) to g_D^2 is indeed of order g_D^4. Secondly, we repeat that g_D may not have a finite meaning at all.

Let us now consider a substitution of the form

$$g_D^2 = g_R^2 + p_1\, g_R^4 + p_2\, g_R^6 + \cdots ,$$

$$m_D = m_R\, (1 + q_1\, g_R^2 + q_2\, g_R^4 + \cdots). \tag{2.8}$$

We will choose p_i and q_i and obtain a preferred set of variables g_R and m_R. We have

$$\frac{\mu d}{d\mu} g_R^2 = \beta^R(g_R^2) = \beta_1 g_R^4 + \beta_2 g_R^6 + \beta_3^R g_R^8 + \ldots \quad ,$$

$$\frac{\mu d}{d\mu} m_R(-1 + \alpha_R(g_R^2));$$

$$\alpha_R = \alpha_1 g_R^2 + \alpha_2^R g_R^4 + \ldots \quad . \qquad (2.9)$$

We find that β_1, β_2 and α_1 are unaffected by the change. These parameters are universal. But

$$\beta_3^R = \beta_3^D + \beta_1(2p_2+p_1^2) + \beta_2 p_1 \quad ,$$

$$\alpha_2^R = \alpha_2^D - \beta_1 q_1 - q_1 p_1 \quad . \qquad (2.10)$$

We now choose $p_1 = o$ and p_2, p_3,... and q_1, q_2,... in such a way that all coefficients β_3^R, β_4^R,... and α_2^R, α_3^R,... are equal to zero. Thus we have obtained parameters g_R and m_R that are equal to g_D and m_D up to computable higher order corrections, and they have known Callan-Symanzik functions α and β. Clearly, if this β function has a zero one should not attach any physical relevance to that.

We can now solve* for $g_R(\mu)$ and $m_R(\mu)$:

$$\frac{1}{|\beta_1| g_R^2(\mu)} + \frac{\beta_2}{\beta_1^2} \log \left(\frac{|\beta_1|}{g_R^2(\mu)} - \beta_2\right) = \log (\mu/\mu_o); \qquad (2.11)$$

$$\frac{\mu m_{fR}(\mu)}{m_{fo}} = \left(\frac{|\beta_1|}{g_R^2(\mu)} - \beta_2\right)^{\alpha_1/\beta_1} \quad . \qquad (2.12)$$

---

*By further redefinition of g_R with counter terms of order g_R^3 one can simplify this solution, but with no particular merit.

Here μ_o and m_{fo} are integration constants. They all
have dimension of a mass and they are the true renormalization group invariant physical parameters of
the theory. The advantage of the above definition is
that it is finite to all orders of the perturbation expansion and even for the summed theory, if it approaches
the asymptotic expansion in the deep Euclidean region.
A similar definition of quark masses has been given in
ref. 5.

As already noted in the Introduction we should not
be tempted to assume that μ_o and m_{fo} are therefore
necessarily the only parameters of the theory. Due to
the "instantons" (classical field configurations in
Euclidean space-time) there are effects that give rise
to amplitudes[6] proportional to

$$g_R^{-C} \exp \ [-8\pi^2/g_R^2 \ \pm \ i\theta][1+\theta(g_R^2)]. \qquad (2.13)$$

For the gauge group SU(3) we have

$$C = 12 \ . \qquad (2.14)$$

(C counts the number of zero eigenmodes of the gauge
fields in the presence of an instanton). θ is a free
parameter, observable if all parameters $m_{fo} \neq o$. Clearly, the effects that depend on θ do not show up in the
usual perturbation expansion in g_R. The mere existence
of phenomena that do not show up in the usual perturbation expansion shows the importance of the questions we
will consider.

3. THE COMPLEX COUPLING CONSTANT PLANE FOR THE MASSLESS
                                    THEORY

The massless theory is defined by

$$m_{fo} = o \ , \ f = 1,\ldots, N_f \leqslant 16 \ . \tag{3.1}$$

In this limit we have a global chiral $SU(N_f) \times SU(N_f)$
$\times U(1)$ symmetry. There is only one known parameter
left, which is μ_o (the parameter θ in this limit is
meaningless because it only fixes the preference coordinate in the broken chiral $U(1)$ group).

Instead of taking the variable μ_o we consider the
theory as a function of $g_R(\mu)$ at some fixed μ. The relation between $g_R(\mu)$ and μ_o is given by the solution
(2.11) of the renormalization group equation. This we
do because the theory is known as a series of integer
powers of $g_R(\mu)$.

First we must give a definition of the Green's
functions which is as accurate as our definition of the
physical variables (that is, the result must be finite
to any order of the perturbation expansion and remain
finite after integration of the renormalization group
equation). One restriction will always be made. We
only consider gauge-invariant amplitudes. A very good
example to work with is the time ordered product of two
(or more) bilinear quark operators: one adds to the
Lagrangean a source term

$$\bar{\psi}(x) \ J(x) \ \psi(x) \ ,$$

where J may contain Dirac and flavor indices but no
color indices, and consider that part of the vacuum-
vacuum amplitude that is linear both in $J(o)$ and $J(x)$.

Let us first renormalize dimensionally,

$$J(x) = J_D(x) + \frac{g_D^2}{\epsilon} J_{D1}(x,g_D^2) + \frac{g_D^4}{\epsilon^2} J_{D2}(x,g_D^2) + \cdots \quad . \tag{3.2}$$

As before, J_{D1}, $J_{D2}, \cdots$ are chosen such that we obtain a perturbatively finite Green's function. In momentum space

$$\Gamma_{pert}^D(k^2,\mu,g_D^2) = \Gamma_0(k^2,\mu) + g_D^2 \Gamma_1(k^2,\mu) + \cdots \quad . \tag{3.3}$$

The renormalization group equation is

$$[\frac{\mu d}{d\mu} + \beta^D(g_D^2) \frac{\partial}{\partial g_D^2} + \gamma_{pert}^D(k^2,\mu,g_D^2) = 0 \quad . \tag{3.4}$$

Here γ also contains the canonical part of the dimension of Γ(depending on the number and type of external lines). The general solution is

$$\Gamma_{pert}(k^2,\mu,g_D^2) = Z(g_D^2) \Gamma(g_D^2(\mu), k^2/\mu^2) \quad , \tag{3.5}$$

with

$$\frac{d}{dg_D^2} \log Z(g_D^2) = - \frac{\gamma^D(g_D^2)}{\beta^D(g_D^2)} \quad , \tag{3.6}$$

and

$$\frac{d}{d\mu} \Gamma_D (g_D^2(\mu), \frac{k^2}{\mu^2}) = 0 \quad . \tag{3.7}$$

In the simplest example, where we look at the σ-channel ($\bar{\psi}J\psi = J(\bar{\psi}\psi)$), we have

$$\gamma^D = \gamma_0 + \gamma_1 g_D^2 + \gamma_2^D g_D^4 \cdots \quad ,$$

with
$$\gamma_o = 2 ,$$

$$\gamma_1 = - 1/\pi^2 . \tag{3.8}$$

The right hand side of eq. (3.6) is then

$$\frac{z_o}{g_D^4} + \frac{z_1}{g_D^2} + z_2 + \ldots, \tag{3.9}$$

with
$$z_o = - \gamma_o/\beta_1 ,$$

$$z_1 = - \gamma_1/\beta_1 + \gamma_o\beta_2\beta_1^2 . \tag{3.10}$$

Thus

$$\log Z(g_D^2) = - \frac{z_o}{g_D^2} + z_1 \log g_D^2 + z_2 g_D^2 + \ldots . \tag{3.11}$$

Not only will we replace g_D here by g_R, but also we will multiply Z with finite corrections of order g_R^2 such that the coefficients z_2, z_3,... all vanish. Again we make use of the fact that the subtraction procedure for Γ_{pert}^D was arbitrary anyhow and thus we obtain Γ_{pert}^R, equal to Γ_{pert}^D up to higher order computable corrections, and

$$Z^R(g_R^2) = g_R^{2z_1} \exp(-z_o/g_R^2) \tag{3.12}$$

exactly.

According to eq. (3.7) we can write

$$\Gamma_D(g_D^2(\mu), k^2/\mu^2) \equiv \Gamma_R(g_R^2(\mu), k^2/\mu^2) =$$

$$= \Gamma(k^2/\mu_o^2) , \tag{3.13}$$

where μ_0 is given by (2.11). And we have

$$\Gamma^R_{pert}(k^2,\mu,g^2_R) = Z^R(g^2_R)\ \Gamma(k^2/\mu^2_0)\ . \qquad (3.14)$$

Notice that we now have one unknown function Γ of one variable $z = k^2/\mu^2_0$. The analytic structure of Γ_{pert} for complex k^2 follows from general physical requirements (fig. 1). In mesonic channels we expect a cut starting at the origin, because massless pions can be produced. In baryonic channels the cut will start at some finite negative value for z. The discontinuity across the cut will show peaks where the resonances are. It is important to consider the second Riemann sheet across the cut. There we expect series of poles due to the resonances. In a simplified version of the theory SU(3) is replaced by SU(N) and the limit $N \rightarrow \infty$, g^2N fixed is taken. Then the poles move to the real axis and the cut disappears[7].

Substituting (2.11) into (3.14) we find the behavior for complex g^2_R as well. The singularities described above occur at

$$\log\ (k^2/\mu^2) + \frac{2}{(-\beta_1)g^2_R(\mu)} + \frac{2\beta_2}{\beta^2_1}\ \log\ (\frac{-\beta_1}{g^2_R(\mu)} - \beta_2) =$$

$$a + (2n+1)\pi i\ , \qquad (3.15)$$

where a is real for the cut and may have a small imaginary part for the poles. If k^2 and μ are kept fixed and positive then (3.15) gives the singularities in complex g^2_R plane (see fig. 2). We assume that β_1 is negative. The origin of g^2_R plane is a density point of cuts (or poles if $N \rightarrow \infty$) both for the meson and for the baryon channels. In the case of a finite gauge group

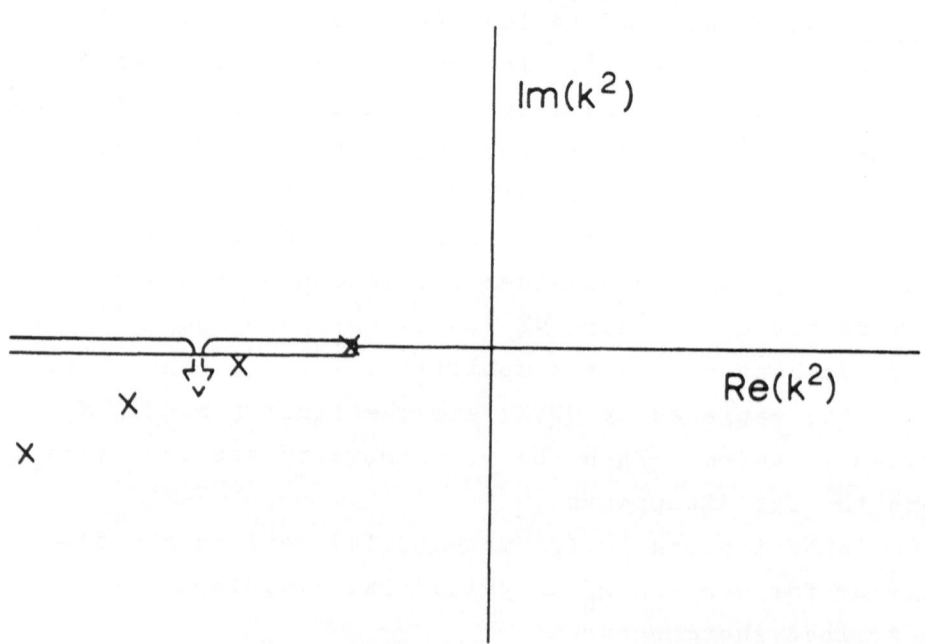

Fig. 1.   Our basic assumption is that Green's functions
are analytic except for a cut along the negative real
axis in k^2 plane.   If we try to continue analytically
beyond the cut (arrow) then we will encounter poles due
to the resonances (crosses).

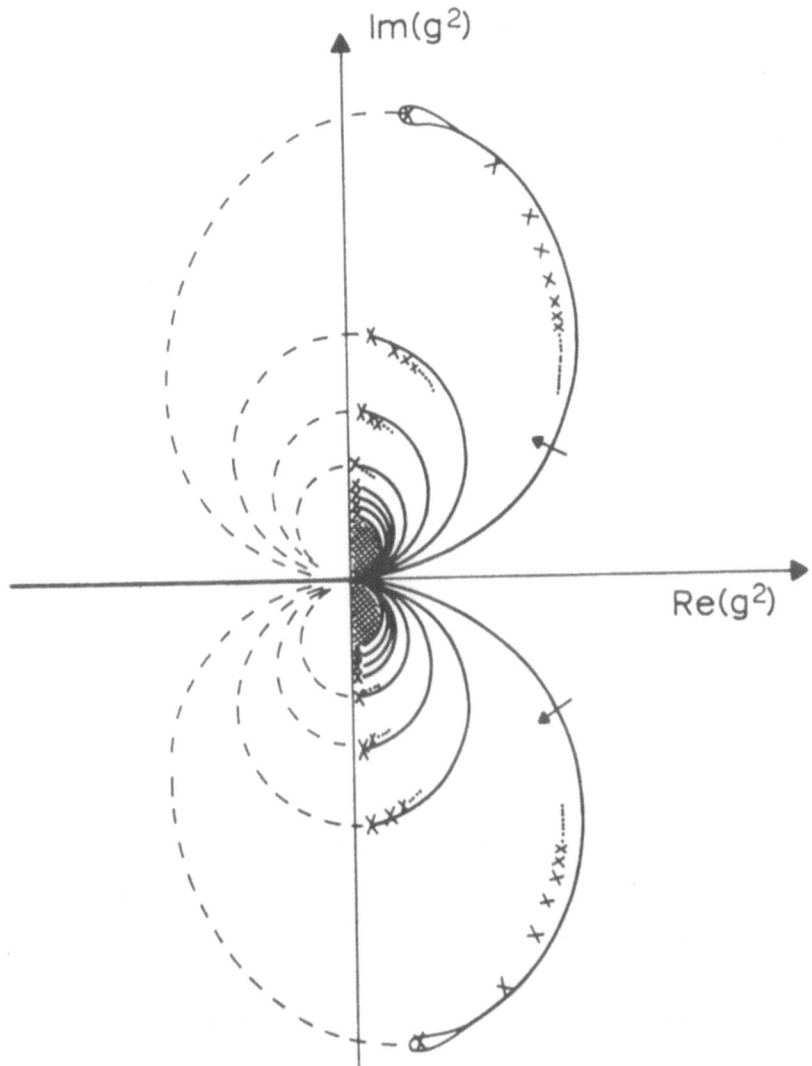

Fig. 2. Singularities in complex $(g^R)^2$ plane. The solid
lines are cuts that will always be present. In the
baryonic channels the cuts stop halfway; in the mesonic
channels they continue to form full circles (dotted lines)
because we have massless mesons. (The circles are slightly deformed due to the β_2 term). If we try to continue
across the cut (arrows) we encounter poles (crosses).

one might hope that continuation across the cut would
be possible but we fear that the resonance poles in the
second Riemann sheet will be unavoidable. The general
expectation is that these series of poles tend to infinity in k^2 plane and thus cluster at the origin of g_R^2
plane. In conclusion, we find that close to the origin
of g_R^2 plane, Γ is analytic in a region of the form

$$|\mathrm{Im}\ g_R^2| < C(\mathrm{Re}\ g_R^2)^2 \quad , \tag{3.16}$$

with $C = \dfrac{-\beta_1}{2}$ or only slightly larger.

In less than four dimensions ($d<4$) the logarithms
of eq. (3.15) are replaced by powers and we get a region
of analyticity of the form

$$|\mathrm{Im}\ g_R^2| < C\mathrm{Re}\ g_R^2 \quad , \tag{3.17}$$

with $C = \tan\frac{\pi}{2}(4-d)$ or only slightly larger. In Γ_{pert}
we get in addition to these singularities also the
singularity of $Z(g_R^2)$ which is a cut along the negative
real axis (usually z_1 is not integer). See fig. 3.

The region (3.16) is too small to have convergence
of the Padé approximation, or even, as we shall see,
the Borel resummation procedure (despite the absence of
instantons in the SU(∞) theory). The region (3.17) is
large enough for Borel resummation to converge only if
$d \leqslant 3$, were it not that there may be other troubles
with that procedure in less than four dimensions (see
chapter 6).

4. BOREL RESUMMATION

The Borel resummation procedure for a function
Γ_{pert} (g^2) consists of writing the Laplace transformation

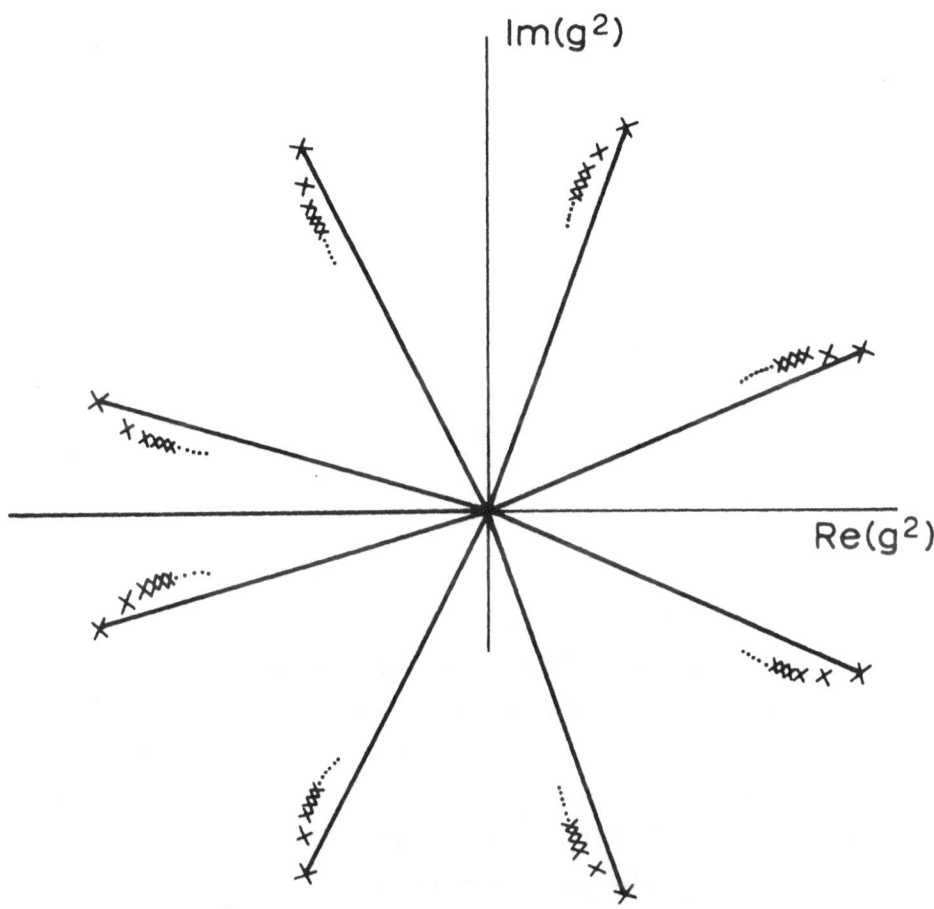

Fig. 3.   Singularities in complex g^2 plane if $\varepsilon = 0.26$.
Thick lines are cuts.

$$\Gamma_{pert}^{\cdot}(g^2) = \int_0^\infty F(z) \, e^{-z/g^2} \, dz/g^2 \ . \qquad (4.1)$$

If Γ_{pert} has the perturbation expansion

$$\Gamma_{pert} = a_o + a_1 g^2 + a_2 g^4 + \dots \quad , \qquad (4.2)$$

then the series for $F(z)$ is

$$F(z) = a_o + \frac{a_1 z}{1!} + \frac{a_2 z^2}{2!} + \dots \quad , \qquad (4.3)$$

which converges better. The singularity of F closest to the origin (if any) will determine the asymptotic behavior of the series $a_o, a_1, \dots$. The question is whether $F(z)$ is determined by its perturbation expansion and whether then $\Gamma(g^2)$ can be determined through (4.1).

Statement:

In a field theory with both an infrared and an ultraviolet cutoff the functions $F(z)$ have a finite radius of convergence. The singularities of $F(z)$ are branch points whose locations are determined by classical solutions of the field equations.

The statement is easy to prove if we take as ultraviolet cutoff a Euclidean space-time lattice and as infrared cutoff a Euclidean space-time box. Then all functional integrals are just finite dimensional integrals.

Let us illustrate the proof for a "field theory" with just one field variable A at just one space-time point x. The action is

$$S(A) = -\frac{1}{2} A^2 - \frac{1}{g^2} V(gA) \quad , \qquad (4.4)$$

with $V(x) = V_3 x^3 + V_4 x^4 + \dots$. Here g^2 is the expansion parameter.

$$W(g^2) = \frac{1}{g} \int dA \, \exp S(A) =$$

$$\frac{1}{g^2} \int dx \, \exp \frac{1}{g^2} [-\frac{1}{2}x^2 - V(x)] \quad . \tag{4.5}$$

The factor $1/g$ in the first line is only for convenience. Leaving it out would give rise to an irrelevant complication.

Obviously the function $F(z)$ of eq. (4.1) is now

$$F(z) = \int dx \, \delta[z - \frac{1}{2}x^2 - V(x)]$$

$$= \sum_i (g^2 S'[A_i])^{-1} \Big/ S(A_i) = z/g^2 \quad . \tag{4.6}$$

Notice the branch point singularities at those points z_i where

$$S'[A_i] = \frac{\partial S}{\partial A_i} = 0 \quad ,$$

$$z_i/g^2 = S(A_i) \quad . \tag{4.7}$$

This result generalizes without many complications to multidimensional integrals. The equations (4.7) then correspond to the classical Lagrange field equations.

Still, however, we are far away from a real field theory. Rather than complete functional integrals as a function of a <u>bare</u> expansion parameter g^2 we are interested in renormalized connected Green's functions as a function of a renormalized parameter g_R^2. Now an interesting feature of the Laplace transformation (4.1)

is that the positions of the branch point singularities
of $F(z)$ are the same as those of the singularities of
the Laplace transforms of

$$\frac{\partial \Gamma}{\partial g^2} \ , \ f(g^2)\Gamma \ , \ \exp \ \Gamma \ , \ \ln(1+g^2\Gamma) \ , \ \text{etc.},$$

as one can easily verify. Thus, by exponentiation and
differentiation it will be possible to obtain the
analytic structure of the Laplace transforms of
connected Green's functions. The location of their
branch points is the same as those of $F(z)$. Also,
finite renormalizations,

$$g^2 \rightarrow g^2 + \theta(g^4)$$

leave the positions of the singularities unaffected.
This is how refs. 1,2 obtained the result that the
singular points z_i of $F(z)$ arise from classical solutions of $\partial S/\partial A_i = 0$ (the Lagrange field equations)
where $S(A_i) = z_i/g^2$, even if Γ_{pert} is a connected Green's
function and if g^2 is renormalized.

Real solutions of the field equations in Euclidean
space-time, with finite total action (instantons) give
singularities in $F(z)$ on the positive real axis. These
singularities correspond to the opening up of new types
of field configurations with increasing action (those
with a nonvanishing total Pontryagin winding number).
Comparison of (4.1) and (4.5) suggests how to integrate
over these singular points, without difficulties. By
doing the perturbation expansion around the instanton
field configurations we find the behavior of $F(z)$ close
to and beyond these singular points. All in all we expect that when all classical solutions of the field

equations with finite total action are known then $F(z)$
can be determined completely by convergent perturba-
tive methods. The instantons give rise to a complication that is not insurmountable[2].

5.   PROBLEMS ASSOCIATED WITH THE BOREL RESUMMATION
     The first problem we encounter if we try to pursue
this program is that the analytic structure of $F(z)$ is
determined by all solutions of the classical equations
that have a finite total action, including the complex
solutions. The solutions where the fields are real
usually have some physical interpretation and can be
categorized. But to determine all nonreal solutions
may be very hard if not impossible. There could be
very many, possibly infinitely many of them with small
or even vanishing total action. There seems to be no
simple positivity principle to rule out such an infinite
class. Not all of these solutions will be equally
important though. Each of them gives rise to a branch
point and thus to additional Riemann sheets in the
z-variable. Many of the branch points will be in those
other Riemann sheets and thus not affect the asymptotic
behavior of the perturbation expansion in z. The
criterion will be provided by connecting the given
solutions continuously to the vacuum configuration
through a series of field configurations $A(x,\lambda)$,
$o < \lambda < 1$. For $\lambda = o$ we have the vacuum, say, and for
$\lambda = 1$ the given solution. The action $S[A(\lambda)]$ will follow
a certain path in the complex plane. If that path has
to go around another solution point S' in the complex
plane then we are in a different Riemann sheet than the
origin. Not $S[A(1)]$ but S' will determine the radius
of convergence of $F(z)$. Even so, we do not know at

present how to find all relevant complex solutions.

We note that the single BPST instanton is an example
of a solution that cannot be connected continuously to
the vacuum because of its nonvanishing winding number.
The instanton anti-instanton pair is the first known
solution that will limit the radius of convergence of
$F(z)$ in QCD to $16\pi^2$ (because the total action of that
pair is $16\pi^2/g^2$). Instantons also give rise to effects
in theories with Higgs mechanism although no finite
classical solution exists. Therefore we expect a
singularity of F at $z = 16\pi^2$ also in these theories.

The second problem we wish to mention may be
equally troublesome. It has been noted that the formal
result can be proven in an alternative way in finite
theories such as the anharmonic oscillator (quantum
field theory in 0+1 dimensions) by counting and estimating the Feynman diagrams[8]. In renormalized theories
however the diagrams must undergo subtractions and a
consequence of these subtractions is a more complicated
dependence of external momenta through powers of
logarithms. At higher orders these logarithms make it
much harder to estimate bounds for these diagrams. Indeed, naive guesses for these diagrams suggest that the
perturbation coefficients diverge so badly that even the
series (4.3) will diverge for all z. This is a new
feature due to renormalization and one therefore wonders
whether perhaps the interchange of the limit $n \to \infty$
(high terms in the perturbation expansion) and the limit
$\Lambda \to \infty$ (renormalization cutoff) do not commute. One can
merely conjecture that interchange is allowed because the
resulting expressions seem to make sense even after renormalization, but there seems to be no way of proving
that.

And here is the third problem. It concerns the integral in eq. (4.1). Does it converge at infinity? Let us consider the one theory where we know the analytic behavior as a function of g_R^2, which is the massless gauge theory, considered in chapter 3. There are cuts and/or poles at

$$- 2/\beta_1 g_R^2 \simeq a_i + (2n+1)\pi i \ , \qquad (5.1)$$

which is a simplification of eq. (3.15). Here a_i is real or nearly real and can become arbitrarely large. Consequently the integral

$$\int_0^\infty F(z) \ [\exp \ (-\beta_1/2) \ (-az)] \ [\exp \ (-\beta_1/2)(2n+1)\pi i z] dz$$

will be infinite, i.e. diverges regardless of how large a is (remember that $-\beta_1$ is positive). Therefore F(z) diverges for z real and large, faster than any ex- ponential of z, and oscillates with periods of the order of $|\frac{4}{\beta_1}|$ and fractions thereof. So, for this theory we answered the question: no, the integral diverges very badly. Attempts to redefine it in terms of some cutoff procedure will be as ambiguous as the original divergent series.

This divergence does not occur in 3-dimensional massless QCD but, as we shall see, there are other problems there.

We think there is no reason to assume that adding masses to the quarks would improve the situation, since we are not able to apply our analysis there.

6. AN INFRARED DIVERGENCE IN EUCLIDEAN SPACE
In four dimensions we have no infrared divergences

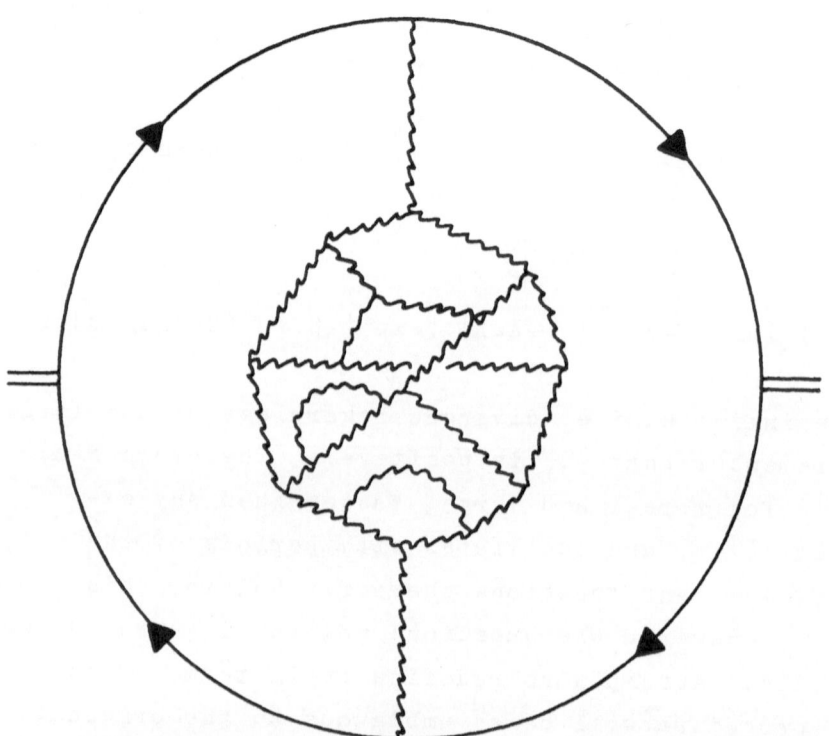

Fig. 4.   Diagrams of this type give the first infrared
divergences in less than four dimensions in Euclidean
space.   In 4 dimensions they might spoil Borel summability.

in Euclidean space. But for renormalization it is convenient to consider the theory in $4-\varepsilon$ dimensions, ε
small and positive. Then, if we look at fixed order in
the perturbation expansion we expect an infrared divergence, from the following arguments. If we consider
the L-loop corrections to the gluon propagator, then,
by power counting, we find the following momentum dependence:

$$\Delta P_{\mu\nu}(k) \propto (\delta_{\mu\nu} - \frac{k_\mu k_\nu}{k^2})\ (k^2 - i\varepsilon)^{-1-\frac{\varepsilon L}{2}} . \qquad (6.1)$$

If we insert this into a larger diagram (fig. 4), we
encounter integrals of the form

$$\int d^{4-\varepsilon} k\ \Delta P_{\mu\nu}(k)\ f_{\mu\nu}(k) . \qquad (6.2)$$

Since we consider only gauge invariant amplitudes, $f_{\mu\nu}$
will satisfy a Ward identity:

$$k_\mu f_{\mu\nu}(k) = 0 \quad ,$$

$$ \qquad\qquad (6.3)$$

$$k_\nu f_{\mu\nu}(k) = 0 \quad .$$

From this we derive that $f_{\mu\nu}$ must have a second order
zero at $k = 0$:

$$f_{\mu\nu}(k) \propto (\delta_{\mu\nu} k^2 - k_\mu k_\nu)\ \text{if}\ k \to o . \qquad (6.4)$$

So if $k \to o$, the integrand in (6.2) behaves as

$$d^{4-\varepsilon} k\ (k^2)^{-\frac{\varepsilon L}{2}} . \qquad (6.5)$$

This starts to diverge if

$$L > \frac{4}{\varepsilon} - 1 . \qquad (6.6)$$

The total diagram is at least of order

$$(g^2)^{4/\varepsilon} \quad . \qquad (6.7)$$

So at finite ε we have an infrared divergence at large but finite order in the perturbation expansion. It is not clear how to give a direct physical interpretation of this divergence. In more than two dimensions there seems to be no reason for expecting cancellations between these divergences for different diagrams at the same order. But, if we sum the gluon selfenergy diagrams to obtain a new gluon propagator, then the "divergent" ineducible parts behave as a mass insertion and at high orders we expect logarithmic dependence of this mass insertion, so then the infinite coefficient for the diagrams of order $(g^2)^{4/\varepsilon}$ is replaced by a term of order

$$(g^2)^{4/\varepsilon} \log g . \qquad (6.8)$$

The Laplace transform will consequently contain terms of order

$$z^{4/\varepsilon} \cdot \log z \quad .$$

Notice that then $F(z)$ must have a cut starting at the origin of z-plane. This is the complication for theories in less than 4 dimensions as we alluded to in the beginning.

7. CONCLUSION

We do think perturbation theory up to two loops is essential to obtain an accurate definition of the theory.

But we were not able to obtain sufficient information on the theory to formulate a mathematically self-consistent procedure for accurate computations.

Our results made us skeptical against approaches that rely on Padé or Borel resummation procedures.

ACKNOWLEDGEMENT

The author wishes to thank the Aspen Center for Theoretical Physics for its hospitality.

REFERENCES

1. L.N. Lipatov, Pisma JETP $\underline{24}$, 179 (1976) and
    Leningrad preprints.

2. E. Brezin, J.C. Le Guillou and J. Zinn Justin,
    Saclay preprints DPh-T/76/102 and DPh-T/76/119.

3. G. 't Hooft, Nucl. Phys. $\underline{B61}$, 455 (1973).

4. D.R.T. Jones, Nucl. Phys. $\underline{B75}$, 531 (1974).
    A.A. Belavin and A.A. Migdal, Gorky State University,
    USSR preprint, Jan. 1974.
    W.E. Caswell, Phys. Rev. Lett. $\underline{33}$, 244 (1974).

5. H. Georgi and H.D. Politzer, Phys. Rev. $\underline{D14}$, 1829
    (1976).

6. G. 't Hooft, Phys. Rev. Lett. $\underline{37}$, 8 (1976).
    G. 't Hooft, Phys. Rev. $\underline{D14}$, 3432 (1976).
    C. Callan, R. Dashen, D. Gross, Princeton preprint.

7. G. 't Hooft, Nucl. Phys. $\underline{B72}$, 461 (1974).

8. C.M. Bender and T.T. Wu, Phys. Rev. Lett. $\underline{37}$, 117
    (1976).
    C.M. Bender, M.I.T., invited talk given at the
    Second Los Alamos Workshop on Mathematics in the
    Natural Sciences (August 1976), to be published in
    Advances in Mathematics.

PARTICIPANTS

Kazuo Abe
University of Michigan

Richard Arnowitt
Northeastern University

Marshall Baker
University of Washington

James Ball
University of Utah

Michael Barnett
Stanford Linear Accelerator
  Center

Asim Barut
University of Colorado

Mirza Bég
Rockefeller University

C. Bouchiat
University of Paris, South

Richard Brandt
New York University

Laurie Brown
Northwestern University

Arthur Broyles
University of Florida

David Campbell
Los Alamos Scientific
  Laboratory

Cyrus Cantrell
Los Alamos Scientific
  Laboratory

Peter Carruthers
Los Alamos Scientific
  Laboratory

Kenneth Case
Rockefeller University

Owen Chamberlain
University of California

M. S. Chen
Center for Theoretical Studies

Ta-Pei Cheng
University of Missouri

Hirsh Cohen
IBM

Charles Conley
University of Wisconsin

John Cornwall
University of California

Stanley Deans
University of Florida

Gary Deem
Bell Laboratories

Robert Diebold
Argonne National Laboratory

P.A.M. Dirac
Florida State University

Bernice Durand
University of Wisconsin

Loyal Durand
University of Wisconsin

Glennys Farrar
California Institute of
  Technology

Richard Feynman
California Institute of
  Technology

727

Thomas Fields
Argonne National Laboratory

Paul Fife
University of Arizona

Paul Fishbane
University of Virginia

Hermann Flaschka
University of Arizona

Kenneth Fox
University of Tennessee

Paul Frampton
University of California

Daniel Freedman
State University of New York
  at Stony Brook

Richard Friedberg
Columbia University

Harald Fritzsch
CERN

Harold Galbraith
Los Alamos Scientific
  Laboratory

Murray Gell-Mann
California Institute of
  Technology

Howard Georgi
Harvard University

Fred Gilman
Stanford Linear Accelerator
  Center

Sheldon Glashow
Harvard University

O. W. Greenberg
University of Maryland

Marcus T. Grisaru
Brandeis University

David Gross
Princeton University

Eugene Gross
Brandeis University

Feza Gürsey
Yale University

C. R. Hagen
University of Rochester

Francis Halzen
University of Wisconsin

Brosl Hasslacher
California Institute of
  Technology

A. W. Hendry
Indiana University

Jarmo Hietarinta
Ohio State University

Bernard Hildebrand
Energy Research and
  Development Administration

Robert Hofstadter
Stanford University

Joseph Hubbard
Center for Theoretical Studies

Antal Jevicki
Princeton University

Kenneth Johnson
Massachusetts Institute of
  Technology

Gordon Kane
University of Michigan

Stuart Kasden
Princeton University

David Kaup
Clarkson College of
  Technology

Boris Kayser
National Science
  Foundation

J. P. Kernevez
Universite de Technologie
  de Compiegne

John Klauder
Bell Laboratories

Abraham Klein
University of Pennsylvania

A. D. Krisch
University of Michigan

Behram Kursunoglu
Center for Theoretical
  Studies

Kenneth Lane
Stanford Linear Accelerator
  Center

Herbert Lashinsky
University of Maryland

Harold Lecar
National Institutes of
  Health

Don Lichtenberg
Indiana University

Harry Lipkin
Fermi Laboratory

Marvin Marshak
University of Minnesota

David McLaughlin
University of Arizona

Sydney Meshkov
National Bureau of Standards

Himrich Meyer
University of Wuppertal

Michel Mille
Center for Theoretical Studies

Peter Minkowski
University of Bern

John Moffat
University of Toronto

Pran Nath
Northeastern University

André Neveu
Institute for Advanced Study

Patrick O'Donnell
University of Toronto

Reinhard Oehme
University of Chicago

Harold O'Gren
Indiana University

Martin Olsson
University of Wisconsin

Heinz Pagels
Aspen Center for Physics

Sandip Pakvasa
University of Hawaii

William Palmer
Ohio State University

Michael Parkinson
University of Florida

Emmanuel Paschos
Brookhaven National
  Laboratory

J. Patera
University of Montreal

Earl Peterson
University of Minnesota

Arnold Perlmutter
Center for Theoretical
  Studies

Chris Quigg
Fermi Laboratory

Pierre Ramond
California Institute of
  Technology

Mario Rasetti
Center for Theoretical
  Studies

L. G. Ratner
Argonne National Laboratory

Charles Rhodes
Stanford Research Institute

Jabus Roberts
Rice University

Rudolf Rodenberg
III. Physikalisches Institut
  der Technischen Hochschule

Joseph Romig
Boulder, Colorado

S. Peter Rosen
Energy Research and
  Development Administration

Carl Rosenzweig
Syracuse University

Ralph Roskies
University of Pittsburgh

Jonathan Rosner
Institute for Advanced
  Study

Hanno Rund
University of Arizona

Robert Sachs
Argonne National Laboratory

Abdus Salam
International Centre for
  Theoretical Physics

Howard Schnitzer
Brandeis University

Jonathan Schonfeld
California Institute of
  Technology

Julian Schwinger
University of California

Alwyn Scott
University of Wisconsin

Gino Segré
University of Pennsylvania

L. M. Simmons, Jr.
Los Alamos Scientific
  Laboratory

Richard Slansky
Los Alamos Scientific
  Laboratory

Charles Sommerfield
Yale University

Vigdor Teplitz
Virginia Polytechnic Institute
  and State University

G. 't Hooft
Stanford Linear Accelerator
  Center

Gerald Thomas
Argonne National Laboratory

Yukio Tomozawa
University of Michigan

T. L. Trueman
Brookhaven National
  Laboratory

Hung-Sheng Tsao
Rockefeller University

P. Van Nieuwenhuizen
State University of New
  York at Stony Brook

Kameshwar Wali
Syracuse University

Jill Wright
Bedford College

Eli Yablonovitch
Harvard University

York-Peng Yao
University of Michigan

G. B. Yodh
University of Maryland

Akihiko Yokosawa
Argonne National Laboratory

Henry Yuen
TRW Systems

Norman Zabusky
University of Pittsburgh

Fredrik Zachariasen
California Institute of
  Technology

Bruno Zumino
CERN

ORBIS SCIENTIAE 1977

MONDAY, JANUARY 17, 1977

LAW SCHOOL AUDITORIUM

OPENING ADDRESS OF WELCOME

SESSION I -

"POLARIZED PARTICLES AND SPIN EFFECTS IN HIGH ENERGY
  PHYSICS"

Moderator:
Alan Krisch, University of Michigan

Dissertators:
P.A.M. Dirac, Florida State University
    "DYNAMICAL METHODS FOR STREAMS OF MATTER"(30 min.)

Behram Kursunoglu, University of Miami
    "ORIGIN OF SPIN" (30 min).

F. Halzen, University of Wisconsin
    "POLARIZATION EXPERIMENTS - A THEORETICAL REVIEW"
    (40 min.)

J. B. Roberts, Rice University
    "MEASUREMENT OF SPIN DEPENDENCE OF $\sigma_{tot}(pp)$"(20 min.)

A. Yokosawa, Argonne
    "pp SCATTERING - AMPLITUDES MEASUREMENTS AND A
    POSSIBLE DIRECT-CHANNEL RESONANCE IN pp SYSTEM"
                                              (20 min.)
K. Abe, University of Michigan
    "LARGE P$_\perp^2$ SPIN DEPENDENCE OF P-P ELASTIC
    SCATTERING(20 min.)

SESSION II -

"POLARIZED PARTICLES AND SPIN EFFECTS IN HIGH ENERGY
  PHYSICS"

Moderator:
Alan Krisch, University of Michigan

Dissertators:
L. Dick, CERN
    "POLARIZATION EXPERIMENTS IN EUROPE"(40 min.)

R. E. Diebold, Argone
    "ELASTIC AND INELASTIC POLARIZATION EFFECTS OBSERVED
    WITH THE ARGONNE EFFECTIVE MASS SPECTROMETER"(20 min.)

E. A. Peterson, University Minnesota
"INCLUSIVE ASYMMETRIES" (20 min.)

Harold Ogren, Indiana University
"PRELIMINARY POLARIZATION RESULTS AT FERMILAB
ENERGIES"(20 min.)

Owen Chamberlain
"POLARIZED TARGET EXPERIMENT AT FERMILAB"(20 min.)

TUESDAY, JANUARY 18, 1977

SESSION III -

"SUPERSYMMETRY AND SUPERGRAVITY"

Moderator:
Daniel Z. Freedman, State University of New York
at Stony Brook

Dissertators:
Richard Arnowitt, Northeastern University
"LOCAL SUPERSYMMETRY AND INTERACTIONS"(30 min.)

Daniel Z. Freedman, SUNY at Stony Brook
"SUPERGRAVITY FIELD THEORIES AND THE ART OF
CONSTRUCTING THEM"(30 min.)

Peter van Nieuwenhuizen, SUNY at Stony Brook
"RENORMALIZABILITY OF SUPERGRAVITY"(30 min.)

Bruno Zumino, CERN
"TOPICS IN SUPERGRAVITY AND SUPERSYMMETRY"(30 min.)

SESSION IV -

"FLAVOR INTERACTIONS"

Moderator:
H. Fritzsch, CERN

Dissertators:
H. Fritzsch, CERN
"QUANTUM FLAVORDYNAMICS"(30 min.)

F. Gilman, SLAC
"PRODUCTION OF NEW PARTICLES IN ELECTRON POSITRON
ANNIHILATION"(30 min.)

H. Meyer, Wuppertal University, Germany
"RECENT RESULTS ON e^+e^- ANNIHILATION FROM PLUTO
AT DORIS"(30 min.)

M. Barnett, SLAC
"THE SEARCH FOR HEAVY PARTICLES"(30 min.)

C. Bouchiat, Ecole Normale Superieure, Paris
  "PARITY VIOLATION IN ATOMIC PHYSICS AND
  NEUTRAL CURRENTS"(30 min.)

WEDNESDAY, JANUARY 19, 1977

SESSION V -

"HADRON PHENOMENOLOGY"

Moderator:
Sydney Meshkov, National Bureau of Standards,
                Washington, D. C.

Dissertators:
Richard P. Feynman, California Institute of Technology
  "CORRELATIONS IN HADRON COLLISIONS AT HIGH
  TRANSVERSE MOMENTUM"(30 min.)

Jonathan L. Rosner, Institute for Advanced Study,
                Princeton
  "FINAL STATES IN CHARMED PARTICLE DECAYS"(30 min.)

Howard Georgi, Harvard University
  "THE WINNER OF THE VECTOR-MODEL LOOK-ALIKE
  CONTEST"(30 min.)

Harry J. Lipkin, Fermi Laboratory and Argonne
  "THE ALEXANDER...ZWEIG RULES AND WHAT IS WRONG
  WITH PSEUDOSCALAR MESONS"(30 min.)

SESSION VI -

"ATTEMPTS TO SOLVE QUANTUM CHROMODYNAMICS"

Moderator:
Murray Gell-Mann, California Institute of Technology

Dissertators:
David Gross, Princeton University
  "A MECHANISM FOR QUARK CONFINEMENT"(30 min.)

H. David Politzer, Harvard University
  "THE STATUS OF ξ-SCALING"(30 min.)

Ta-Pei Cheng, University of Missouri
  "MUON NUMBER NONCONSERVATION IN GAUGE THEORIES"(30 min.)

J. M. Cornwall, UCLA
  "SEMICLASSICAL PHYSICS AND CONFINEMENT"(30 min.)

G. t'Hooft, SLAC and the University of Utrecht
  "SOME OBSERVATIONS IN QUANTUM CHROMODYNAMICS"(30 min.)

THURSDAY, JANUARY 20, 1977

SESSION VII -

"SOLITONS AND NONLINEAR PARTICLE THEORY"

Moderator:
David K. Campbell, Los Alamos Scientific Laboratory

Dissertators:
Andre Neveu, Institute for Advanced Study, Princeton
    "SOME RECENT DEVELOPMENTS ON SOLITONS IN
    TWO-DIMENSIONAL FIELD THEORIES"(45 min.)

Antal Jevicki, Institute for Advanced Study, Princeton
    "PATH INTEGRAL QUANTIZATION OF SOLITONS"(30 min.)

Richard Friedberg, Barnard College
    "NONTOPOLOGICAL SOLITONS"(30 min.)

Paul Frampton, University of California
    "VACUUM BUBBLE INSTANTONS"(20 min.)

SESSION VIII -

"TOPICS IN NONLINEAR MATHEMATICS WITH APPLICATIONS TO
 PHYSICS"

Moderator:
Hermann Flaschka, University of Arizona

Dissertators:
Henry Yuen, TRW, Redondo Beach, California
    "NONLINEAR DEEP WATER WAVES: A PHYSICAL
    TESTING GROUND FOR SOLITONS AND RECURRENCE"(30 min.)

David Kaup, Clarkson College, Potsdam, N.Y.
    "SOLITONS AS PARTICLES, AND THE EFFECTS OF
    PERTURBATIONS"(30 min.)

N. J. Zabusky, University of Pittsburgh
    "COHERENT STRUCTURES IN FLUID DYNAMICS"(30 min.)

FRIDAY, JANUARY 21, 1977

SESSION IX -

"NONLINEAR MOLECULAR PROCESSES"

Moderator:
Cyrus Cantrell, Los Alamos Scientific Laboratory

Dissertators:
Eli Yablonovitch, Harvard University
      "COLLISIONLESS MULTIPHOTON DISSOCIATION OF SF_6:
      A STATISTICAL THERMODYNAMIC PROCESS"(30 min.)

H. W. Galbraith, Los Alamos Scientific Laboratory
      "STRUCTURE OF THE VIBRATIONAL STATES IN THE ν_3-
      FUNDAMENTAL AND ITS OVERTONES IN SF_6 AND
      MULTIPHOTON ABSORPTION EFFECTS"(30 min.)

Kenneth Fox, University of Tennessee
      "REVIEW OF ROTATIONAL STRUCTURE IN EXCITED
      VIBRATIONAL STATES OF SPHERICAL-TOP MOLECULES"
      (30 min.)

C. K. Rhodes, Stanford Research Institute
      "ISOTOPE EFFECTS IN MOLECULAR MULTIQUANTUM"(30 min.)

SESSION X:

"NONLINEAR STRUCTURES AND OSCILLATIONS IN BIOCHEMICAL
MEDIA"

Moderator:
Norman J. Zabusky, University of Pittsburgh

Dissertators:
J. P. Kernevez, Universite de Technologie de Compiegne
      "SPATIO-TEMPORAL STRUCTURATION IN IMMOBOLIZED
      ENZYME SYSTEMS"(30 min.)

Paul C. Fife, University of Arizona
      "THEORIES AND CONJECTURES ON MEMBRANE-SUPPORTED
      WAVES AND PATTERNS"(30 min.)

Harold Lecar, National Institutes of Health
      "PHYSICAL MECHANISMS OF NERVE EXCITABILITY"(30 min.)

Hirsh Cohen, International Business Machines Corp.
      "MATHEMATICAL DEVELOPMENT IN HODGKIN-HUXLEY
      THEORY AND ITS APPROXIMATION"(30 min.)

SUBJECT INDEX